A PLANT SCIENCE MONOGRAPH
Edited by Professor Nicholas Polunin, M.S.(Yale), M.A., D.Phil., D.Sc.(Oxon.)

CROP PRODUCTION IN DRY REGIONS

Volume I: Background and Principles

I. ARNON

Ingénieur Agronome (Gembloux); Ph.D.(Jerusalem). Associate Professor of Agronomy, Hebrew University. Former Director of the Volcani Institute of Agricultural Research, Israel.

LEONARD HILL
LONDON 1972

An Intertext Publisher

Published by Leonard Hill Books
a division of
International Textbook Company Limited
158 Buckingham Palace Road
London SW1

First published 1972

ISBN 0 249 44043 1

Printed in Great Britain by Cox & Wyman Ltd.

FOREWORD

The impending world food shortage has, in recent years, directed the attention of international and other agricultural bodies to the dire need for developing more effectively the agricultural potential of the dry regions. These are the areas of meagre and undependable rainfall, in which the average precipitation is deficient in relation to evaporation. The dry regions include the arid zones—in which arable crop production is not possible without irrigation, and also their semi-arid fringes, in which rainfall, though precarious, is sufficient for certain types of crops, requiring special management techniques.

World interest in the development of the dry regions is evidenced by the Arid Zone Programme of UNESCO, started in 1951 with the object of promoting and stimulating research that has a bearing on the problems of arid lands; as a result of such research, numerous publications have appeared. Various bodies in the United States, U.S.S.R., India, and elsewhere, have also organized symposia, sponsored numerous publications, and amassed a huge store of important data on the arid regions and the problems of their development. Research activities in these regions, though lagging far behind those of the temperate regions, are still considerable in the aggregate, as evidenced by the numerous papers that are being published on them. What appears to be lacking is the condensation of that which is most relevant in this mass of information to the problems of crop production in the arid regions and their semi-arid fringes.

The author was therefore happy to accept the suggestion of Professor N. Polunin, the Editor of Plant Science Monographs, to write a book that would attempt to fill this void. Everywhere in the dry regions of the world, traditional land-use patterns are breaking down as national and international resources are being devoted increasingly to the development of agriculture. In the light of these developments, it was felt that such a book would be timely and would fulfil a real need by contributing to an understanding of crop requirements under arid and semi-arid conditions while also stimulating improved production practices.

The subject as a whole is so vast that it has only been possible to cover it in reasonable detail in two volumes. These are discrete to the extent that the first surveys general principles and practices whereas the second deals with specific crops and their cultivation, etc.

In the first five chapters of Volume I, background information on the history, environment, and traditional land-use, of the drylands and their semi-arid fringes is given. Chapters VI to XVIII are concerned with the basic principles of arid-land crop production, grouped in three main sections: the first is devoted to the rational use of water, the second to basic agronomic practices, and the third to plant protection problems.

Whilst the basic principles of agricultural practice are universal, an effort has been made in these sections to stress what is specific or unique to crop production under the conditions of the dry regions. In a book of this nature, the author is continuously faced with the problem of deciding what general information should be included in what is destined to be a specialized account of dryland crop production. In the absence of an objective yardstick, the dividing line between what is essential and relevant to the subject treated, and what can be omitted, is never well defined, and therefore depends largely on the subjective feelings of the author. Nor was a full, exhaustive regional treatment found to be practicable, though an effort was made to indicate the relevance of the subjects treated to the very diverse regions which constitute the world's arid lands.

Chapter XIX, the last of Volume I, is concerned with agricultural development in the dry regions. Much has been written recently on agricultural development by specialists in the many economic, social, and political, aspects involved in the transition from traditional to modern agriculture. An attempt by a non-specialist to condense this extremely wide and complex field into a single chapter, in a book devoted to crop production, may therefore appear presumptuous. And yet, the basic problems of agricultural development are germane to the application of the agricultural techniques described in this work. Without some understanding of the interactions between the application of improved technology and other facets of development, and in particular of the economic and social preconditions required before improved practices can be adopted by farmers, attempts to apply such improved practices may be irrelevant, wasteful, and even occasionally harmful. This appears, to the author, as sufficient justification for the inclusion of a chapter on the problems involved in the transition from traditional to modern agriculture.

Coming to Volume II we may note that a very wide range of rain-fed and irrigated field-crops can be grown in the arid and semi-arid regions. Where mild winters prevail, most temperate crops can be grown; during the hot summer that is typical of drylands, the production of many subtropical and even certain tropical crops is possible. However, production of wheat, for example, is entirely different in a semi-arid region from the usual practice in temperate climates; similarly, rice or cotton production in the dry regions requires cultural techniques that are very different from those in use in the humid tropics. Volume II is accordingly devoted to a description of the principal crops produced in the dry regions, their economic significance, and the cultural techniques used there. It was not the intention of the author to give detailed instructions for the growing of these crops, but rather to demonstrate how the basic principles described in Volume I are applied in practice to the main crops of the dry regions, meanwhile dealing with the specific problems to be encountered.

The overall objective of the author has been to give an up-to-date, comprehensive, and concise, review of the present knowledge on crop production in dry regions. A vast array of literature was perused in the preparation of the text, and a synthesis of the results of research and available information was attempted in the light of the author's own experience in the field. Only a small proportion of the literature consulted has actually been cited in the text. The references that are quoted are either in support of the general principles enunciated, or are chosen as appropriate examples to illustrate these principles.

In a work of this scope, it is almost inevitable that errors of omission or commission should intrude. Quite possibly some important aspects may have been overlooked, and there will assuredly be justification for criticism and comments. These will be welcomed by the author, *inter alia* against reprintings or possible further editions.

ACKNOWLEDGEMENTS

The author has been active in agricultural research and development in an arid and semi-arid environment for some forty years. He is not, and could not possibly be, an authority on all the special fields which the work encompasses. A number of chapters have therefore been reviewed by several colleagues, each in his special field, whose critical and constructive comments the author gratefully acknowledges: Dr D. Drori, Dr Y. Ephrat, Dr M. Horowitz, Dr U. Kafkafi, Mr D. Lachover, Mr N. Rosenman, Dr Y. Rotem, Dr Y. Shalhevet, Dr N. Tadmor, Prof. Y. Vaadia, Dr M. Vermesh, and Dr D. Yoel.

It is a particular pleasure to acknowledge the great help I received from Mrs V. Priel, who read and corrected the manuscript with painstaking care, and to thank Mrs M. Ben-Chaim for the arduous work of deciphering my handwriting and typing the manuscript.

Finally, I wish to express my gratitude and warm thanks to the Editor,

Professor N. Polunin, who suggested the subject of the book, then proposed a tentative outline, and has since and always been ready with advice on which I have relied throughout the preparation of the manuscript.

The sources of the illustrations have been acknowledged in the appropriate captions.

November, 1971
Bet Dagan, Israel

I. Arnon

TABLE OF CONTENTS

LIST OF PLATES

FIGURES IN TEXT

LIST OF TABLES

INTRODUCTION

It is an interesting fact of history that in most of the arid zones in the Near East, India, North Africa, and Central America, very advanced agricultural civilizations developed, flourished, and eventually disappeared.

Whatever the reasons that prompted early Man to choose the arid and semi-arid lands in preference to the more favourable conditions for crop production of the temperate regions with a reliable rainfall, it is clear that he chose an area in which the balance between his own activities and the environment is extremely precarious.

The central problem of sustained land-use in the dry regions has always been to find and maintain a balance between man's requirements and the productive capability of the land. This has rarely been achieved.

Throughout history Man has, by over-use, consistently reduced the productive capacity of the arid lands. Originally, a large proportion of the more or less scanty vegetative cover consisted of palatable and nutritious grasses and shrubs. Overgrazing caused an overall reduction of the plant cover, and the large-scale replacement of useful plants by unpalatable species. The reduced cover intensified erosion by wind and rain, starting a process which is frequently irreversible. Gullies cut through alluvial valleys, water run-off increased, and storage reservoirs filled with silt.

Where the original plant-cover of areas on the fringes of the arid regions was ploughed up for crop production, the consequences were still more disastrous. The classic, though by no means isolated, example is that of the 'dust-bowl' of the Great Plains of the United States. In most regions in which crop production was based on irrigation, the consequences of cultivation were in the long run no less catastrophic; salinization and waterlogging have reduced once-fertile regions to barren wastes. Economic deterioration was accompanied by political instability and the inability to withstand the invasions of nomadic tribes. Wide areas were gradually abandoned, and the erstwhile cradles of civilization became man-made deserts.

With the explosive increase in world population, there is increasing pressure on the drylands, which, with their semi-arid fringes, constitute fully one-third of the land area of the globe. This increased pressure applies both to the areas with primitive agriculture and low standards of living, as in the arid regions of Asia and North Africa, and to technologically advanced regions with a high standard of living, such as the western part of the United States.

The arid parts of a great number of countries are now usually vast empty areas. A large proportion is desert, and its contribution to food production is minimal. Other vast areas on the fringes of the deserts in North Africa and Asia, with rainfall that is just sufficient or even adequate to grow a single adapted crop during a few months of the year, are still cultivated by primitive methods that have not changed since biblical days—with resultant low yields. In other vast areas with similar climatic conditions—in the U.S.S.R., North America, and Australia—though cultivation is with the most modern equipment and production is extensive, the yields are low in comparison with those that are common in regions of temperate climate. And yet the combination of high insolation, favourable temperatures, and long growing periods, in these dry regions makes possible levels of crop production that can exceed considerably those achieved in the temperate regions—if only an adequate supply of water can be provided.

Relatively small areas of former desert, such as the Imperial Valley in California and the Jordan Valley in Israel, produce a large variety of crops with record yields—an indication of the potential of these regions when water is no longer a limiting factor and modern agricultural technology is applied. The development of other areas along the same lines could make substantial contributions to the economies of their countries and to the increasing food requirements of the world as a whole. It is hoped that the present work will stimulate and help to guide overdue improvements.

I

ARID CLIMATES AND THEIR WORLD DISTRIBUTION

Definition and Classification of Arid Areas

The word arid (from 'arere', to be dry) implies a deficiency of rainfall as the basic characteristic of the arid climate. However, the amount of rainfall cannot serve as an adequate measure of aridity unless this amount is related to its effectiveness.

The effectiveness of rainfall depends on its seasonal distribution, the rate of evaporation, the nature of the soil, and the vegetative cover. The rate of evaporation is determined by temperature, air humidity, and wind movement. An arid climate is defined as one in which, for the greater part of the year, precipitation is less than the potential evapotranspiration—that is, evaporation plus water-loss from plants.

A number of attempts have been made to classify the different types of dry climates, generally based on the *amount* of rainfall and its *effectiveness*. None of the classifications that have been developed is ideal in the sense that it is simple and complete, that the criteria are really objective, and that climatic factors relate satisfactorily to natural vegetation, to soil types, or to land-use (Koeppe & Long, 1958).

The Köppen system of climates (Köppen & Geiger, 1936) is the most generally accepted one, being based mainly on the relation between rainfall (amount and distribution) and temperature. It assumes that:

(a) the higher the temperature, the greater the amount of precipitation that will still result in the same degree of aridity;

(b) relatively less precipitation is needed if it is concentrated during the cool season, more is required if distribution throughout the year is uniform, and the highest requirement is associated with precipitation that is concentrated in the warm season.

Köppen first defines five major categories of world climate, of which one only—comprising those regions in which the average precipitation is deficient in relation to evaporation—is termed dry.

The dry climates are subdivided into *deserts* (or arid climates), in which rainfall is insufficient for arable crop production, and *steppes* (semi-arid climates), in which rainfall is sufficient for certain types of crops when adequate management techniques are adopted and grasses constitute an important element of the native vegetation. A further subdivision is based on seasonal distribution of rainfall, according to whether the greater part of the annual precipitation occurs during the six cooler months of the year (winter precipitation) or during the six warmer months of the year (summer precipitation). Otherwise, it is considered that there is no distinct season of rainfall.

TABLE I

The Köppen Classification of Dry Climates (Köppen & Geiger, 1936)

Boundary	*Winter rainfall*	*No clearly-defined period of rainfall*	*Summer rainfall*
Desert/Steppe	$P=T$	$P=T+7$	$P=T+14$
Steppe/humid	$P=2T$	$P=2(T+7)$	$P=2(T+14)$

$P=$ annual precipitation (in cm)
$T=$ annual mean temperature (°C)

A final subdivision is based on temperature: *hot* dry climates and *cool* dry climates, according to whether the mean annual temperature is above or below 18°C, respectively.

Since latitude is the principal determinant of temperature, the drylands are classified as:

low-latitude (or hot) deserts and low-latitude (or hot) steppes, and middle-latitude (or cool) deserts and middle-latitude (or cool) steppes.

The following works should also be considered in this connection:

Martonne (1962) attempted to find an empirical relationship between precipitation and temperature, that would enable a more accurate and general definition of dry climates. This relationship is expressed in his '*aridity-index*', which can refer to a selected period of a few days, a month, a season, or an entire year.

$$I = \frac{np}{t + 10}$$

where I = aridity index,
n = number of rainy days,
p = mean precipitation per day, and
t = mean temperature of the selected period in °C.

When calculated on an annual basis, $I < 20$ is characteristic of an arid climate and $20 < I < 30$ defines a semi-arid climate.

Gaussen (1954), from studies of natural vegetation, developed a classification of climates based on the number of arid months in a year. A month is considered arid if $p < 2t$, where p = mean monthly precipitation (cm) and t = mean monthly temperature (°C). This classification gave good results when applied to Mediterranean climates, but was found less satisfactory when applied on a world-wide scale (Wallen, 1966).

Thornthwaite (1948) bases his classification of climate on two elements: (a) water supply in the form of precipitation, and (b) water needs resulting from evapotranspiration. Precipitation and evapotranspiration are due to different meteorological causes and may be markedly different from each other both in quantity and in seasonal distribution. In principle, Thornthwaite defines a climate as moist when precipitation exceeds evapotranspiration; a *dry climate* occurs where evapotranspiration is markedly in excess of precipitation.

Thornthwaite further differentiates between *actual evapotranspiration*, which in an arid climate may be very low simply because the moisture supply is limited, and *potential evapotranspiration*, or 'water need', which is the amount of water that would be transpired and evaporated under ideal conditions of soil moisture and vegetative cover. The amount of potential evapotranspiration being dependent only on climate, by comparing it with precipitation Thornthwaite developed a classification system for arid climates. In Thornthwaite's method, potential evaporation is computed by an empirical formula involving mean monthly temperature and average day-length. When the monthly curves of potential evapotranspiration and rainfall are plotted as smooth curves, and the two curves are compared, it is possible to determine if and when precipitation exceeds evapotranspiration (Figure 1).

When precipitation is adequate to supply the water that is needed for potential evaporation, the 'moisture index' is considered as equal to zero; moist climates have positive values and arid climates have negative values of this index.

The values of other important factors influencing evaporation—such as wind, humidity, and solar radiation—are considered to be linked with the variations in temperature and were therefore omitted from the formula, in order to simplify the method without affecting its usefulness.

Thornthwaite's method of calculating potential evapotranspiration was developed in a climate of summer rainfall and has not proved entirely satisfactory in other types of dry climates.

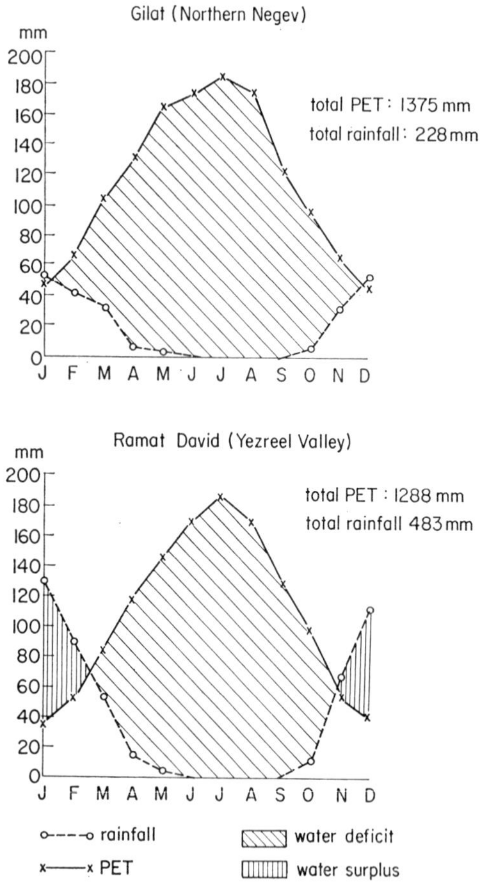

Figure 1. Comparison of seasonal precipitation and potential evapotranspiration: annual march of rainfall and potential evapotransipation (PET) through the year at two contrasting regions of Israel: Gilat (Northern Negev, above) and Ramat David (Yezreel Valley, below). PET was calculated according to Penman's formula (see p. 192).

Selyaninov (cf. Davitaya, 1964) proposed the principal arid-zone climatic index used in the U.S.S.R., namely the heat and moisture ratio:

$$C = \frac{\Sigma P}{0{\cdot}10.\Sigma \mathrm{H}}$$

where ΣP= total amount of precipitation (in cm) for the period with an average temperature above 10°C,

$\Sigma \mathrm{H}$= accumulated temperatures for the same period,

Isolines* of this coefficient with values of C=0·5 constitute the boundaries

* Isolines are imaginary lines connecting places that have the same coefficient.

of the desert, values of $C=0{\cdot}7$ indicate the boundaries between dry steppe and typical steppe, and values of $C=1{\cdot}0$ constitute the boundaries between steppe and forest steppe.

The main advantage claimed for the hydrothermic coefficient of Selyaninov is its simplicity and the general availability of the meteorological data required for its calculation.

Meigs (1953) prepared a basic map of the distribution of semi-arid and arid zones in the world, based on certain refinements of the Thornthwaite system. Major climatic divisions are based upon degree of aridity (extremely arid, arid, and semi-arid) using Thornthwaite's 'moisture index', as follows:

	Moisture index
extreme arid climate	–57 or outer limit
arid climate	between –40 and –57
semi-arid climate	between –20 and –40

The outer margin of the semi-arid region, –20, is approximately the outside limit beyond which inadequate precipitation is not the main agricultural risk; the dividing line between semi-arid and arid, –40, separates lands where rain-fed arable cropping can still be attempted with appropriate dry-farming techniques, from those in which the risk of aridity is too great. In the extreme arid climate, at least twelve consecutive months without rainfall have been recorded and there is no regular seasonal rhythm to the rainfall.

Subdivisions are based on temperature, using the means of the coldest and warmest months, in the following ranges: below 0°C; between 0 and 10°C; between 10 and 20°C; between 20 and 30°C; and above 30°C.

A further subdivision is based on season of precipitation, as in Köppens' classification.

The *cold deserts* of Alaska, Siberia, and the Western part of the Tibetan plateau, are excluded by Meigs. These are the areas where the mean temperatures in the warmest months are less than 10°C, and where even irrigation cannot ensure a reasonable level of productivity.

A semi-arid climate is not necessarily *intermediate* between dry and humid, but can be a mixed climate, in which a completely dry season alternates with a fairly moist season. For example, Jerusalem, with its semi-arid climate, has an average annual rainfall of 600 mm, which is practically identical to that of London (620 mm); yet the former has a completely dry season of six to seven months, whilst the latter has a humid climate practically throughout the year (Figure 2).

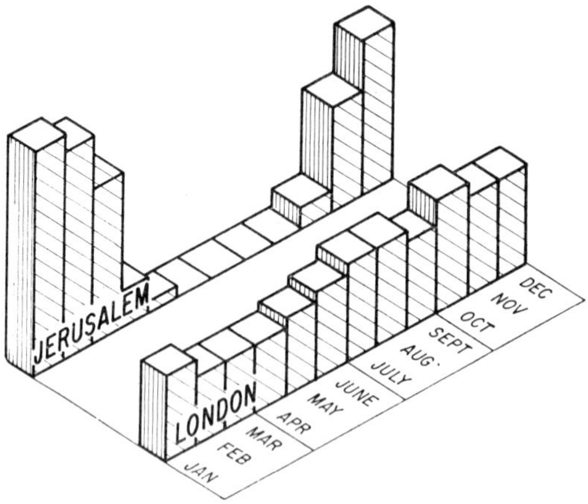

Figure 2. Typical seasonal patterns of rainfall distribution in semi-arid (Jerusalem) and humid (London) climates with approximately the same annual precipitation rates (averaging 600 and 620 mm respectively).

Causes of Aridity

Rainfall generally occurs when the moisture-laden air rises, expands, and cools. As a result, vapour condenses and precipitation occurs.

The distribution of world rainfall is influenced by a great number of factors, of which the principal ones are: the characteristics of the general circulation of the atmosphere, relief, and the distribution of land and sea (Butzer, 1961).

General circulation of the atmosphere

The earth's moisture is involved in a cycle of continuous circulation (Figure 3) called the 'water cycle' or 'hydrologic cycle'. Water is evaporated from the

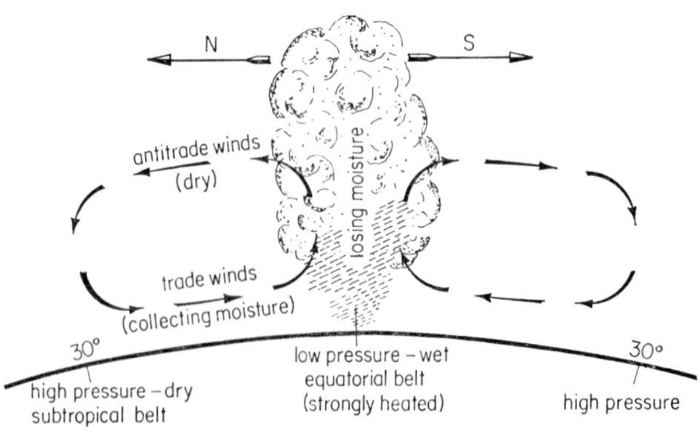

Figure 3. Schematic presentation of the circulation pattern of the atmosphere between the equator and 30° latitude north and south.

soil and from the water surfaces by solar energy, and transported as vapour by the general circulation of the atmosphere. Eventually the moist air is cooled to the point of condensation and may fall as precipitation.

The basic cause for the general circulation of the atmosphere is the unequal heating of the earth by the sun. The polar regions have an annual heat deficit, whilst the equatorial belt receives more heat from the sun than is lost by radiation into space. This temperature gradient between the equatorial belt and the polar caps provides the energy to set in movement a general circulation of the atmosphere, transporting heat from low to higher latitudes, and thereby mitigating the disparities in heat distribution over the earth.

Expressed in its most simple terms, the air heated over the equatorial belt would rise, flow polewards in the upper levels of the atmosphere while progressively losing heat by radiation, and then subside over the poles. The cooled air would then return to the Equator near the earth's surface.

This theoretical circulation pattern would be along the meridians of longitude, as its basic causes are the temperature gradient along the meridians. However, experience shows that there is not a mere single circulation system conforming to the above oversimplified pattern; instead there are mainly three systems, essentially zonal in character, that result in a consistent pattern of winds and precipitation the functioning of which is still not fully understood.

The breakdown of the expected meridional circulation system into a number of zonal systems is due to the combined effects of: (a) The earth's rotation, which causes a deflection of all moving air—clockwise in the northern hemisphere, and counter-clockwise in the southern hemisphere, (b) The existence of a number of *distinctive pressure belts*:

1. In the *equatorial region* there is a belt of relatively low and very uniform pressure, called the Doldrums.

2. In the *subtropical latitudes* there is a broad belt of high uniform pressure. In the northern hemisphere this high-pressure belt lies approximately between latitudes 25 and 35°N, and is confined mainly to the oceans. In the southern hemisphere, the belt is approximately along the 30°S latitude, and encircles the entire globe (Hare, 1961).

3. In the vicinity of the *Arctic and Antarctic Circles*, there are belts of low pressure.

4. Relatively high pressure occurs at the *two geographical poles*.

The combined effect of the meridional temperature gradient, the earth's rotation, and the pressure belts, results in three main circulation systems of the earth's atmosphere:

1. *Equatorial Low-pressure Belt*

The trough of low pressure in the equatorial belt causes surface air to move in from the subtropical high-pressure belts. As this air flows over warm oceans and tropical jungles which evaporate huge quantities of water, it becomes moisture-laden. The heated land-mass of the equatorial region in turn heats the damp surface air, which expands and therefore becomes lighter and rises. As the heated air rises it continues to expand, thereby using up some of its energy and becoming progressively cooler. When the air has been cooled to below dew-point, cloud formation occurs. Continued cooling then causes precipitation, resulting in heavy rainfall in the afternoon or evening. This occurs practically all the year around, and is typical for the equatorial belt.

2. *Subtropical High-pressure Belts*

The air that has ascended from the equatorial belt, after losing much of its moisture, flows poleward north and south in the upper regions of the atmosphere. By the time it is above latitude 30°N or S, the airstream is deflected by the earth's rotation. This causes an accumulation of air above these latitudes (called the Horse Latitudes) and produces belts of high pressure on the earth's surface.

Those parts of the airstream that have lost their heat descend slowly over the high-pressure subtropical belts. The slowly descending air is so dry that there is no condensation, the sky remains clear, and practically no rainfall occurs. Thus large arid and semi-arid zones are created in these latitudes. In winter, storms from the higher latitudes occur occasionally, resulting in moderate winter rainfall.

The Trade-winds. In addition to air circulation in the upper atmosphere, surface winds are also involved in determining the rainfall régime of the dry region. Because of the subtropical high-pressure, surface winds blow out of the subtropical highs to the equatorial low, moving in to replace the air that has ascended aloft at the Equator in the manner described above. These surface winds, instead of being northerly and southerly in the northern and southern hemispheres, respectively, are deflected by the earth's rotation—to the right in the northern hemisphere and to the left in the southern hemisphere, thereby becoming north-east and south-east winds. They are called the 'trade-winds' or, simply '*trades*'.

These winds are steady and moderate, carrying moisture away from the subtropical belt. As the air moves from the subtropical belt towards the Equator, its temperature rises and its relative humidity remains low. The skies are clear, the weather is warm, and only light rainfall occurs. Wherever mountain ranges force the trade-winds to rise, rainfall occurs on the windward side, whilst the leeside remains arid. As the trades approach the Equator they become progressively more humid and unstable; however, they do not actually reach the Equator, but rise at some distance from it. As a result, a narrow low-pressure belt is formed between the two belts of trade-winds—the so-called doldrum or equatorial calms, in which the air is generally calm or the winds are very light.

The general circulation of the atmosphere, involving the equatorial low-pressure belt and the subtropical high-pressure belts, explains why, on a world-wide scale, the major arid and semi-arid regions are generally found in the zones between latitudes 20° and 30° in each hemisphere—in particular in the western parts of the continents, where upper-level subsidence is particularly strong. The aridity even affects rainfall over the oceans in these latitudes, proving that a process effectively preventing precipitation is involved.

In these zones are found the great tropical and subtropical deserts. The eastern sides of the continents are the windward sides; they are under the influence of maritime air currents and so their air masses are relatively warm and humid, causing relatively abundant rainfall. These regions may therefore be considered as a poleward extension of the rainy tropics (Trewartha, 1954).

3. *The Polar Circle Low-pressure Belts*

Air moves polewards from the subtropical high-pressure belts to the sub-polar lows mainly at the surface. As a result of deflection by the earth's rotation, the winds develop into the so-called 'mid-latitude westerlies' or, simply, '*westerlies*'. These are neither constant nor steady in direction, as they are affected by various secondary circulation factors. The regions of prevailing westerlies lie between 35° and 60° in each hemisphere, and comprise the cool, humid climates of the earth.

Briefly it can be said that the circulation systems of the atmosphere, which tend to be zonal in character, are the basic determinant of world climates. These climates accordingly tend to occur in latitudinal zones, as was already observed by ancient civilizations.

RELIEF

When rising ground, or mountains running at right-angles to the prevailing winds, deflect moisture-laden air upwards, the air cools, resulting in condensation and subsequent precipitation. This is called 'orographic' rain (Figure 4). When the air moves down the leeside of the mountain range, it has already lost most of its moisture. During subsidence, the air again becomes warmer and its relative humidity is decreased further, which explains why the leeside of a mountain has little rainfall. Examples of this so-called rain-shadow effect are the Patagonian Desert, which lies in the rain-shadow of the Andes, and much of the Australian Desert behind the Queensland coast, which is situated in the rain-shadow of the mountains that lie across the path of south-easterly winds.

High plateaux also tend to be dry, because here the air has lost most of its moisture before reaching these high altitudes.

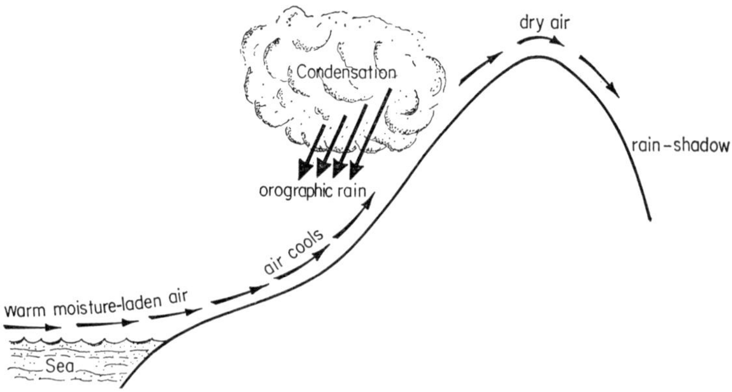

Figure 4. Effect of relief on rainfall 'orographic' rain.

DISTRIBUTION OF LAND AND SEA

Winds may also arise as a result of the seasonal heating and cooling of the continents, which warm up rapidly in summer and cool rapidly in winter. Such winds are the so-called '*monsoons*', which in summer are moist and blow landwards whereas in winter the direction is reversed. This latter outflow of dry air is characterized by lack of precipitation.

The greater part of the earth is covered by oceans; the distribution of land and sea will, therefore, have a marked effect on climatic patterns only in regions in which large land-masses are involved—such as Asia and, to a lesser degree, Africa, the Americas, and Australia—and in which the zonal circulation pattern may be entirely obscured by the great thermal differences between land and sea.

Air moving over warm oceans becomes moisture-laden. When it moves inland, any factor that causes this air to ascend, such as rising ground or a wedge-shaped mass of cold air, will result in cooling, condensation, and precipitation (Figure 5). Land areas near warm ocean waters are characterized by high temperatures, high air humidity, and heavy rainfall. However, the moisture-laden winds progressively lose most of their moisture whilst passing over the land-mass, so that the continental interiors, far removed from the warm oceans, are generally dry. The Gobi Desert is an example of aridity due to sheer distance from oceanic moisture. A characteristic of the continental arid regions is that precipitation may occur at all seasons, with a tendency for there to be more rainfall in the summer than in the winter. In winter, the low temperatures reduce the water-vapour capacity of the air, whilst in summer there is an indraft of warm, humid air from the oceans, with the surface-heating of the air causing convective activity which occasionally results in precipitation.

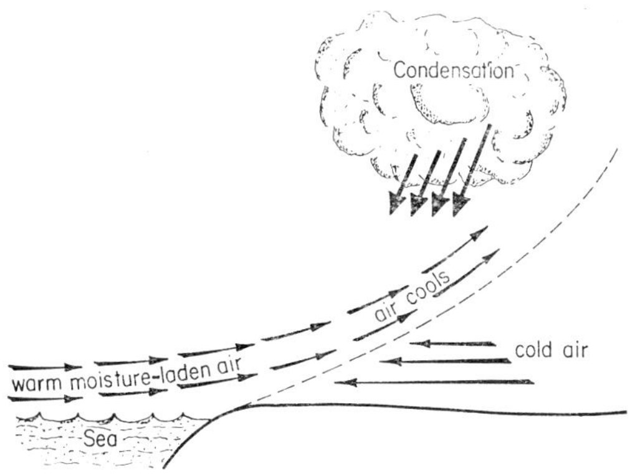

Figure 5. Effect of warm moisture-laden air encountering a cold air-front.

However, coastal areas may also be arid if they border on oceans with cold currents flowing parallel to the coast—especially if the normal temperatures of the region are high. This is a regular characteristic of the tropical and sub-tropical west coasts along which currents, which have originated over the poles, flow towards the Equator. Warm air moving landwards from far out in the ocean is cooled when it moves over the cold water of the ocean near the shores; as a result of the drop in temperature, heavy fog may be formed. As the air which has lost much of its humidity moves inland, it is warmed by the land surface, and its relative humidity is also lowered. Coastal regions with cold water along their shores are therefore arid or semi-arid, notwithstanding the frequent morning fogs near the coast. The deserts of Chile, Peru, and Southwest Africa, are typical examples of aridity caused by cold oceanic currents.

SUMMARY

Of all the factors which lead to aridity, the two most important (McDonald, 1959) are the low frequency of occurrence of an adiabatic rise of large bodies of air, and the remoteness from an oceanic moisture source in the prevailing wind direction. The principal arid regions of the world occur where one or both of these factors operate:

1. In the flat trade-wind deserts of Africa (Sahara), Arabia, and Australia. The primary reason for the aridity of these regions is the chronic lack of lifting processes. This is indicated by the fact that within these regions are isolated 'humid islands'—e.g. isolated mountains, such as the Atlas of North Africa, which, by causing orographic lifting, have more precipitation than the surrounding plains (McDonald, 1959).

The reasons for the scarcity of dynamic lifting processes in these low-latitude arid regions are:

(a) They are too far equatorward to be strongly influenced by the extra-tropical wave cyclones that increase precipitation in the regions lying polewards of the deserts.

(b) On the other hand, they are too far poleward to benefit from the summer poleward movements of the humid equatorial air.

(c) They are also too far westward in their land-masses to be affected by the hurricane activity of low-latitude east-coast littorals.

2. In the interior of continents, especially in areas enclosed within mountain-barriers (Mongolia, southwestern Siberia, the Great Plains of the U.S.A.).

3. On the tropical and subtropical west coasts of continents (Peru, northern Chile).

4. On leeward mountain slopes (Patagonia, Australian desert).

5. On high plateaux (Tibet).

The reason for aridity in these regions is the remoteness from an oceanic moisture source. This remoteness may be due either simply to distance, or to the so-called 'hydrometeorological remoteness' caused by mountain-barriers that block the inflow of moisture-bearing winds (rain-shadow deserts) (cf. McDonald, 1959).

Types of Dry Climates

The dry regions constitute about one third of the land surface of the world. They are divided almost equally into arid and semi-arid regions (Koeppe & Long, 1958). The dry regions occur in five large zones, separated from each other by oceans or wet equatorial zones (Meigs, 1953). The largest of the five is the North African–Eurasian dry zone, extending from the west coast of Africa and reaching into India and farther north-east. The others are in South Africa, North America, South America, and Australia.*

ARID HOT CLIMATES

These lie in the vicinity of latitudes 20–25° in both hemispheres, with extreme margins generally at 15° and 35°. They are located on the western, leeward sides of the continents. The Sahara, an apparent exception, is really an extension of the south-western Asian desert region.

SEMI-ARID HOT CLIMATES

On the poleward fringes of the hot deserts are the transitional zones of precipitation whose amounts lie between those of arid and humid climates.

* *See* the 'continental' maps later in this work, indicating the arid and semi-arid areas in Africa (Fig. 6, p. 14), Eurasia (Fig. 7, p. 18), North America (Fig. 8, p. 21), South America (Fig. 9, p. 24), and Australia (Fig. 10, p. 25), and their accompanying accounts—Ed.

Thus on the west coasts are regions with dry summers and cool, moist winters (Mediterranean-type climate—*see below*), while on the desert fringes on the equatorial side are the semi-arid regions with high temperatures but enjoying some of the hot-season rainfall of the tropics.

MEDITERRANEAN-TYPE CLIMATES

At latitudes of around 30° to 40°, on the west sides of the continents, are located the areas with a Mediterranean-type climate. In terms of latitudinal locations and temperatures, this is a subtropical climate, being dry in the summer and cool and wet in the winter. In summer, all the winds and high-pressure belts 'move polewards with the sun', and as a result the dry regions extend to the higher latitudes, up to 40°—hence the dry summer which is characteristic of these lands. In winter, when the sun 'retreats towards the Equator', the high-pressure and wind belts receive occasional rain—hence the moderate winter rainfall.

Mediterranean climates are found in the regions bordering on the Mediterranean Sea, in central California, in central Chile, along a narrow coastal plain near the south-west tip of Africa, and in small areas in western and southern Australia.

The Mediterranean climate is characterized by mild winters and hot, but not oppressive, summers. Frosts and snow are rare. Rainfall is light to moderate, and generally occurs in heavy downpours. Sunshine is abundant even during the winter months.

MIDDLE-LATITUDE DRY CLIMATES

Eastwards of the above, in the middle latitudes, lie the dry regions with cold winters and essentially summer rainfall. The margins lying towards the summer-dry subtropics have winter rainfall (Critchfield, 1966), which is characterized by heavy showers separated by long dry periods. Rainfall efficiency is generally low, because of the low air humidity, intense sunshine, and strong winds

Summer temperatures may be as high as in the tropical deserts (Critchfield, 1966). Surface wind-speeds are generally high because of the width of the plains and plateaux and the lack of obstructions.

The distribution of these dry regions is related to their distance from the sea and to the presence of mountain barriers.

THE DRY REGIONS OF THE WORLD

AFRICA

Northern Africa is characterized by the unusual longitudinal breadth of the dry regions and their extension towards the Equator, in contrast to the situation south of the Equator. As a consequence, in Africa (Figure 6), the transitional wet-and-dry climates are much narrower on the northern side of the Equator than on the southern side (Trewartha, 1961).

The Sahara and Its Margins

The Sahara is the largest desert of the world and extends practically across the African continent between approximately 15° and 33°N, covering some 9

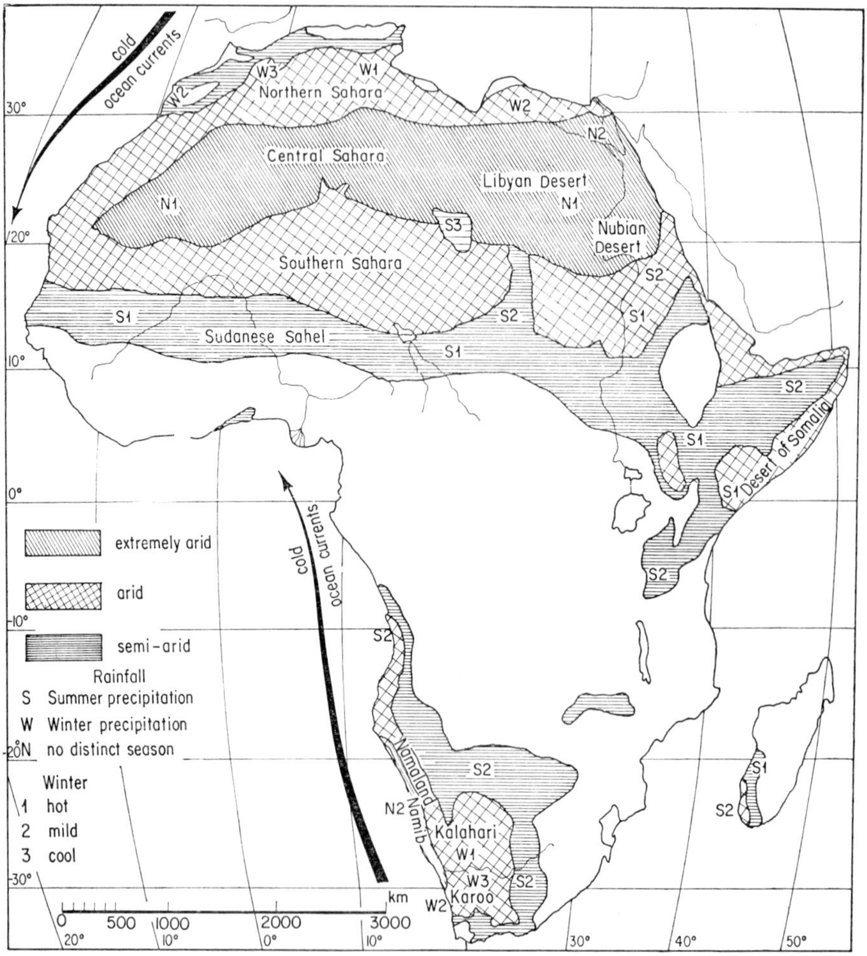

Figure 6. Map of the arid and semi-arid regions of Africa. (After Meigs (1953): by courtesy of the United Nations Educational, Scientific and Cultural Organization.)

million square kilometres. In reality the Sahara is only a part of a still larger area of dry tropical–subtropical climate, extending from the Atlantic Ocean across northern Africa, Arabia, and western Asia, to northwestern India–Pakistan (Trewartha, 1961) (cf. our Figures 6 and 7).

The Sahara is in the trade-wind belt: dry north-east winds prevail over major sectors, becoming hotter and drier as they cross the land-mass. In the Sahara are large areas of barren, shifting sand, as well as stony wastes, hills and mountains, valleys and depressions. There are also a number of oases, the largest of which is the narrow valley of the Nile.

There is a great range in diurnal as well as annual temperatures, and frosts may occur in the winter months. The intense heating by day, contrasted with the cooler nights, results in strong winds during the day and generally calm nights; dust storms are frequent. The greater part of the Sahara has less than 100 mm of annual rainfall, mostly occurring in heavy showers. The central, extremely arid core, comprises the Central Sahara, the Libyan Desert, and the Nubian Desert. These are bordered by the less dry deserts: the northern Sahara, with a rainfall not exceeding 200 mm and occuring in winter, and the southern Sahara, with rainfall occurring in summer.

On the northern fringe of the Sahara is a narrow strip of steppe, stretching across Morocco and Algeria, and marking the transition between the desert on one hand and the coastal strip with a Mediterranean climate on the other. South of the Sahara is a tropical area, the Sudanese Sahel, which is a narrow region of steppe separating the desert from the tropical savannas bordering on the rain forests. It has a semi-arid climate, with a short rainy season during the summer months (June to September) and the rain falling mainly in thunderstorms. Its most striking characteristic is the lack of important relief and, therefore, of consequent weather modification.

The northern boundary of the Sudanese Sahel is the isohyete of about 250 mm of mean annual rainfall and a mean 'moist period'* of 55 days; it has the climatic characteristics corresponding to settled millet cultivation.

There is no precise boundary of the semi-arid zone to the south. It gradually merges into the humid savannas and forests, characterized by a wet season that exceeds the dry season in length.

This Sudanese Sahel includes portions of the territories of Senegal, Mali, Upper Volta, Mauritania, Ghana, Niger, Togo, Dahomey, Nigeria, Cameroun, Chad, and Sudan (Cochemé & Franquin, 1967).

Along the east coast of tropical-equatorial Africa, between about 12°N and 2–3°S of the Equator, is a dry region, comprising the desert of Somalia, eastern Ethiopia, and northeastern Kenya (Figure 6). The aridity of this region

* 'Moist period or 'moist period plus reserves' refers to the season during which an annual crop can be grown with dry farming.

is considered by Trewartha (1961) to be a 'climatic abnormality', because an equatorial dry climate along an eastern littoral is unusual. In such a latitudinal and geographical location a tropical humid climate generally prevails. The situation of this region on the windward side of the continent, with highlands rising abruptly interior to the coast, would also appear to warrant greater precipitation than is actually received.

South of the Equator in East Africa the water deficiency is generally not as severe as it is in corresponding regions to the north. Only a small area in eastern Kenya is really desert. However, semi-arid climates with unreliable rainfall are widespread, both along the coast and in the interior.

No satisfactory explanation for the rainfall deficiency over East Africa is yet available (Trewartha, 1961).

Southern Africa

According to Trewartha (1961), the most striking climatic peculiarity in Southern Africa is the large proportion of the total area in which dry and semi-arid climates predominate (Figure 6).

Desert conditions prevail along the coast of southwestern Africa, from Angola to the Cape Province. Whatever rain does occur falls in summer in the north, and throughout the year in the south.

This region comprises the extremely arid Namib Desert, and the less dry Namaland, Kalahari, and Karoo, covering some 550,000 square kilometres in all.

The Namib Coastal Desert belongs to the class of deserts which are extremely dry as a result of being located along the tropical and subtropical western littorals bordered by cool ocean currents and colder upwelled water (Trewartha, 1961). The cold currents off the shore cause frequent fogs, high air humidity, and relatively mild temperatures, and yet the Namib Desert is characterized by intense aridity. In a narrow coastal strip some 30 to 50 km in width extending over a latitudinal stretch of about 15°, the annual rainfall is under 125 mm. The lowest precipitation coincides with the coastline and there is a slow increase inland (Trewartha, 1961).

Two-thirds of the Republic of South Africa are in the arid or semi-arid area consisting of four main plateaux (Dyer, 1955):

The Coastal Plateau—ranging from sea-level to 300 m elevation, with a winter rainfall of up to 120 mm. A cold ocean current flowing up the coast causes fogs and heavy dews.

The Second Plateau—with an elevation of 300 to 600 m, is surrounded by mountains. The western part is the Little Karoo, with a rainfall of 150 to 200 mm.

The Third Plateau—to the north of the former, has an elevation of 600 to 900 m and is called the Great Karoo. It has 120 to 200 mm of rainfall, mostly in winter.

The Central Plateau—is a large inland area with an elevation of 900 to 1500 m, called the Upper Karoo. The rainfall averages 125 to 500 mm, mostly in summer precipation. Winters are cold, with frequent night frosts. Grazing is the predominant land-use (Walter, 1962).

To the north, the central plateau descends into the Kalahari sand veld.

EURASIA

South-west Asia has vast areas of hot desert extending from Arabia to western India (Figure 7). Most of Arabia is a hot and barren desert plateau covering approximately 2·5 million square kilometres; it has very little rainfall and what there is occurs mainly in the winter months. The northern part is a stony or sandy desert producing some pasturage; in the centre are mountain chains with alluvial cultivated valleys, whilst the south is in the rain-shadow of these mountains and is therefore extremely arid desert (the Rub el Khali). In the region between the lower Tigris and Euphrates is a hot desert which was once widely irrigated. The scanty precipitation occurs in winter. The region has rich alluvial soil, and was extensively cultivated in the past. Only a small part is cultivated at present; thc reasons for the decline that has occured are described in chapter V.

The Desert Zone of India and Pakistan

The arid zone of India and Pakistan covers an area of some 750,000 square kilometres. In the north it takes in the southern part of the Punjab, on the east it is bounded by the Aravalli Mountains, in the south it reaches the coast, and in the west it reaches the western border of West Pakistan (Bharadwaj, 1961). The regions comprising this arid zone include Sind, Baluchistan, all Rajasthan west of the Aravalli Mountains, the south-western tip of Punjab (India), south-western Punjab (Pakistan), and most of the North-west Frontier Province.

The most important source of irrigation water in this region is the Indus River system, so that most of the cultivated land is restricted to the area through which the Indus River and its tributaries flow, or in which their waters are carried by canals. The Indus plains have one of the world's largest and oldest irrigation systems, including some 12 million hectares of irrigated land, with thousands of kilometres of canals. The annual water-flow of the catchment area of the Indus and its tributaries is more than double that of the Nile (Kaser & Allison, 1967).

Winters are cool, with occasional frosts. The hot season begins in early April; June is the hottest month, with day-time temperatures reaching 50°C or

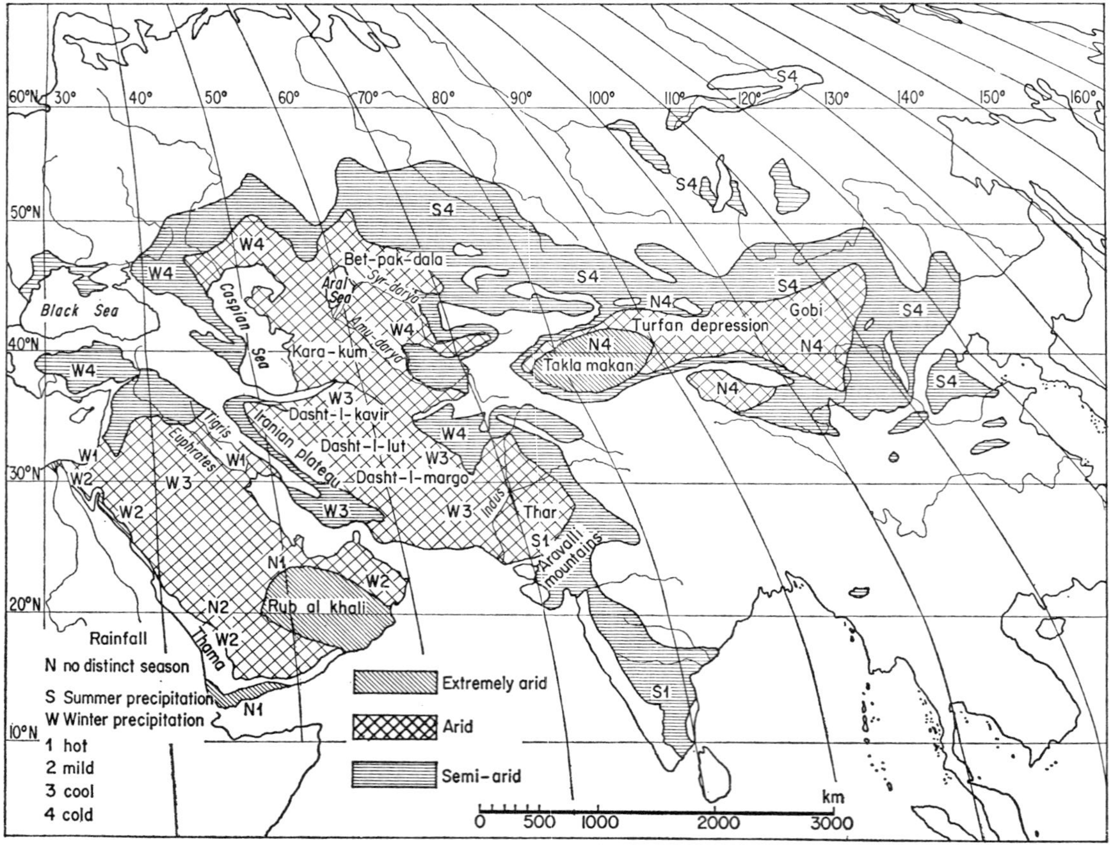

Figure 7. Map of the arid and semi-arid regions of Eurasia. (After Meigs (1953): by courtesy of the United Nations Educational Scientific and Cultural Organization.)

even higher during heat-waves. Relative humidity decreases with the rise in temperature; it is generally 35–60% during the morning hours and 10–30% at noon. Monsoon showers, accompanied by a drop in temperature, occur in September.

The length of the rainy season and the annual amount of rainfall decrease from east to west; average annual precipitation varies from 150 mm in Sind to 250 mm in south-western Punjab and 275 mm in western Rajasthan. More than 90% of the precipitation occurs during the hot monsoon period and is therefore not very efficient. Winter rainfall, which occurs mostly in the north-western tip of the desert, ranges from 60 to 125 mm and decreases gradually towards the south, where it is less than 25 mm.

In the southern half of Arabia is a highland region bordering on the Red Sea coast that has a warm, semi-arid climate with prevailing summer rains. A second highland strip in south-east Arabia, near the Gulf of Oman, has a similar climate.

Middle-latitude Dry Climates of Eurasia

A highland area including northern Iraq and Syria and most of Asiatic Turkey extends from Anatolia to the Iranian plateau. The area has short dry summers, long cold winters, and scanty rainfall—most of which falls in winter (Figure 7).

The Iranian plateau itself, lying between the Indus and the Tigris valleys, is surrounded by mountain ranges with scanty precipitation—mostly falling in the winter months and including a fair amount of snow.

In south-western Afghanistan and north-western Baluchistan are desert plateaux and hilly steppe regions. Temperatures are very high in summer and very low in winter; humidity is low and sunshine is plentiful.

Farther to the east and north, in those parts of Eurasia which are the farthest away from oceans, between latitudes 35° and 50°, is a huge arid and semi-arid plain, stretching from the Black Sea to Manchuria. Over two-and-a-half million square kilometres of the Soviet Union are steppe and desert, of which the central Asian deserts account for almost two million square kilometres (Manton, 1954). Annual rainfall ranges from zero to 150 mm in the desert, and from 150 to 300 mm in the semi-desert areas (Rode, 1955). The region comprises the extremely arid great deserts of Central Asia, including the Aral-Caspian Depression—an immense lowland, barren, cold desert—the Takla Maklan desert of Sinkiang, and the Gobi Desert of Mongolia (Figure 7). These are deserts of drifting sand, with cold winters, and very hot summers having scanty precipitation.

Only two great rivers traverse the central Asian deserts: the Syr-Darya, the waters of which are used for irrigation in Uzbekistan, and the Amu-Darya, which is used for watering land in Turkmenistan (Manton, 1954).

In the desert zone within the U.S.S.R. (Serozem soil region), average July temperatures reach 26–30°C. The combination of high summer temperatures and low rainfall results in very low relative humidity (20–30% on the average) and high potential evapotranspiration, which may reach 2000 mm per year (Rode, 1955). Winter temperatures are low, but increase from north to south ranging from −12°C average minimum temperature in January in the north to −2°C in the south. However, very low minimum temperatures occur throughout the zone, and may reach −30°C even in the south.

Surrounding these deserts are extensive steppe regions comprising much of the Black Sea area, of the Volga region, of Azerbaijan, of Turkmenistan, of Uzbekistan, of Kazakhstan, of Mongolia, and of Manchuria. These regions have sparse rainfall throughout the year, with hot summers and warm dry autumns. The climate of the northern belt becomes more severe from west to east, with winters becoming longer and more severe and the snow-cover thinner. Summers are too cool in the west, and too short in the east, to justify intensive irrigation. The main characteristic of the greater part of these regions is the very cold winter, with minimum temperatures that may fall as low as 25–30°C below zero, resulting in the soil freezing to various depths. Winters are milder in the south-western part—the Turanian depression, the Caucasus, and southern Ukraine. A considerable proportion of the Eurasian steppe region consists of highly fertile, vast alluvial plains. These result from age-old deposits from the large number of great rivers traversing the region with their tributaries, and include black earth zones and deep brown soils with a thick grass cover.

In the Russian steppes (Chernozem zone) the average July temperature fluctuates between 20 and 23·5°C. Average annual precipitation is between 300 and 450 mm, with a maximum in early summer (June) and with fairly long dry periods between rains. Another summer characteristic is the frequency of the dry and strong east or south-east winds, which may cause the relative humidity to drop very rapidly to 15–20 per cent and the temperatures to rise to 40°C (Rode, 1955).

In the drier steppes (chestnut and brown soils zone) the average precipitation is 250–350 mm, with one-quarter to one-third falling in summer. The main sources for water storage in the soil are the autumn rains and the spring snow-thaw. Most of the summer precipitation evaporates, and very little run-off occurs (Vilenskii, 1957). Average July temperatures range from 20–25°C, and January temperatures from −7°C to −17°C. In eastern Kazakhstan, minimum temperatures of −40°C have been recorded (ibid.).

NORTH AMERICA

The arid and semi-arid areas of North America lie almost wholly west of the 100th meridian (cf. Figure 8). There are two rainfall régimes: winter rains prevailing in the west and summer rains in the east. Between the two zones

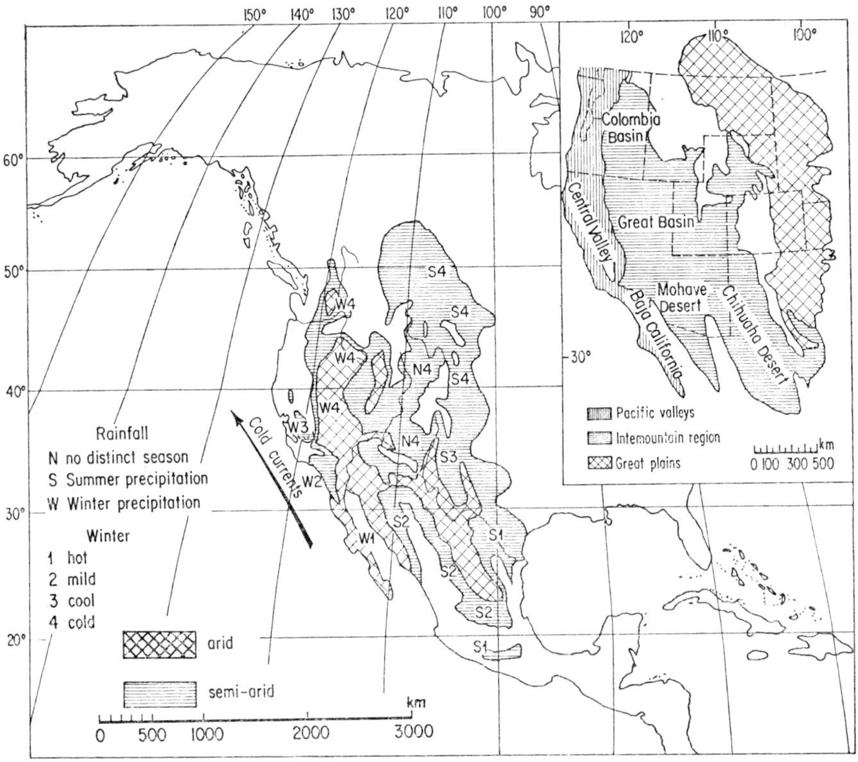

Figure 8. Map of the arid and semi-arid regions of North America. (After Meigs (1953): by courtesy of the United Nations Educational, Scientific and Cultural Organization.)

lies a large area, in south-east California and south-west Arizona, which seldom receives either winter or summer rainfall except in small amounts, and is therefore the driest part of the region (Logan, 1961).

The dry regions of North America can be divided into three major groups: the Pacific valleys, the Intermountain region, and the Great Plains. The Sierra Nevada Mountain Range separates the Pacific valleys from the Intermountain region, whilst between the latter and the Great Plains are the Rocky Mountains.

The Pacific Valleys

The Central Valley of California has a climate that varies from semi-arid in the Sacramento Valley to the north, to arid in the more southern San Joaquin Valley. Almost all of the rainfall occurs in winter. The extreme heat and low relative humidity are the reasons for the very high evapotranspiration rates that characterize this area, which is mostly irrigated.

In the south, there are relatively small areas of hot deserts, comprising the Imperial and Death Valleys of southern California and the Salt and Gila Valleys of Arizona. Still farther south is the coastal desert of the Mexican peninsula of Baja California, with hot summers and near-tropical winters. Rainfall in the north is of the winter Mediterranean type, while in the south it is mainly of tropical origin and concentrated in summer. The driest part of Baja California is in the centre.

The Intermountain Region

The Intermountain region lies in the rain-shadow of the Sierra Nevada mountain range. Summer temperatures are high; winters, whilst cold, have temperatures that are somewhat milder than might be expected at this latitude. A dominant climatic feature is the biannual rainfall variation: one maximum occurring in winter, the second during the warm season. The biannual variation in rainfall is most marked in central Arizona, with a decrease in amplitude in all directions from this centre (Trewartha, 1961).

The Intermountain region is divided into two main sectors: the northern part is a cool-desert (the so-called 'sage-brush zone') and the southern part is a warm desert (the creosote-brush zone). Within the sage-brush zone are a number of distinct sub-regions, such as the Columbia Basin in the North and the Great Basin in the centre of the region.

The Columbia Basin is a high-altitude, dry plateau, though on the fringes of its arid core is a semi-arid belt that is largely cultivated with winter wheat. The Columbia River skirts around the northern and western margins of this sub-region. However, because this river flows through a very deep canyon and therefore lies far below the plateau surface, the great volume of flowing water is not easily available for irrigation.

The Great Basin is typical range country; grazing is seasonal and extensive, and settlement is extremely sparse. The western part, lying in the rain-shadow of the Sierra Nevada, gets the least rainfall, with annual averages of about 125 mm or less. Summer showers become more frequent with increasing distance from the coast (MacPhail, 1963).

The warm desert, or creosote-brush zone, is characterized by very low rainfall and very hot summers. In the south-west are the Mohave and Sonoran Deserts. The western part of the Mohave Desert gets winter rainfall. The Sonoran Desert has mild winters and hot summers, and a frost-free growing period of not less than 300 days, permitting the production of 'off-season' winter food-crops.

Eastward lies the high-altitude Chihuahua Desert, with snow in winter and occasional torrential summer rains.

The Great Plains

The Great Plains are the climatic analogues of the Russian steppes, stretching in a belt from about 22° to 54°N, and thus reaching into Canada. They form the largest area of grassland of North America. Whilst winter temperatures vary widely, averaging from 10–20°C in the south to below 0°C in the north, the differences in summer temperature are far less marked. The amount of insolation and rainfall, and the relative humidity of the air, vary very little from south to north. In general, precipitation is greatest along the eastern margin of the Southern Plains. Most of the rain occurs in summer. High winds and the long hours of sunshine contribute to high evapotranspiration rates and reduce the effectiveness of precipitation, so that light rains are of little value. The winters are dry, with clear skies and the ground mostly frozen. Soil-water recharge is therefore confined mostly to the period in the autumn before freezing, and to early spring after thawing (Shaw, 1965). Crop production is precarious and the rural population very sparse.

On the northern fringe of the Great Plains is the Canadian dry belt, comprising southern Alberta and Saskatchewan, in which rainfall is still lower than normal for the general region, and falls to as low as 300 mm. This part of the region also has a higher degree of rainfall variability, a longer duration of droughts, and a higher ratio of evapotranspiration to precipitation than the general region (Trewartha, 1961).

SOUTH AMERICA

The arid zone of South America forms a long narrow strip along the western coast between the sea and the Andes, reaching almost to the Equator (Figure 9). Included is the Atacama Desert, in Peru and northern Chile, which is an almost rainless region considered by some to be the driest area on earth (McDonald, 1959). From northern Peru (approximately 5° to 6°S) to northern Chile (approximately 26°S) the highest average rainfall in this zone is under 50 mm, while in parts no rainfall has been recorded over periods of one or more decades. At best, effective rains occur at intervals of several years. The extreme aridity is surprising, in that no part of this desert is far from a tropical ocean.

A unique feature of this arid region is that occasionally, once in several years, frequent torrential rains may occur during one or more summer months, so that, practically overnight the desert takes on the characteristics of an area of tropical wet climate. These heavy rains are extremely destructive, as they flood the irrigated valleys, destroying crops, roads, bridges, and even entire villages (Trewartha, 1961). Despite its rainlessness, the coastal strip is extremely foggy, because of a cold ocean current that parallels the coastal desert.

In the rain-shadow of the Andes is a semi-arid region extending inland to the east; this includes the pampas plains of the Gran Chaco, with mild

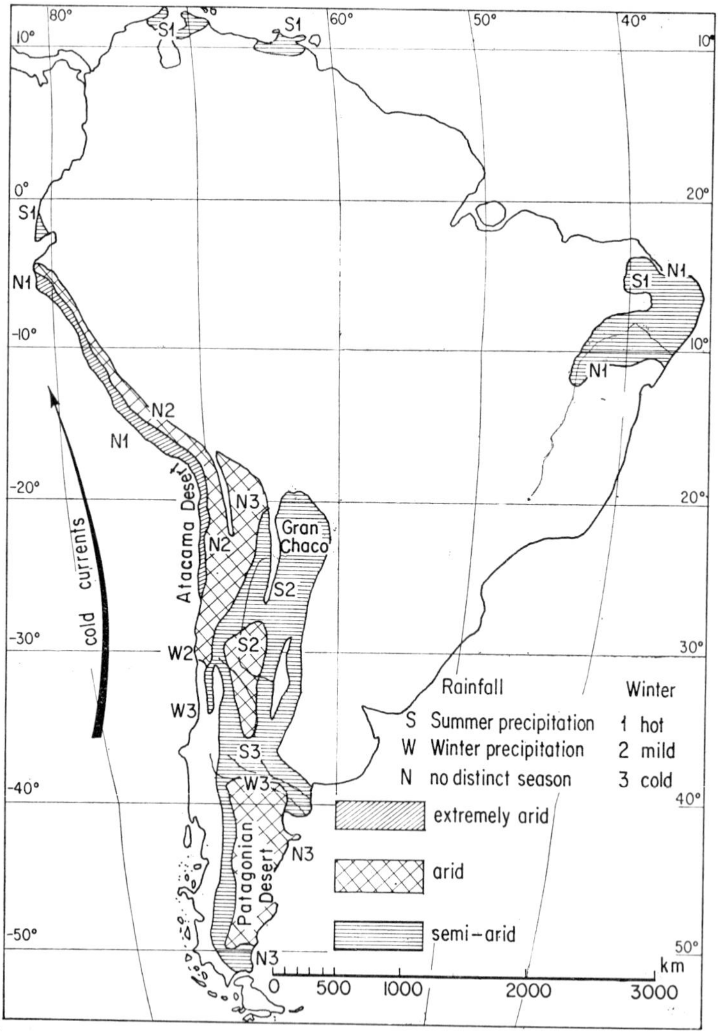

Figure 9. Map of the arid and semi-arid regions of South America. (After Meigs (1953): by courtesy of the United Nations Educational, Scientific and Cultural Organization.)

winters and moderate winter rainfall. Western and southern Argentina have a dry, mid-latitude climate (Critchfield, 1966). To the south is the Patagonian Desert which covers some 500,000 square kilometres, with cool winters and rainfall occurring throughout the year (Figure 9). Precipitation is low, however, totalling about 125 mm annually along the coast, but increasing somewhat inland, towards the base of the Andes (Trewartha, 1961). The Patagonian Desert is not barren, but produces pastures on which sheep are raised.

Much farther north, along the Colombian and Venezuelan coasts, are found arid and semi-arid areas with most of the rainfall occurring in summer and autumn, and with relatively few rainy days.

In northern Brazil there occurs an area of dry climate which is atypically situated, occupying an intermediate position between the winter-rainfall area to the east and the summer-rainfall area to the south. The area of drought varies considerably in extent and location from year to year; fairly extensive areas have under 600 mm of annual precipitation while there are restricted areas of less than 400 mm. The rains occur in the form of heavy showers extending over a small area at a time (Trewartha, 1961).

AUSTRALIA

Dry climates occupy the entire mainland of the continent of Australia except in its northern, eastern, and part of its southern, margins (Figure 10).

About 40 per cent of Australia, or some 3·2 million square kilometres,

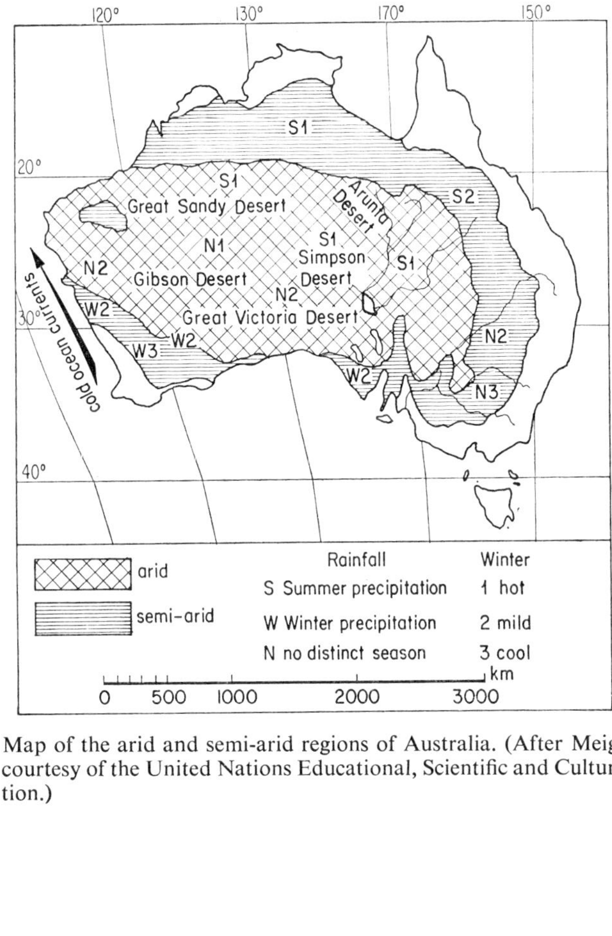

Figure 10. Map of the arid and semi-arid regions of Australia. (After Meigs (1953): by courtesy of the United Nations Educational, Scientific and Cultural Organization.)

consists of hot desert, with mild to hot winters. This desert occurs not only in the region of the subtropical belts of the trade-winds, but also in the rain-shadow of the mountain ranges of the eastern coast.

Although the Australian desert is one of the most extensive in the world, precipitation in it is higher than in other low-latitude deserts. Thus the desert parts of Australia average about 125 mm of rain annually (McDonald, 1959). Whilst in most other low-latitude deserts the annual rainfall increases inland, the opposite is true in Australia (Trewartha, 1961). The rains occur in the north mainly in summer, and in the south mainly in winter. Between these two regions lies a zone without a well-defined rainy season; the occasional precipitation can occur in any month of the year (Walter, 1962). The driest parts of the Australian desert average about 125 mm of rain annually (McDonald, 1959).

Around the central desert core of Australia, with the exception of parts of the west and south coasts, is a fairly broad ring of semi-arid lands with erratic summer rainfall and tropical temperatures, which separates the desert from the humid climates on the margin of the continent. Grazing is practised in the more favourable areas of these semi-arid lands.

References

BHARADWAJ, O.P. (1961). The arid zone of India and Pakistan. *Arid Zone Res.*, **17**, 143–73.

BUTZER, K.W. (1961). Climatic change in arid regions since the Pliocene. *Arid Zone Res.*, **17**, 31–56.

COCHEME, J. & FRANQUIN, P. (1967). A study of the agroclimatology of the semi-arid area south of the Sahara in West Africa. *Tech. Rep., F.A.O., Rome*, vii + 325 pp., illustr. (mimeographed).

CRITCHFIELD, H. J. (1966). *General Climatology*. Prentice-Hall, Englewood Cliffs, New Jersey, xi + 420 pp., illustr.

DAVITAYA, F. F. (1964). Climatic resources of arid zones and their agricultural use in the USSR. *Arid Zone Res.*, **26**, 31–36.

DE MARTONNE, E. *See* MARTONNE, E. DE.

DYER, R. A. (1955). Structural and physiological features of the vegetation of arid and semi-arid areas of the Union of South Africa. *Arid Zone Res.*, **5**, 20–24.

GAUSSEN, H. (1954). Théorie et classification des climats et des microclimats. *VIII Congr. Int. Bot., Paris*, Section 7 et 8, pp. 125–30.

HARE, F. K. (1961). The causation of the arid zone. *Arid Zone Res.*, **17**, 25–30.

KASER, R. F. & ALLISON, S. V. (1967). Use of modern equipment and techniques in the planning and development of water resources. pp. 509–24 in *Water for Peace*, U.S. Government Printing Office, Washington D.C., vol. 8, v + 827 pp., illustr.

KOEPPE, C. E. & LONG, G. C. DE (1958). *Weather and Climate*. McGraw-Hill, New York, vii + 449 pp., illustr.

KÖPPEN, W. & GEIGER, R. (1936). *Handbuch der Klimatologie*. Gebr. Borntraeger, Berlin.

LOGAN, R. L. (1961). Land utilization in the arid regions of southern Africa. Part II: South West Africa. *Arid Zone Res.*, **17**, 331–8.

McDONALD, J. E. (1959). Climatology of arid lands. *Arid Lands Colloquia, 1958–59,* University of Arizona, Tempe, pp. 3–13.

MACPHAIL, D. D. (1963). Regional setting. pp. 21–53 in *Aridity and Man* (Ed. C. Hodge). Amer. Assoc. for Adv. of Science, Washington D.C., xviii + 584 pp., illustr.

MANTON, S. M. (1954). Biological research and the productive transformation of steppe and desert in the Soviet Union. pp. 148–55 in *Biology of Deserts* (Ed. J. L. Cloudsley-Thompson). Inst. of Biology, London, iv + 224 pp., illustr.

MARTONNE, E. DE (1962). Une nouvelle fonction climatologique: l'indice d'aridité. *La Météorologie Nouvelle*, **2**, 449–58.

MEIGS, P. (1953). World distribution of arid and semi-arid homoclimates. *Arid Zone Res.*, **2**, 203–10.

RODE, A. A. (1955). *Soil Science.* Publishing House for the Wood and Paper Industry, Moscow: Goslesbumizdat, 517 pp., illustr. (Translated from Russian by 'Israel Program for Scientific Translation, Jerusalem'.)

SHAW, R. H. (1965). The prediction of soil moisture for the winter period in Iowa. *Iowa St. J. Sci.*, **39**, 337–44.

THORNTHWAITE, C. W. (1948). An approach toward a rational classification of climate. *Geog. Rev.*, **38(1)**, 55–94.

TREWARTHA, G. T. (1954). *An Introduction to Climate.* McGraw-Hill, New York, vii + 402 pp., illustr.

——(1961). *The Earth's Problem Climates.* University of Wisconsin Press, Madison. vi + 334 pp., illustr.

VILENSKII, D. G. (1957). *Soil Science.* Ministry of Education of the USSR, Moscow, 448 pp., illustr. (Translated from Russian by 'Israel Program for Scientific Translation, Jerusalem'.)

WALLEN, C. C. (1966). Arid zone meteorology. pp. 31–50 in *Arid Lands: A Geographical Appraisal.* (Ed. E. S. Halls). UNESCO, Paris, xviii + 461 pp., illustr.

WALTER, H. (1962). *Die Vegetation der Erde in Okologischer Betrachtung.* Gustav Fisher, Jena, Band I, xiii + 538 pp., illustr.

II

CLIMATIC FACTORS AND THEIR EFFECT ON CROP PRODUCTION

Atmospheric Moisture

RAIN

The distinctive trait of the arid lands is, of course, insufficient precipitation. In the deserts the average amount of rainfall is never great, though heavy showers may occur occasionally. The mean annual precipitation totals at a few typical desert locations are: Yuma, 75 mm; Baghdad, 175 mm; Khartoum, 150 mm. In most parts of the Egyptian Desert, the annual rainfall does not exceed 10 mm, while in parts of the Libyan Desert several years may pass without any rainfall whatsoever (Migahid, 1962). Precipitation in some desert areas is sometimes too small to be measurable.

In connection with conditions under which light rainfalls only wet the soil surface and usually evaporate without having any appreciable effect, the concept of effective rainfall has been proposed; this is the minimum amount of rainfall, occurring on a single occasion, that ensures at least some water storage under favourable soil conditions. To be reasonably safe from evaporation, the rain must penetrate to a depth of at least 10–12 cm (Staple, 1964). The amount needed for this has been estimated as 15–20 mm for a single rainfall. The annual effective rainfall commonly amounts to about one-third of the total rainfall (Bagnold, 1954).

Another concept, of great biological significance in the desert, is the mean period between effective storms (ibid.).

The desert areas are usually surrounded by a fringe that receives more precipitation than they do; these are the steppes or semi-arid areas. However, these latter are generally seasonally dry areas—receiving practically all their precipitation during one season of the year, whilst the rest of the year is more or less dry. Excluding the monsoonal deserts of Asia, as a general rule the semi-arid fringe bordering on the temperate deserts gets winter precipitation, whilst the similar fringe bordering on tropical deserts gets summer rainfall (Bagnold, 1954).

The season of the year at which precipitation occurs has a considerable bearing on rainfall efficiency and on the amount of run-off. It has already been pointed out that cool-season rainfall is more efficient than that which is concentrated during the warm season, when evapotranspiration is at its maximum. Depending on the season, precipitation may occur mainly as rain or as snow. If the ground is frozen, infiltration will be low or nil. If the vegetation is in full leaf, a larger proportion of the rainfall than otherwise will be intercepted and lost by evaporation before reaching the soil. If most of the vegetation is dormant, the soil will become saturated sooner than when the vegetation is active, thereby also increasing run-off.

The dividing line between arid and semi-arid areas is usually a minimum of 250–350 mm (Koeppe & Long, 1958) in winter rainfall areas, and of 500 mm in summer rainfall areas, though local factors may cause deviations from these figures.

The following cycle of water-supply and water-use usually occurs in regions with alternating wet and dry seasons (Thornthwaite, 1956): (a) a period during which precipitation is in excess of water requirements and water accumulates in the soil and in reservoirs; (b) a dry period, during which a water deficiency occurs, as stored water is used for evapotranspiration and actual evapotranspiration falls below potential evapotranspiration.

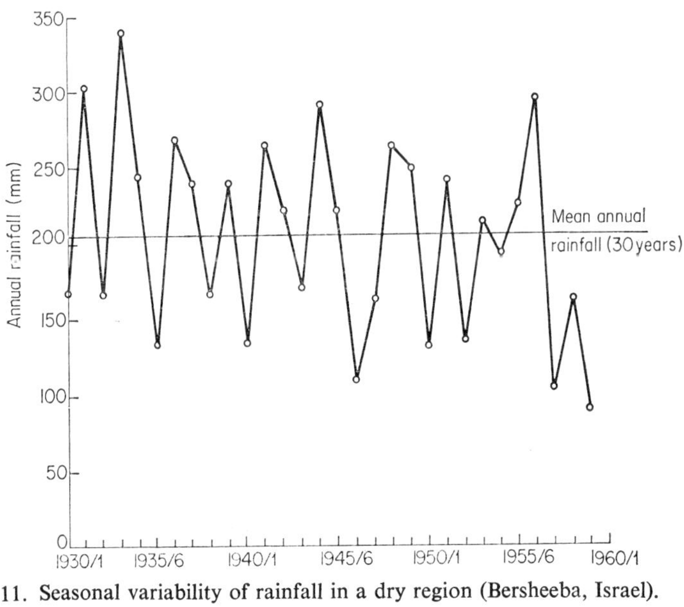

Figure 11. Seasonal variability of rainfall in a dry region (Bersheeba, Israel).

Dependability of Rainfall

It is a general rule that the drier the climate, the less reliable is the rainfall (Koeppe & Long, 1958). The arid regions are therefore characterized by considerable variability in the annual amounts of rainfall and its distribution in time (Figure 11). Because of this great variability, the dividing lines between arid, semi-arid, and humid, areas shift considerably from year to year.

Variability of rainfall is the greatest hazard to crop production in the semi-arid regions. In above-average rainfall seasons, cultivators may be tempted to plough and sow large areas on the drier fringe of a semi-arid region; then in the following below-average seasons, which are usually accompanied by higher-than-average temperatures, disastrous crop failures may result.

In a study covering a period of 68 years, from 1878 to 1946, on the effect of rainfall variability on the prosperity of agriculture in eastern Montana, the following picture emerged (Lommasson, 1947):

TABLE II

Rainfall Variability in Eastern Montana

	No. of years	Percentage of total No. of years
Abundant rainfall (double the average)	3	4·4
Average or above-average rainfall	31	45·5
Dry years, below average	24	35·3
'Killer' years, with extreme droughts	10	14·8

The probability of obtaining a bumper crop in this 'wheat region' is not more than one year out of twenty-three whilst complete crop failures, as a result of drought, can be expected in one year out of seven.

Frequency of Rainfall and Its Intensity

A further characteristic of rainfall in the arid regions is that rain frequently comes in a few heavy showers of short duration, which engender considerable run-off. It is not uncommon for 50 per cent of the annual precipitation to occur on 10–15 per cent of the rainy days (White, 1966).

Relation between Rainfall and Yields

Notwithstanding the major influence of rainfall on yield levels, yields are not always directly proportional to the amount of precipitation. Generally, yield levels are determined by the amount of precipitation above the basic minimum required to enable the crop to achieve maturity. If, for instance, under given circumstances 250 mm is the necessary minimum precipitation for a grain crop, 225 mm, or a reduction of only 25 mm, may result in complete crop failure. Conversely, 50 mm above the minimum requirement may double the yields.

It is accordingly important in dryland farming to have even a relatively small amount of water stored in the soil, prior to sowing the crop. This is the main justification for maintaining a period of fallow—even if this method is generally wasteful of land and of water. The response to additional rainfall will also depend on soil fertility.

Rainfall may also be in excess of the optimum and thereby cause reduced yields. This may sound paradoxical in relation to semi-arid climates. However, it will be remembered that the total rainfall in a semi-arid region may equal that characterizing a humid climate and yet be concentrated in a very short period. (The mean annual rainfall of Jerusalem is approximately equal to that of London, as was indicated on p. [5].) When this rainfall is concentrated in 4–5 months of the year, there may easily be periods when the rate of precipitation exceeds the capacity of the soil to absorb the rainfall. As a result, considerable run-off occurs from the waterlogged soil, plant nutrients are leached out of the root-zone, and crops are adversely affected by anaerobic conditions —especially if the excess precipitation occurs during the cool season.

The distribution of the rainfall and the coincidence or otherwise of a plentiful supply of soil moisture at periods of maximum crop requirement, also have a bearing on yield levels. A period of drought at the time of earing of cereals will have a more adverse effect on yields than if it occurs earlier or later in the growing period.

Excess moisture after sowing may reduce germination; heavy rainfall during earing may cause lodging and impair pollination.

SNOW

As snow melts slowly in the higher altitudes, it is an important source of water-supply to many arid regions. Thus it provides a sustained stream-flow that may continue into the summer, and is a valuable source of recharge of the ground-water basins. However, for this important water resource to be fully beneficial, efficient flood control measures are essential.

DEW

When surfaces, such as those of leaves or stones, lose heat rapidly by radiation, the air in contact with these surfaces is also cooled. When the temperature of the air falls below a critical point called the 'dew-point', dew condenses on the cold surfaces.

The immediate source of dew is the condensation of water vapour from the atmosphere. This vapour is replaced from the free atmosphere (dewfall) and from the soil (distillation), providing a moisture supply for continuous condensation throughout the night. If absorbed, dewfall is a real addition of moisture to the plant; distillation merely transfers available moisture from the soil to the atmosphere.

The formation of dew occurs mainly at night because it depends on radiational cooling of leaf and soil surfaces until they reach dew-point temperatures. According to Slatyer (1967), the main factors contributing to dew formation are:

(a) a clear sky—cloudiness reduces radiation from the earth and plants;
(b) low wind-speed—otherwise turbulent mixing in the lower atmosphere will prevent the temperature from falling;
(c) high air-humidity—so that the dew-point is reached with relatively little cooling; and
(d) vegetation of low heat-capacity—open-branched and with leaf-spread over a considerable height, so as to facilitate radiation loss.

The total amount of dewfall is unlikely to exceed 1 mm nightly (Slatyer, 1967).

Effects of Dew on Plant Growth

Dew absorption has been found to vary widely with plant species; it is largely dependent on the intensity and duration of dewfall as well as on the soil moisture régime, being higher under dryland conditions than under irrigation (Waisel, 1958).

Dew can only provide a very small proportion of the water requirements of a normally transpiring plant, but may be of some importance to plants under water-stress. Occurring as it does at night, it may accelerate the restoration of leaf turgor, and in the morning can delay the onset of renewed stress (Slatyer, 1967). In arid regions, dew may therefore be beneficial to plant growth and make a positive contribution to the water balance of the plant. However, Monteith (1963) estimates that these benefits are probably of little significance, as in cloudless summer weather, in an arid climate, the ratio of potential condensation to evaporation is approximately 1:4. Moreover, the severest stress for plants that are inadequately supplied with water occurs during the hottest hours of the day, long after the dew has completely evaporated.

Increasing Dew Condensation

In the Negev region of Israel, many hill sides are covered with regularly spaced heaps of stones (Plate I) that were probably built by the Nabateans, around the first centuries B.C. or A.D.; in Arabic, their name is Teleilat el Einab (grapevine mounds). This suggests that they were built to supply dew to individual plants, whose roots were beneath the stones. Monteith (1963) calculates that dew condensation from these mounds may have amounted to as much as 2–3 litres per night. Considerable doubts have, however, been expressed as to whether dew collection was really the function of these mounds.

In the Crimea, much bigger stone-piles have been found on the hills above the port of Theodosia. The potential calculated yield from condensation on them is 300 litres per night (Monteith, 1963).

GUTTATION

Water that is exuded by plants, usually on overcast nights, when the soil is warm and moist, may be mistaken for dew. Such water is commonly in the form of drops and the phenomenon is known as 'guttation', being of a physiological and not a meteorological nature (Long, 1958). Guttation typically occurs most actively through special organs termed hydathodes.

FOG

When a warm, saturated air mass rapidly replaces a cool, dry air mass over a cool surface, fog or mist results. During overcast nights, this may cause the deposition of relatively large amounts of water on plants and on the soil surface (Long, 1958).

Fog is of very frequent occurrence along rather narrow coastal regions which border on oceans with cold currents that run parallel to the coast. It affects plant growth through the high air-humidity, through wetting of the aerial parts of plants, and through humidification of the surface of the soil (Emberger & Lemée, 1962). There are no accurate measurements of the amount of moisture supplied by fog to the hydric balance of plants, but both type and density of vegetation appear in places to be influenced by the frequency of fog occurrence.

Frequent fogs or mists, by reducing evapotranspiration, may considerably increase the effective humidity of a region without this being reflected in the various formulae devised for defining aridity (Troll, 1958). In certain deserts, in which no precipitation may fall for several consecutive years, vegetation types occur which derive their entire water requirements from fogs or mists. Typical regions of this nature are the Atacama desert in northern Chile and Peru (cf. Figure 9), and the Namib desert in southern Africa (cf. Figure 6).

RELATIVE HUMIDITY

Relative humidity has a considerable effect on evapotranspiration, and hence on the water requirements of crops. In the arid regions relative humidity tends to be low, a mere 12–30 per cent being common around mid-day (Trewartha, 1954). The low humidity combined with the high temperatures increases the difficulties of maintaining an adequate supply of water to the plants.

Absolute atmospheric humidity is not always low in arid regions; because of the high temperatures, desert air may contain a considerable quantity of water vapour. Dews that occur during the cooler nights may, however, attenuate atmospheric dryness—at least if they are substantially absorbed by plants. A high relative humidity may partially compensate for low rainfall in regions with prevailing onshore winds from the sea.

In the semi-arid areas, there may be considerable variations in relative humidity, which tends to be dependent on the direction of the prevailing winds.

Solar Radiation

The arid regions are characterized by predominantly clear skies during both day and night, permitting a large amount of solar energy to reach the earth. However, a large proportion of this energy is lost: by radiation back to the atmosphere and to space, by being used for warming the soil, and by evapotranspiration.

Solar energy provides two essential needs of plants: (a) light, required for photosynthesis and for many other functions of the plant—including seed germination, leaf expansion, growth of stem and shoot, flowering, fruiting, and even dormancy (Stoughton, 1955); and (b) thermal conditions required for the normal physiological functions of the plant.

LIGHT

The three characteristics of light which affect plant growth and development are duration, intensity, and quality (wavelength).

Duration and Length of Day

The duration of light is of major importance to the growth and development of plants. The effect of photoperiodism—the relative length of daily light and dark periods—on the vegetative and reproductive stages of development is well known. Plants have been classified on the basis of their photoperiodic requirements for floral initiation into long-day (usually more than 14 hours), short-day (less than 10 hours), day-neutral, and indeterminate (requiring 12–14 hours of light daily to initiate flowering). Actually, the duration of the night or of complete darkness is often more important than the length of daylight.

Most typical tropical crops are short-day plants (requiring long-nights for flowering), whilst those of the higher latitudes are generally long-day plants.

Some plants have different photoperiodic requirements for different developmental stages: wheat, for example, is day-neutral for floral initiation, but long-day for fruiting.

Control of the time of flowering is extremely important for the adaptability of a crop to a given environment. Of all the recurrent climatic conditions, photoperiod is the most predictable, changing in a regular manner with latitude. Therefore, the ability of the plant to respond to photoperiod provides it with a reliable mechanism for assuring that seed formation occurs at a time of favourable conditions of temperature, moisture, and light, which are required for this function (Calder, 1966).

Length of day is largely a function of latitude. At the time of the equinoxes, the length of day is practically the same at all latitudes. At the beginning of summer, day-length between latitudes 10 and 30° is 13–14 hours; between 30 and 50°, in the middle latitudes, it is 14–16½ hours; at 60° it is approximately 19 hours (Figure 12).

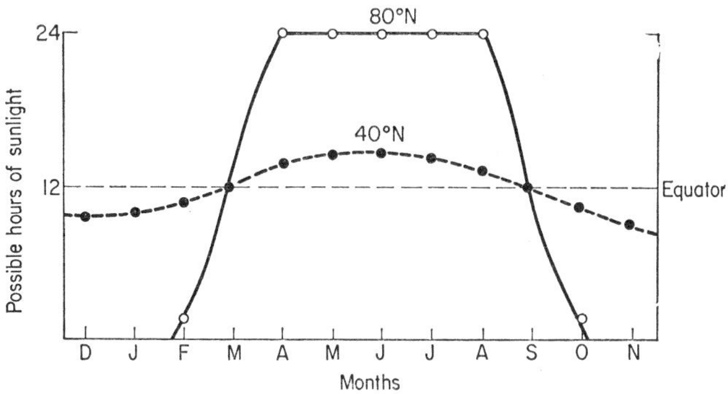

Figure 12. Variation of day-length throughout the year at the Equator, 40°N and 80°N. (By courtesy of the World Meteorological Organization.)

Whilst for optimal photosynthesis high light intensities are essential, the formative processes under the influence of photoperiodism are usually influenced by very low light intensities (Wassinck, 1954).

The large amount of sunshine is the greatest potential asset of the desert regions for agriculture, and may reach 75–90 per cent of the possible sunshine. Even during the winter months in the semi-arid areas, with winter rainfall, sunshine is plentiful. The hot and mild-arid and semi-arid regions of the world are better endowed with long hours of bright sunshine, combined with moderately high temperatures, than are any other parts of the world.

To a large extent irrigation can control the moisture régime of extensive areas, but the energy régime cannot be modified economically, except under the very limited conditions of greenhouses (Slatyer, 1967). Therefore, if an ample water supply can be provided by irrigation, yield levels can be achieved that are higher than in any other region (Davidson & Philip. 1958). This higher potential productivity is reflected by the figures in Table III.

TABLE III

Potential Productivity of Different Regions of the Earth
(adapted from Wit, 1967)

North latitude (degrees)	Number of months with mean temperature above 10°C	Carbohydrate/ha/year (in kg × 10^3)
70	1	12
60	2	21
50	6	59
40	9	91
30	11	113
20	12	124
10	12	124
0	12	116
−10	12	117
−20	12	123
−30	12	121
−40	8	89
−50	1	12

The major arid and semi-arid regions of the world are generally found between latitudes 20 and 30° in each hemisphere, with extremes often lying far outside these limits (cf. Figures 6–10). The figures in Table III reflect the great agricultural productive potential, per unit area, of these regions.

In Figure 13, the seasonal changes in the rate of net dry-matter gain per unit

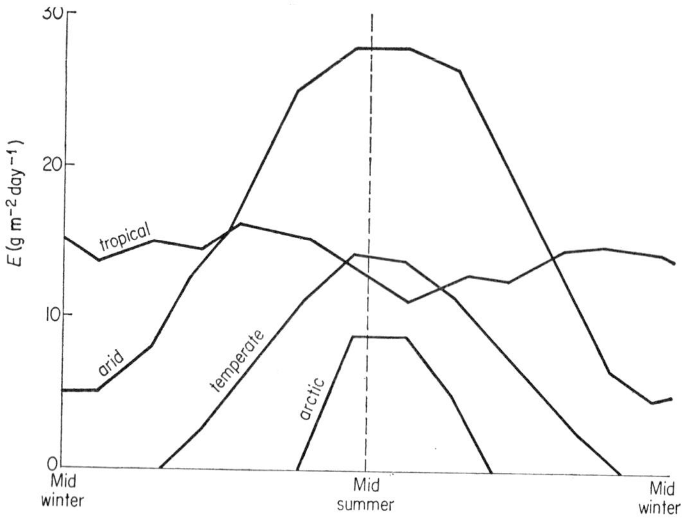

2

Figure 13. Seasonal changes in the rate of net dry-matter gain per unit leaf area (*E*) for sunflower plants, predicted from multiple regression on light and temperature, for four climates. (Wilson (1969): by permission of the author.)

leaf area (E) for sunflower plants grown in four different climates, have been predicted from multiple regression on light and temperature (Wilson, 1969). That author has calculated that the annual dry-matter production of widely spaced sunflower plants, when nutrient and water-supply are non-limiting (expressed on a leaf-area basis), are as follows:

Arctic (Resolute, N.W.T., Canada)	670 g m^{-2} yr^{-1}
Temperate (Kew, England)	2170 g m^{-2} yr^{-1}
Tropical (Benin City, Nigeria)	5110 g m^{-2} yr^{-1}
Arid (Deniliquin, N.S.W., Australia)	5522 g m^{-2} yr^{-1}

Extremely high values of net dry-matter gain per unit leaf area have been found to occur in plants growing in the arid zones around latitude 30° in summer, when clear skies, fairly long days, and a sun rising high in the sky, combine to give more light than in any other climate. Provided soil nutrients and water are not limiting, it is in the arid regions that the greatest annual crop production is possible, as the mean annual light receipt in the 20–30° latitudinal zone is slightly higher than at any other latitude.

Light Intensity and Crop Yields

Light is the most important factor influencing the photosynthetic activity of a crop, as it is the energy source for the whole process. Light utilization by the crop has two limitations: the maximum quantum yield at low intensities, and light saturation at high intensities (Loomis *et al.*, 1967).

Minimum Light Requirements

It has generally been found that at low light intensities, for the individual leaf, there is a linear relation between light intensities and rate of photosynthesis. Theoretically, photosynthesis is possible at any light intensity, however low; practically, respiration dominates when light intensity is too low. Net assimilation will be zero at a light intensity of 500 candles, and a minimum of 500 to 1000 foot candle is required for effective rates of photosynthesis, at which the photosynthesic gas exchange is greater than the respiratory gas exchange (Blackman & Black, 1959).

Light Saturation

With increasing light intensity, photosynthesis of the single leaf obeys the law of diminishing returns. Extremely high light intensities even have an inhibitory effect on photosynthesis—a phenomenon called solarization (Wang, 1963).

In most crop plants, light saturation for single leaves is reached at a light intensity of 0·2 cal/cm^2 minute, which is typical of the light intensity on an overcast day with the sun at its zenith. On a clear day, light intensity may be four times as great (de Wit, 1967), so that a large proportion of the light is far in excess of what fully exposed leaves can utilize. For this reason, it was previously assumed that light could not generally become limiting for crop production.

However, in the field, light is not spread evenly over the photosynthetic surface but commonly passed by reflection and transmission through several layers of leaves, its intensity falling off exponentially with the path-length through absorbing layers. The fully exposed leaves in the canopy may absorb several times the amount of light needed for saturation, thereby absorbing a large proportion of the available light energy without increasing photosynthesis. The intensity of the light falling on the leaves in the lower layers depends on the initial light intensity, on the ratio between direct and diffuse light, on the number and size of the leaves, their angle of incidence, and distribution, and on the transmission and reflection of the leaves (Verhagen *et al.*, 1963). At ground-level, the final light intensity is usually below compensation point, i.e. that light intensity at which the gas exchange resulting from photosynthesis is equal to that resulting from respiration (Bonner, 1962).

Very high light intensities are therefore rarely capable of saturating the whole canopy in the field, and the photosynthetic rate of the foliage *as a whole* continues to increase up to very high light intensities, as more and more leaves in the shaded parts of the canopy reach light saturation (Figure 14).

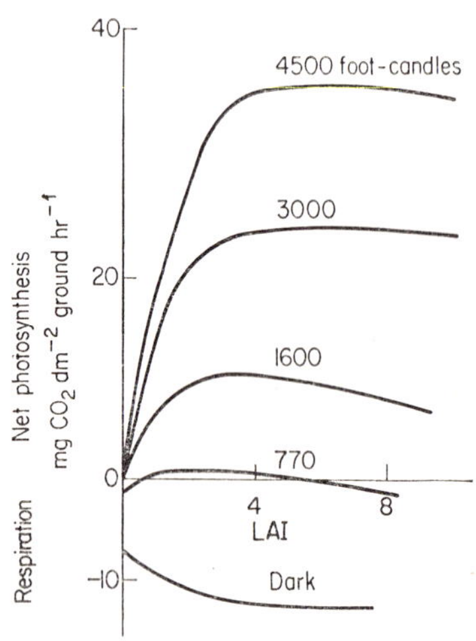

Figure 14. Growth of cotton communities in controlled environments at various light levels (Loomis *et al*, 1967). The 'dark' curve shows the curvilinear nature of dark respiration with increasing leaf-area index (LAI). (Based on data of J. L. Ludwig, T. Saeki and Evans: by courtesy of Academic Press, New York, and Professor R. S. Loomis.)

On a world-wide basis, photosynthetic efficiency is less than 1 per cent, generally as a result of deficiencies in nutrition, limited moisture supply, inadequate crop protection, etc. In agriculturally advanced areas, efficiencies of 2–2·5 per cent are common (Army & Greer, 1967). When growing conditions are optimal, and after a closed crop surface is reached, a number of agricultural crops have achieved an efficiency of light utilization of 5–6 per cent (Alberda, 1962).

Much of the difference between actual yields in the field, and the calculated potential yield based on maximum possible conversion of light energy by the photosynthetic process (estimated at 77 grammes of dry matter per square metre per day [Loomis & Williams, 1963]), appears to be due to light saturation of the upper part of the canopy only, whilst the lower leaves may be contributing very little on clear days, and possibly show a negative net assimilation on cloudy or rainy days. Therefore, as long as light saturation of the canopy as a whole does not occur, any increase in light intensity will increase productivity. Even in environments with very high peak intensities, productivity was found to show a linear relationship with the logarithm of relative light intensity (Loomis *et al.*, 1967). For example, the growth-rate of Subterranean Clover was found to be determined solely by the amount of light-energy received, and to be independent of temperature. However, this is only part of the picture. A leaf subjected to such radiation heat-loads must be able to dissipate rapidly the greater part of the absorbed energy, or its temperature could rise to as much as 100°C over air temperature within less than a minute (Idso *et al.*, 1966). This dissipation of heat occurs in three ways: by thermal radiation from the leaf, by removal of heat via convection currents, and by transpiration.

Solar radiation can therefore affect plant productivity in two different ways simultaneously: directly by its intensity, and indirectly by its effect on leaf temperature.

On a relatively cool day, the rate of photosynthesis follows the incidence of solar radiation closely, for both shaded and unshaded leaves. By contrast, on a warm day, the sunlit leaf undergoes a midday slump at which it stops photosynthesizing, resuming the process only just before sunset. The shaded leaf, on the other hand, shows only a slight reduction in its rate of photosynthesis during the midday period, but keeps on photosynthetizing throughout the day, so that its total daily photosynthesis is three times as great as that of the sunlit leaf—in sharp contrast to what occurs on a cool day (Idso *et al.*, 1966). The midday slump is only partially due to an internal water deficit, the major reason being the effects of the energy environment on leaf temperatures.

Light Quality

Many factors, such as the amount and kind of cloud cover, fog, air pollution, and the colour of the foliage interrupting light, influence the quality of the incident light.

Radiation up to 0·25 micron (ultra-violet spectrum) is harmful to most plants; from 0·30 to 0·55 micron it has a photoperiodic effect; from 0·40 to 0·69 micron it is most effective in photosynthesis. Above 0·74 micron (the infra-red spectrum), 'light' has practically no effect on photosynthesis; its main effect is thermal, and respiration is encouraged (Wang, 1963).

TEMPERATURE

Temperatures in the hot, arid climates are the highest in the world (Critchfield, 1966). The generally clear skies facilitate maximum radiation by day, and rapid loss of heat by night—resulting in high diurnal ranges of temperature, which may reach 30° to 40°C in the hot deserts during the summer.

The sparse vegetative cover permits the soil surface to become intensively heated by day, with temperatures at the soil surface reaching 93°C (Trewartha, 1954). The higher the temperature of a body, the more rapidly it loses heat by radiation. Deserts, therefore, both acquire and lose heat very rapidly. At night, about 90 per cent of the heat accumulated during the day is radiated into space, as compared with 50 per cent in humid climates (Cloudsley-Thompson, 1954).

The annual ranges of temperature are less marked in hot deserts than in many other regions, but are far greater than in the humid tropics. Daily average ranges during the hottest months in the hot desert area are 27–35°C, with daily maxima of 37–43°C. The coolest months in these areas average around 10°C, with occasional night frosts. Temperatures in the semi-arid hot climates are generally similar to those of the more arid regions (Critchfield, 1966).

Effects of Temperature on Crop Production

Plants can grow only within certain limits of temperature. For each species and variety there are not only optimal temperature limits, but also optimal temperatures for different growth-stages and functions, as well as lower and upper lethal limits.

We have seen that light has an overriding effect on the rate of photosynthesis. However, it is involved in only one aspect of the photosynthetic process, namely, as a source of energy for the conversion of carbon dioxide and water to carbohydrates.

There are certain biochemical processes preceding and following the reduction of carbon dioxide, which are affected mainly by temperature (Gaastra, 1963). As long as light is limiting, temperature has little effect on the rate of photosynthesis. However, when light is not limiting, as is generally the case in dry regions, the biochemical processes associated with photosynthesis become the limiting factor, so that the effect of favourable temperatures on the rate of photosynthesis is increased.

In general, high temperatures accelerate growth processes. Rarely are high temperatures *per se* the direct cause of death of plants, provided the water supply is adequate. However, beyond a certain limit, which depends on the crop, the stage of development, and the physiological process involved, high temperatures may have detrimental effects on crop production.

The photosynthetic process appears to become heat-inactivated at excessively high temperatures which, however, do not inhibit respiration, so that apparent photosynthesis declines rapidly under these conditions (Moss *et al.*, 1961).

Retardation of growth and difficulties in fertilization, even in heat-loving crops such as maize and sorghum, occur at temperatures that are often well below the lethal limit. The harmful effects of excessive temperatures are usually aggravated by lack of available moisture. Hot, dry winds will further increase the damage.

In the Great Plains of the U.S.A., it was found that high summer temperatures may even override the effects of summer rainfall on the yields of crops. Thus variability in yields of spring wheat was clearly associated with variability in average maximum temperatures in June, and variability in maize yields was most closely associated with variability in average maximum temperatures in July (Moldenhauer & Westin, 1959). With increasing temperatures, evapotranspiration increases. Hence, a moisture supply that is adequate in a cooler climate may be deficient in a warmer region—a situation which may be aggravated by hot winds. As already pointed out, the daily alternation of high and low temperatures is a characteristic feature of arid climates. This may also affect plant production. For example, certain weeds, such as Johnson Grass (*Sorghum halepense*) and Bermuda Grass (*Cynodon dactylon*), germinate better under conditions of temperature alternations than under a constant temperature.

Frosts

The clear skies at night permit considerable radiation from the earth in arid regions. The rapid cooling of the land at night causes an inversion of temperature, the air near the cool soil becomes heavy and does not rise, so that night temperatures at soil level are typically low in the desert and many reach freezing-point, even after a relatively hot day. Cold air will tend to flow into low places, so that localized pockets of frost may occur in an area which is generally frost-free. In the middle-latitude dry zones, the incidence of killing frosts determines the length of the growing-season, which is practically limited to the time-interval between the last killing frost of spring and the first killing frost of autumn. The shorter this period is, the earlier maturing will be the varieties that can be grown and the lower their potential yielding ability. In the arid regions with mild winters and no or rare frosts, it is possible to produce two, and sometimes three, crops a year. The annual number of cuttings of a perennial forage crop, such as Lucerne (Alfalfa), varies from three (in a region with killing frosts) to ten (in a region with mild winters).

In warm regions, occasional and exceptional frosts may do considerable damage because the crops grown in such regions are usually very susceptible to low temperatures.

CLOUDINESS AND ITS EFFECT ON LIGHT INTENSITY

In desert regions the skies are generally clear. This favours high insolation by day and rapid radiation by night. As a result of the heating of the land during the day, rising air currents are induced, producing the typical 'dust-devils'. The warmed air often rises sufficiently high to be cooled to dew-point and form clouds; these may result in precipitation which, however, rarely reaches the ground. If and when it does reach the ground, the rain is usually local, torrential, and brief (Debenham, 1954).

When the sun shines through a gap in the clouds, the reflection from the clouds may, for a short period, increase and even double the intensity of incoming radiation, whilst a light cloud-cover may completely intercept direct radiation, reducing it to zero (Slatyer, 1967).

Whereas overcast skies may reduce direct radiation, and diminish light intensity to about one-fifth of that of a clear sky, photosynthesis will not be reduced proportionally, because of the even distribution of the light (Wit, 1967).

WINDS

The arid regions are characterized by frequent and strong winds, which are due partly to considerable convection during the day. The usually sparse vegetation is not capable of slowing down air movement, so that dust storms are a frequent concomitant of wind movement. Much of the dust is carried for considerable distances, to form loess deposits in neighbouring regions. The heavier sand-particles carried by the wind scour the soil surface.

Winds also affect growth mechanically and physiologically. The sand and dust particles carried by the wind may damage plant tissues. Emerging seedlings may be completely covered or, alternatively, the roots of young plants may be exposed by strong winds. Winds may also cause considerable losses by inducing lodging, the breaking of stalks, and the shedding of grain.

The physiological effect of winds consists mainly in increasing transpiration as well as evaporation from the soil. Hot dry winds may, however, adversely affect photosynthesis, and hence productivity, by causing closure of the stomata even when soil moisture is adequate (Crafts *et al.*, 1949).

The arid regions are under the influence of winds which originate either over the sea or over deserts. In summer, the desert regions become heated to high temperatures; the heated air rises, causing an indraught of winds from neighbouring regions. If these winds pass over an ocean, they become moisture-laden. These are the monsoon winds, such as those that invade the Sahara from the south-west during the northern summer.

During the winter, the opposite effect occurs: the desert areas are cooler than the adjacent tropical areas, and the winds blow from the deserts towards the tropical areas. These winds are relatively cool and dry at the beginning of the growing season, but become progressively hotter and drier as the spring advances. They are called Harmattan in most of tropical West Africa north of the Equator, and Khamsin in the Near East. The total number of days in spring during which these dry, hot winds blow may be relatively small, but their effect on the crops and their yields may be considerable. Overnight, temperatures may soar from 25 to 46°C, and the relative humidity may drop to 5–10 per cent.

If the cereal crops are earing when a Khamsin strikes, the ears may be actually 'blasted', turning white, without any seed formation following. When the Khamsin occurs at a later stage, seeds may shrivel and mature ahead of time.

The hot, dry winds may also occasionally have beneficial effects. The main reason for the low incidence of the very dangerous Yellow Rust of Wheat (*Puccinia glumarum*) in the Near East, is the susceptibility of the pathogen to the high temperatures that occur when the sirocco winds blow. In the rare years when no Khamsins occur in the spring season, this Rust may cause devastating reductions in yield.

Moderate winds have a beneficial effect on photosynthesis by continuously replacing the carbon dioxide absorbed by the leaf surfaces. When turbulence was low in a dense maize canopy, the maximum observed photosynthetic efficiency in the utilization of radiation was found to be 2 per cent; efficiency was doubled to 4 per cent, under conditions of high radiation, when turbulence provided sufficient carbon dioxide (Lemon, 1963). It was also found that photosynthetic efficiences increased with increasing wind—even when the maize was visibly wilting as a result of water-stress, and notwithstanding stomatal closure.

The favourable effect of wind on photosynthesis was most marked at the lower leaf-layers; increasing wind-strength permitted ever deeper leaf-layers to approach the maximum rate of photosynthesis possible under the amounts of radiation received (Lemon, 1963).

SHELTER-BELTS

Much work has been done on the effect on crop production in the semi-arid steppes of the U.S.S.R. of planting shelter-belts of trees (Albenskii & Nikitin, 1956). Many shelter-belts were also planted in the semi-arid plains of the United States. Whilst some conflicting results have been recorded, the vast majority of reports indicate beneficial effects due to windbreaks, and their value in the semi-arid regions of the world is well documented (Rosenberg, 1967).

Windbreaks have been very successful in areas where an appreciable part of the total precipitation is in the form of snow, as they aid in uniform distribution of the snow, and thus improve the supply of moisture to spring-grown crops (McGinnies *et al.*, 1963). While all reasons for the beneficial effects of windbreaks are not known, the favourable distribution of soil moisture, and the favourable influence on the internal water régime of sheltered plants, are considered by Rosenberg (1967) as probably the most important. In semi-arid windswept regions where irrigation is practised, the frequent dry winds with relatively high velocities and high temperatures increase the horizontal advection of energy. Under these conditions windbreaks are effective in reducing evapotranspiration and decreasing plant water-stress. By preventing turgor loss and stomatal closure, photosynthetic potential is increased..

As a result of these effects, windbreaks increase the efficiency of water use not only in semi-arid regions, but also in deserts where oasis-type irrigation is practised.

In the U.S.S.R. it was found that the yields of grain-crops were increased by 20–30 per cent, and those of some grasses by 100–200 per cent, in the sheltered zone. The advantages of the shelter-belts were most marked in years of drought, as utilization of the limited water available for transpiration was more efficient by the sheltered crops than by others (Albenskii & Nikitin, 1956).

On the leeward side of a shelter-belt, the reduction in wind velocity is felt for a distance that equals, on the average, 30 times the height of the protective stand, and on the windward side 5 times this height. The zone of influence on crop yield was generally found to extend to the leeward for a distance of approximately 20 times the height of the trees, with the maximum effect occurring within a distance of 10 times this height (Stoeckler, 1962).

Experience has shown that windbreak efficiency is increased by increasing permeability. A dense windbreak causes air to descend abruptly behind it, producing turbulent eddying. If about 50 per cent of the wind passes through the barrier, this air-flow will prevent the air from descending abruptly to the ground (Albe, 1958) (Figure 15). A permeable windbreak reduces wind-speed less, but the effect persists for a greater distance downwind. The shelter-belts

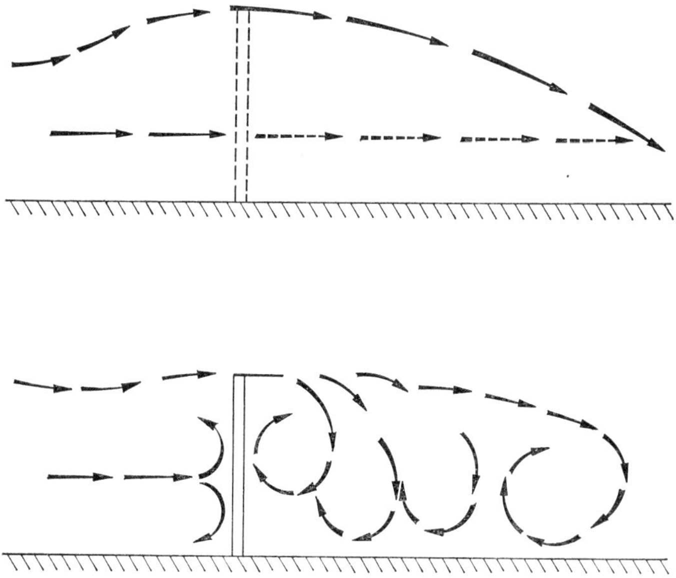

Figure 15. Effectiveness of a windbreak in relation to its permeability.

planted during the dust-bowl years in the Great Plains were of the dense type and were for this reason of low effectiveness (Rosenberg, 1967).

Trees are usually planted as shelter-belts, but tall plants such as bamboos, castor beans, maize, and elephant grass, can make quite effective windbreaks in the warm dry regions.

Failures with shelter-belts have usually occurred when establishment was poor and early care was negligent or the moisture supply was inadequate.

Potential Evapotranspiration

In the arid regions the rate of potential evapotranspiration, as contrasted with precipitation, varies little from year to year. As a general rule, conditions that cause a reduction in the amount of precipitation will at the same time favour potential evapotranspiration. Therefore, the discrepancy between water requirements and water supply in arid regions is still further aggravated.

Potential evapotranspiration may greatly exceed precipitation, as indicated by the following examples: at Las Cruces, New Mexico, average annual precipitation was found to be 216 mm, as compared with 2 337 mm for potential evapotranspiration (Hardy & Derraugh, 1947). At Yuma, Arizona, potential evapotranspiration is 55 times as great as rainfall (Koeppe & Long, 1958).

Carbon Dioxide Content of the Atmosphere

The carbon dioxide content of the atmosphere is of major importance to plant growth. Photosynthesis is approximately proportional to the concentration of CO_2 in the air surrounding the foliage of the crop (Monteith, 1965). An important source of CO_2 is the decomposition of organic residues by micro-organisms, though 'soil respiration' is not sufficient to enable rapid rates of photosynthesis to take place (Moss *et al.*, 1961).

The concentration of carbon dioxide in the atmosphere may vary locally very considerably, ranging from half to several times the normal value, which is around 0·03 per cent by volume.* In fields of dense, growing crops, when there is no wind, levels may fall well below 0·03 per cent. However, decreases in photosynthetic rate due to a local deficit of CO_2 rarely exceed 20 per cent (Monteith, 1965).

Increasing carbon dioxide concentration, if it could be economically justified, might have a considerable effect on crop yields, especially when high light intensities and favourable temperatures prevail at the same time.

Microclimatology

Microclimatology deals with the pertinent factors of the environment in the zone lying between the highest level affected by plants locally and the lowest depth to which air penetrates into the soil (Albe, 1958).

MICROCLIMATE AT PLANT LEVEL

The microclimate just above a crop and under its canopy (eco-climate) is influenced by the particular type of crop, and may be strikingly different from the climate of the surrounding environment. Even emerging seedlings will alter the climate near the soil surface, by reducing air movement and by shading the ground. As the plants grow, the extremes in temperatures of the soil surface are reduced; even so, the range of temperatures at the soil surface may be double that of the temperatures recorded in the standard meteorological screen.

The eco-climate may be more humid and cooler than the atmosphere above

* According to Dr Francis S Johnson ('The balance of atmospheric oxygen and carbon dioxide', *Biol. Conserv.*, **2**(2), 83–89, January 1970). 'The present concentrations [of CO_2 in the atmosphere] are about 6 per cent greater than values that existed near the turn of the century . . . the widely-quoted figure of 10 per cent [being] based on questionable data.' This increase has largely been due to the unprecedented almost universal burning of fossil fuels, which at the present rate of adding 0·7 p.p.m. of CO_2 to the atmosphere per annum would take about 40 years to produce a further 10 per cent increase. However, the rate of usage of fossil fuels is increasing so rapidly that a 10 per cent increase above present CO_2 levels seems more likely to be reached in 20 or fewer years; this could lead to substantially increased photosynthesis.—Ed.

the crop, which circumstances may create favourable conditions for the spread of certain diseases. Leaf temperatures in the sunlight may be higher than the air temperature by day and lower than the air temperature by night. The intensity and quality of light change as the rays pass through the plant canopy.

SOIL MICROCLIMATE

The climate in the soil may be strikingly different from that of the atmosphere above it. Only part of the radiation that reaches the soil's surface is absorbed, the amount reflected depending on certain characteristics of the soil's surface. The percentage of the incoming radiation that is reflected is called the 'albedo', which varies according to plant cover, soil type, and moisture content, etc., as shown by the figures in Table IV.

TABLE IV

Albedo of Different Soil Surfaces (Thornthwaite, 1958)

Surface	*Albedo* (percentage of incoming radiation)
Wet sand	9
Dry land	18
Grass cover, green crops	15–30
Forest	15–18

Among other moisture-saving characteristics of desert vegetation, is the commonly light-grey colour of the foliage, which has a high albedo and is considered by Thornthwaite (1958) to be the most important moisture-conserving property of such plants.

As a result of differences in albedo, different thermal régimes will exist on different types of soil surface layers and the air above them—even when the amounts of incoming radiation are identical. For example, a mulch on the surface of the soil absorbs most of the incoming radiation; the soil remains cool, whilst the surface of the mulch itself and the overlying air become very hot during the day-time and cool rapidly during the night hours.

Wet soils absorb more radiant energy than dry soils. Most of the energy is used for the evaporation of the soil-water; as water is also a good conductor of heat, the remaining radiant energy will warm deeper soil layers. By contrast, the surface of dry soils, and the air layer above them, will be at higher temperatures than is the case with moist soil. Variations in temperature of a desert soil at a depth of 0·4 cm from 17·0°C in the early morning hours to 71·5°C at midday were recorded at Tucson, Arizona, U.S.A. The maximum temperature was 21·1°C higher than at a depth of two centimetres in the soil.

In research carried out at the Laboratory of Climatology in the Great Plains region of the United States, it was found that, when a soil is at field-capacity—i.e. holding all the water it can without being waterlogged—80–85 per cent of the incoming energy was used for evapotranspiration; as soil moisture was depleted, this proportion decreased.

In arid regions, as a result of the low moisture content of the soil and the sparse vegetation, actual evapotranspiration is low and most of the incoming radiation serves to heat the soil and the air above it. When the surface of a moist soil is covered by vegetation, the amount of evapotranspiration is determined mainly by the amount of incoming radiation, and is otherwise mainly affected by differences in albedo between different plant covers. However, as the soil dries, the ability of different types of vegetation to use soil moisture will make itself felt, and thereby affect actual evapotranspiration.

The temperature of the top layer of the soil is usually higher than that of the overlying atmosphere, while daily fluctuations in temperature decrease with increasing depth in the soil. At a depth of 50 cm, daily variations of temperature in the soil practically cease (Migahid, 1962). At a depth of 1–3 m, temperatures are fairly stable throughout the year.

Soil temperature has considerable influence on plant growth and on the soil micro-organisms. Warm (20–30°C), moist soils are a favourable medium for most of the cultivated crops that are grown in arid regions.

As a result of the heating of the soil surface, which may be considerable in arid climates, a steep temperature gradient occurs, which may cause a considerable movement of water through the soil.

There is a continuous exchange of air between soil and atmosphere. When water penetrates into the soil, air is expelled; as the soil dries, fresh air is drawn into the soil. The air in the soil is usually moisture-saturated, excepting under conditions of extreme dryness.

References

ALBE, E. M. Fournier D' (1958). The modification of microclimates. *Arid Zone Res.*, **10**, 126–46.

ALBENSKII, A. V. & NIKITIN, P. D. (1956). *Handbook of Afforestation and Soil Amelioration.* State Publishing House for Agriculture, Moscow (Translated from Russian by Israel Program for Scientific Translation, Jerusalem).

ALBERDA, T. (1962). Actual and potential production of agricultural crops. *Neth. J. Agric. Sci.*, **10**, 325–33.

ARMY, T. J. & GREER, F. A. (1967). Photosynthesis and crop production system. pp. 321–32 in *Harvesting the Sun – Photosynthesis in Plant Life* (Ed. A. San Pietro, F. A. Greer, and T. J. Army). Academic Press, New York. ix + 342 pp., illustr.

BAGNOLD, R. A. (1954). The physical aspects of dry deserts. pp. 7–12 in *Biology of Deserts* (Ed. J. L. Cloudsley-Thompson). Institute of Biology, London, iv + 224 pp., illustr.

BLACKMAN, G. E. & BLACK, J. N. (1959). Physiological and ecological studies in the analysis of plant environment. XII: The role of light as a limiting factor. *Ann. Bot.* (*N. S.*), **23**, 131–45.

BONNER, J. (1962). The upper limit of crop yield. *Science*, **137** (3532), 11–15.

CALDER, D. M. (1966) Inflorescence, induction and initiation in the Graminae. pp. 59–73 in *The Growth of Cereals and Grasses* (Ed. F. L. Milthorpe and J. D. Ivins) Butterworths, London, xii + 359 pp., illustr.

CLOUDSLEY-THOMPSON, J. L. Ed. (1954). *Biology of Deserts*. Institute of Biology, London, iv + 224 pp., illustr.

CRAFTS, A. S., CURRIER, H. R. and STOCKING, C. R. (1949). *Water in the Physiology of Plants*. Ronald Press, New York, xxi + 240 pp., illustr.

CRITCHFIELD, H. J. (1966). *General Climatology*. Prentice-Hall, Englewood Cliffs, New Jersey, xi + 420 pp., illustr.

D'ALBE, E. M. FOURNIER. *See* ALBE, E. M. FOURNIER D'.

DAVIDSON, J. L. and PHILIP, J. R. (1958). Light and pasture growth. *Arid Zone Res.*, **11**, 181–7.

DEBENHAM, F. (1954). The geography of deserts. pp. 1–6 in *Biology of Deserts* (Ed. J. L. Cloudsley–Thompson) Institute of Biology, London, iv + 224 pp., illustr.

EMBERGER, L. & LEMÉE, G. (1962). Plant ecology. *Arid Zone Res.*, **18**, 197–212.

FOURNIER D'ALBE, E. M. (1958). *See* ALBE, E. M. FOURNIER D'

GAASTRA, P. (1963). Climatic control of photosynthesis and respiration. pp. 113–18 in *Environmental Control of Plant Growth* (Ed. L. T. Evans) Academic Press, New York, xvii + 449 pp., illustr.

HARDY, F. & DERRAUGH, L. F. (1947). The water and air relations of some Trinidad sugar-cane seeds. *Trop. Agric. Trin.*, **24**, 76–87.

IDSO, S. B., BAKER, D. G. & GATES, D. M. (1966). The energy environment of plants. *Adv. Agron.*, **18**, 171–218.

KOÉPPE, C. E. & LONG, G. C. DE (1958). *Weather and Climate*. McGraw-Hill, New York, xvii + 341 pp., illustr.

LEMON, E. (1963), Energy and water balance of plant communities. pp. 55–79 in *Environmental Control of Plant Growth* (Ed. L. T. Evans) Academic Press, New York, viii + 449 pp., illustr.

LOMMASSON, T. (1947). The influence of rainfall on prosperity in eastern Montana, 1878–1946. *Mimeographed Rep.*, 7, Region I, U. S. Forest Service.

LONG, I. F. (1958). Some observations on dew. *Met. Mag.*, **87**, 161–8.

LOOMIS, R. S. & WILLIAMS, W. A. (1963). Maximum crop productivity. *Crop Sci.*, **3**, 67–72.

———, ——— & DUNCAN, W. G. (1967). Community architecture and the productivity of terrestrial plant communities. pp. 291–308 in *Harvesting the Sun—Photosynthesis in Plant Life* (Ed. A. San Pietro, F. A. Greer and T. J. Army) Academic Press, New York, ix + 342 pp., illustr.

LUDWIG, J. L., SAEKI, T. & EVANS L. T. (1965). Photosynthesis in artificial communities of cotton plants in relation to leaf area. *Austr. J. Biol. Sci.*, **18**, 1103–18.

MCGINNIES, W. G., MCCOMB, A. L. & FLETCHER, J. E. (1963). Role of watersheds and forests in the arid west. pp. 227–307 in *Aridity and Man* (Ed. C. Hodge) Amer. Ass. Adv. Sci., Washington, D.C., xviii + 584 pp., illustr.

MIGAHID, A. M. (1962). The drought resistance of Egyptian desert plants. *Arid Zone Res.*, **18**, 213–23.

MOLDENHAUER, W. C. & WESTIN, F. C. (1959). Some relationships between climate and yields of corn and wheat in Sprink County, S. D., and yields of milo and cotton at Big Spring, Tex. *Agron. J.*, **51**, 373–6.

MONTEITH, J. L. (1963). Dew: facts and fallacies. pp. 37–56 in *The Water Relations of Plants* (Ed. A. J. Rutter & F. H. Whitehead) Blackwell, Oxford, 394 pp., illustr.

——, (1965). Light distribution and photosynthesis in field crops. *Ann. Bot.*, **29**, 17–37.

MOSS, D. N., MUSGRAVE, R. B. & LEMON, E. R. (1961). Photosynthesis under field conditions. III: Some effects of light, carbon dioxide, temperature and soil moisture on photosynthesis, respiration and transpiration of corn. *Crop Sci.*, **1**, 83–87.

ROSENBERG, N. J. (1967). The influence and implications of windbreaks on agriculture in dry regions. pp. 327–49. in *Ground Level Climatology* (Ed. R. H. Shaw) Amer. Ass. Adv. Sci., Washington, D.C., xii + 395 pp., illustr.

SLATYER, R. O. (1967). *Plant-Water Relationships.* Academic Press, London, xii + 366 pp., illustr.

STAPLE, W. J. (1964). Dry land agriculture and water conservation. pp. 15–30 in *Research on Water.* ASA Spec. Publ. Soil Sci. Soc. Amer., Madison, Wisconsin, vi + 121 pp., illustr.

STOECKLER, J. H. (1962). Shelter belt influence on Great Plains field environment and crops. *Prod. Res. Rep. U.S. Dep. Agric.*, **62**, 26 pp., illustr.

STOUGHTON, R. H. (1955). Light and plant growth. *J. Roy. Hort. Soc.*, **80**, 454–66.

THORNTHWAITE, C. W. (1956). Climatology in arid zone research. pp. 67–84 in *The Future of Arid Lands* (Ed. G. F. White) Amer. Assoc. Adv. Sci., Washington D.C., ix + 453 pp., illustr.

—— (1958) Introduction to arid zone climatology. *Arid Zone Res.*, **11**, 15–22.

TREWARTHA, J. T. (1954). *An Introduction to Climate.* McGraw-Hill, New York, vii + 402 pp., illustr.

TROLL, C. (1958). Climatic seasons and climatic classification. *Orient. Geogr.*, **2**(2), 141–65.

VERHAGEN, A. M. W., WILSON, J. H. & BRITTEN, E. J. (1963). Plant production in relation to foliage illumination. *Ann. Bot.* (N.S.)., **27**, 627–40.

WAISEL, Y. (1958). Dew absorption by plants of arid zones. *Bull. Res. Coun. Israel*, **6D**, 180–6.

WANG, J. Y. (1963). *Agricultural Meteorology.* Pacemaker Press, Milwaukee, Wisconsin, 693 pp., illustr.

WASSINCK, E. C. (1954). Remarks on energy relations in photosynthesis processes. *Proc. Ist. Int. Photobiol. Congr.*, Amsterdam, Biology Sect. V, pp. 307–22.

WHITE, G. F. (1966). The world's arid areas, pp. 15–30 in *Arid Lands: A Geographical Appraisal* (Ed. E. S. Hills) UNESCO, Paris, xviii + 461 pp., illustr.

WILSON, J. WARREN (1969). *Maximum Yield Potential.* Paper presented at Colloquium on the Transition from Extensive to Intensive Agriculture, Israel, 33 pp., illustr. (mimeographed).

WIT, C. T. DE (1967). Photosynthesis: its relationship to overpopulation. pp. 315–20 in *Harvesting the sun—Photosynthesis in Plant Life* (Ed. A. San Pietro, F. A. Greer, & T. J. Army) Academic Press, New York, ix + 342 pp., illustr.

III

SOILS OF THE ARID ZONES

Factors of Soil Formation in Arid Regions

The principal factors of soil formation are the nature of the parent material, climatic conditions, vegetation, soil organisms, and relief—all of which interact with one another, thereby intensifying or mitigating the effects of each individual factor.

The soil-forming processes, under the combined effects of climatic and biological factors, proceed in a number of well-defined stages until the soil reaches 'maturity'—at which stage it is more or less in equilibrium with its environment.

PARENT MATERIAL

The mineral content of arid soils depends mainly on the nature of the parent rock. In sierozems of central Asia, it was found that kaolinite predominates in the clay fractions of certain soils, and montmorillonite in others (Rozanov, 1951). The influence of the parent material on soil characteristics is most apparent in young, imperfectly developed soils.

However, as many rocks give rise to often essentially similar final products of disintegration and chemical change, though in different proportions, the influence of the parent material may become largely obliterated in mature soils by the various soil-forming processes.

CLIMATE

Though the nature of the parent rock has some influence on soil properties, climatic factors tend to have a greater influence in this respect, especially on soil properties of agricultural significance. This is borne out by the fact that mature soils which have developed in different regions but under similar climatic conditions, have certain common characteristics, which are modified only by the nature of the parent rock and other specific local conditions such as slope, run-off, drainage, etc.

The soil-forming factors in the dry regions of the world are essentially the same as those influencing soil formation in more humid regions; their relative intensities are, however, different, and the soils formed are characteristic of the arid regions (Buol, 1965). In arid climates, physical weathering processes predominate whilst the chemical weathering processes, that are very effective in humid climates, can operate effectively only during the relatively short periods during which the soils are wet. The most important processes of soil formation in arid regions are therefore erosion and redeposition, together with leaching of soluble materials and their redeposition. Clay formation, organic matter production, oxidation and reduction, etc., are processes of lesser importance.

The two main climatic factors influencing soil formation are *rainfall* and *temperature* (Figure 16), while *winds* can also be important.

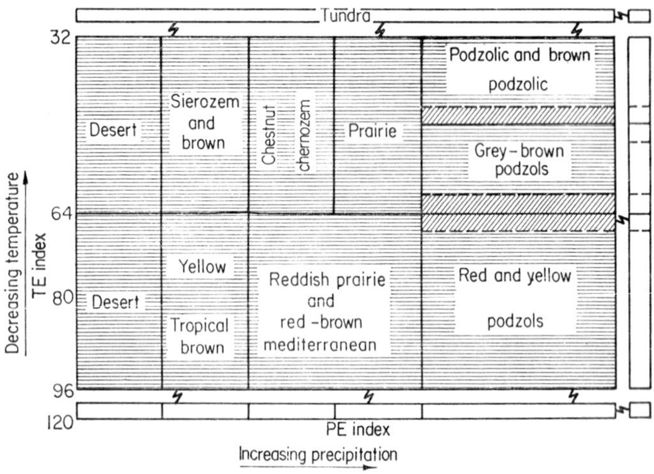

Figure 16. Relation between climates and the great soil groups.

Rainfall

Water is essential for all the processes involved in chemical weathering of the parent rock: hydrolysis, hydration, solution, leaching, eluviation, oxidation, and carbonation. When the soil dries, dehydration also has important effects.

The alternate wetting and drying of the parent materials, which is particularly marked in the arid regions, also contributes to physical weathering—as a result of the stresses and strains that occur with each cycle of wetting and drying. The limited precipitation and rapid evaporation characterizing the arid regions, limit the depth of penetration of the rain, and result in long periods during which most of the soil is dry. Soluble materials are carried downwards in solution, and solid particles by eluviation, to the depth of penetration of the water. This depth depends on the nature of the soil, and on the amount and intensity of rainfall. It is only very rarely that the entire soil profile is leached. As a result, soluble salts and calcium carbonate accumulate at a relatively shallow depth; the carbonates frequently form a hard-pan, called 'caliche', that is typical of many soils in the dry regions. Silica may contribute to the cementation of soil layers. The calcium carbonate may also appear as soft or hard nodules in the 'calcium layer'. A continuous process of soil erosion may expose this calcium layer, which is then called 'croûte calcaire'. These calcium crusts cover large areas of desert and semi-desert soil (Kelley, 1951).

The fact that a horizon of calcium carbonate is characteristic of relatively low-rainfall lands, has been used to classify all soils into two major groups: the *pedocals* of the arid and semi-arid regions, in which the characteristic constituent is calcium carbonate; and the *pedalfers* of the humid regions, in which alumino-silicate and iron compounds are more characteristic (Marbut, 1921).

The subsoil of many soils in the dry regions shows a higher *clay content* than the top-soil (Buol, 1965). It is assumed that conditions are more favourable for clay formation in the subsoil, because it is moist for longer periods than is the top-soil. Notwithstanding the low rate of clay formation in arid regions, many of the soils are rich in clay (Kelley, 1951).

In the semi-arid regions, on the fringes of the deserts, precipitation is more plentiful, moisture penetrates more deeply, and it is retained for longer periods. This makes possible a better-developed vegetative cover. The soils of these regions, especially in the cooler, temperate semi-arid lands, are therefore characterized by a dark-coloured top-soil, with a relatively high content of organic matter, and a subsoil that is rich in soluble elements and in calcium.

The Soil Profile. As a result of the leaching etc. processes described above, it is accepted practice to distinguish between a number of horizons of the soil profile:

A: is the alluvial horizon from which material has been leached out;
B: is the eluvial horizon in which the clay carried down from A has accumulated; and
C: represents the disintegrated but less-weathered parent material.

Concentrations of carbonates, sulphates, and soluble salts, may be found in the master B horizon (and indicated as Bca, Bs, and Bsa, respectively) or in the C horizon (indicated as Cca, Cs, and Csa, respectively).

Temperature

The intense heating during the day and rapid cooling at night, that are characteristic of arid regions, are conducive to alternating expansion and contraction of the parent rock which cause cracking of the rocks and the flaking off of rock fragments; however, the importance of this process appears to have been overestimated (Hills *et al.*, 1966).

High temperatures also intensify the chemical weathering processes in the relatively short periods during which the soils are moist, the speed of chemical reactions doubling, approximately, with every increase of 10°C in temperature.

Winds

Winds intensify the drying effects of aridity and promote erosion. The sparser the vegetation, the greater the erosive power of the wind. In the desert and semi-arid regions, the high-velocity winds carry sand particles and cause a considerable amount of abrasion and scouring of rock surfaces. They also transport and finally deposit loess and sands.

The desert winds are the main causal agent of the typical desert 'pavements'—the 'regs' and 'serirs', also called 'desert armour' in North America (Plate 2). The desert pavement is a residual layer of rock fragments, 2–5 cm thick, which remains after all the fine particles have been carried away. It can form only on soils which consist of a mixture of fine and coarse particles.

Winds are also the causal agents of great depressions which appear in many deserts. A water-table close to the surface transforms many such depressions into oases (Hills *et al.*, 1966).

BIOLOGICAL SOIL-FORMING AGENTS

Vegetation generally plays a far less important role in the soil formation processes of desert soils than in those of the semi-arid regions.

Soils in the dry regions are characterized by a low content of organic matter, as a result of the sparse vegetation producing but little residue. The organic matter content of some desert soils may be much less than 1 per cent. The high temperatures favour the predominance of oxidizing processes, resulting in limited formation of humic acids (Marbut, 1929). The C/N ratio is relatively low, possibly as a result of the activity of nitrogen-fixing blue-green Algae, which are frequently present on the soil surface. Nitrifying bacteria have also been shown to be active in relatively dry soils (*see* p. 337).

When the soil is poor in calcium and contains sodic alkali, the humus is dispersed, producing the typical black colour of these soils. While the typical soil fauna of humid soils, especially earthworms, are commonly absent from desert soils, many other animals burrow in the soil to protect themselves from the hot sun and dryness during the day, and thereby have some influence on soil characteristics.

RELIEF

Relief affects soil formation by influencing run-off, erosion, sedimentation, and drainage. The run-off from slopes may collect in depressions, creating relatively humid 'micro-regions' which result in the formation very locally of soils that are characteristic of regions of higher rainfall.

There exists a close relationship among the factors affecting soil formation—especially relief, climate, and soil cover. In the extremely arid areas, the lack of plant cover favours the removal by wind or water of the fine soil particles, which can be redeposited only in closed basins. In somewhat less arid conditions, the ecological environment in depressions or on the cooler slopes may enable development of a dense vegetation cover that is sufficient to collect and hold fine eroded particles. With increasing precipitation and on approach to semi-arid conditions, most of the area that is of a moderate relief will be covered with fine soil.

Classification of Soils of Arid and Allied Zones

The main soil groups of the arid and allied zones are:

(a) *The zonal soils*, which are the so-called 'normal soils', whose principal characteristics reflect the influence of climate and vegetation on their formation. They can be divided according to whether the area is desert, semi-arid, or sub-humid, each of these categories having its typical soils (Aubert, 1962).

(b) *The azonal soils*, which are similar to the parent materials from which they are formed, being too young to have well-developed characteristics.

(c) *The intrazonal soils*, which are those in which a local factor, such as relief, parent material, or age, have had the dominant influence in forming the soil. Solonchak and solonetz soils belong to this group.

As a result of the high rates of erosion which characterize the arid and allied zones, these have a lower proportion of 'normal' soils than do the humid regions, but a high proportion of intrazonal and, especially, azonal soils (Kellogg, 1953).

ZONAL SOILS

Extremely Arid Conditions

Desert Soils. Desert soils are so little weathered that many soil scientists consider them to be still rock, and not soil at all. The physical condition of these media is usually favourable for plant growth, while the nutrient supply in them is commonly ample and well balanced, moisture being the yield-limiting factor.

These soils have an AC or ABC profile. The A horizon is light-coloured, often vesicular, and weakly platey, with a low organic-matter content (less than 0·5 per cent). The B horizon, if present, is finer-textured than the A.

The younger soils are usually calcareous throughout, show accumulation of soluble salts and gypsum, and are shallow (20–40 cm). The older soils, on residual landscapes, are usually deeper and less calcareous than the younger ones. Desert soils are dry most of the time but may occasionally be wetted to a considerable extent by infrequent but heavy rains or surface floods (F.A.O., 1964). About half the total area of these soils is virtually devoid of vegetation but much of the rest provides some browse. With irrigation, these soils may become very productive.

Arid Areas

These are areas with an average annual rainfall of commonly 100–250 mm though it may be up to double the latter figure—especially when mostly in summer (cf. p. 29). The soils are somewhat more developed than are full desert soils, but as their development is confined to very short periods every year, it is extremely slow. Such soils have been subjected to very little leaching and, because of the sparse vegetation, are low in organic matter.

Sierozems. These are soils with an AC or an A(B)C profile. The A horizon is weakly developed, light greyish-brown or light-grey, vesicular, and weakly platey, with a low to moderate content of organic matter (less than 2 per cent). The B horizon, if present, is finer-textured than the A and has a prismatic structure.

These soils are commonly weakly calcareous in the surface horizons, become highly calcareous in depth, and the zone of lime accumulation may be present in the (B) while a gypsiferous zone is usually present in the subsoil. These soils occur in climates with cool to cold winters, and rainfall during the cool season. The soils are dry for most of the year and are only occasionally moistened throughout their depth (F.A.O., 1964).

Brown Soils. These are soils with an A(B)C profile. The A horizon (10–25 cm) has a distinct crumb structure, with a moderate content of organic matter. It grades into a brown to dark-brown, coarse sub-angular blocky (B) horizon.

Brown soils are generally calcareous throughout and frequently the (B) horizon overlies a zone of lime accumulation. The silicate clays are dominantly of the illite and montmorillonite groups. These soils, which vary in depth from 30 to 50 cm, occur in areas with cool to cold winters, and with rainfall during the warm season. The soils are dry for most of the year and are only occasionally moistened throughout (F.A.O., 1964).

Tropical (Arid) Brown Soils. These are soils with an AC or an A(B)C profile. They have a well-developed crumb-structured A horizon (20–30 cm); the organic matter content is low (about 1 per cent) but evenly distributed in depth. A weakly platey structure may develop at the surface. The (B) horizon has a weak, sub-angular and blocky structure. Its colour is brown to reddish-brown, reflecting a segregation of iron in the profile.

Tropical brown soils may be weakly calcareous and show accumulation of $CaCO_3$ in the deeper soil layers. The silicate clays are composed of a mixture of kaolinite, illite, and montmorillonite. The depth of these soils ranges from 50 to 80 cm. They occur in tropical climates in which the soil is dry for most of the year. The short rainy season occurs in the hot period of the year, during which the soil is occasionally moistened throughout (F.A.O., 1964).

Agricultural Characteristics of Soils of Arid Areas. Most soils of the arid areas, as a result of limited leaching by rainfall, are usually fairly well supplied with mineral nutrients, though these may be in poor balance with one another. Potassium is usually abundant, the content of phosphorus is relatively low, and deficiencies of micronutrients are quite common. Because of the sparse vegetation, these soils are low in organic matter and nitrogen.

Without irrigation, the sierozems are used mainly for range-land for sheep. However, these soils have good physical characteristics and can be productive under irrigation. Soils with caliche crusts at or near the surface have little agricultural value (Dudal *et al.*, 1966).

The least arid of these soil groups—the brown soils—are frequently used for dry-farming; yields on them tend to be low and erratic, and the risks of failure are high. A common characteristic of the arid soils is their lack of stability, and man's intervention usually results in erosion. When irrigated, they are easily leached if the drainage is good; but if drainage is poor, they are easily salinized.

Semi-Arid Areas

The semi-arid areas normally have an average annual rainfall of 250–350 mm, or more than 500 mm if mainly in summer (cf p. 29), but show great variations from year to year. The following soils are particularly characteristic of these areas.

Yellow Soils. These are soils with an A(B)C profile. The A horizon is moderately high in its content of organic matter. The (B) horizon is brown to yellowish-brown, sub-angular and blocky, and has more clay than the C horizon but shows little or no evidence of clay movement. The sub-angular structure is sometimes combined into prisms. The (B) horizon may overlie a layer of $CaCO_3$ accumulation.

Yellow soils range from 40 to 80 cm in depth and are usually calcareous.

They are found in regions of warm Mediterranean climate in which the soil is dry for several months of the year. Rainfall occurs in the cool season, during which the soil is moistened throughout its depth for a brief period (F.A.O., 1964).

Chestnut Soils. These are soils with an A(B)C or ABC profile. The A horizon is relatively thick (30–50 cm) and granular, with a moderately high content of organic matter which is mull.* The A horizon may grade into a dark-brown to reddish-brown B horizon with a prismatic structure and often showing evidence of clay movement. The base-saturation is generally higher than 80 per cent, while the ratio of bivalent to monovalent cations decreases with depth. The B horizon may overlie a horizon of lime accumulation which is often hardened, for these soils are moderately calcareous and the content of calcium carbonates increases in the lower horizons. The silicate clays are mainly of the illite group. Chestnut soils occur under semi-arid climates in which the soil is dry for several months of the year and is only occasionally moistened throughout its depth during the rainy season (F.A.O., 1964). They may also occur in areas of sub-humid climate and so link up with the next category.

Semi-arid to Sub-humid Areas

Chernozems. These are soils with an AC or A(B)C profile. The A horizon is deep (40–100 cm), granular, and dark-coloured to nearly black (hence the derivation of the name from the Russian) because of the high content of organic matter. The A horizon generally shows a high degree of biological activity. It grades into the C or into a light-brown to brown (B) horizon. The (B) horizons are usually transitional layers between the A horizons, which are rich in organic matter, and the C horizons consisting almost entirely of parent material (Winters & Simonson, 1951).

Chernozems may be calcareous throughout or have a layer of lime accumulation of which the depth varies with the $CaCO_3$ content of the parent material and with climatic conditions. Chernozems have a high base-saturation. They develop in temperate to cool, sub-humid climates in which the soil is dry for several months of the year but may be moistened throughout its depth during the rainy season (F.A.O., 1964); their typical vegetation is grassland.

The chernozems are the great wheat-producing soils of the world. They have high nutrient status and good soil structure, and occur in areas of smooth topography; they are therefore suitable for a great variety of crops. If the moisture régime is favourable, crops on these soils respond well to applications of nitrogen and phosphorus. Deficiencies of trace elements may occur *inter alia* because of the high pH (Dudal *et al.*, 1966).

* Mull is the humus of well-aerated, moist soils.

Vertisols. These are soils with an AC profile. The A horizon is thick and, although dark in colour, relatively low in organic matter content. The upper part of the A horizon is generally granular when dry and sometimes develops a crust, whereas the lower part usually develops a prismatic structure.

The vertisols are heavy-textured and contain clay minerals which cause the soil to contract and expand with alternating wet and dry conditions. Many of these soils are self-mulching and develop a typical micro-relief known as 'Gilgai'. They often contain lime concretions at variable depths, and salt may be found in the lower part of the profile. Vertisols are found in a wide variety of climates, though they usually have a marked dry season (F.A.O., 1964).

Red-brown Mediterranean Soils. These include the non-calcic brown soils and are soils with an ABC profile. The A_1 horizon is well developed and has a moderately high organic-matter content which is mull; this horizon merges into a red or brown, angular-blocky B horizon. A weakly developed A_2 horizon may be present. The clays are of the illite and kaolin groups. The base-saturation is higher than 35 per cent and increases with depth. During the dry period of the year, the A and B horizons become hard and in some soils hard-pans have been reported to occur below the B horizon. The boundaries between the main horizons are distinct.

The red-brown Mediterranean soils range in depth from 40 cm to 100 cm or occasionally more. They occur in humid and sub-humid climates with a marked dry season. The soil is dry for several months of the year but is moistened throughout its depth during the cool rainy season.

These are agricultural soils, suitable for a number of crops—especially irrigated ones. Their productivity depends on their depth, as they are easily eroded. These soils are generally poor in phosphate, and respond well to fertilization—especially under irrigation (Hoore, 1964).

AZONAL SOILS

Rock and Rock Debris. These are virtually landforms covered with consolidated or unconsolidated mineral materials showing no soil development. Included in this category are barren deserts, moving dunes, salt plains, and highly gypsiferous deposits (F.A.O., 1964).

Lithosols. These are soils with an (A)C profile. The (A) horizon is weakly developed, shallow, mineral, generally stony, and merges into or rests upon consolidated rock or rock debris (F.A.O., 1964). The soils of this group cover about 21 per cent of the world's land surface, but account for only 2·5 per cent of the arable land. They are usually too shallow or stony for crop production. Occasionally, they may have some agricultural potential; in North Africa they are used for growing vines and in South Africa for extensive grazing (Hoore, 1964).

Regosols. These are again soils with an (A)C profile in which the (A) horizon is weakly developed. They differ from lithosols in that they are formed from unconsolidated or soft materials that are predominantly mineral but not stony (F.A.O., 1964). Unlike the lithosols, they are deep soils with uniform soil characteristics from the surface downwards to a depth of 150 cm or more. Most of these soils are sands.

Sandy Soils. In arid regions, sandy soils are derived from the breakdown of sandy sedimentary rocks (Kellogg, 1953). Because of the sparse vegetation, the dryness, and the absence of clay particles, the components of these sandy soils are easily moved by the wind, to form shifting sand-dunes.

For desert vegetation, the best soils are undoubtedly the sandy soils, and the most favourable sites are along the lower, gentle slopes of dunes (Bagnold, 1954). With their high permeability, sandy soils rapidly absorb rainfall even from relatively light precipitation, and run-off is minimal even after heavy storms. At a depth of 20 to 30 cm, water may remain to form a moist if unsaturated zone between dry layers of sand, both above and below for extremely protracted periods.

It has long been recognized that sandy soils, when irrigated, frequently form the most productive oases of the North African deserts (Hilgard, 1930). Their productivity may remain unimpaired even after long periods of irrigation with relatively saline water (Plate 3).

INTRAZONAL SOILS

Saline Soils

Many arid soils have a high content of soluble salts, which may have been derived from a variety of sources:

(a) they may have resulted from the weathering of the parent rock without having been removed by the scanty precipitation;

(b) they may have been brought in by the wind, originating in sea-spray or from dry efflorescences of salt on the surface of other saline soils; or

(c) they may have risen to the surface, by capillarity, from mineralized ground-waters (the higher the water-table, the greater the concentration of salts). If the water-table approaches to within 2 to 2·5 m of the surface, the situation begins to become critical (Aubert & Kovda, 1967).

Because in arid regions the rainfall is not sufficient to leach them away, these salts tend to accumulate in the soil, while the high evaporation-rate concentrates them near the soil surface.

Arid soils usually have a pH above 7·0, and are frequently highly alkaline. In calcareous soils, the alkalinity is mainly due to calcium saturated clays, the pH of which is less than 8·5. In soils in which sodium saturated clays predominate, sodium hydroxide forms by hydrolysis, and, after reacting with CO_2, produces sodium carbonate. These soils have a pH that is usually higher than 8·5.

A high proportion of the saline soils of the world are alluvial in origin. As a rule, they are of high potential fertility, which can be realized if the excess salts are removed by appropriate procedures (Kelley, 1951). Without such reclamation, crop production is either impossible or at a very low level of productivity.

Saline Soil Types

There are three types of soils with excessive concentration of salts: saline soils, with excessive concentrations of soluble salts; sodic soils,* with an excess of exchangeable sodium; and saline-sodic soils, which have a high concentration both of soluble salts and exchangeable sodium.

Saline Soils (Solonchaks). These are soils in which the total concentration of salts has reached a level that interferes with the normal development of most crop plants. This usually occurs at levels exceeding 0·1 per cent of soluble salts (Kearney & Schofield, 1936). A generally adopted criterion for estimating soil salinity is the electrical conductivity, at 25°C, of the solution extracted from a saturated soil paste. A soil is considered to be saline if the electrical conductivity of the saturation extract is 4 mmhos/cm† or more, and the percentage of exchangeable sodium is less than 15 (Richards, 1954).‡

Sodium salts constitute only a relatively small proportion of the total salt content of these soils, and sodium is consequently not adsorbed in significant quantities. The other salts that may be present in various proportions consist mainly of the cations calcium, magnesium, and potassium, and the anions chloride and sulphate—with, to a lesser degree, bicarbonate, carbonate, borate, and nitrate. The exchangeable ions are mainly Ca^{++} and Mg^{++}, the exchange complex being almost entirely saturated with these two ions.

In these soils, white salt-crusts appear on the surface and the pH is usually less than 8·5. Because of the large excess of soluble salts over exchangeable sodium, these soils are generally flocculated; soil structure is usually satisfactory and water-stable. The main cause of the negative effect on plant growth is the increased osmotic potential, with its concomitant reduction in availability of soil-water.

* Formerly called 'alkali' soils.

† Electrical conductance is expressed in 'mhos', i.e. reciprocal ohms, while electrical conductivity has dimensions of mho/cm. As the standard unit mho/cm is large, the unit millimhos/cm, or mmhos/cm, ($EC \times 10^3$), is generally used in salinity work.

‡ The limits indicated for salinity and sodicity are not rigid, but depend also on the plants concerned (Bernstein, 1962).

Sodic (Alkali) Soils (Solonetz). These are soils in which the total concentration of salts is not excessively high (less than 4 mmhos/cm), but the proportion of exchangeable sodium is sufficiently high to affect plant growth adversely. When the ratio of soluble Na^+ to soluble $Ca^{++} + Mg^{++}$ exceeds 2, the sodium is capable of displacing calcium and magnesium in the exchange complex, and sodium is adsorbed on the surface of clay particles.

$$\boxed{\text{Clay}}\,Ca^{++} + 2Na^+ \rightleftharpoons \boxed{\text{Clay}}\begin{matrix} Na^+ \\ \\ Na^+ \end{matrix} + Ca^{++}$$

The adsorbed sodium is not easily leached from the soil, unless it is in turn displaced from the exchange complex by sufficiently large additions of calcium to the soil.

When the clay calcium has been almost completely replaced by sodium, some of the excess sodium combines with soil CO_2 to form Na_2CO_3. This increases soil pH, which may reach values of 8·5 to 10·0 (Richards, 1954). The sodium carbonate causes the dispersion of part of the soil organic matter, which becomes dissolved in the soil solution. The water rising by capillarity to the surface of such soils leaves behind a dark residue, called 'black alkali'.

As a result of the displacement of calcium and magnesium from the exchange complex, the clay particles also tend to become dispersed and soil structure deteriorates, with a resulting decrease in percolation rates. The dispersed clay particles tend to move downwards with rain or with the irrigation water; they are thus removed for the surface soil, which in due course may become coarse-textured. The eluviated clay accumulates at a lower depth, forming a dense, impermeable layer. As a result of alternate wetting and drying, typical morphological structures develop which have the form of columns, prisms, or blocks, when dry (Kelley, 1951).

Whilst the replacement of calcium by sodium in the exchange complex is a reversible process, the eluviation of the clay particles and the formation of zones of clay accumulation is irreversible—which may be the main obstacle to the reclamation of these soils.

When wet, alkali soils become plastic and sticky; after drying, hard clods are thrown up from them by tillage operations. Consequently it becomes very difficult to prepare a good seed-bed, or to produce good tillage, where such soils are concerned. Moreover, plant growth suffers from inadequate aeration of the soil, low infiltration rates, and the formation of hard crusts on the soil surface after rains or irrigation.

Saline-Sodic Soils. Saline-sodic soils form as the result of combined processes of salinization and alkalization. They are soils in which a high content of total salts (greater than 4 mmhos) is combined with a high proportion of exchangeable sodium (more than 15 per cent); soil pH is usually less than 8·5. The high concentrations of soluble salts have a flocculating effect, which offsets the adverse tendency of adsorbed sodium to cause dispersion of the clay particles. Therefore, as long as the soil solution contains a high proportion of soluble salts, a more or less favourable soil structure can be maintained, and infiltration rates will remain relatively high.

However, when the soluble salts are leached out, in the absence of an abundance of calcium, the sodium clays become hydrolized, forming free sodium hydroxide. This deflocculates the clay particles, so that the saline-sodic soil is transformed into a sodic soil with its undesirable characteristics. If, however, sufficient gypsum is present in the soil, the leaching water dissolves the calcium, which replaces the exchangeable sodium concurrently with the leaching of the salts, effectively preventing the deterioration of soil structure.

Factors Influencing the Formation of Saline Soils

Climate. Saline soils occur most frequently in regions of hot and dry climate. The aridity, high temperature, and high evaporation rates, are preconditions for a preponderance of evaporation over drainage, resulting in an adverse salt-balance. Soil salinity may attain 75 per cent, and the soils may be covered by a salt crust up to 50 cm thick in extremely arid deserts, such as those of Arabia and Chile (Kovda *et al.*, 1967).

Soils in deserts with a cold winter may also be highly saline. Here, the low solubility at low temperatures of certain salts is an important factor contributing to salinity.

Topography. A hot, dry climate is not sufficient in itself to cause the formation of saline soils. When the ground-water table is deep and evaporation from the soil is equal to precipitation, salts do not accumulate.

When an arid region lies in a deep depression, surrounded by mountain-chains, underground water-flow carries dissolved salts into the dry area. As these waters are under considerable pressure, an 'artesian' rise of salts results in salt accumulation at or near the surface. These accumulations usually occur in secondary depressions, and give rise to *soluble salt-pans*—the so-called 'sabkhas' or 'shotts'.

Irrigation and Drainage. Irrigation waters always contain salts, albeit of different kinds and in different concentrations. Water losses by evapotranspiration inevitably cause an increase in concentration of the salts in the root-zone. The crop itself removes only infinitesimal amounts of salt: a cotton crop producing 4 tonnes of cottonseed per hectare, removes barely 10 kg of sodium.

The rapidity with which salts accumulate in the root-zone depends on a number of factors; these include the quality and quantity of irrigation water applied, the method of irrigation, and the adequacy of drainage.

A layer of sand overlying a heavier soil type is a highly effective mulch for moisture preservation. With overhead irrigation, and with proper management (organic manuring, fertilizers, etc.), these soils may be extremely productive.*

Alluvial Soils

These are soils with an (A)C profile formed on recent deposits of fluvial, marine, or lacustrine origin. The A horizon, when present, is weakly developed and merges into undifferentiated, wet, mineral material that constitutes the C horizon and often shows evidence of reduction (gley). The total thickness and organic matter content vary sharply with drainage and climate (F.A.O., 1964).

Alluvial soils occur on bottom-lands, floodplains, and low terraces along streams, the term 'alluvial' applying to recently deposited water-laid materials which have been little affected by the local environment. Their characteristics are determined by the nature of the parent materials, and the manner in which they have been transported, sorted, and deposited.

Alluvial soils are characteristic of the principal irrigated areas of the world, such as the Nile Valley, the basins of the Tigris and Euphrates, the basin of the Indus and its tributaries, etc.

The fertility of alluvial soils depends on their origin: if they developed from poor rocks or from laterite, their fertility will be extremely low; if they developed from the erosion of nutrient-rich soils and rocks such as limestone and basalt, they can be very fertile (Kellogg, 1953).

REFERENCES

AUBERT, G. (1962). Arid zone soils. A study of their formation, characteristics, utilization and conservation. *Arid Zone Res.*, **18**, 115–38.

—— & KOVDA, V. A. (1967). Soils in relation to salinity, irrigation and drainage. pp. 92–122 in *Irrigation and Drainage of Arid Lands*. F.A.O./UNESCO, Paris, 663 pp., illustr.

BAGNOLD, R. A. (1954). The physical aspects of dry seasons. pp. 7–12 in *Biology of Deserts* (Ed. J. L. Cloudsley-Thompson) Institute of Biology, London, iv + 224 pp., illustr.

BERNSTEIN, L. (1962). Salt-affected soils and plants. *Arid Zone Res.*, **18**, 139–74.

BUOL, S. W. (1965). Present soil-forming factors and processes in arid and semi-arid regions. *Soil Sci.*, **99**, 45–49.

* On the other hand, in many desert areas as for example in southern Iraq, surface salt accumulation—often in the form of extensive white crusts—may be sufficient to prevent all agriculture over considerable areas, which thus become almost totally unproductive with prolonged irrigation in the absence of adequate drainage.—Ed.

D'HOORE, J. L. *See* HOORE, J. L. D'
DUDAL, R., TAVERNIER, R. & OSMOND, D. (1966). *Soil Map of Europe*. F.A.O., Rome. 120 pp., 6 maps.
F.A.O. (1964). Preliminary definitions, legend and correlation table for the soil map of the world. *World Soil Res. Repts* No. 12, 17 pp. (mimeographed).
HILGARD, E. W. (1930). *Soils: Their Formation, Properties, Composition and Relations to Climate and Plant Growth in the Humid and Arid Regions.* Macmillan, New York. xxvii + 593 pp., illustr.
HILLS, E. S., OLLIER, C. D. & TWIDALE, C. R. (1966). Geomorphology. pp. 53–76 in *Arid Lands: A Geographical Appraisal.* (Ed. E. S. Hills) UNESCO, Paris, xviii pp., illustr.
HOORE, J. L. D' (1964). *Soil Map of Africa.* Publ. Comm. for Technical Cooperation in Africa, Lagos, 205 pp., 6 maps.
KEARNEY, T. H. & SCHOFIELD, C.S. (1936). The choice of crops for saline land. *Circ. U.S. Dep. Agric.*, **404**,
KELLEY, W. P. (1951). *Alkali Soils.* Reinhold, New York. vi + 176 pp., illustr.
KELLOGG, C. E. (1953). Potentialities and problems of arid soils. pp. 19–42 in *Desert Research Proceedings* Research Council of Israel, Jerusalem, 641 pp., illustr.
KOVDA, V. A., EGOROV, V. V. & HILLS, N. S. (1967). Landscapes in relation to irrigation, drainage and salinity. pp. 218–45 in *Irrigation and Drainage of Arid Lands*, F.A.O./UNESCO, Paris, 663 pp., illustr.
MARBUT, C. F. (1921). The contribution of soil surveys to soil science. *Soc. Prom. Agr. Sci. Proc.*, **41**, 116–42.
—— (1929). The relation of soil type to organic matter. *J. Am. Soc. Agron.*, **21**, 943–50.
RICHARDS, L. A. (Ed.) (1954). Diagnosis and improvement of saline and alkali soils. *Handb. U.S. Dep. Agric.*, **60**, U.S. Govt. Printing office, Washington D. C., vii + 160 pp., illustr.
ROZANOV, A. N. (1951). *The Serozems of Central Asia.* Translated from Russian by the Israel Program of Scientific Translations, Jerusalem (1961), 550 pp., illustr.
WINTERS, E. & SIMONSON, R. W. (1951). The subsoil. *Adv. Agron.*, **3**, 2–86.

IV

VEGETATIONAL FEATURES OF THE ARID ZONES

The natural vegetation reflects all the influence of climatic factors, and it is therefore not surprising that vegetation zones coincide closely with climatic zones (Figures 17 and 18).

Though vast expanses of the arid lands are not cultivated, it would be a mistake to assume that the types of 'natural' vegetation encountered on them have not been profoundly affected by Man. This is especially true for desert environments in which the equilibrium between plant cover and environment is most precarious. Once this equilibrium is destroyed, the process may well be irreversible—especially if soil erosion is increased as a result of man's intervention. Overgrazing, burning during the dry season, and cutting of shrubs and trees for fuel, have caused great changes in the composition of the natural vegetation of most of the world's deserts and semi-arid regions.

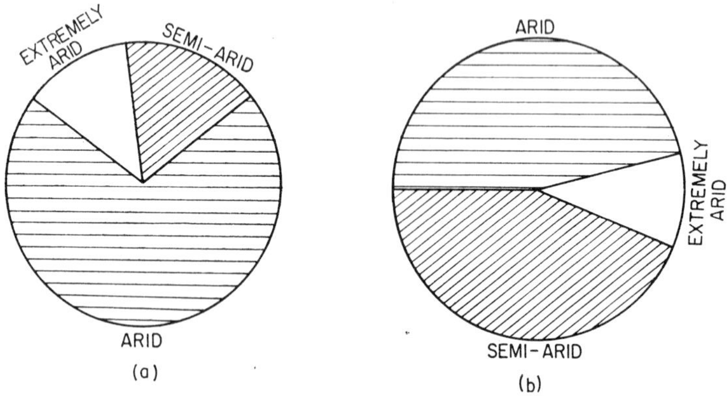

Figure 17. World areas of arid and semi-arid land based on (a) Vegetation and (b) Climate. (Data from Shantz (1956).)

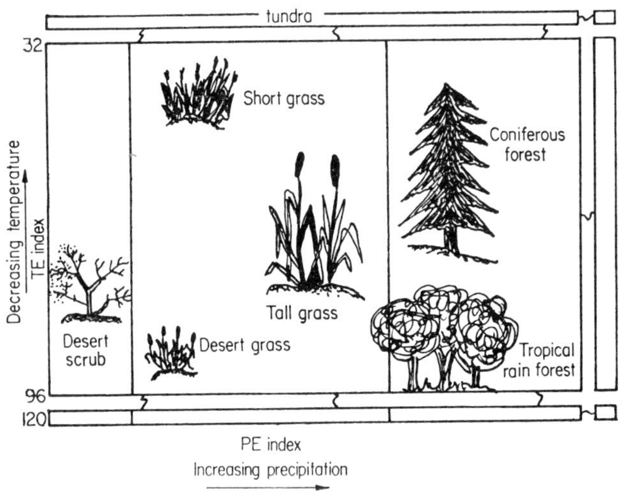

Figure 18. Relation between climates and the main vegetation types.

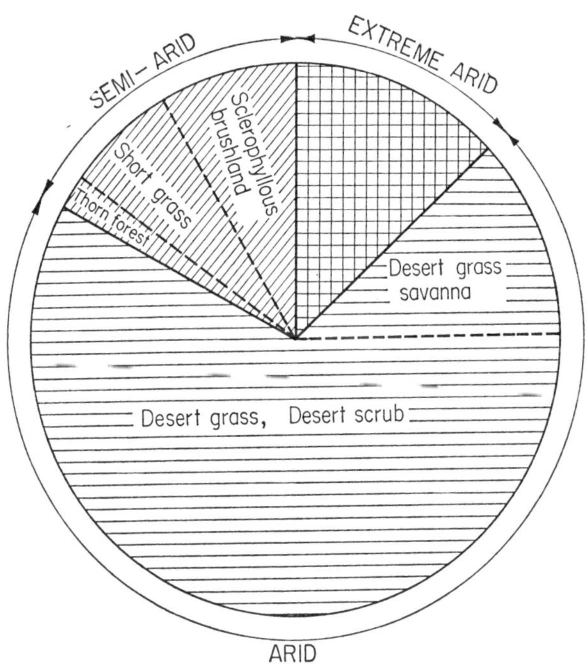

Figure 19. Relative areas of different vegetation types in arid regions. (Data from Shantz (1956).)

Principal Types of Desert Vascular Plants

Only a relatively limited number of types of plants are adapted to living in the very arid regions, but they may be found over immense areas (Figure 19). There are two basic groups of desert vascular plants: ephemerals and perennials (Zohary, 1962).

Ephemerals are annual, short-lived plants, able to achieve maturity within a few weeks after germination. The ephemerals occur in considerable numbers in rainy periods, but do not appear during drought years.

Single heavy rainstorms allow the water to penetrate the soil to a sufficient depth, and the soil to maintain its moisture long enough, to enable seeds that have lain dormant for long periods to germinate and the resulting plants to grow and produce seeds. This ephemeral vegetation is usually the main pasture resource on which nomadic flocks depend in deserts.

Some typical species of ephemerals are *Mesembryanthemum forskalei*, *Zygophyllum simplex*, *Plantago ovata*, *Trigonella stellata*, *Matthiola livida*, and *Stipa cupensis* (Migahid, 1962).

Perennials are the permanent and dominant species of the desert. They maintain themselves by successful drought evasion or drought endurance, or by their physiological, morphological, or anatomical adaptations to reduce water-loss, and hence their ability to survive prolonged periods of saturation deficits.

The two main groups are the succulent perennials, such as the cacti of the American deserts and the *Euphorbia* spp. of the African and Asian deserts, and the non-succulent perennials comprising woody 'herbs', grasses, shrubs, and trees: e.g. *Hammada* spp., *Anabasis* spp., *Acacia* spp., *Artemisia* spp. Also prevalent in many deserts are various lichens and other non-vascular cryptogams.

In years of prolonged drought, a large proportion of the vegetative growth of the non-succulent perennials may dry up completely; conversely, in rainy years, they respond quickly to the favourable moisture conditions by renewed growth.

Effects of Climate and Physiographic Features on Desert Vegetation

PRECIPITATION

Under arid conditions, the density of vegetation decreases with decreasing precipitation, giving a 'diffuse type of vegetation' (Monod, 1954) which is characterized by a very shallow root-system with considerable horizontal growth (Walter, 1961).

In many deserts the vegetative 'cover' is so sparse that there is commonly no competition between species or even between individual plants (Emberger & Lemée, 1962). The struggle for survival is, rather, with the environment. The absence of competition is the direct result of the ability of the already established plants to prevent competitors from gaining a roothold in their vicinity. In certain cases, such as *Parthenium argentatum* (Guayule), the roots of the plant excrete inhibitors into the surrounding soil, which effectively suppress young *Parthenium* seedlings. In other cases, such as *Encelia farinosa*, it is the plant's own dead leaves and detritus that contain the inhibitors. All these inhibitors are water-soluble, so that the greater the precipitation is, the more effective will be the leaching and the closer together the plants can grow. Conversely, with insufficient precipitation, the plants remain widely and regularly spaced—a characteristic feature of most desert shrubs (Went, 1955).

Desert areas where rainfall is sufficient to maintain permanently such 'diffuse', sparse vegetation are called 'rain-deserts' (Zohary, 1962).

TOPOGRAPHY

Topography plays a major role in the water-supply of desert vegetation. Water runs off slopes and collects at lower levels, which may therefore receive moisture far in excess of the actual rainfall. The amount received will depend on the size of the catchment area, and its effectiveness will depend on the uniformity of water distribution in the receiving area. Frequently, the run-off will concentrate in the narrow beds of *wadis** and the receiving areas will then be very small in relation to the catchment area. Wadis and other depressions may contain fine alluvial soil which has better water relationships and higher fertility than the surrounding rocky desert slopes.

In regions with less than 100 mm of rainfall annually, soil moisture is usually insufficient to maintain even a diffuse type of vegetation. In such an environment, the so-called 'run-off deserts' (Zohary, 1961), most of the vegetation is concentrated in depressions. This gives rise to the paradoxical situation that in deserts with less than the minimum precipitation needed to maintain plant life, the dominant type of vegetation (the 'restricted type' of Monod, 1954) is less xeromorphic than the diffuse vegetation occurring under conditions of higher precipitation. This restricted vegetation has a deeper root-system, which is not surprising if one realizes that a total annual precipitation of not more than 50 mm may provide the equivalent of 500 mm or more in the restricted areas in which the run-off is concentrated.

Most of the Syrian Desert and southern Iraq are 'rain-deserts', whilst the Egyptian Desert, large parts of Sinai, and the southern Negev, with still less

* Watercourses that are dry except during and soon after rainy periods. Such intermittent streams in the desert regions of south-western North America are called 'arroyos' —Ed.

rainfall are 'run-off deserts'. The critical rainfall above which diffuse vegetation occurs is conditioned by relative humidity, temperature, and soil type, but usually lies between 50 and 100 mm. However, where a sandy soil overlies a fine-textured soil (such as loess), the lower limit of 50 mm may still be sufficient to maintain diffuse vegetation (Zohary, 1962) (cf. our Plate **4**).

VEGETATION-TYPES IN THE ARID CLIMATES

THE LOW LATITUDES

The principal vegetation-types in regions of arid climate in the low latitudes depend largely on whether the precipitation occurs predominantly in winter or in summer.

The winter-rainfall vegetation belt stretches from Morocco, through North Africa and the Near East, up to Baluchistan, with corresponding areas in California, South Africa, northern Chile, and western Australia.

The summer-rainfall vegetation belts occur in North America from Mexico to Arizona, in South America from Gran Chaco to western Argentina, and in South Africa from the Kalahari to the eastern Karoo.

The dominant vegetation-types of the warm winter-rainfall-region deserts of Eurasia tend to consist of such grasses as Halfa Grass (*Stipa tenacisssima*), such shrubs as *Artemisia herba-alba,* and such dwarf-shrubs as *Astragalus* spp. The dominant trees are commonly *Acacia* spp. and *Zizyphus* spp. (Troll, 1956).

In the arid pastoral areas of southern Australia, with a mean annual rainfall of less than 250 mm, the dominant species are commonly perennial shrubs of *Atriplex* (salt-bushes) and *Kochia*. Both are palatable to sheep. Overgrazing has resulted in marked damage to these associations, and recovery has generally been limited (Trumble & Woodroffe, 1954).

The warm regions with summer rainfall in North America have annual precipitation that generally varies between 75 mm and 275 mm. Evaporation rates are extremely high. The main components of the vegetation of such areas include Creosote-bush (*Larrea tridentata*), Catclan Acacia (*Acacia greggii*), Whitethorn (*A. constricta*), Palo Verde (*Cercidium floridum* and *C. microphyllum*), Bur-sage (*Franseria deltoidea* and *F. dumosa*), varieties of cacti (*Opuntia* spp. and others), Ocotillo (*Fouquiera splendens*), and Tarbush (*Flourensia cernua*) (Pase, 1965) (cf. our Plate 5).

The Karoos of South Africa have a desert scrub vegetation with a large proportion of succulents.

The grazing value and carrying capacity of the above types of vegetation are extremely low. When sufficient rainfall occurs, however, winter or summer ephemerals occasionally provide improved grazing.

SEMI-DESERT SCRUBS

On the fringes of the low-latitude deserts, on the equatorial side, there is usually a region of semi-desert scrub—a transitional type of vegetation between the true deserts and the tropical savannas (Polunin, 1960). These occur in Africa on the southern fringes of the Sahara, in Arabia, in the south-western United States, and in Mexico. The rainfall in these regions ranges from 200 to 500 mm; the semi-desert scrubs consist of a more or less dense growth of thorny bushes, dwarfed wiry grasses growing in scattered patches, and fleshy-leafed herbs with underground storage organs. In regions in which the rainfall is divided into two short seasons, succulent plants play a greater role: on the semi-arid fringe, thorny shrubs are still present and grow in association with the succulents. The lower the rainfall is, the greater the proportion of succulents, forming the so-called 'cactus-desert'. In the south African Karoo, the numerous types of succulents that are present characterize the transition region from winter to summer rainfall.

Typical succulents in these semi-desert scrubs include Euphorbias in the Old World and cacti in the New World—as well as Agaves, Aloes, and Yuccas (Polunin, 1960). In Australia, the succulents are absent but the peculiar porcupine-grasses (*Triodia* and *Plectrachne* spp.) occur (Walter, 1962).

MAQUIS AND CHAPARRAL

Polewards, on the fringes of the low-latitude deserts, are the regions with a Mediterranean-type climate. The typical vegetation is a man-made disclimax, due to overgrazing and fire, called maquis (Mediterranean) or chaparral (California). The maquis consists of a more or less dense growth of hard-leafed shrubs, many of which are evergreen, associated with xerophilous grasses and dicotyledonous herbs among which short-lived annuals predominate in numbers. The chaparral consists mainly of a thick scrub, composed of plants of many diverse families, associated with succulent and bulbous or tuberous herbs (Polunin, 1960) (cf. our Plate 6).

THE MIDDLE LATITUDES

There is a great similarity in the vegetation-types of the dry regions of Eurasia and North America. In both, the two dominant vegetation-types are: scrub vegetation in the more arid parts, bordered on their poleward side by semi-arid steppes (Polunin, 1960). Each of these vegetation belts is subdivided into regions having a number of often well-defined types of vegetation.

Desert Scrub Vegetation

In these areas in the middle latitudes the amount of precipitation is roughly between 150 mm and 225 mm. The winters are cold, and winter precipitation comes mostly as snow. In North America, Big Sage-brush (*Artemisia trident-*

ata) is widely the dominant plant. It is associated with a number of perennial grasses: Blue-gramma (*Bouteloua gracilis*), Western Wheat-grass (*Agropyron smithii*), and Galleta (*Hilaria jamesii*), though these tend to require better moisture conditions. On the more saline soils *Atriplex confertifolia* is common. The grazing value of this association depends on the ratio of sage-brush to grasses, the former having a very low grazing-value (Pase, 1965).

The vegetation of this belt is matched in Eurasia by open stands of *Artemisia* spp., mainly *A. maritima* and *A. incana,* which occur in the dry regions stretching from the Caspian depression to the Gobi desert. Other prominent scrubby plants include species of *Kochia*, *Eurotia*, *Atriplex*, *Pyrethrum*, and *Ephedra*; on solonetz soils species of *Salsola* and *Salicornia*, and on the solonchaks many other chenopods abound and may dominate.

Semi-arid Grasslands or Steppes

The semi-arid grasslands consist of an almost continuous sward of short grasses, interspersed with shrubs which dominate mainly where the steppes border on desert areas. On their moister margins the steppes merge into the prairies.

Precipitation in the areas of these semi-arid grasslands varies from 250 mm to 400 mm, with most of it falling as rain in summer. In North America the dominant grasses include gramas (*Bouteloua* spp.), Red Three-awn (*Aristida longiseta*), and needle-grasses (*Stipa* spp.), on the heavier and more mature soils, whilst on bottom areas, with soils varying from loams to slightly developed sandy soils, there is a shift towards Western Wheat-grass (*Agropyron smithii*) and Buffalo Grass (*Buchloe dactyloides*) (McGinnies, 1955). Overgrazing has led to the spread of a number of undesirable types of shrubs. The carrying capacity ranges from 8 to 24 ha per animal unit* (Pase, 1965).

The corresponding vegetation in Eurasia consists largely of other needle-grasses (*Stipa* spp.) and a large number of herbs (Troll, 1956).

HALOPHYTES

Characteristics

Halophytes consist of two main types (Walter, 1962):

(a) *Facultative halophytes*, which are relatively salt-tolerant but grow better on non-saline soils.

(b) *Euhalophytes*, which grow better on soils with an appreciable salt-content than on non-saline soils.

* *Animal unit:* The grazing area required to support a mature cow for twelve months and her suckling calf for six months. Five sheep or goats are rated as one animal unit.

Halophytes have at least three characteristics which facilitate their survival on saline soils (Hayward & Bernstein, 1949):

(1) the ability to increase the osmotic values of their tissue fluids in response to increased osmotic pressure of the soil solution;
(2) the ability to accumulate considerable quantities of salts in their tissue fluids, and
(3) protoplasm that is not easily damaged by accumulations of sodium salts in the cell-sap.

Recent investigations have shown that the problem of halophytes is less the ability to adapt to the increased suction-pressure of the soil than the ability to regulate the intake of salts and consequently resist their adverse effect on the plant protoplasm (Walter, 1961).

If the plasma's resistance to salinity is sufficiently high, the problem of water uptake is alleviated: by the taking up and storing of salts in the cells, the suction-pressures developed by the plants, and their power to extract water from a saline soil, are increased. However, if the salt-resistance of the plant plasma is low, the suction-pressure is insufficient and the plants are incapable of taking up water and nutrients from a saline soil. The level at which the internal salt-concentration becomes critical is what differentiates the halophytes from the non-halophytes (Repp, 1961). The internal salt-content of halophytes may be quite high: ash contents of 25–50 per cent of the dry matter have been found, of which 20–50 per cent was sodium (ibid.).

Plants take up salts along with the water which they absorb. As this water is transpired, salts accumulate in the plant tissues. Some halophytes are able, by the following mechanisms, to overcome the problem of an excess accumulation of salts (Walter, 1961):

(a) *Salt excretion* through glandular cells in the leaves: this is an active process which only a small proportion of halophytes are capable of performing.
(b) *Increasing succulence:* as the salt-concentration of the cells increases, morphological modifications occur; among other changes, the cuticle becomes thicker and the stomata become sunken. Also, water-storage organs tend to become more massive and leaves to thicken. These changes are generally accompanied by a reduction in growth in length (Peterson, 1961). The water content of the tissues increases, with a concomitant dilution of the potential salt-concentration.
(c) *Abccission:* when the internal salt-concentration becomes excessive, the older leaves die prematurely and are shed, e.g. in *Atriplex halimus*. The later this occurs, the better the plant is adapted to saline conditions (Repp, 1961). Plants that lack a regulatory mechanism simply die when a certain level of salt-concentration is reached.

Though halophytes may be resistant to drought, salt-resistance is not identical with drought resistance, as is frequently assumed (Repp, 1961).

Habitats

In general, the habitats of halophytes in arid regions are wet; soils that are both dry and saline are usually bare of macroscopic vegetation (Walter, 1961), though occasionally xero-halophytes may be found in these habitats. The principal plant associations of saline areas consist of Chenopodiaceae, e.g. of the genera *Arthrocnemum* and *Salicornia*, as well as Juncaceae and certain Compositae. Species also occur which, although not true halophytes, are more or less tolerant of salinity; examples are found in the grass genera *Spartina*, *Glyceria*, and *Atropis*, and in the species *Cynodon dactylon* (Grillot, 1956).

PHREATOPHYTES*

It may appear nonsensical to include this category of plants which transpire rapidly among the plants of the dry regions. However, in most arid regions there are occasional habitats with adequate moisture throughout the year: such are the banks of streams and rivers, or places where ground-water lies within reach of deep-rooting plants. Phreatophytes or luxury consumers of water in these moist habitats can cause considerable losses to the water economy of these regions. Tamarisks (e.g. *Tamarix gallica*), poplars or cottonwoods (*Populus* spp.), willows (*Salix* spp.), and Mesquite (*Prosopis juliflora*), are typical examples of this category. Mesquite, for example, with an extensive root-system spreading for more than 30 metres, was found to have a water requirement from three to four times as great as that of the grasses with which it competed (McGinnies, 1955).

The shallower the water-table, the greater the water use by phreatophytes. These plants thrive best when the water-table, or its capillary fringe, is 2 to 3 metres below the surface (Fletcher & Elmendorf, 1955). However, certain species are capable of withdrawing water from a depth of 30 metres or more.

In the vicinity of concentrations of phreatophytes, such as Mesquite or cottonwoods, the water levels of wells fluctuate daily: the water drops in the day-time, when the plants are transpiring actively, and rises again during the night, when transpiration is at its lowest. The size of the daily fluctuation depends on the stage, vigour, and density, of the plant growth. When killing frosts cause the cessation of growth, the fluctuation in water level ceases (Fletcher & Elmendorf, 1955). In the western United States, it has been estimated that as much water is consumed by phreatophytes growing in and along streams and irrigation canals as is consumed by irrigated crops (Dixey, 1966).

* Phreatophytes are deep-rooted plants capable of tapping the underground water reserves.

The removal of phreatophytes, and their replacement by shallow-rooted species, could therefore achieve considerable savings of water.

Adaptation of Plants to Drought

There are three principal ways in which plants can adapt to drought:

(1) Evading the drought periods;
(2) Maintaining a favourable internal water-balance, thereby postponing the negative effects of drought; and
(3) Surviving long periods of drought.

DROUGHT EVADENCE

The evadence of drought is the simplest means of adaptation of plants to desert conditions. Many desert plants germinate at the beginning of the rainy season and have an extremely short growing-period, which is confined to the rainy period. Between germination and seed maturity, as few as five (Polunin, 1960) to six (Kassas, 1966) weeks may suffice. In rainy years these 'ephemerals' may occur in considerable profusion, but they do not appear in severe drought years. Instead, they remain dormant—usually in the form of seeds or fruits.

Dormancy and Germination

The seeds of many species that grow in regions with hot, dry seasons, are not able to germinate at high temperatures; they are thereby protected from destruction (Evenari, 1962). In many desert plants, the process of germination can be stopped and restarted at different stages between inhibition and the appearance of the seedling rootlet and shoot. Seeds may remain soaked and return to a state of secondary dormancy without losing their viability, or they may dry out completely after wetting without losing their germinability (ibid.).

The presence of germination inhibitors in such seeds serves as an 'internal raingauge', which permits germination only after an amount of precipitation has fallen that will be sufficient to remove the inhibitor (Went, 1952).

As seeds germinate, vacuoles are formed in the cells and their drought-resistance decreases very markedly before emergence (Iljin, 1957).

Early Maturity

In areas with alternate rainy and dry seasons, the ability of a cultivar to achieve maturity before the soil dries out is the main adaptation to drought of cultivated crops. Therefore, early maturity of varieties of wheat and barley is a breeding objective in the Mediterranean region, and in the semi-arid regions of India and Australia. However, a reduction of the growing period usually signifies a lower potential yielding ability—so that early maturity, as an insurance against early drought, is a sacrifice of potential yield.

MAINTAINING A FAVOURABLE INTERNAL WATER-BALANCE

With the exception of succulent plants, xerophytes may transpire as much as, or even more, per unit area of transpiring surface, than mesophytes, *when the water supply is adequate*. It is only during periods of dryness that the morphological characteristics of the xerophytes give them a distinct advantage in reducing water-losses (Killian & Lemée, 1956).

Further, the efficiency of water-use, as expressed by the ratio of dry matter produced per unit of water used in transpiration, is in general not greater in xerophytes than in mesophytes. This also holds true for drought-resistant varieties of crop plants. In short, dryness is not favourable to xerophytes any more than it is to mesophytes; the xerophytes are merely more resistant to an adverse water-supply (Killian & Lemée, 1956).

Efficient Water Uptake

Root Systems. The root systems of xerophytes are characterized by a wide variety of apparent adaptations. These include responses to such predominant soil conditions as the duration of soil dryness and the depth that is normally wetted, the presence and nature of soil constituents, and the degree of soil salinity.

Plants become adapted to dry conditions mainly by developing an extended root-system, rather than by structural modifications of the roots (Shields, 1958). The root systems of many desert plants are usually shallow, but they are able to adapt themselves, in respect of branching, extent, and orientation, to regional and seasonal variations of the soil-depth at which moisture becomes available, and to the physical characteristics of the soil.

Artemisia herba-alba, which is common in the dry regions of North Africa, at the arid limit of its habitat has roots that grow to a depth of only 5 cm, because moisture does not usually penetrate beyond this depth in its common habitat. In less extremely dry regions, its root penetration is deeper (Killian & Lemée, 1956).

However, certain plants of the semi-arid areas have extremely deep root-systems, enabling them to reach the water-table and thereby maintain active growth even after the upper soil has dried out completely. Typical examples are two common and widespread leguminous plants, *Alhagi maurorum* and *Prosopis stephaniana*, which may have roots penetrating to 15 metres or deeper.

There are considerable differences between cultivated plants in the extent, depth, and efficiency, of their root systems. Sorghum has nearly twice as many small branch-roots per unit length of main roots as has maize and it is assumed that this is one of the main reasons for its greater drought-resistance. An interesting adaptation of the roots of some xerophytes, such as certain *Aristida* and *Pennisetum* species of the Sahara, and *Lygeum spartum* (another gramineous plant) of the arid parts of the Mediterranean region, consists of a dense growth of root-hairs which excrete a mucilage that agglutinates the soil particles around the root. Thus is formed a sheath that helps to protect the root from desiccation.

Another adaptation of some xerophytes is their speedy response to a re-wetting of the soil by the rapid formation of secondary rootlets ('rain-roots') which may appear only two to three hours after a fall of rain (Evenari, 1962).

Plants growing on 'hamadas' and rocky desert soils in rain-deserts usually respond quickly to fluctuations in moisture content, by their ability to grow rapidly towards sources of available moisture. This characteristic is also apparent in certain cultivated plants. It has been shown experimentally that drought-resistance of varieties of spring wheat is related to the rapidity of growth of the primary roots (Aamodt & Johnston, 1936).

Root-to-top Ratio R/T. A high root-to-top ratio is a very effective means of adaptation of plants to dry conditions (Killian & Lemée, 1956), as under such conditions the growth-rate of the roots considerably exceeds that of the shoots. The transpiring surface is thereby reduced, whilst the root system of the individual plant obtains its water from a large volume of soil.

R/T commonly increases with increasing aridity—except in the case of succulents (Walter, 1962). The seedlings of desert perennials may produce a few leaves, after which top-growth ceases for a few months, whilst the root system continues to develop (Kassas, 1966). One-year-old seedlings of *Erodium hirtum* that have germinated under dry conditions have roots that are 20–30 cm deep, whilst the aerial growth is not more than 1–3 cm high: under moist conditions, the R/T ratio of this plant is much smaller (Evenari, 1962). Another example is that of *Aristida pennata*, which, 20 to 25 days after germination, may have roots one metre long but only a minuscule top-growth (Khachkarov & Korovine, 1942).

Reduced Water-loss Mechanisms for Water-loss Control

Reduced leaf Area. The principal means of reducing water-loss of xeromorphic plants is their ability to reduce their transpiring surface. Apart from the common means of keeping the aerial parts small, perhaps the simplest form of this reduction of the transpiring surface is the *rolling* or curling of leaves at times of water stress, a characteristic exhibited by many grasses. The rolling of leaves has been shown to reduce transpiration by almost 55 per cent in semi-arid conditions, and by 75 per cent in desert xerophytes (Stålfelt, 1956).

A more extreme adaptation to dry conditions is the *shedding* of leaves, as in *Fouquiera splendens* (Shields, 1958) and species of *Euphorbia* (Kramer, 1959). Many desert species, such as *Artemisia herba-alba*, have two kinds of leaves—large winter leaves that are shed at the end of the rainy season, and very small summer leaves which replace the large ones (Zohary, 1961). In other species, e.g. *Cerdicium microphyllum* (Palo Verde) of the Sonora Desert, the leaves are reduced to vestigial structures, and photosynthesis is restricted to the green stems (Polunin, 1960). Some desert species, such as *Hammada salicornica* and *Retama cactum*, are leafless, except for structures that are present for very short periods (Shields, 1958; Migahid, 1962).

Leaf Structure. Various morphological characteristics of leaves help to reduce the transpiration rate and may affect survival of plants under drought conditions: leaves with thick cuticles and a waxy surface, sunken stomata, the presence of spines, etc., are common and effective in this.

Plants in dry habitats usually have thicker cuticles than those growing under moist conditions. The cuticle consists of a hydrophilic framework in which wax platelets are embedded, and under conditions of high humidity this framework absorbs water and swells, thereby separating the wax platelets and increasing the permeability of the cuticle (Overbeek, 1956). This may explain why the leaves of plants in arid regions can absorb dew and water vapour.

The smaller leaf size and increased leaf thickness of drought-resistant plants is associated with certain modifications of leaf structure (Shields, 1958); thus the palisade tissue is more developed and the spongy mesophyll is proportionally reduced, the cells in general are smaller and have thicker walls, and stomata are commonly more numerous per unit area except in the case of many succulents.

In many instances the principal effect of leaf adaptation to dry conditions, such as thick cell-walls, is to prevent wilting or collapse of the leaf, rather than to ensure reduced water-loss (Kramer, 1959). Indeed the palisade type of mesophyll actually favours higher transpiration (Shields, 1958). This explains why xeromorphic types of leaves frequently transpire *more*, when amply supplied with water, than mesomorphic types, as mentioned above (Shields, 1958; Kramer, 1959), and thus make possible an efficient use of water during the short periods of wet conditions which are typical for desert environments. As soon as the soil moisture-level again becomes critical, the stomata close and the morphological adaptations to reduce water-loss, mentioned above, become operative.

Drought resistance thus comprises the ability of the plants both to endure water-loss and to reduce water-loss.

In certain species, drought stimulates the production of epidermal hairs (Shields, 1958). The mere presence of pubescence on the leaves is no longer considered as an advantage in reducing transpiration (Kramer, 1959), unless it be through increasing their albedo (*see* p. 47). However, at high air-humidity some desert plants are capable of absorbing water through the hairs on their leaves (Shields, 1958).

Succulence. Succulent plants, such as the cacti, tend to have a very thick cuticle and relatively few stomata. Some desert plants have morphological characteristics which reduce the absorption of radiation within the visible as well as the infra-red wave-lengths, in addition to characteristics associated with reduction of water-loss by transpiration (Billings & Morris, 1951).

Stomatal Frequency and Location. A small number of stomata, or stomata that react quickly to water-deficits by closing, can retard the development of water-deficits. Some desert plants have less than 100 stomata per cm^2, as compared with averages of 10 000 per cm^2 in mesophytes (Killian & Lemée, 1956). In many species (e.g. *Retama raetam* and *Nerium oleander*), the stomata are located in depressions or cavities in the leaves, which is a feature that can further reduce transpiration by limiting the impingement of air currents.

The stomata of desert Euphorbias and cacti are commonly closed during the day and open at night, so that transpiration is low when the demands of evaporation are high. Under these conditions, the carbon dioxide required for photosynthesis is derived partly from organic acids produced in anaerobic respiration (Kassas, 1966).

The drought-resistance of sorghum and certain varieties of maize and groundnuts is attributed, at least partly, to the ability of the stomata to resume their normal functioning rapidly after a period of water-stress (May & Milthorpe, 1962).

ABILITY TO SURVIVE LONG PERIODS OF DROUGHT

Osmotic Pressure

Plants growing in temperate regions commonly have an average osmotic pressure of the order of 10 atmospheres, whilst in the Arizona desert the average osmotic pressure of the plants tested was 20 atm (Iljin, 1957). Together with the morphological adaptations already mentioned, the higher internal osmotic pressure helps to prevent dangerous internal desiccation, and apparently makes the plasma itself more drought-enduring (Vaadia & Waisel, 1967).

When plants are grown in their natural environments, their osmotic potentials tend to be characteristic for each ecological group: most crop plants have values of 10–20 atm, xerophytic plants commonly have values of 30–40 atm, halophytes may reach values of 100 atm or more, and plants growing on hamadas may show even higher values. These differences become far less marked when plants of the different ecological groups are grown under the same conditions of water-stress (Slatyer, 1963). Table V gives examples of the osmotic potentials of plants growing in different habitats, and indicates how much higher these tend to be in summer than in winter.

TABLE V

Osmotic Pressure in Selected Desert Plants Growing in Different Habitats
(Zohary, 1962)

Habitat	*Species* (examples)	*Osmotic pressure* (atm)	
		*winter value**	*summer value*
Sand-dunes	*Zilla spinosa, Artemisia monosperma, Calligonum comosum*	7–14	15–16
Run-on habitats	*Acacia raddiana, Ochradenus baccatus*	13	17–22
Salines	*Arthrocnemum glaucum, Suaeda monoica, S. tetrandra*	28–50	89–113
Hamadas	*Reaumuria palaestina*	39	203

* The differences between winter and summer values are small under favourable conditions and very large under unfavourably dry conditions.

Ability to Maintain Reserves of Water in Tissues

Certain xerophytes have water reserves in hypertrophic underground organs which are often greatly swollen. In *Pachypodium bispinosum*, a South African plant, the tubers account for 95 per cent of the total weight of the plant, and contain 91 per cent of its water (Walter, 1962).

Succulents are a special group, frequent in the deserts of North and South America and South Africa, rare in the deserts of Asia and the Sahara, and virtually absent from the deserts of Australia. They tend to be preponderant in arid regions which have short rainy seasons annually (Walter, 1962), being able to survive periods of absolute drought during which they are almost completely unaffected by their environment, both aerial and edaphic. This they manage by living very frugally on reserves of water which they accumulate in their storage organs during periods when water is available in the soil. The amount of water conserved, which can be quite considerable, is estimated at 2 000 to 3 000 litres for a single plant of *Carnegiea gigantea* (Killian & Lemée, 1956).

Protoplasmic Dehydration

There are great differences between plant species in their ability to endure desiccation, in relation to both the duration and the degree of dehydration. The more water the plant can hold in its tissues under stress conditions, the greater will be the ability of the protoplasm to withstand permanent injury during drought. The 'bound water'† prevents the protein molecules of active

† Bound water: Due to the presence of hydrophilic colloids, a fraction of the water in the tissues is so tightly held, that the energy of the water molecules is apparently reduced to such a low level that other properties are also changed. This fraction is called 'bound water' (Currier, 1967).

sites from coagulating (Vaadia & Waisel, 1967). As the tissues become desiccated, the protoplasm becomes increasingly dense and its viscosity gradually increases; when dehydration is severe, complete gelation of the protoplasm occurs and finally it may become rigid to the point of brittleness (Levitt, 1956).

There are also great differences between different plant organs in their ability to withstand desiccation. Seeds that are filled with food reserves but have no vacuoles can survive very long dry periods, often of many years. Buds of higher plants also have no vacuoles, and are very resistant to desiccation. Certain desert plants survive a degree of desiccation that causes the death of their leaves; but the buds on the stems remain alive and resume growth when the drought ends.

In many species of perennial plants, the above-ground parts die off at the onset of the hot, dry season and the underground parts, such as rhizomes, bulbs, corms, and tubers, remain alive but dormant (Vegis, 1963).

Drought-hardening

Drought-resistance of many plants can be increased by subjecting them for several days to moderate dehydration, to low temperatures, or to treatment with growth-retarding substances. Plants that recover from a period of drought are also usually less affected by a renewed period of water-stress (May & Milthorpe, 1962).

When hardened plants are subjected to drought, their protoplasm has a lower viscosity than that of similar but non-hardened plants, and a higher permeability to water. The adapted protoplasm does not become rigid and brittle as soon as that of unadapted plants, and is better able to hold water against dehydrating forces (Levitt, 1956). Higher rates of photosynthesis, lower rates of respiration, and a higher root/shoot ratio, also characterize hardened plants (May & Milthorpe, 1962). Drought-hardened plants are, moreover better able to survive exposure to heat than are unhardened plants (Levitt, 1956).

Pre-sowing Hardening of Plants to Drought. Very young seeds, which are still in the milky stage, and have attained only one-eighth of the volume of the mature seed, can be completely air-dried without losing their vitality (Hubac, 1962). The excised embryo can be dried to a moisture content of 5 per cent, and will resume normal development and germinate if re-moistened and placed on a suitable substrate.

As germination proceeds, the sprouting embryo rapidly loses its ability to withstand desiccation without loss of viability. The critical period occurs at the time of emergence of the radicle (or first seminal-root). It is the radicle that dies first under dehydration, but, before it reaches this point through desiccation, the seed or embryo can be submitted to a number of alternate hydrations and desiccations without harm. Indeed a succession of alternate hydrations and desiccations of the seeds was found to increase the resistance of the seedlings to desiccation. As long as the hydration and desiccation have only a purely physical effect on the colloids of the embryo, its vitality is not affected. By contrast, biological dehydration, involving death of the radicle, is irreversible.

Based on these findings, a method for increasing the drought-resistance of plants by treatment before sowing was proposed and developed by Russian researchers (Genkel & Henkel, 1961). However, their results were not confirmed by experiments carried out in the United States (Waisel, 1962) and in Israel (Evenari, 1962).

HEAT RESISTANCE

In arid lands, dryness and high temperatures are generally associated with each other. Because of the high net radiation and the little water that is available for evapotranspiration, *resistance to heat* by the plant plays a very large role (Emberger & Lemée, 1962). The resistance of plants to heat is accordingly just as relevant to their adaptation to arid conditions as is their resistance to drought; when irrigation is practised, heat-resistance becomes the main consideration. When the plant has an ample supply of water, its metabolism may be adversely affected by excessively high air or soil-surface temperatures, that may eventually cause the death of the more susceptible species.

It is assumed that at certain temperatures, specific essential enzymatic reactions are completely inhibited. This limits or prevents the elaboration of certain metabolic products that are essential to the growth of plants. If, for each plant, it were possible to ascertain what specific metabolite is affected by high temperatures, it might be possible, by an appropriate intervention, to improve plant productivity under high temperatures (Kurtz, 1958).

According to Soviet workers (Petinov & Molotkovsky, 1962), the protective mechanism of heat-resistant plants is based on the regulation of the respiration process. Under the influence of excessive heat, protein formation is seriously affected, leading to excessive formation and accumulation of NH_3, which is toxic to plant tissues. Heat-resistant plants react by the accelerated production of organic acids during the respiratory process; these organic acids bind NH_3 in the form of salts, thereby minimizing the effects of heat. The researchers found that foliar zinc sprays increased the heat-resistance of plants, because the zinc activates certain enzymes and stimulates oxidation reactions in vegetative cells.

Economic Plants of the Desert

DESERT PLANTS USED BY DESERT INHABITANTS

Since time immemorial, desert plants have supplied many needs of the local human inhabitants. Foods and spices, fibres, medicines and drugs, shelter materials, and weapons, are obtained from a large variety of desert plants.

Fruits. The *Opuntia* and *Agave* spp. are the principal fruit-producers of the North American deserts (Burgess, 1965). The succulent Euphorbias of the African deserts, however, contain large amounts of very toxic alkaloids. Berries of many species are additional sources of nutrition.

Starch-filled Storage Organs. Edible roots, rhizomes, tubers, bulbs, etc., are also found in a large variety of desert plants. The 'hearts' or tender, inner parts of many species (e.g. *Agave*, *Yucca*, etc.), after roasting, serve as a staple food of many American Indian tribes.

Seeds and Nuts. Seeds and nuts, legume pods and acorns, are important food-sources. The seeds of many of the annual species—grasses, composites, legumes, chenopods, mustards, and others—are gathered and used fresh, dried, parched, or ground. Some plants, such as various *Prosopis* species, produce an abundance of large and nutritious seeds (Burgess, 1965).

Greens and Herbs. The young shoots or leaves of many desert plants are used as salads or condiments.

Drugs and Medicines. Many desert plants have biochemical constituents that have a therapeutic value or serve as narcotic drugs or stimulants, etc. Dried or ground parts of the plant, as well as decoctions, have traditionally been used for these purposes.

Fibres. Many desert plants supply fibres that have a wide variety of uses, *Yucca* and *Agave* are the principal genera involved (Bender, 1963).

Miscellaneous Uses. In addition to the above, desert plants supply building materials, weapons, tools, ornaments, dyes, gums, fuel, soap, perfumes, etc.

DESERT PLANTS OF POTENTIAL COMMERCIAL SIGNIFICANCE FOR ARID GROWING-CONDITIONS

The potentially most interesting arid-zone plants are those which contain constituents whose manufacture by the plant is not adversely affected by aridity or may even benefit from dry growing-conditions, and whose commercial value per unit weight is very high. Plants containing alkaloids, essential oils, and mucilaginous materials, clearly belong to this category.

Plant species producing constituents of economic value, but whose yield is proportional to the vegetative growth of the plant—such as fibres or food reserves—and have a relatively low intrinsic commercial value, are, *a priori*, less promising.

The number of desert plants containing *alkaloids* are relatively few, but their alkaloid content can be very high (Paris & Dillemann, 1960), with members of the Solanaceae commonly the richest source. Thus the most important alkaloids of desert plants are the hyoscine and hyoscyamine groups found in *Datura inoxia* and *D. stramonium*, *Hyoscyamus muticus* and *H. albus*, and *Physochlaina praelata*, while steroidal alkaloids have been isolated from various species of *Solanum*, e.g. *S. carolinense* and *S. xanthocarpum*. Ephedrine is contained in a few species of *Ephedra*, and a variety of other alkaloids, too numerous to mention, are found in other desert plants (Chopra *et al.*, 1960).

The second important group of constituents contained in many desert plants consists of *essential oils*; the secretion of essential oils appears to be a xerophytic character conferring a degree of protection against drought (Fluck, 1955). The main sources are plants of the families Labiatae: *Salvia officinalis*, *Lavandula officinalis*, *L. latifolia*, *Rosmarinus officinalis*; and Umbelliferae: *Foeniculum vulgare*, *Pimpinella anisum*, *Ferula alliacea*, *F. asafoetida*, *F. foetida*, and *F. galbaniflua* (Chopra *et al.*, 1960). Other important sources of aromatic oils are members of the genus *Artemisia* (Compositae).

The third group is composed of the mucilaginous plants, producing medicinal *gums*, which are contained in various species of *Acacia* and in *Butea monosperma* (Leguminosae). The mucilage plays a part in water-retention and is therefore an important xerophytic character; it has also been established that gum production is favoured by low atmospheric and soil humidity (Paris & Dillemann, 1960).

Examples of Desert Plants used Commercially

Various species of *Hyoscyamus* are specific to certain desert areas. Thus *H. muticus* is indigenous to Egypt, the Near East, Arabia, Iran, and India. It is the only wild plant in the Egyptian Desert that is of sufficient commercial value to justify its export (Drar, 1954). Attempts to cultivate *H. muticus* in various countries other than Egypt have not met with success (Chopra *et al.*, 1960). Even in Egypt, the plants under cultivation had a lower alkaloid content than those growing wild, while the percentage of alkaloids was greatest in plants growing under the most arid conditions (Paris & Dillemann, 1960).

Citrullus colocynthis (Cucurbitaceae) is native to Arabia, Syria, Egypt, and India. The pulp of the unripe fruit yields the commercial drug colocynth, which is used as a cathartic. *C. colocynthis* used to be collected and exported in large quantities from Egypt; it is now grown in Cyprus and Spain (Chopra *et al.*, 1960).

Ephedra (Gnetales) species are found mostly in the arid regions of the temperate zone. The dried young branches are a source of the alkaloid ephedrine, which has an action similar to that of adrenalin. *Ephedra* species have been cultivated successfully in the U.S.A., England, Kenya, and Australia (Chopra *et al.*, 1960).

Various species of *Agave* (Amaryllidaceae) are being grown for their sapogenins in Italy, Sicily, and Malta (Chopra *et al.*, 1960). One species, *A. roseana*, yields 2·5 per cent by dry-weight of hecogenin, a steroidal sapogenin which is a precursor of cortisone and other steroidal hormones. Another species, *A. siselana*, has been grown commercially on a large scale in East Africa and India, for the production of fibres for the sisal industry. Agaves are extensively cultivated for fibres in the Pueblo Desert of Mexico (Chopra *et al.*, 1960), but attempts to grow *Agave siselana* in the arid parts of Israel (200–250 mm rainfall) have not been economically successful.

Stipa tenacissima (Gramineae) is collected in North Africa for the production of paper pulp.

Yucca (Liliaceae) is one of the genera with the greatest abundance of steroid saponosides. Its species are plants of the desert and semi-desert regions of Mexico and the south-western United States, and are exploited commercially for their long textile fibres; after extraction of the fibres, the residue serves as a source of saponosides. The plants are extremely slow-growing, and long periods must elapse before it is possible to return to 'harvested' areas (Paris & Dillemann, 1960).

Aloe (Liliaceae) is a genus of xerophytic plants that are indigenous to East and South Africa. Aloes can be grown on very poor soils. The concentrated juice provides the drug aloes, which is a mixture of glucosides and is largely used as a cathartic. Some species are grown as a crop in South Africa, Arabia, and India (Chopra *et al.*, 1960), while *Aloe vera* is cultivated in the Dominican Republic, along the Venezuelan coast, and on the Island of Arruba—all for the commercial production of the drug aloes (Paris & Dillemann, 1960).

Artemisia (Compositae) species are abundant in arid regions of the Asiatic steppes and in the western United States, South Africa, and South America (Chopra *et al.*, 1960). Santonin-bearing *Artemisia* species (mainly *A. cina*) are being grown in Russian and Iranian Turkestan. So far, attempts to grow *A.*

cina outside these regions have not been successful (Paris & Dillemann, 1960). The santonin is an anthelmintic. *A. herba-alba* is widespread in North African deserts. Though it contains no santonin it has vermifuge properties. These plants are collected and marketed as Barbary santonica (ibid.).

Wormwood (*Artemisia absinthium*), a native of northern Asia and Afghanistan, is cultivated in the United States. The essential oil extracted from it has a tonic and stimulating effect on the digestive organs (Chopra *et al.*, 1960).

Canaigre (*Rumex hymenosepalus*) occurs in the arid regions of Mexico and the U.S.A., where wild stands were used as a source of tannin (Bender, 1963). After these stands had been depleted, attempts were made to grow canaigre as a commercial crop, but it could not compete economically in the long run with other sources of tannin.

Guayule (*Parthenium argentatum*) is a small Mexican semi-desert shrub, containing 10–18 per cent of high-quality rubber (Tysdal & Rands, 1953). Though considered economically very promising, and notwithstanding a considerable research effort and some production in the immediate post-war period, guayule rubber production has not developed much in recent years (Polhamus, 1962).

The rubber-bearing *Scorzonera tau-saghys* and *Taraxacum kok-saghys*, dandelion-like plants that were found growing wild in the Tien Shan Mountains of Asia, have been developed into cultivated plants, providing the major native rubber crop of the U.S.S.R. (Manton, 1954).

Euphorbia antisyphilitica, a source of candelilla wax which has a great variety of industrial uses (Bender, 1963), has been in commercial use for many years.

Protein and Oil Contents of Desert Plants. The seeds of many species of desert plants have been found to have above-average protein and oil contents, while some species contain oils of unusual fatty-acid composition (Bender, 1963). Of these, *Lesquerella fendleri*, indigenous to semi-arid North America, has seeds with an oil content ranging from 20 to 40 per cent, and the oil has potential industrial uses.

PROSPECTS OF COMMERCIAL EXPLOITATION OF DESERT PLANTS

Harvesting from Natural Stands

The natural stands of most potentially valuable desert plants are too diffuse to permit harvesting to be economical, except possibly for people with a very low standard of living. Even plants of medicinal value that are relatively abundant in the desert may encounter difficulties when it comes to marketing, due to competition from cheaper sources. In other cases, the plants are not sufficiently abundant to provide an adequate and continuous supply. In the few cases of a seemingly abundant supply, such as that of candelilla plants in Mexico (Pultz, 1956) or of *Hyoscyamus muticus* in Egypt, which latter is collected for its alkaloid constituent (Drar, 1954), excessive harvesting has rapidly led to a depletion of the natural stands.

Domestication

In the cases of a number of desert plants of potential economic value, a very large effort in plant breeding, management practices, and methods of processing, has been expended as a preliminary step to domestication. Examples of this are Guayule for rubber production and Canaigre for tannin. Commercial production in the truly arid conditions that are characteristic of the native habitat of the plants concerned have usually failed to be economical. It is probably unrealistic to expect commercial cultivation in areas with less than 200 mm of annual rainfall.

If the desert plants are grown under more favourable conditions of precipitation, and still more so if they are irrigated, the content of the valuable constituents is markedly reduced in most cases. This is not surprising in view of the fact that the production of these constituents by the plant is a protective measure against aridity.

Up to the present, success in a few limited cases, such as the cultivation of Guayule, has been dependent on war-time shortages or other special economic considerations—such as the desire to be independent of outside sources, which has led to the extensive cultivation of the rubber-producing Kok-saghys in the U.S.S.R. Once the situation returns to normal, however, cultivation has usually ceased.

The potentialities of the world's flora seem sufficiently well known to enable the conclusion to be drawn that the likelihood of finding crops of economic significance that can be grown commercially on a large scale in the arid regions, without irrigation, is only slight (Whaley, 1952). Whatever specific advantage plants that are native to arid lands may have, these are attenuated when the plants are grown under irrigation. With an assured water-supply, there is a great choice of mesophytic crops, with high-yielding potentialities and commercial value; the native desert plants can rarely compete under these conditions.

References

AAMODT, O. S. & JOHNSTON, W. A. (1936). Studies on drought resistance in spring wheat. *Cand. J. Res.*, **14**(c), 122–52.

BENDER, G. L. (1963). Native animals and plants as resources. pp. 309–37 in *Aridity and Man*. Amer. Assoc. Adv. Sci., Washington, D.C., xviii + 584 pp., illustr.

BILLINGS, W. D. & MORRIS, R. J. (1951). Reflection of visible and infra-red radiation from leaves of different ecological groups. *Amer. J. Bot.*, **38**, 327–31.

BURGESS, R. L. (1965). Utilization of desert plants by native peoples: an overview of southwestern North America. pp. 6–21 in *Native Plants and Animals as Resources in Arid Lands of the Southwest United States* (Ed. L. J. Gardner) Arizona State College, tempe, v + 80 pp., illustr.

CHOPRA, I. C., ABROL, B. K. & HANDA, K. L. (1960). Medicinal plants of the arid zones, with particular reference to the botanical aspects. *Arid Zone Res.*, **13**, 11–53.

CURRIER, H. B. (1967). Nature of plant water. pp. 307–19 in *Irrigation of Agricultural Lands* (Ed. R. M. Hagan, H. R. Haise & T. W. Edminster) Agronomy series, *Am. Soc. Agron.* Madison, Wisconsin, **11**, xxxi + 1180 pp., illustr.

DIXEY, F. (1966). Water supply, use and management. pp. 77–100 in *Arid Lands: A Geographical Appraisal* (Ed. E. S. Halls) UNESCO, Paris, xviii + 461 pp., illustr.

DRAR, M. (1954). Plants for raw material in the deserts of Egypt. pp. 70–76 in Proc. Symposium on *Scientific Problems of Land Use in Arid Regions*, Heliopolis 219 pp., illustr.

EMBERGER, L. & LEMÉE, G. (1962). Plant ecology. *Arid Zone Res.*, **18**, 197–212.

EVENARI, M. (1962). Plant physiology and arid zone research. *Arid Zone Res.*, **18**, 175–96.

FLETCHER, H. C. & ELMENDORF, H. B. (1955). Phreatophytes, a serious problem in the West. pp. 423–9 in *Water*, Yb. Agric., U.S. Dept. Agric., xiii + 751 pp., illustr.

FLUCK, H. (1955). The influence of climate on the active principles in medicinal plants. *J. Pharm. Lond.*, **7**, 361–83.

GENKEL, P. A. & HENKEL, P. A. (1961). Drought resistance of plants: methods of recognition and of intensification. *Arid Zone Res.*, **16**, 167–74.

GRILLOT, G. (1956). The biological and agricultural problems presented by plants tolerant of saline or brackish water and the employment of such water for irrigation. *Arid Zone Res.*, **4**, 9–35.

HAYWARD, H. E. & BERNSTEIN, L. (1949). Plant-growth relationships on saline and alkaline soils. *Adv. Agron.*, **1**, 1–38.

HUBAC, C. (1962). La reviviscence en aptitude à l'hydrobiose et ses variations naturelles et experimentales chez les embryos et les plantules. *Arid Zone Res.*, **16**, 271–4.

ILJIN, W. S. (1957). Drought resistance in plants and physiological processes. *A. Rev. Pl. Physiol.*, **8**, 257–74.

KASSAS, M. (1966). Plant life in deserts. pp. 145–80 in *Arid Lands: A Geographical Appraisal* (Ed. E. Hills) Methuen, London, viii + 461 pp., illustr.

KHACHKAROV, D. N. & KOROVINE, E. P. (1942). *La Vie dans les Deserts*. Payot, Paris, 361 pp., illustr.

KILLIAN, C. & LEMÉE, G. (1956). Les xérophytes: leur économie d'eau. pp. 787–824 in *Handbuch der Pflanzenphysiologie* (Ed. W. Ruhland) Springer Verlag, Berlin, Band III, xxii + 1073 pp., illustr.

KRAMER, P. J. (1959). Transpiration and the water economy of plants. pp. 607–726 in *Plant Physiology, a Treatise* (Ed. F. C. Steward) Academic Press, New York, Vol. II, viii + 758 pp., illustr.

KURTZ, E. B. (1958). A chemical basis for the adaptation of plants. pp. 23–29 in *Bioecology of the Arid and Semiarid Lands of the Southwest* (Ed. Lora M. Shields & L. J. Gardner) New Mexico Highlands University Bulletin, 69 pp., illustr.

LEVITT, J. (1956). Significance of hydration of the state of protoplasm. pp. 650–1 in *Handbuch der Pflanzenphysiologie* (Ed. W. Ruhland) Springer Verlag, Berlin Band III, xxii + 1073 pp., illustr.

MCGINNIES, W. G. (1955). Characteristics and distribution of arid-zone vegetation as related to land management in the United States. *Arid Zone Res.*, **5**, 25–28.

MANTON, S. M. (1954). Biological research and the productive transformation of steppe and desert in the Soviet Union. pp. 148–55 in *Biology of Deserts* (Ed. J. L. Cloudsley-Thompson) Institute of Biology, London, iv + 224 pp., illustr.

MAY, L. H. & MILTHORPE, F. L. (1962). Drought resistance of crop plants. *Fld. Crop Abstr.*, **15**, 171–9.

MIGAHID, A. M. (1962). The drought resistance of Egyptian desert plants. *Arid Zone Res.*, **16**, 213–33.

MONOD, T. (1954). Modes contracté et diffus de la végétation saharienne. pp. 35–44 in *Biology of Deserts* (Ed. J. L. Cloudsley-Thompson) Institute of Biology, London, iv + 224 pp., illustr.

OVERBEEK, J. VAN (1956). Studies on the relation between molecular structure and penetration of growth regulators into plants. pp. 205–10 in *The Chemistry and Mode of Action of Plant Growth Substances* (Ed. R. L. Wain & F. Wightman) Academic Press, New York, xi + 312 pp., illustr.

PARIS, R. & DILLEMANN, G. (1960). Medicinal plants of the arid zones with particular reference to the pharmacological aspects. *Arid Zone Res.*, **13**, 55–91.

PASE, C. P. (1965). Grazing and watershed values of native plants. pp. 31–40 in *Native Plants and Animals as Resources in Arid Lands of the Southwest United States* (Ed. L. J. Gardner) Arizona State College, tempe, v + 80 pp., illustr.

PETERSON, H. B. (1961). Some effects on plants of salt and sodium from saline and sodic soils. *Arid Zone Res.*, **14**, 163–8.

PETINOV, N. S. & MOLOTKOVSKY, U. G. (1962). The protective processes of heat resistant plants. *Arid Zone Res.*, **16**, 275–83.

POLHAMUS, L. G. (1962). *Rubber: Botany, Production, and Utilization.* World Crops Books (Ed. N. Polunin) Leonard Hill, London, & Interscience, New York, xvii + 449 pp., illustr.

POLUNIN, N. (1960). *Introduction to Plant Geography and Some Related Sciences.* Longmans Green, London, and Barnes & Noble, New York, xix + 640 pp., illustr.

PULTZ, L. M. (1956). Problems in the development and utilization of arid lands plants. pp. 414–8 in *The Future of Arid Lands* (Ed. G. F. White) Amer. Assoc. Adv. Sci., Washington D.C., ix + 435 pp., illustr.

REPP, G. (1961). The salt tolerance of plants: basic research and tests. *Arid Zone Res.*, **14**, 153–9.

SHANTZ, H. L. (1956). History and problems of arid lands development. pp. 3–25 in *The Future of Arid Lands* (Ed. G. F. White) *Amer. Assoc. Adv. Sci.* Washington D.C., ix + 453 pp., illustr.

SHIELDS, L. M. (1958). Morphology in relation to xerophytism. pp. 15–22 in *Bioecology of the Arid and Semi-arid Lands of the Southwest* (Ed. Lora M. Shields & L. J. Gardner) New Mexico Highlands University Bulletin, 69 pp., illustr.

SLATYER, R. O. (1963). Climatic control of plant water relations. pp. 32–52 in *Environmental Control of Plant Growth* (Ed. L. T. Evans) Academic Press, New York, viii + 449 pp., illustr.

STÅLFELT, M. G. (1956). Die Stomatare Transpiration und die Physiologie der Spaltöffnungen. pp. 351–426 in *Handbuch der Pflanzenphysiologie* (Ed. W. Ruhland) Springer Verlag, Berlin, Band III, xxii + 1073 pp., illustr.

TROLL, C. (1956). Das Wasser als Pflanzengeographisher Faktor. pp. 750–86 in *Handbuch der Pflanzenphysiologie* (Ed. W. Ruhland) Springer Verlag, Berlin, Band III, xxii + 1073 pp., illustr.

TRUMBLE, H. C. & WOODROFFE, K. (1954). The influence of climatic factors on the reaction of desert shrubs to grazing by sheep. pp. 129–47 in *Biology of Deserts* (Ed. J. L. Cloudsley-Thompson) Institute of Biology, London, iv + 224 pp., illustr.

TYSDAL, H. M. & RANDS, R. D. (1953). Breeding for disease resistance and higher yield in Hevea, Guayule and Kok-saghyz. *Agron. J.*, **45**, 234–42.

VAADIA, Y. & WAISEL, Y. (1967). Physiological processes as affected by water balance. pp. 354–68 in *Irrigation of Agricultural Lands* (Ed. R. M. Hagan, R. H. Haise & T. W. Edminster) Agronomy Series; Amer. Soc. Agron., Madison, Wisconsin, II, xxxi + 1180 pp., illustr.

VAN OVERBEEK, J. *See* OVERBEEK, J. VAN.

VEGIS, A. (1963). Climatic control of germination, bud break and dormancy. pp. 265–88 in *Environmental Control of Plant Growth* (Ed. L. T. Evans) Academic Press, New York, viii + 449 pp., illustr.

WAISEL, Y. (1962). Presowing treatments and their relation to growth and to drought, frost and heat resistance. *Physiol. Plant.*, **15**, 43–66.

WALTER, H. (1961). The adaptation of plants to saline soils. *Arid Zone Res.*, **14**, 129–34.

—— (1962). *Die Vegetation der Erde in Okologisher Betrachtung*. Gustav Fischer Verlag, Jena, Band I, xii + 538 pp., illustr.

WENT, F. W. (1952). The effects of rain and temperature on plant distribution in the desert. pp. 232–7 in *Desert Research*, Research Council of Israel, 641 pp., illustr.

—— (1955). The ecology of desert plants. *Scient. Am.*, **192**(4), 68–75.

WHALEY, W. G. (1952). Arid lands and plant research. *Scient. Monthly*, **75**, 228–33.

ZOHARY, M. (1961). On hydro-ecological relations of the Near East desert vegetation. *Arid Zone Res.*, **16**, 199–209.

—— (1962) *Plant Life of Palestine, Israel and Jordan*. Ronald Press, New York, vi + 262 pp., illustr.,

V

TRADITIONAL LAND-USE IN ARID ZONES AND ITS EVOLUTION

THE BEGINNINGS OF AGRICULTURE IN ARID LANDS

IN THE OLD WORLD

For about half-a-million years, Man subsisted by gathering 'wild' food, by hunting, and by fishing. About 10 000 years ago, Mediterranean Man started producing food in the Near East, first in the highlands and plateaux which flank this region and subsequently in the drainage basin of the Euphrates and Tigris.

As a result, three great cultural-historical events occurred in this region: the earliest appearance (i) of the use of blade-tools, (ii) of settled village-farming communities based on cereal agriculture and animal domestication (about 10 000 years ago), and (iii) of urban civilization based on irrigation agriculture (about 5 000 years ago) (Braidwood, 1952). All the available evidence indicates that the first efforts at domestication were devoted to the growing of cereal grains (Braidwood & Reed, 1957).

Helbaek (1959) postulates that domestication of a wild plant must have taken place within the region of its natural distribution. In the Near East* both wheat and barley were grown concurrently for bread. The probable progenitor (*Hordeum spontaneum*) of domesticated barley is found over a wide area, practically from Turkestan to Morocco, whilst that of domesticated wheat (*Triticum dicoccoides*) is restricted to a much smaller area, situated within the wider range of wild barley. It can therefore be assumed that plant domestication in the Old World originated in this restricted area, situated between the

* Near East. This term was formerly used to denote the Balkan States and the area of the Ottoman Empire. Since the Second World War, it is commonly used as an alternative for 'Middle East', encompassing the region of S.W. Asia and N.E. Africa stretching from Turkey through Iran, Iraq, and Arabia to the Sudan and countries bordering the E. Mediterranean. It is in the latter sense that the term Near East will be used in this work.

Zagros Mountains (Iraq, Iran), the Taurus Mountains in southern Turkey, and the uplands of Galilee (Peterson, 1965).

Evidence of the earliest appearance of kernels of partly domesticated forms of wheat, barley, and several pulses, was found in the excavations at Jarno, a village in Iraqi Kurdistan. These findings indicate that partly domesticated grains were already used for human consumption in the seventh millennium B.C., although actual cultivation at that time has not yet been proven. However, at the village site of Matarrah, at a lower elevation than Jarno, only *cultivated* Emmer Wheat was found (Helbaek, 1959); these findings belong to the sixth millennium B.C. The Natufian culture of Palestine (approximately 7th millennium B.C.) used flint sickle-blades with stone hafts, as well as hoes for loosening the soil before sowing—indicating that Natufian Man was not only a food gatherer but also a food producer (Albright, 1956). Though the Natufians were cultivators of grain, they apparently continued to gather wild food, to fish, and to hunt (Plate 7).

It is generally agreed that the domestication of food animals occurred at a somewhat later date than incipient plant cultivation (Reed, 1959).

In the early days of crop domestication, the cultivation of cereal grains was practised in the higher elevations, which have relatively ample winter rainfall. However, as crop cultivation involved tilling the soil, it became imperative for the early cultivators to move down from the natural habitat of wheat on the mountain slopes, to lower and more easily-tilled land, searching for sites with adequate rainfall for the crops and a good water-supply for domestic needs and the establishment of the village farming communities which characterized the agrarian culture (Reed, 1959). During the Neolithic period (sixth to fifth millennium B.C.) settlements developed wherever springs made this possible—usually on foothills near fertile plains.

With the increasing population, the initially small area of crop domestication expanded, and it was necessary to move from the open-forested slopes of the hills of Mesopotamia, Syria, and Palestine, to the alluvial plains. This movement, from the fringes of the semi-arid area to the arid areas of the region, was possible only by resorting to irrigation. In the sixth millennium B.C., there is evidence of large-scale cultivation already of Emmer Wheat under irrigation in the alluvial plain of lower Iraq—beginning what Albright (1956) calls 'irrigation culture', after people had learned to dig canals and build dams in order to divert water for irrigating large areas.

From the Near East, this agrarian culture spread throughout Asia Minor, and to Egypt, the Indus Valley (late third to early second millennium B.C.), and the Far East (Helbaek, 1959). Crossing to Europe also came relatively early; at the end of the fifth millennium B.C., Einkorn and Emmer Wheats were already cultivated on the large loess plains ranging from the Danube Delta to the mouth of the Rhine (ibid.).

Irrigation farming has considerable cultural implications. It made possible the growing of food for large populations, not all of whom had to be occupied in agriculture. This gave impetus to technical inventions, such as loom weaving and the wheel, and to cultural achievements such as writing, the development of mathematics, and artistic advances.

With the expansion of the irrigation networks, a well-developed social organization became a prerequisite for the proper maintenance of irrigation dams, channels, and structures; also needed were the enactment of laws safeguarding water-rights, and the supervision of their proper distribution. A central authority, usually based on a theocracy, became established over a large area which gradually evolved into an irrigation state. This occurred in Mesopotamia and Egypt (4000–3000 B.C.), and in China (around 2000 B.C.).

When the irrigated areas had reached their full expansion and were no longer able to feed the growing population of the State, the stage was set for expansion by conquest of neighbouring territories and transformation of the State into an empire exacting tribute from the conquered peoples. Central authority passed from the theocracy to a military establishment.

Empires which developed out of the irrigation states began with the Sumerian Dynasty in Mesopotamia, the Early Dynasty in Egypt, and the Chou periods in China.

These empires subsequently disintegrated; their decline can be attributed to conquest by stronger forces, to revolts of the dominated peoples, to a weakening of central authority and the consequent neglect of the complex infrastructure needed for maintaining irrigation farming, or alternatively to the loss of soil fertility due to salinization and waterlogging of the soil.

In most cases there was probably an interplay of two or more of these different factors, of which loss of soil fertility on large tracts may well have been the initial one. After a relatively long period of decline in production and in population, the whole cycle might begin again a number of times, as was the case in Mesopotamia, Egypt, and China.

IN THE NEW WORLD

Available evidence indicates that Man entered the New World from the north-west about 10 000 years ago and spread slowly southwards (Willey, 1960). There is reason to assume that the traditional agriculture of the American Indians developed independently of that of the Old World and was based on food plants that had been domesticated from indigenous wild plants (Armillas, 1956). And yet, this development occurred on lines very similar to those that characterize Neolithic agriculture in the Old World.

The domestication of plants and other important cultural developments in the New World occurred in the arid regions of Central and South America (Woodbury, 1963). The two main zones in which the most important New

World crops originated are in the Mexico-Guatemala region, and the Andean region in the south (Armillas, 1956). Hunting and food-gathering continued for a long time to provide the main sources of food—even after plant domestication made incipient farming possible—and were still being practised up to the 19th century by the Indians in Utah and Nevada, indicating an extremely successful subsistence adjustment to their arid climate (Woodbury, 1963).

The earliest evidence for incipient agriculture in the New World (7000–5000 B.C.) comes from the semi-arid hill country of Tamaulipas, north-eastern Mexico, and consists of traces of domesticated squash and possibly domesticates of peppers and small beans. Considerably later (between 3000 and 2000 B.C.), a primitive small-eared maize appeared in the same area, where fully-settled village farming is evidenced from about 1500 B.C. From thereabouts village farming, based on cultivated food plants, spread northwards to Mexico and southwards as far as Peru, with the maize plant as the vital element, complemented by beans and pumpkins (Willey, 1960). By 700 B.C., a relatively well-developed type of maize appeared also at Hauca Prieta in Peru, indicating that links had been established between the farming villages of Middle America and Peru (ibid.).

The oldest villages with archaeological evidence of farming are found at elevations at which rainfall was sufficient for growing rain-fed crops. There is also evidence of early farming on the Gila River in Arizona, where there is an annual precipitation of only around 250 mm; it is assumed that the local Indians planted their crops in the wet margins of the river after the flood-water had receded (Woodbury, 1963).

As in the Old World, the expansion of irrigation farming made necessary the establishment of a strong central authority. This occurred in the New World in Peru and Mexico between 500 B.C. and 500 A.D. Out of these irrigation states developed the empires of the Toltic and Aztec periods in Mexico and of the Tiahuanacan period in the Andes (Steward, 1956) (cf. our Plate 8).

General Patterns of Land-use in the Dry Regions

Throughout the historical period, certain patterns of land-use have evolved that characterize the agriculture of the dry regions of the world. These patterns are extraordinarily similar over very wide areas, notwithstanding the great diversity due to various combinations of topography, soil type, and climate. Land-use has also shown a surprising amount of stability, having remained practically unchanged in the Old World throughout the historical period—even to the present day.

From the agricultural standpoint, arid areas are those in which rainfall is not sufficient for crop production, whilst in semi-arid areas rainfall is sufficient for adapted crops. In the desert areas as such, the only agricultural use is grazing of the sparse plant cover by livestock; pastoral agriculture is therefore the dominant form of land-use of vast expanses of desertic terrain.

Crop production in deserts becomes possible only with irrigation, whether by flash-floods, annual inundation, or a year-round water supply. In the semi-arid regions, rain-fed crops can be grown with a reasonable degree of security —the risk of complete failure of the harvest is in general not greater than two years in ten—but the low and erratic rainfall limits the number and variety of crops that can be grown, as well as their yields. Cereal cultivation is the main source of livelihood.

The amount of rainfall that has been set, somewhat arbitrarily, as the limit of regular dry-land farming, is around 250–300 mm in areas with predominantly winter-rainfall, and approximately 500 mm for areas with summer rainfall (Wallen, 1966). The closer arable crop production is practised to the fringes of the desert area, the more precarious and uncertain it becomes.

SEDENTARY AGRICULTURE

Sedentary agriculture established itself early in large, contiguous areas along great rivers such as the Nile, the Tigris, the Euphrates, and the Indus. It also became established in isolated pockets of well-watered oases of the deserts, more precariously on the floodplains of the stream-beds in the deserts, and in places on the semi-arid fringes of the deserts where rain-fed crops were produced.

This pattern of land-use has led to a virtual separation of arable cropping from animal husbandry. Conflicts between the nomadic pastoralist and the sedentary cultivator are a recurrent theme in the Old World throughout the historical period, as characterized by the biblical strife between Cain and Abel. This theme repeated itself in the New World, when the arid regions of the American West were first settled by the White Man.

The lot of the sedentary cultivator in the dry regions was always precarious and hard. Even when the water-supply was fairly adequate and assured, irrigated crop production was associated with back-breaking labour in a difficult climate and under unsanitary conditions.

The primitive irrigation farmer, in addition to typical agricultural operations—such as soil preparation, sowing, weeding, and harvesting—was burdened with additional back-breaking chores, of which the actual watering of the crops was the most onerous. The maintenance of the irrigation system—canals, ditches, levees—also involved unending drudgery. Weakened by malaria and bilharzia and other diseases endemic to the irrigated areas, the cultivator was under constant pressure from the nomads. His life was also made difficult and bitter by a social environment that was based on the exploitation of his labour. Most of the available arable land in the oases and water-spread areas of the Sahara and the Sahel, for example, was, and still is, owned by Arab or Berber nomads. The land is worked by share-croppers of Negro slave origin (Haratin) or impoverished Berbers; for their work they receive about one-fifth of the crop (Monod & Toupet, 1961).

The maintenance of the irrigation system required an enormous amount of manual labour. Right up to the end of the 19th century, the fellaheen in Egypt were required to clear, without pay, the mud that had accumulated in the irrigation canals (corvée). This they did by scooping the mud from the canal bed with their hands and dumping it on the high banks of the canal (Addison, 1961). Even the agriculture which they practised was beset by difficulties: irrigation increased the hazards of pests and favoured the proliferation of diseases; the fragile soils typical of the arid regions broke down easily under irrigation; and rising water-tables and salt accumulation imperilled the very basis of their livelihood. Paradoxically, the cultivators who were dependent on the erratic rainfall of the region, whether in the production of rain-fed crops or by utilizing flash-floods, were generally better off than those practising typical irrigation agriculture. In Biblical times, the Israelites considered it a special blessing that they were not compelled to irrigate as in Egypt, but could depend on rain and dew for growing their crops (Reifenberg, 1953).

NOMAD PASTORAL AGRICULTURE

By contrast, the nomad pastoralist appears to have achieved a far better balance with his environment than did the sedentary cultivator. His basic livelihood and indeed his whole way of life, was dependent on his flocks. Camels, sheep, and goats, are able to subsist on the sparse vegetation of the desert. They are able to travel relatively large distances to water; they can also use water of high salinity that is unfit for human use or even for irrigation.

His flocks provided the nomad with drink, protein foods, and the materials needed for clothing and shelter. They were mobile, and so the nomad could adjust to the vagaries of the arid climate by seasonal movement of the livestock in search of pasture and water. He also had additional sources of income: he had gained a virtual monopoly for transportation over the traditional desert routes—such as the trans-Sahara slave trade from tropical Africa to the Mediterranean, and the ancient 'Silk road' between China and the Near East. Levies exacted from pilgrims and merchants travelling on the desert routes, tribute from cultivators working the nomad-owned land in the oases on a share-cropping basis, plunder from slave-collecting razzias and other raids, augmented the income of the nomad tribes, even though they were not equally divided among all its members.

With the aura of romanticism that usually surrounds the nomad way of life in the desert, it is tempting to overestimate the degree of adaptation of the nomad pastoralist to his environment. A balance was indeed achieved, but at a very low standard of living for the vast majority of the tribal members.

Though the Bedouin does not generally engage in the same back-breaking labour as the fellah, his life is arduous and knows many discomforts and dangers. The balance between livestock and the forage resources of the desert is also very precarious, notwithstanding the mobility of the pastoralist and his flock. A longer-than-usual period of drought might wipe out a large proportion of his herds, especially the young stock. After a period of relative abundance, the numbers of his stock might grow out of all proportion to the normally available forage resources, and overgrazing might follow before a balance was again achieved by an increased mortality rate. The nomads did not normally slaughter their animals for food, the herds being considered as non-expendable capital—thereby preventing any planned adjustment of livestock numbers to available forage. The nomads also suffered, and still suffer, from many diseases.

The precarious balance between the nomad and his environment, achieved and maintained throughout centuries, has been disrupted in recent times. His essential role in the transport of merchandise and travel along the desert routes has become obsolete. Law and order have practically abolished his income derived from raiding and tribute.

CONQUESTS AND THEIR AFTERMATH

On the basic theme of conflict between nomads and sedentary farmers, was superimposed a sequence of conquests and their aftermath, which is repeated throughout the history of the arid regions. The names of the invaders and the dates may differ in different regions, but the basic sequences are very similar. The invaders were of two types: nomads—such as, at various periods, the Seljuks, Turkomen, Mongols, Uzbeks, and Arabs—and well-established empires such as those of Babylonia, Persia, Greece, Rome, Byzantium, etc.

After a preliminary phase of destruction during the conquest, the cultivable land was subsequently allocated as spoils of war to warriors in whose interest it therefore was to further agricultural production. However, the agricultural population, already decimated by the conquest, was submitted to heavy taxation and oppression, generally by corrupt administrations. Heavy levies, needed to maintain the armies, were exacted. In order to counteract the tendency of the agricultural population to abandon the land and concentrate in the towns, laws were enacted binding the cultivators to the land, transforming them into semi-serfs and serfs. Another solution was to employ slave labour on the large estates of the big landowners.

A typical example of the consequences of large-scale invasion by nomads is that of Cyrenaica, which up to the 11th century had developed a fairly prosperous agriculture. The invasion of Arab nomads in the 11th and 12th centuries resulted in the almost complete extinction of sedentary farming (Despois, 1961). By contrast, after the Roman conquest of the Levant and of North

Africa, the Romans established fortifications (limes) which protected the cultivated areas from the depredations of the nomadic tribes, and introduced improved agricultural practices such as the use of fallow, crop rotation, and organic manuring. Irrigation systems and aqueducts for the transport of water were constructed. Large states were established.

Periods of strong central government favoured sedentary cultivation. Law and order provided security from the invasions of the nomads, and made possible the execution of large-scale terracing in the hills and of irrigation projects in the valleys. Conscript labour or slave labour were probably used for these purposes.

The relative prosperity of agriculture does not necessarily mean that the lot of the cultivator was enviable. The social gulf between chieftains and their families on one hand, and of those who did the actual tilling of the soil on the other, was enormous; the former lived in mansions, the latter in hovels (Albright, 1956). Every weakening of the central authority was inevitably accompanied by incursions of the nomads, who at the best extracted tribute but more commonly pushed back the limits of the cultivated areas by ravage, plunder, and destruction. In the Maghrib, under the pressure of the nomads, many villages disappeared and cultivation became restricted to the mountain regions and the vicinity of the towns (Despois, 1961).

Slave labour has been abolished, but where farming is still practised in the arid regions by traditional methods—whether arable farming or nomadic herding—living standards are extremely low and life is difficult and hazardous to this day.

Desert Pastoral Agriculture

THE NEAR EAST AND NORTH AFRICA

In the arid areas of the Near East and North Africa, as well as on the land that is too steep or shallow or stony for cultivation in the semi-arid fringes of these areas, the sole traditional form of land-use is grazing for livestock production. The stubble of cultivated crops also provides a welcome addition to the grazing areas in summer, when natural grazing has completely dried out.

Sheep and goats predominate through the region; in the deserts there are also numerous camel flocks, whilst a small, hardy breed of dual-purpose cattle range the foothills and graze the stubble of cultivated crops and the spontaneous vegetation on fallow land in the areas of higher rainfall. Adjustment to varying decrees of aridity is achieved by various forms of seasonal movement of the stock.

In parts of North Africa, the flocks spend the summer months on the high mountain ranges and are brought down to the plains in winter. Stock movement may also occur according to the distribution and amount of rainfall—

e.g. from the low-lying desert in winter, or to the maritime plains in summer.

In parts of Iraq, Jordan, and Syria, flocks of sheep and camels are completely nomadic and graze in the steppes, returning to cultivated areas only in years of exceptional drought. A form of semi-nomadism is also practised in the semi-desert parts of this region, the livestock being moved about over relatively short distances.

Whilst regular cultivation is alien to the mentality of the nomad pastoralist, the occasional sowing of patches of wheat or barley, in areas that have been wetted by a flash-flood, or in low-lying lands to which he can divert some water by constructing a small dam on a wadi, is an important, if irregular, source of food supply (Plate 9).

In recent years, as a result of improved veterinary services and improved security conditions, the number of grazing animals has increased considerably, so that overgrazing has become even more accentuated than in the past. The better grazing lands have also been taken under the plough, so that overstocking has led to deterioration of the remaining areas that are left available for grazing.

The combination of overstocking and variability in rainfall leads to considerable fluctuations in the numbers of livestock. For example in Algeria, a livestock population of 8 million head was built up during a succession of good years, to be reduced in 1945, after a few drought years, to 2 millions (Droughin, 1962).

WESTERN EURASIA AND CENTRAL ASIA

Nomadic herding has been the main form of land-use from time immemorial in the deserts of Western Eurasia and Central Asia, where appropriate breeds of stock, mainly camels and sheep, were able to subsist on xerophytic and halophytic vegetation and on saline water unfit for human consumption (Kovda, 1961).

The most important plants of the cold desert pastures are such perennials as wormwoods (*Artemisia terrae-alba* and *A. herba-alba*) and *Salsola laricifolia*, which supply year-round grazing. In the southern, warmer parts of the Central Asian Desert, numerous winter and spring ephemerals predominate, which, after rains, provide large amounts of nutritious forage. The length of the grazing season depends largely on winter temperatures. In the coldest parts of the arid zone, animals have to be sheltered for 150–180 days each year, during which time they have to be fed with specially-prepared forage reserves. This period is shortened to 80–90 days in Kazakhstan and Mongolia, and to only 30–40 days in the deserts of Asia Central.

The frequency of droughts and of fodder shortages varies considerably: from 10–20 years out of 100 on the semi-arid fringes of the area, to 50–70 years out of 100 in the fully desert zone.

Transhumance is generally practised, with the animals moving to higher altitudes during the hot months, but descending into the flat desert plains and valleys during the colder months.

After centuries of haphazard grazing, much of the natural vegetation has deteriorated, as a result of over-use and of cutting for fuel. Dust-storms are frequent and the movement of sand-dunes is practically ubiquitous. Modern developments consist mainly of improved social and cultural amenities for the pastoral populations of the region, improved communications, and the establishment of a network of canals, wells, and borings. Central camps have been set up, with shelters and reserves of forage. Veterinary services are being provided, better breeds of stock are being developed, and wherever possible animal husbandry and arable cropping are being integrated on large collective and state farms.

SOUTH AFRICA

The arid regions of South Africa comprise an area with predominantly winter precipitation—the Karoo—and an open savanna with predominantly summer precipitation. The name Karoo is derived from a Hottentot word signifying parched or desiccated (Talbot, 1961); it also designates the characteristic desert-shrub vegetation of the region. In the savanna, grasses dominate, with scattered thorn-trees. These regions, despite their aridity, originally supported a fairly large and varied fauna of herbivores and the carnivores that preyed upon them. The native Bushmen subsisted on hunting, fishing, and food-collecting, and were able to survive for fairly long periods in areas that were lacking in surface water.

In the 14th century, the pastoralist Hottentots moved into the area with their flocks (Shapera, 1936). Each tribe claimed as its territory the area in which it grazed its herds throughout the year; this was regarded as the inalienable communal property of the whole tribe (Talbot, 1961). Their cattle were of the Zebu type, and served as the parent stock from which the white settlers subsequently bred the modern Afrikander cattle. The sheep were non-woolled, extremely hardy breeds, well adapted to grazing the range. The native goats were of very mixed stock—hardy, large-sized, and very prolific.

The arid interior was first used by white farmers for grazing after 1730. The great demand for meat arising out of the war between Britain and France, which brought numerous warships and troop-ships into Table Bay, as well as the increasing number of merchant ships travelling around the Cape, provided a practically unlimited market for fresh meat, and for the tail-fat, soap, and candles, derived from the sheep. With such economic incentives, livestock production expanded rapidly.

Land tenure was conferred on the basis of periodically renewed grazing licences for vaguely-defined areas. This favoured nomadism, and the graziers would move to new pastures whenever the need arose. At a later stage, grazing rights referred to specified localities, which could be identified—usually by the farmbuildings and corrals of the homestead. An individual holding would comprise the area within the radius of half-an-hour's walk from the farm-house. Homesteads were established wherever water made permanent settlements possible.

The Hottentot flocks were completely displaced from the range; many of the Hottentots themselves were employed by the whites as herdsmen, shepherds, wagon drivers, etc., so that land-use of the range continued to be based mainly on the experience and traditions of the Hottentots.

Transhumance was generally practised. After the winter rains, abundant and nutritious annuals appear, and the grazing is sufficiently succulent to enable the stock to subsist in areas without surface water. In summer, large areas in the region of summer rainfall, called the 'trekveld', served for grazing on a communal basis, after summer rainstorms had filled the water-pools. The movements of the stock were generally parallel to those of the game, so that the summer trek of the graziers provided not only fresh pastures but also new hunting-grounds. The regular seasonal movements of the stock were frequently upset by drought, invasions of Brown Locust (*Locustana pardalina*), and the appearance of enormous herds of migratory Springbok (*Antidorcas marsupialis*), which together could completely destroy the grazing. When this occurred, the graziers had no choice but to search for alternative grazing areas. Both seasonal and emergency migrations were an essential element of the land-use of the area.

Towards the end of the 18th century, Merino sheep were introduced into the Cape. By cross-breeding with selected Cape sheep, a dual-purpose Merino-type sheep was evolved, which was a better meat-producer than the Cape sheep and at the same time produced high-quality wool. In the middle of the 18th century, some fat-rumped Hedjaz sheep were introduced, from which, by selective breeding, a distinctive South African breed was developed—the Blackhead Persian. This breed is extremely hardy and disease-resistant, relatively prolific, and well-adapted to grazing under arid conditions. In 1909, Karakuls were introduced, which are raised for the pelts of their unborn and new-born lambs. The fact that the ewes do not have to nurse their young, gives them an advantage over other breeds of sheep, so that Karakuls can thrive where other breeds would starve.

Goats played a complementary role to sheep, as they could graze on vegetation that was unsuitable for or inaccessible to sheep.

In addition to the traditional farm animals, an original source of income was provided by ostrich farming in the Karoo. At first, the prime feathers were the main product; subsequently, feathers for dusters, ostrich-skin leather, and ostrich biltong, found a ready market. Ostrich raising also provided the first impetus for irrigating small areas of the desert. Because of the need to provide an adequate source of proteins for the birds, Lucerne (alfalfa) was grown under irrigation as a supplement to the natural range.

The Kalahari Desert was opened to settlement after the First World War, with the Government establishing bore-holes before allocating any usable land to the settlers. The grazing was found to be better suited to cattle; sheep suffered from internal parasites, diseases, and depredations by jackals. However, cattle also, failed to thrive, and suffered from a disease called 'lamsiekte'. Only as late as 1919 was it shown that this disease was due to a deficiency of phosphate, reflecting the extremely low phosphate-content of the soils. After this problem had been overcome by a supply of bone-meal or phosphate licks, cattle ranching, based mainly on Afrikander stock, was able to develop. This took two forms: (a) a fairly primitive form of combined beef and dairy production, whereby the calves were kept kraaled, whilst the cows grazed the range within a limited radius; and (b) extensive beef ranches, on units of 4 000 ha and more. Beef production was a far more rational form of land-use than the primitive dairy ranching. The steers required less water and less labour, could graze much farther from water, and could make better use of the feed.

Range Improvement and Deterioration

The first and major improvement to be adopted was the fencing of the grazing land into paddocks, each with a drinking place. This made possible rational pasture management, which increased carrying capacity by at least 30 per cent during normal years. By periodic resting of parts of the veld, reserves of forage were established, to be used during drought years. The need for transhumance was reduced, and seasonal trekking became the exception. Fencing also helped to prevent the spread of stock diseases, and made continuous herding by day and kraaling by night unnecessary.

Locusts, migrating Springbok, and predatory carnivores such as jackals and leopards, were brought under control. The improved conditions made it possible to increase the overall number of sheep and to replace many of the hardier Afrikander and Blackhead Persian sheep by the more remunerative wool-producing Merino and, to a lesser degree, mutton-and-wool crossbreeds. These possibilities were, however, not always fully exploited, and land-use was not always judicious: large areas of veld deteriorated as a result of overgrazing, as stocking rates were frequently based on the carrying capacity of good years. The decline of trekking brought in its wake increased pressure on grazing lands, on which rational grazing was not effectively carried out. Perennial grasses disappeared and were largely replaced by toxic and unpalatable species. Run-off and soil erosion increased as a result of the reduced plant cover and exposed soil. The introduction of exotic plants such as the Prickly-pear (*Opuntia megacantha*) and Jointed Cactus (*O. aurantiaca*), which spread rapidly over enormous areas, was extremely destructive to grazing, and also caused the death by poisoning of thousands of animals annually. After many costly efforts, these cacti were brought under effective control by a combination of mechanical, biological, and chemical, methods (King, 1966).

Nevertheless, as a result of these various vicissitudes, most of the original vegetation of the range has changed considerably—usually for the worse, the deterioration being most marked in the more arid parts.

THE AMERICAN DESERT

The first advent of the White Man, in the middle of the 18th century, in this region was mainly concerned either with missionary activities or with the search for gold, and had little direct impact on land-use (Armillas, 1961). But in the second half of the 19th century, there began an influx of settlers from the East, mainly graziers. This was the era of the 'open range', a settler would lay a claim to a watering place, and thereby have effective control over vast areas of adjacent grazing-lands.

Conservation methods were practically unknown; nor did the ranchers, lacking legal rights to the range, have any incentive to make basic improvements on their holdings. The first procedures for establishing title to the lands, and the system of land division itself, were based on those prevalent in the humid areas of the United States, without regard to the differences in climate, water resources, soil type, and topography. As a result, land was allotted without any appraisal of its potentialities, and the size of the allocations remained constant whatever the potential productive capacity of the land; the country was also divided into rectangular holdings without concern for erosion problems. Misuse and overstocking reduced the carrying capacity of the range by over fifty per cent in the course of several decades (Logan, 1961).

It was only in 1934, during the periods of droughts and depression, that an Act was passed to change this situation. A land-classification authority was established, and the grazing land was brought under the control and management of the federal government.

Under the primitive and rigorous conditions prevailing on the range, only a very hardy breed of cattle, capable of travelling large distances and requiring a minimum of water, could exist. The typical breed of the period, adapted to these conditions, was the Texas Longhorn—whose productivity, however, was low.

Though the ranches were usually based on a reliable water-supply, the sparse vegetation made it necessary to graze huge expanses of land with widely-spaced water-holes, so that the cattlemen actually led a semi-nomadic existence. A form of transhumance was also practised, with the animals grazing in the valley bottoms in winter and on the higher mountain areas in summer.

On the range, sheep have certain disadvantages over cattle. They tend to uproot the grass, thereby denuding the range and increasing the hazards of erosion. They pasture in compact flocks, trampling everything in their path, with their small hooves doing more damage to young seedlings than the larger hooves of the cattle. On the other hand, sheep are better adapted to grazing the sparse vegetation of desert range.

Sheep and cattle are not incompatible, and can graze the same range, as is the practice in the Utah desert range. However, on the whole, cattle-production is predominant on the American range; this may be due largely to the preference of the American consumer for beef rather than mutton.

The Conflict between Grazing and Farming

It is interesting to note that the traditional conflict between the pastoralist and the crop farmer, characterizing the dry regions of the Old World, was repeated in the drylands of North America in the 19th century.

Homesteaders were required to clear for cultivation and grow a crop on at least part of their land in order to obtain a legal title to it. In order to keep out the cattle, the land had to be fenced. This, in many cases, prevented free access of the cattle to water, and thereby made large areas of adjacent range virtually unusable. It was only after a long period of bitter conflict that stockmen and farmers found that their production could be complementary—that the development of the full potentialities of the range, with its unreliable and fluctuating forage supply, was dependent on a balancing source of animal fodder, which in turn required assured reserves of forage. Forage reserves, in the form of hay or silage, were usually produced under irrigation. Thus stock production on a high level was dependent on the integration of animal husbandry and forage production.

Whilst irrigated agriculture was highly successful in the desert areas, attempts at dry-farming, based on the methods developed in more humid regions, generally ended in failure. Occasionally, the sowing of dry-land crops of grain coincided with years of more or less adequate rainfall; this encouraged putting larger areas under the plough. The results, in the long run, were not only financially disastrous; this practice also destroyed, sometimes permanently, the native vegetation and laid the land bare to erosion by wind and rain.

The production of forage reserves under irrigation made it possible to reduce the areas required by the individual rancher, and also to fence the range. Thereafter, the rancher had the incentive to conserve and even improve his holdings. The extension of the railways made the long drives to the markets unnecessary. An assured forage supply, improved range, and smaller distances to be covered by the cattle, made it possible to replace the Longhorn by less hardy but more productive breeds, which produced a higher quality of meat. The Hereford from Britain became the dominant breed of the range; at a later stage a new breed, the Santa Gertrudis, was produced from a cross between a Zebu-type animal from India with the British Shorthorn, to withstand heat and fly-borne tropical diseases. Technological improvements were also introduced into the range: windmills, well-pumps, water-storage tanks, and pipe-lines, improved the supply of water; scientific range management replaced overgrazing while improved veterinary techniques, better-bred animals, and supplementary feeding, increased productivity and quality. It became possible to replace migratory herding by different forms of controlled grazing. Where transhumance continued to be practised, the cattle were transported by trucks from the winter to the summer range and *vice versa*.

A family-sized ranch requires at least 150 head to maintain an acceptable standard of living according to American standards. The average size of a ranch in the American desert regions is about 8 000 ha (Logan, 1961). Most of the land is leased from the Government; the fees paid by the rancher are usually invested by the Government in improving the range.

Three types of ranching economies are practised on the desert ranges: (a) young steers are purchased outside the ranch, to be raised on natural grazing until they achieve the most economic weight; (b) a calf-crop is produced on the ranch, to be sold shortly after weaning; and (c) two-year-old steers are raised on the ranch (or older steers, according to the type of range).

ARID AUSTRALIA

The aboriginal population of the arid areas of Australia was extremely sparse. They were nomadic and lived from hunting and food-gathering only, making no attempt to cultivate plants or to domesticate animals; therefore they had no disturbing influence on plant cover and did not cause degradation of the soil, as was the case in other regions (Wadham, 1961).

White settlement in the dry interior started around 1820 and was prompted by the excellent market in Britain for Australian wool, produced in the humid coastal strip. The first attempts at sheep-raising in the dry region were generally unsuccessful; it was only far later that it was demonstrated that the basic reasons for these failures were the low phosphorus content of the soils and, in certain cases, the lack of specific trace elements. However, occupation developed rapidly, and within a relatively short period most of the accessible areas were being used for sheep-grazing. Cattle were better adapted to the arid range than were sheep, mainly because they could graze in a far greater radius around the watering points, but they were economically far inferior: markets for beef were limited and the other principal products, hides and tallow, could not compare with wool in economic value.

Towards the end of the 19th century, two developments occurred which had a far-reaching effect on land-use in the arid areas of Australia: (a) a large number of bores were developed, many of them free-flowing and others requiring pumping, which supplied semi-saline water for the stock and made possible a considerable increase in the intensity of stocking; and (b) there was an epidemic spread of rabbits, which competed with the stock for the available grazing. As a result of both these developments, overgrazing caused the pastures to deteriorate rapidly.

As in all desert areas, there were great annual fluctuations in production, due to recurrent droughts; considerable fluctuations in the price of wool also created conditions of economic instability.

In Australia as elsewhere, the legalizing of individual claims to the grazing areas was followed by fencing, improvement of water supplies, construction of buildings, and attempts to improve pasture management.

The size of the individual holding was based on the principle that the landowner should be assured of an income roughly comparable to the basic wages prevailing in the cities. However, estimates based on the price of wool, with its considerable fluctuations, could not be very accurate or lasting. In the 1930s, the number of sheep estimated as necessary for a satisfactory family income was set at 3 000; the grazing area required for an efficient farm unit varied from 1 200 hectares in south-western New South Wales, to 20 000 hectares in the more arid areas situated farther north in New South Wales.

Earlier, a number of economic setbacks had occurred, due to the epidemic spread of sheep diseases and pests, to droughts, to price fluctuations of wool, and to the depredations of rabbits. These had culminated in a near-catastrophe in 1891, when several successive years of drought, which caused the death of 50 million sheep coincided with a sharp drop in the price of wool and a worldwide financial crisis. Sheep populations declined by 35 to 60 per cent, according to the regions, with the more arid areas incurring the greatest losses. Recovery was gradual, but the flock population never again attained the levels reached before 1891.

Grazing in Australia is pushed farther beyond the dry side of the 250 mm isohyet than in America. Grazing density of the 'sparse-lands' of Western Australia is about one cow per 160 ha, whilst in Arizona, ranges with less carrying capacity than one cow per 20 ha, are not in use (McDonald, 1959).

In Australia, nomadic land-use does not exist (Christian, 1964). There is some movement of stock during drought periods to less-affected pasture areas, and also movement of stock from breeding to fattening areas—the nearest approach to nomadic grazing. Stock routes over which the herds travel are provided with water-bores, at distances of from one to two days' travel.

RECAPITULATION

Pastoral Systems

Undoubtedly the best use that can be made of the desert areas, as such, is extensive grazing by livestock. *Nomadism* and *transhumant pastoralism* are very ancient natural adaptations to the sparse vegetation, seasonal growth of the vegetation, and drying up of watering places, that are characteristic of arid lands; they are therefore the traditional forms of land-use of enormous desert tracts in practically all the dry regions of the world. Even wild animals migrate spontaneously under these conditions, and, like them, the nomad pastoralist follows the rain according to seasonal patterns, seeking new pastures.

Seasonal pastures are generally of two kinds: winter and early-spring grazing on the low-lying, warmer and drier areas, and summer grazing in the higher hill and mountain ranges. In addition, grazing on the stubble of cultivated crops may occupy the transition period between these two types of grazing. Seasonal pastures can be used in two ways (Barth, 1962):

(a) transhumance: herds that are resident in one area, are moved to an adjacent area for a few months, during the latter's productive period;
(b) long-range nomadism: this involves the seasonal migration of the whole human population with their herds over long distances along seasonal routes, alternating between different types of grazing areas, or 'moving with the rain', over vast expanses of the desert. This implies continuous tent-dwelling and a completely nomadic way of life for the whole tribe.

Primitive pastoral methods can, at best, provide a very low standard of living to the pastoralists, who have traditionally always sought supplementary sources of income (Plate 10).

Ranching Systems

In zones with a somewhat higher and more reliable rainfall, sedentary systems become possible—notwithstanding the long dry season that follows the rains. The two salient features of such 'ranching' are:

(1) The fencing of the area, making possible planned grazing procedures, and

(2) The establishment of adequate watering points, which by proper spacing make it possible to reduce localized overgrazing. Self-feeding salt installations at the right locations also induce the livestock to graze the range more uniformly. With modern methods, a large capital investment, and very large areas available per farming unit, a satisfactory standard of living can be achieved.

RAIN-FED ARABLE CROPPING IN SEMI-ARID AREAS

CROPPING SEASONS

It has already been pointed out that a semi-arid climate is usually a mixed climate, in which a completely dry season alternates with a fairly moist season.

The humid season of the semi-arid regions can be subdivided into three principal periods (*see* Figure 20; also Cochemé & Franquin, 1967).

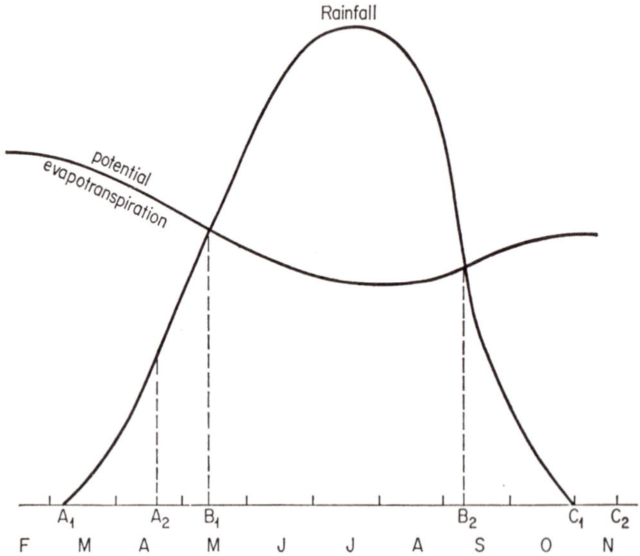

Figure 20. Subdivisions of the cropping season in a semi-arid climate.
A_1: Beginning of rainy season;
A_2: Beginning of transition period (precipitation = half of P.E.T.);
B_1: End of transition period, beginning of humid period;
B_2: End of humid period;
C_1: End of rainy season;
C_2: Exhaustion of moisture reserves in the soil.
(From Cochemé & Franquin (1967): by courtesy of Food and Agriculture Organization of the United Nations.)

(a) *The pre-humid period,* during which precipitation is markedly inferior to potential evapotranspiration.

This period begins with the first rain (point A_1 in the figure) and ends when precipitation equals potential evapotranspiration (point B_1). This period is normally the time for sowing, which may be in dry soil if it precedes the first effective rainfall, or in moist soil if unusually early rains occur. Sowing in dry soil of course involves the risk that rainfall, while sufficing to cause germination, may fail to provide sufficient moisture for emergence and continued growth. Only crops that can effectively withstand a number of periods of desiccation between the beginning of germination and full emergence—such as wheat, barley, and pencillaria—can be sown under these conditions. The risk involved in sowing in dry soil is considered to be justified by the possibility of increased yields, due to the likelihood of a longer growing-season and more effective use of precipitation early in the season.

Where the risk of sowing in dry soil is too great, or in regions in which the soil is too dry and hard for tillage with primitive equipment, the pre-humid period is mostly devoted to the preparation of a seed-bed. The tillage operations in this case become easier as one approaches point B_1; but once this point is passed, they become increasingly difficult and even impossible in heavy soils, as the soils become too wet for effective tillage.

At point A_2 on the graph (Figure 20), precipitation equals one-third to one-half of potential evapotranspiration (depending on albedo). This is the normal mean evaporation rate of a bare soil (or one with a sparse vegetative cover) which is only occasionally moistened by rainfall. In principle, this is the time when germination and emergence become possible, and from this period onwards real evapotranspiration will increase exponentially, together with leaf area, until the crop shades the soil entirely. Ideally, this should coincide with point B_1 on the graph.

(b) *The humid period,* during which total precipitation is in excess of potential evapotranspiration, and therefore potential and real evapotranspiration are equal.

The 'humid' period does not coincide with the rainy season: it begins later and ends earlier. During the humid period, the excess of precipitation over real evapotranspiration, provided it is not lost as run-off, is stored in the soil. If precipitation is sufficient, the entire root-zone is saturated to field-capacity, and excellent moisture conditions for crop production and especially for crop maturation during the third (c) period will prevail. Excess precipitation will either cause run-off or drain beyond the root-zone, or both.

The relative positions of B_2 on the graph (Figure 20), and the dates of flowering (or earing for cereals), have a considerable impact on potential yield levels. The longer the humid period, and the later it ends, the greater will be the prospects for the maturation under favourable conditions of a larger number of fruits per unit area. However, the greater the coincidence of flowering period and peak moisture-supply, the greater will be the danger of epidemic diseases.

(c) *The post-humid period,* in which total precipitation is inferior to potential evapotranspiration.

At the end of the rainy period, real evapotranspiration continues until the reserves of water in the soil formed during the humid period are exhausted.

Crop production may also be possible during the completely dry period, provided the land remains fallow during the humid season, and precipitation has been sufficient to fill the soil reservoir to the depth of the root-zone. Adapted crops can then be sown during the dry season, and even produce high yields without receiving a drop of rainwater throughout their growing period.

THE NEAR EAST AND NORTH AFRICA

In the semi-arid areas of the Near East and North Africa, traditional agriculture is almost exclusively based on winter cereal monoculture—of wheat in the higher-rainfall areas, and of barley in the less favoured ones. These small-grain cereals have proven, through long experience, to be the food-crops best adapted to the semi-arid climate. Their growth-patterns coincide more or less closely with the distribution of rainfall, and they are able to survive periods of drought. The winter cereals are sown mainly after uncultivated fallow; the lower the rainfall, the longer will be the period of fallow.

The soil cultivation is superficial and does not completely eradicate weeds, so that the plant cover is apt to be rapidly restored during the fallow period. Until the beginning of this century, only the best lands were cultivated, and marginal areas were left for grazing. Yields were generally low and irregular. But in more recent years, population pressure has led to the cultivation of marginal lands, coupled with the introduction of mechanical ploughing; this has frequently led to soil erosion and crop failures.

Some annual forage crops, such as vetch (*Vicia sativa*), Berseem (*Trifolium alexandrinum*), and Subterranean clover (*T. subterraneum*), are well adapted to the semi-arid climate with rainfall in the mild winters. They are not used to a large extent in the traditional agriculture of these regions, but have proven their value, both for forage production and for improving the general fertility-level, in the modern agriculture of Australia, for instance.

Brichambaut & Wallen (1963) state that evapotranspiration is very much the same in most parts of the semi-arid region of the Near East, and that, as a result, the limit of dry-land cropping is closely related to annual rainfall.They have therefore attempted to determine the lowest amount of rainfall needed for regular dry-land farming in the Near East, taking into account the variability or reliability of annual precipitation in different areas. They found that a mean annual rainfall of 240 mm, with a variability of 37 per cent, was the normal minimum requirement for regular dry-land farming in the region. Where variability was much greater, as on the slopes of Zagros in northern Iraq, a much higher annual average was required.

Allowing for crop failure due to drought in two years out of ten, it was found that the minimum amount in an individual year varied from 180 mm in areas with regular rainfall, such as Jordan, south-western Syria, and the so-called Fertile Crescent,* up to 230 mm in the north of Iran—where the low winter temperatures extend the dry-farming season into the summer, with consequent high evapotranspiration. The borderline zone for dry-land farming, and hence the limit between semi-arid and arid conditions, was thus found to be between 200 and 300 mm of annual rainfall in these regions.

SUBTROPICAL AFRICA

The Sudanese Sahel consists, from the agricultural viewpoint, of two distinct regions (Cochemé & Franquin, 1967):

* Fertile Crescent is the area of fertile land extending in a semi-circle from Israel to the Persian Gulf.

(a) *a northern region*, bordering on the Sahara. Like the semi-arid region to the north of the Sahara, this region is one of cereal monoculture, with an alternating cereal—fallow sequence. However, wheat and barley are replaced by pencillaria, which is the only grain-crop that is successfully grown. Sorghum is also grown on inundated lands, after the waters recede.

The northern boundary of this region is difficult to define precisely, as it moves in the direction of the Sahara or southwards according to the prevailing cycles of wetter- or drier-than-average years, respectively. For a long period of years, the northern limit approximates the isohyet of 250–300 mm.

(b) *the southern region*, of 'extensive polyculture', in which millet, sorghum, niebé, groundnuts, cotton, and maize, can be grown, under a rotation scheme in which an annual crop alternates with fallow.

This region comprises a number of sub-regions, in order of increasing precipitation:

Sub-region A. Principal crops: pencillaria (or sorghum) and niebé.
Sub-region B. Principal crops: pencillaria (or sorghum) and groundnuts; secondary crop: niebé.
Sub-region C. Principal crops: groundnuts, cotton; secondary crops: pencillaria, niebé, maize.
Sub-region D. Principal crops: maize (or sorghum or pencillaria), groundnuts, cotton; secondary crops: tuber crops (manioc), rice, fonio.

SEMI-ARID REGIONS OF THE RUSSIAN PLAIN, CAUCASUS, AND CENTRAL ASIA

The soils of the semi-arid zones of Eurasia are characterized by their high natural fertility, including, as they do, extensive areas of black and chestnut soils (Kovda, 1961). The earliest attempts at farming, during the Neolithic period, were of necessity on a very small scale: the thick, tough steppe sod could not be penetrated by the simple and primitive implements available at the time. It was only after the invention of bronze and iron tools, in the middle of the first millennium B.C., and the start of the use of draught animals, that the virgin black soil could be cultivated on a relatively extensive scale.

Burning the plant cover and sod was an essential part of the primitive farming system practised. Hard spring wheat and millet were the principal crops, though barley, hemp, and flax, were also grown. By the 8th–9th centuries A.D., the three-field rotation system (winter crops, spring crops, fallow) predominated in Kiev-Rus, the large feudal state. A wide variety of crops were grown, including winter and spring wheats, rye and millet, barley, peas, hemp, flax, poppies, buckwheat, and lentils. The limit of dryland farming was roughly the isohyet for 350 mm annual rainfall.

The Mongolian invasion by Genghis Khan in the 8th century destroyed the agricultural economy of Kiev-Rus. The cultivated land was largely abandoned, and arable farming was replaced by the nomadic stock-raising of the invaders. Farming in the area was resumed after the expulsion of the Mongols, and again became based on the previous system of the three-course rotation. In the 18th and 19th centuries, arable cropping gradually spread from the black-earth steppes of the Ukraine and central Russia to progressively drier regions in the Crimea, Caucasus, and the Volga area, in which large areas of black earth as well as chestnut soils were brought under cultivation down to the 250–mm isohyet. In addition to the three-course system in the more humid parts of the region, long-term fallow was practised in the drier parts. Here, droughts were relatively frequent, 23 being recorded during the 19th and first half of the 20th centuries.

Improved techniques initiated in recent years consist of mechanized operations aimed at completing all farming activities during the seasonally optimal periods—apart from deep autumn ploughing, snow retention on the fields, and the use of fallow.

In the semi-arid Soviet central Asian region, most of the annual precipitation (200–250 mm) occurs during the cool season; summer precipitation is either sparse or entirely absent. Spring rains here bring the soil to field-capacity to a depth of 100–150 cm, and make possible a form of dry-farming called 'bogara'. The main crop is winter wheat. Below a depth of 200–300 cm, however, soil moisture is always at the wilting point or below. By autumn, a wheat crop will have exhausted most or all the available moisture, but approximately 70 mm of water will remain available for the following wheat-crop; this water, together with later precipitation, makes it possible to obtain a reasonable yield. The summer soil-moisture deficit coincides with the maturation of the cereal.

SEMI-ARID AUSTRALIA

Arable cropping in Australia is practised in semi-arid regions with rainfall ranging from 250 to 500 mm (Wadham, 1961). Originally, the areas for cropping had to be cleared of trees and scrub, and then levelled. Yields were low, and very large areas had to be allocated for the individual farmstead: 259 hectares was the standard for the more favoured areas, and increasingly larger amounts were allowed for the drier regions.

Progress in the early years consisted mainly of mechanical innovations and improvements—tree and stump pullers, scrub beaters, land levellers, and mechanical harvesters. Average yields actually declined, most probably because of the gradual depletion of soil nutrients that had been initially low in any case (Stephens & Donald, 1958), and exceeded 680 kg/ha only once in the period from 1890 to 1905. In South Australia yields averaged only 255 kg/ha and crop failures were frequent.

The first improvement came with the breeding, by William Farrer and his colleagues, of wheat varieties that were adapted to the short Australian growing-season and to mechanical harvesting. The use of fertilizers, and the introduction of the combined seed- and fertilizer-drill, also contributed to higher yields. The high prices and considerable demand for wheat during the years of the first World War favoured an extension of arable cropping into the arid zone. In the beginning, continuous cropping to wheat was general; later, the system of fallowing was introduced gradually and became common by 1920.

World crop surpluses, recurrent droughts, and low prices, subsequently brought financial disaster to many of the wheat-growing farms; the payment of a government subsidy to wheat farmers became necessary to alleviate their critical economical situation. Meanwhile the continual wheat cropping, even in a crop-fallow system, caused a gradual deterioration of soil structure: soil tillage became progressively more difficult and had to be restricted to an increasingly narrower soil-moisture range.

The problem of deterioration was solved by introducing a ley-wheat rotation, including specially adapted, self-seeding annual leguminous species—subterranean clovers in the more humid areas, and annual species of *Medicago* in the drier regions, which were also fertilized with superphosphate. These leguminous plants occupied the land for four consecutive years, to be followed by three years of wheat cropping. Shorter rotations, consisting of two years of clover ley followed by a single cereal crop, have also been proposed (Watson, 1963). The legumes increase soil nitrogen (originally 0·01 to 0·05 per cent in the drier soils) to levels suitable for wheat production without additional nitrogenous fertilizers (Stephens & Donald, 1958).

Subsequently, the addition of certain trace-elements to soils on which legumes had not been successful with superphosphate alone, enabled a considerable extension of the cropping areas into areas that had previously been considered sterile.

It was found that for each ton of superphosphate used on legumes, increments of 747 kg of nitrogen (Donald & Williams, 1954), 58 kg of exchangeable magnesium, and 230 kg of exchangeable calcium (Williams & Donald, 1957), were obtained. The leys served as pastures for sheep, and in due course about one-third of the total sheep population of Australia was maintained on the wheat farms.

The sheep fed on the stubble of wheat crops and on the naturally regenerating or sown pastures—a system of rotational land-use that proved to be very well adapted to the semi-arid agriculture in Australia. It benefits crop and animal, economizes on fertilizer use, and helps to control weeds as well as to reduce erosion on cultivated fields.

DRY-FARMING IN THE GREAT PLAINS REGION OF THE U.S.A.

The Great Plains are a region of predominantly summer rainfall and constitute the principal wheat-producing area of the U.S.A. Here 'dry-farming,' as a scientific method for producing crops under conditions of limited moisture, was developed.

The beginnings of dry-farming were greatly influenced by mistaken theories —in particular, that maintaining a 'dust mulch' on the soil surface by frequent cultivations, would break up the soil capillaries and thereby prevent loss of water by evaporation. European-type tillage equipment was used that pulverized the soil.

It was further assumed that very deep ploughing and subsoiling would increase water-uptake by the soil, and that plants would obtain their water by capillary movement from a water-table. Dry-farming practices were accordingly used on a large scale that was made possible by advances in agricultural machinery. By 1930, practically all the land of the Great Plains had been ploughed and the native short-grass cover was almost extinct. The maintenance of a fully pulverized soil-surface increased wind-erosion and reduced moisture infiltration into the soil. From 1931, a succession of droughts struck the region; the results were not only financially catastrophic but caused widespread soil destruction as a result of wind-erosion.

The economic repercussions of an alternation of periods of favourable years with a series of drought years are vividly illustrated by the following, earlier example: in eastern Montana during a period of twelve years (1905–16), rainfall was 26 per cent above average. In 1916, dry-land farming reached its maximum extension. Then started a period of below-average precipitation, which was to continue for 23 years. After four successive years of drought in eastern Montana, during the period 1917–20, 38 per cent of the State banks and 30 per cent of the National banks in Montana went bankrupt; 20 per cent of the farms were abandoned, 50 per cent of the manufacturing concerns ceased work, 50 per cent of the retail businesses closed, and 15 per cent of the population migrated elsewhere (Lomasson, 1947). Yet this was in a period of high wheat prices and demands elsewhere (*see* p. 113).

The magnitude of this catastrophe has since served as the classical example of the danger of adapting land-use to the conditions pertaining during favourable periods, instead of developing a stable agricultural economy that is adapted to the entire range of possible climatic conditions. Following the disastrous effects of the 'dry-farming' system, new tillage methods and equipment were developed to reduce as far as possible the dangers of wind-erosion. Clean-tillage and dust mulches were avoided; when the soil was not covered by a growing crop, it was left in a cloddy condition and was covered by crop residues. Special practices for improving infiltration rates of precipitation and reducing soil-erosion were adopted, such as contour listing and contour cultivation and terracing.

Crop rotations were developed that provided a vegetative cover during the critical periods of the year. As the land still retained much of the fertility that had been built up during the period when it had been under grass, the maintenance of fertility was less of a consideration in selecting crops for the rotation and the place assigned to them in the crop sequence, than was their effect on moisture relations. A common rotation adopted was maize–wheat, because the maize leaves a residue of moisture that is available to the wheat (Mathews & Cole, 1938). However, the commonest rotations are based on wheat and fallow, with the fallow occurring once in a number of years.

It is the residue of moisture in the soil, available at the time of seeding the wheat, which serves as a guide for predicting the yields of grain that may be reasonably expected, and whether wheat should be sown or the land left fallow. If the soil is wet to a depth of one metre or more at seeding time, yields are generally satisfactory (Cole & Mathews, 1940). When the soil is wet only to a very shallow depth, it is considered preferable to leave the land fallow for a year.

RECAPITULATION AND CONCLUSIONS

In all the semi-arid regions of the globe, rain-fed arable cropping is based mainly on production of wheat, which has proven to be the crop best adapted to these conditions. It is outstandingly more valuable than any other crop that can be grown under these conditions: hence the difficulty of planning rotations which include less remunerative crops.

The main problem of these regions is to adjust the farming system to the extremely variable rainfall, so that the maximum benefit can be derived from years with good precipitation, and that drought years need not cause economic breakdown.

It is on the fringes of the arid regions that arable cropping is most precarious and undesirable methods of land-use have led to catastrophes. These regions are characterized by the very irregular pattern of annual precipitation, which varies within wide limits. In years of sufficient precipitation, very good grain yields can be obtained, because of the build-up of fertility during the low-rainfall years. This serves as a temptation to plough up land for cropping—especially when years of high precipitation coincide with favourable marketing conditions for grain. Inevitably, years of under-average rainfall follow, and crops fail more or less completely. In addition to the economic distress involved, the misuse of the land leads to severe erosion, mainly by wind.

As one moves from the desert fringe towards regions with higher annual rainfall, the prospects for satisfactory crop production improve. However, even under these circumstances, occasional years of drought and crop failures are almost inevitable.

The conclusions to be drawn from examples of rain-fed arable cropping in various semi-arid regions of the world can be summarized as follows:

(a) The ploughing up of range-land in high-risk regions with too scanty precipitation to assure reasonable prospects of cropping, should be avoided.

(b) Where the amount of rainfall justifies arable cropping, cropping systems based on the integration of animal husbandry and arable farming appear to give the best prospects for sustained yields at a reasonable level.

(c) Appropriate tillage methods must be adopted for maximum moisture conservation and soil protection.

These aspects are treated in detail in Chapters XIII and XV.

Irrigation Agriculture in the Dry Regions

FLOOD-WATER FARMING

A very crude form of irrigation, called flood-water farming, has been practised since the earliest times in the deserts of the Old and New Worlds. By this means, the insufficient precipitation was supplemented by run-off water from higher-lying areas.

Nabatean Flood-water Farming

The Nabatean Kingdom flourished from the 2nd century B.C. to the 7th century A.D. It was essentially a desert kingdom, extending from Damascus to what is now Saudi Arabia (Evenari & Koller, 1956). The Nabatean economy was based mainly on the commerce between the Near and the Far East. Petra, its capital in the middle of the desert, was at the cross-roads of the caravan routes. A number of important towns, also in the desert, were set up to protect these routes. By developing an elaborate system of rain-catchment, including dams, spillways, and terraces, the Nabateans were able to produce in the desert itself, with an annual rainfall of less than 100 mm, the food required for their towns and for the caravans travelling through their territories (Plate 11).

Rainfall in the desert usually occurs in short, violent downpours, which the hard-baked and relatively impermeable soil is not capable of absorbing. The water runs off the stony, bare hill-sides and even a relatively light rainfall is capable of producing a significant 'water crop', carrying a fair amount of silt (Plate 12). A rainfall of 10 mm on a watershed may cause a flood of 30 000 m^3 during some four to five hours (Evenari & Koller, 1956). There are two possible means of saving this water: by storing the run-off behind dams, or by storing it underground in the beds of wadis* and in level plots on their banks. The

* *See* footnote on p. 69[0].—Ed.

Nabatean farmers chose the latter alternative which is more effective and enduring than the former. A rainfall of a few millimetres on the catchment area may permit the storage, in a restricted area, of the equivalent of several hundred millimetres of rainfall.

Crop production was carried out either in tributary valleys, or in the main valleys:

(*a*) *Cultivation in Tributary Valleys*. In wadis which were not too narrow or of which the banks were not too steep, the flow of water from the hills was slowed down by a series of stone levees. Gradually the areas between the levees became level as a result of deposition of silt, so that terraces were created which could be sown after the flood—mainly with cereal grains. Once the whole wadi was so terraced, the flow of water was completely intercepted and absorbed. The Bedouin cultivate these patches up to the present day, obtaining yields even in dry years (Zohary, 1954).

(*b*) *Cultivation in Main Wadis*. When flash-floods occurred, these were apt to be destructive torrents. The Nabateans realized that they could not control or harness the whole flow of the main wadis. Their approach was therefore to leave the watercourse proper of the wadi undisturbed, whilst practising cultivation on the flat strips of land along both sides of the wadi. Relatively strong dams were built across the wadi, whose purpose was not to create reservoirs, but to raise the level of the water so that it could be diverted to terraces alongside the wadi, through special diversion conduits built upstream from the dam. Between the wadi and the fields, strong walls were built to protect the cultivated plots against erosion. In addition, the fields were divided into level terraces by stone walls built at right-angles to the direction of flow in the wadi.

Flood-water Cultivation in the New World

A very simple and primitive form of flood-water cultivation was practised in what is now the south-western U.S.A., by a number of Indian tribes along the Gila and Colorado Rivers (Gulhati & Smith, 1967). Flood-water farming was also practised by the Hopi Indians and other tribes for whom it is still the dominant form of cultivation (Armillas, 1956).

The preferred sites for flood-water farming are the arroyo (wadi) mouths, where the silt deposited by the floods has built up a shallow fan. The ratio of cultivated-field area to the arroyo watershed varies from 3 to 6:100. The water is distributed over the fields by the building of low earthen dams. During floods, water backs up behind the dam and saturates the soil. After the floods subside, the Indians plant maize, squash, and other crops.

Other sites are located on the floodplains of the main valleys and/or low flood-terraces of large arroyos. The floods are sometimes diverted by the construction of simple earthen dams, about 30 cm high, which last for only a single season; a more permanent type of check-dam is built of brush and loosely-piled stones.

Where seasonal floods were very variable, with years of no floods interspersed with years of floods of destructive force, the agricultural economy was extremely unstable and subsistence had to depend to a large extent on food-gathering and fishing. Where floods were more dependable, and occurred in both summer and winter, two crops a year could be grown and agriculture was the main source of subsistence.

Conclusion. Summing up it may be said that, under certain conditions, flood-water farming was the only, and essential, method for producing crops in the desert. The intensity and number of floods determine the size of the area that can be sown. The large fluctuations from year to year make such farming a very precarious source of livelihood; in certain years, the floods may be rare or short-lived, and the soil will not be moistened to a depth that is sufficient to justify sowing.

It therefore appears doubtful whether a viable agricultural economy can be based on this form of crop production, especially as cultivation is possible only in relatively small and scattered strips and patches, making mechanization uneconomic. The method may, however, have its uses for creating local reserves of forage. Modern methods of using run-off water are also being developed, and will be discussed in Chapter VI.

IRRIGATION FARMING

Four great valley-civilizations flourished in early times in the Old World: those of the Nile, the Tigris-Euphrates, the Indus, and the Hwang-Ho. All four regions are characterized by extremely light precipitation and high summer temperatures, and were therefore dependent on irrigated agriculture.

Very little is known of the early history of the civilizations of the valleys of the Indus and the Hwang-Ho (Drower, 1955); by contrast, considerable information is available on those of Egypt and Mesopotamia—to which regions we will accordingly confine ourselves.

Evolution of Irrigation Agriculture in Egypt

The rainfall equivalent of the annual irrigation supply of Egyptian agriculture works out at approximately 2 000 mm. Yet with only a narrow coastal strip receiving an average of as much as 125 mm of rainfall, Egypt belongs to a typical arid zone (Hamdan, 1961).

Prehistoric. According to Butzer (1958), the Nile Valley was occupied by Man as a hunter or food-gatherer for at least 10 000 years before agriculture was attempted. During the period before the Nile was contained by embankments, only the areas lying above the high-water mark could be inhabited. It is conjectured that the inhabitants descended with their herds into the valley every year, for grazing of the lands after the floods had receded, and that only later in history was the sowing of cereals practised.

The Early Stage: Basin Irrigation. The first to undertake the control of the Nile is said to have been Menes (3400 B.C.). A basin system of irrigation was gradually developed; the floodplain, which lay about 9 m above the river-bed, was divided into basins whose area varied from some 100 sq m to several square km. When the river flooded in July, August or September, feeder canals led water into the basins whose bottoms were then submerged to a depth of one metre or more. After the ground had become thoroughly soaked to the full depth of the root systems, the excess water was drained off—either into an adjacent basin downstream or into the bed of the subsiding river.

The annual flood brought with it some 100 million tons of suspended materials (Addison, 1961), and therefore usually left a thin layer of fertilizing material on the soil surface of the basin. The crop was sown as soon as possible, without any preliminary ploughing. Essentially, a single winter crop, mainly a cereal (barley or wheat), was sown each season, so that this was basically a subsistence economy (Hamdan, 1961). Other crops to be sown were pulses (beans and lentils), onions, and flax.

The basin system is technically a very primitive and expensive form of agriculture, similar in many respects to rain-fed crop production on the semi-arid fringe of the desert. In both cases, only a single winter crop can be grown annually. On the other hand, the system showed a degree of permanency which is most rare for irrigated agriculture: the fertility of the soil and its productive capacity have been maintained under this system for thousands of years.

Several factors contributed to the biological soundness of the system: the thorough annual flushing of the soil, combined with excellent natural drainage, prevented salinization—the foremost problem of irrigated agriculture. The silt deposited annually provided plant nutrients in quantities that were adequate in view of the extensive form of cropping practised, thereby preventing soil exhaustion. The annual fallow period, which followed the harvesting of the cereal, took place during the hottest months of the year; the soil was thereby completely dried out, and as a result the weeds were destroyed and the population of pathogens in the soil was considerably reduced. Under this régime, the soil which had been packed by sowing, weeding, and harvesting operations, usually developed deep cracks whilst drying out, so that air was able to penetrate to a considerable depth and the soil structure was improved on rewetting.

The system was, however, not without its dangers. Successful cropping depended on the time of flooding and on its extent. If the flooding was too late, the growing-season of the winter crops was shortened, with resultant reduced yields. If the flood was too low, the cultivable area was correspondingly reduced; if too high, the basins would be transformed into marshes and epidemics would occur. Hamdan (1961) estimated that during the period from the 14th to the 18th centuries, no fewer than 50 plagues and major epidemics occurred in Egypt.

Nor was the basin system entirely free from the danger of salinization. Probably as a result of neglect of the canal network, normal drainage was hindered in the Berari region of the northern Delta, and some time in the late Roman and early Arab period over half-a-million hectares were lost to Egyptian agriculture following salinization (Hamdan, 1961).

Perennial Irrigation. The basin system made possible subsistence farming only. Many cash-crops, suited to the Egyptian climate, such as sugar-cane, indigo, cotton, vegetables, and rice, require a regular water-supply during the period from March to July, when the water level in the Nile is at its lowest and its flow is greatly reduced. These crops could be grown only by controlled irrigation lasting practically throughout the year. Such perennial irrigation also made it possible to grow more than a single crop annually.

The beginnings of perennial irrigation can be traced to the ponds and small lakes left behind after the flood-waters had retreated (Hurst, 1952). Summer crops could be grown in the neighbouring fields, and watered for many months after the flood, by using buckets filled by hand. Gradually, this method was replaced by primitive mechanical devices for lifting water.

A subsoil water reservoir, whose level ranges from 4–7 m below the surface, according to the time of the year, is fed underground from the Nile. This aquifer was tapped by wells, whose water complemented the river floods and made year-long irrigation possible. However, the water was not silt-laden, and the continuous cropping, made possible by this system, created serious problems of impaired soil fertility. Animal manure, and soil from long-extinct settlements which was rich in phosphorus and potash, had to be applied regularly.

With perennial irrigation, a greater diversity of crops could be grown, and a more complex crop-rotation established, than with basin irrigation. However, for perennial irrigation to achieve the scope required for growing food and export crops on a scale commensurate with the needs of a rapidly increasing population—from 2½ millions at the beginning of the 19th century to about 7 millions before the end of that century (Addison, 1961)—more effective and large-scale methods were required. Around the beginning of the 19th century, it was proposed to build a barrage on each of the two branches of the Nile to the north of Cairo, and thereby raise the water-level in the river, so as to keep water flowing into the canals after the flood had receded. However, this first attempt to construct large-scale river control structures ended in failure. The necessary technical know-how was simply not available at the time, and the barrages that were constructed did not hold back the waters. A more modest and technically simple effort was more successful: an inundation canal—the Ibrahimie Canal—which diverted water from the Nile at Assiat, was dug some 320 km upstream from the Delta.

In the meantime, technical knowledge on the construction of barrages was increasing; by 1864 the Delta barrages had been strengthened and made serviceable and by 1902 a dam had been constructed at Aswan—the first of its kind in the world (Addison, 1961). This, together with a network of main carrier-canals, branch canals, distributaries, and field ditches, made possible the perennial irrigation of two million hectares. In 1912, the Aswan Dam was made higher, and its capacity was more than doubled. Additional diversion barrages were also built—such as the Mohammed Ali Barrages, put in service in 1940, which replaced the by now obsolete Delta barrages.

With the expansion of the irrigated areas, certain dangers became manifest. The almost complete cessation of flow of the Nile River into the sea increased the danger of sea-water invading the Delta; to prevent this from happening, a special barrage was built and completed in 1952. Silting of the canals increased in amount and made it necessary to maintain an army of 400 000 men for four months out of each year to clear the canals. Perennial irrigation can also lead to a rapid rise in the water-table, with resultant waterlogging and salinization of the soils. A network of drainage canals and ditches had to be developed in order to maintain the water-table at an appropriate level and to remove excess salts.

The larger the areas under irrigation, the greater the danger of a catastrophe whenever the Nile flood does not reach its normal proportions. The first step taken to obviate this danger was to transform Lake Victoria at the headwaters of the Nile into an enormous storage reservoir, by controlling the flow from the Lake. The second step was aimed at reducing the considerable evaporation losses resulting from the sluggish flow of the Nile through the swamps in the Sud region of southern Sudan. This was achieved by developing an artificial by-pass channel—the Jonglei Channel—which shortened and quickened the flow of the Nile waters through this region. The final step is the current construction of the giant Aswan High Dam, about 110 metres high above the river-bed, with a capacity of 130 milliards m^3 of water (as compared with the present storage-capacity of 5 milliards (Addison, 1961)).

In 1925, the ratio of perennial to basin irrigation in Egypt was 70 per cent; by 1960 it had increased to 80 per cent, the basins being confined to Upper Egypt (Hamdan, 1961).

Crops and Cropping Systems in Egypt

Under Basin Irrigation. In early times, barley was the principal crop and was gradually replaced by wheat. Pulses (including beans and lentils), as well as onions, were grown widely. Flax was also cultivated extensively, and provided the principal fibre for clothing.

Egypt has always been poor in pasture, and the number of domesticated animals has been relatively small. Therefore, for the livestock, and in particular draught and transport animals, forage crops have had to be grown, the most important of which is Berseem (*Trifolium alexandrinum*). A two-course rotation is practised, with half of the area devoted to the cereal grain and the other half to assorted other crops.

Under Perennial Irrigation. With the beginnings of perennial irrigation it became possible to diversify cropping, especially by the addition of summer and perennial crops. Among the most important cash-crops to be grown were sugar-cane, rice, tobacco, and cotton—also the dye-crops indigo and saffron, which were grown mainly in Middle Egypt. Egypt became an exporter of sugar in the 12th and 13th centuries, but this crop required heavy irrigation. Rice, also introduced in the Middle Ages, was grown on a large scale in the regions in which there was an abundance of easily lifted water. A perennial tree-type cotton, *Gossypium arboreum*, was grown on a small scale.

Cropping also became more intensive with perennial irrigation, and two or three crops could be produced annually. However, yields showed a tendency to decline, mainly as a result of impaired soil fertility. This resulted from several factors: the lack of silt in the irrigation water, the exhaustive cropping schedule, and the increased incidence of pests; also waterlogging and progressive salinization. In the clay soils of the Delta, some of the best cotton soils lost their fertility in this way. The decline in fertility has been halted in recent times by very heavy fertilization with chemicals—mainly nitrogenous and phosphatic fertilizers—as well as by the establishment of an artificial drainage system. The relatively large area devoted to Berseem in the crop rotation is also a major contribution to maintaining soil fertility.

At present a wide range of crops, including tropical, subtropical, and temperate crops, are grown. Not only have new crops been introduced, but traditional crops have been considerably expanded—though some of them, such as saffron and indigo, have practically disappeared, and others, such as flax, have declined considerably.

Cotton is the main crop, the old perennial types having being replaced by the high-quality, long-staple *Gossypium barbadense* of the New World. Maize, introduced relatively recently, has also become an important crop. Indeed maize, Berseem, cotton, and wheat, together occupy 75 per cent of the crop area, and are grown all over Egypt. Maize is the staple food of the fellah,

Berseem is the principal forage crop, and wheat is supplied to the town-dwellers. Cotton is, of course, the main cash-crop, and is especially popular with the larger landowners. Cotton areas are limited by the necessities of crop rotation, but of every four or five crops that are grown in each two-year period, one is cotton (Addison, 1961).

Of the secondary crops, rice and sugar-cane are increasing in importance, whilst beans and barley are declining. Relatively small areas are sown to lupins, lentils, fenugreek, chickpeas, onion, sesame, and groundnuts, which together account for about 3 per cent of the cropped area (Hamdan, 1961). These secondary and minor crops are apt to be highly localized, being grown in certain specialized areas which have proved to be their optimum habitats. Thus sugar-cane is concentrated in Upper Egypt, where the high temperatures are favourable to the crop, whilst flax is grown exclusively in the cooler and wetter north. Rice serves as a reclamation crop on salt-affected land, and hence is grown mainly in the northern half of the delta, which has the most severe salinization problems.

Irrigation water is not sold to farmers, but is delivered as a national service. The taxable value of the land is assessed according to the irrigation facilities provided. The volume delivered to the individual farmer is in proportion to the amount of water in the river at a given time. For example, the extra water required for rice is supplied only if the Nile's flow exceeds a certain minimum figure (Addison, 1961).

Evolution of Irrigation Agriculture in Mesopotamia

The Tigris and Euphrates both originate in the snow-covered high mountains of Turkey, whilst the main tributaries of the Tigris (the Greater and Lesser Zab) have their origin in the highlands of Iran, which are also snow-covered. The Tigris-Euphrates basin lies mostly in the modern country of Iraq.

The origins of irrigation in the Tigris-Euphrates basin are obscure, but it is known that an elaborate system of canals was already in existence before 4000 B.C. Following the conquest of the valley of the Tigris and part of that of the Euphrates, Hammurabi established a strong, centralized government which exercised effective control over irrigation. Several of the laws of the Hammurabi code are concerned with irrigation, the land-tenure system, and various other aspects of agricultural transactions.

The conditions prevailing in Mesopotamia for irrigated agriculture differ significantly in a number of ways from those of the Nile valley, and therefore the evolution of irrigated agriculture in the two regions, and their present-day situation, are entirely different—particularly in the following ways:

(1) The Tigris-Euphrates basin is in a dry steppe; a limited amount of rain-fed crop production is possible in the region. By contrast, the Nile flows largely in a desert, in which no crop production is possible without irrigation.

(2) The inundation by the Nile is regular and predictable. In Mesopotamia, floods occur in spring as a result of melting snows. The actual period of inundation is irregular, and cannot be forecast with accuracy. If the snows in the highlands melt too rapidly, destructive floods occur. The flood peak generally occurs late in spring and is therefore not well synchronized with the water requirements of the major crops. The floods recede about June, at the beginning of the hot season, so that a single flooding is not sufficient, especially as the principal winter crops—wheat, barley, and beans—are sown in November and December, and the peak requirements of the summer crops, such as cotton, sesame, and millet, occur after the floods have subsided. The floods are therefore always potentially dangerous and have to be kept in check. By contrast, the Nile flood reaches Egypt in late summer and early autumn, and the waters subside at exactly the right time for the sowing of the winter cereals.

(3) The Euphrates carries about five times as great a load of silt as the Nile (Whyte, 1961). The annual increment of silt has in due course raised the level of the river-bed, so that it is above the Plain. The Tigris, by contrast, has a deeply-cut bed, with a very low water-level, requiring barrages to make the water available for irrigation.

The heavy load of silt was not an unmixed blessing. It necessitated the constant clearing of the irrigation system, and raised the level of the fields. It is estimated that in the course of 5 000 years, 10 metres' thickness of silt have been deposited in the fields (Jacobsen & Adams, 1958). As a result of this raising of the level of the fields, water has to be raised to a higher level for irrigation, causing a gradual reduction in the flow from the canals. The yearly deposition of silt therefore results in a decline in productivity, whilst the opposite was true in the deep-basin irrigation of Egypt (Gulhati & Smith, 1967). Meanwhile, in Mesopotamia, inadequate maintenance contributed to the siltation of river and canal beds, with a resultant reduced flow and reduced head of water.

After 1 000–1 500 years of successful irrigation agriculture during the Sumarian civilization, during which period most irrigation possibly consisted of uncontrolled flooding through breaches cut in the levees of the water courses, canal irrigation was introduced and a severe salinity problem began to develop, in addition to the difficulties caused by silting. The earliest and most serious manifestations were in southern Mesopotamia, and occurred about 2400–1700 B.C. (Jacobsen & Adams, 1958). The less salt-tolerant wheat was gradually replaced by more salt-resistant barley. By 1700 B.C., wheat cultivation had practically disappeared in the south, and the yields of barley had fallen considerably—from 2500 kg/ha prior to 2400 B.C., to 900 kg/ha by 1700 B.C. (ibid.). This latter figure is approximately the same yield as that obtained from rain-fed crops in the more favoured parts of the region. Whilst the region was not completely abandoned, it never regained its former prosperity.

The great Sumarian cities declined into villages or fell into ruin, causing a breakdown of the Sumarian civilization.

Salinization was milder, and occurred later, in central and northern Mesopotamia; agriculture in these regions therefore continued to prosper. However, wave after wave of invasions took place, each resulting in a period of neglect, when parts of the irrigation system were allowed to go to ruin—to be followed by new upsurges of relative prosperity.

Devastation reached its peak during the Mongol invasions in the mid-13th century; during their occupation of the country, chaos prevailed, irrigation networks broke down, and farming was not encouraged. Neglect continued during the Ottoman era. However, towards the end of Turkish rule, in 1913, the first barrage on the Euphrates, the Hindiye Barrage, was built. This was followed by the much larger Kut barrage on the Tigris, which was completed in 1939.

The large natural depressions which are characteristic of the plain were also used for diverting excess flood-water. They served the dual purpose of helping to keep the floods in check, and of creating reserves which could serve for regulating the water-supply.

In 1959 the third and most important stage of development of the region was completed. The high dam across the Greater Zab at Dokan, built in a deep gorge, made it possible to impound 7 milliards m^3 of water. At present, approximately three million hectares are irrigated in Iraq, half of them by gravity and half by lift-irrigation (Addison, 1961). But the salinity problem over considerable areas is extreme and tending to heighten, to the derogation of agricultural productivity as a whole.

Irrigation Agriculture of the American Deserts

Irrigation agriculture was already practised in prehistoric times in various parts of the American deserts by certain Indian tribes. At the time of the Spanish conquest, large diversion canals were already being used, in the Gila and Salt River Valleys of Arizona, to transport water to the fields—sometimes for distances of 30 kilometres and more (Logan, 1961). These canals followed a regular gradient, though they had been built without any engineering instruments.

In the Rio Grande Valley, Pueblo Indians to this day practise an irrigation agriculture that is very similar in all respects to that practised before the advent of the White Man. Cultivation methods are primitive, yields are low, and the traditional crops—maize, beans, squash, and melons—are produced. Meanwhile the Spanish missions had introduced Mediterranean crops and methods of irrigation, as well as potatoes and tomatoes from Mexico and Peru.

The first large-scale attempts at irrigation agriculture in the desert by Anglo-Saxons were carried out by Mormons, who settled in the Great Salt Lake area of Utah. Religious motivation facilitated the establishment of a communal irrigation system. The Mormons introduced the crops of the northern temperate regions of Europe, with which they were familiar; these included wheat, barley, root-crops, and forage and pasture crops, in addition to the typical European deciduous fruit-trees.

At the beginning of the present century the so-called 'Reclamation Era' was initiated by President Theodore Roosevelt, which aimed at integrating flood-control, power-production, and water-supply, for irrigation and other purposes. All the major streams of the west were harnessed, and it became possible to irrigate the enormous areas of hitherto unproductive land of the western desert. The Grand Coulee Dam in Washington, the Boulder Dam on the lower Colorado River, and the Shasta and Friant Dams of California, are some of the major projects involved.

In the warm-winter desert areas, such as the Imperial-loachella Valley of California, with their long growing-season and practically frostless winters, a vast array of crops can be produced: subtropical crops such as cotton and fruits such as dates and citrus, out-of-season vegetable crops as well as the typical temperate-climate crops such as wheat, barley, flax, etc.

In the cold-winter deserts, such as occur in Nevada, the choice of crops is much more limited, and the major emphasis is on the production of forage and pasture crops, which make possible a type of transhumance livestock industry. Cattle and sheep graze in summer at the higher altitudes, and on the irrigated pastures in autumn and early spring. In winter the animals are stabled and subsist on reserves of alfalfa hay. Other important crops of this area are sugar-beets and celery, which are grown in summer.

Because the best lands for irrigation are usually in the lower-lying areas, where the soils are heavy and have little natural drainage, problems soon arise as a result of the raising of water-tables and the accumulation of salts. Many of the irrigated areas are abandoned after a few years of intensive farming. Elsewhere, master drainage-systems have had to be laid down, so that much of the irrigated land of the western desert area is now adequately drained. At present, further development of irrigation agriculture in the region is dependent on the exploitation of unconventional sources of water-supply and on the large-scale diversion of water from the humid regions. Underground basins and streams are already being exploited to the maximum (Logan, 1961).

SUMMARY

This review of irrigation agriculture in a few regions of the Old and New Worlds indicates the difficulties and dangers with which civilizations based on irrigation are faced. Control of the moisture régime of the soil may appear at first view to make farming practically independent of the vagaries of climate, but actually it introduces a host of problems—social, political, and technical—

with which, in the long run, primitive civilizations were not able to cope. An adequate solution to these problems is essential if the permanency of irrigated agriculture is to be assured; these problems will be discussed in detail in other parts of this work.

References

Addison, H. (1961). *Land, Water and Food. A topical commentary on the past, present and future of irrigation, land reclamation and the food supplies they yield.* Chapman & Hall, London, xii + 284 pp., illustr.

Albright, W. F. (1956). *The Archaeology of Palestine.* Pelican, London, 271 pp., illustr.

Armillas, P. (1956). Land reclamation and soil conservation in Indian America. pp. 245–9 in *The Future of Arid Lands* (Ed. G. F. White) Amer. Assoc. Adv. Sci., Washington D.C., ix + 453 pp., illustr.

—— (1961). Land use in pre-Columbian America. *Arid Zone Res.*, **17**, 225–76.

Barth, F. (1962). Nomadism in the mountain and plateau areas of South West Asia. *Arid Zone Res.*, **18**, 341–56.

Braidwood, R. J. (1952). From cave to village: birth of agriculture and animal husbandry. *Am. Sci.*, **187**(4), 62–66.

—— & Reed, C. A. (1957). The achievement and early consequences of food-production: a consideration of the archaeological and natural-historical evidence. *Cold Spring Harbor Symposia on Quantitative Biology*, **12**, 19–31.

Brichambaut, G., Perrin de & Wallen, C. C. (1963). A study of agroclimatology in semi-arid and arid zones of the Near East. *Tech. Bull. World Meteorol. Organ.*, **56.** 64 pp.

Butzer, K. W. (1958). Der Umweltfaktor in der Grossen Arabischen Expansion. *Saeculum*, **8**, 359–71.

Christian, C.S. (1964). Methods of land use in Australia's arid and semi-arid areas. *Arid Zone Res.*, **26**, 117–22.

Cochemé, J. & Franquin, P. (1967). *A study of the Agroclimatology of the Semi-arid Area South of the Sahara in West Africa.* Tech. Rep. F.A.O., Rome, vii + 325 pp., illustr. (mimeographed).

Cole, J. & Mathews, O. R. (1940). Relation of the depth to which the soil is wet at seeding time to the yield of spring wheat on the Great Plains. *Circ. U.S. Dep. Agric.*, **563**, 20 pp., illustr.

de Brichambaut, G. Perrin. *See* Brichambaut, G. Perrin de

Despois, J. (1961). Development of land use in northern Africa (with reference to Spain) *Arid Zone Res.*, **17**, 219–37.

Donald, C. M. & Williams, C. H. (1954). Fertility and productivity of a podzolic soil as influenced by Subterranean Clover (*Trifolium subterraneum* L.) and superphosphate. *Aust. J. Agric. Res.*, **5**, 644–87.

Droughin, G. (1962). The possibliity of using minor water resources in Algeria. *Arid Zone Res.*, **18**, 371–80.

Drower, M. S. (1955). Water-supply, irrigation and agriculture. pp. 520–57 in *A History of Technology* (Ed. C. Singer *et al.*) Clarendon Press, Oxford, vol. I, 827 pp., illustr.

Evenari, M. & Koller, D. (1956). Ancient masters of the desert. *Am. Sci.*, **194,** 39–45.

GULHATI, N. D. & SMITH, W. C. (1967). Irrigated agriculture, an historical review. pp. 3–11 in *Irrigation of Agricultural Lands* (Ed. R. M. Hagan, R. H. Haise & T. W. Edminster) Agronomy Series; Amer. Soc. Agron., Madison, Wisconsin, II, xxxi + 1180 pp., illustr.

HAMDAN, G. (1961). Evaluation of irrigation agriculture in Egypt. *Arid Zone Res.*, **17**, 119–42.

HELBAEK, H. (1959). Domestication of food plants in the Old World. *Science*, **130**, 365.

HURST, H. E. (1952). *The Nile: A General Account of the River and the Utilization of its Waters.* Constable, London, 326 pp.

JACOBSEN, T. & ADAMS, R. M. (1958). Salt and silt in ancient Mesopotamian agriculture. *Science*, **128**, 1251–8.

KING, L. J. (1966). *Weeds of the World: Biology and Control.* Plant Science Monographs (Ed. N. Polunin). Leonard Hill, London, & Interscience, New York, xxxii + 526 pp., illustr.

KOVDA, V. A. (1961). Land use development in the arid regions of the Russian plain, the Caucasus and Central Asia. *Arid Zone Res.* **17**, 175–218.

LOGAN, R. F. (1961). Post-Columbian developments in the arid regions of the United States of America. *Arid Zone Res.*, **17**, 277–97.

LOMASSON, T. (1947). *The Influence of Rainfall on the Prosperity of Eastern Montana, 1878–1946.* Mimeo Rep. 7, Region I, U.S. Forest Serv.

MCDONALD, J. E. (1959). Climatology of arid lands. pp. 3–13 in *Arid Lands Colloquia*, 1958–59. University of Arizona, Tucson, 92 pp.

MATHEWS, O. R. & COLE, J. S. (1938). Special dry-farming problems. pp. 679–92 in *Soils and Men.* Yb. Agric., U.S. Dep. Agric., xiii + 1232 pp., illustr.

MONOD, T. & TOUPET, C. (1961). Land use in the Sahara-Sahel region, *Arid Zone Res.*, **17**, 239–54.

PERRIN DE BRICHAMBAUT, G. *See* BRICHAMBAUT, G. PERRIN DE

PETERSON, R. F. (1965). *Wheat, Botany, Cultivation, and Utilization.* World Crops Books (Ed. N. Polunin) Leonard Hill, London, & Interscience, New York, xxiv + 422 pp., illustr.

REED, C. (1959). Animal domestication in the prehistoric Near East. *Science*, **130**, 1629–39.

REIFENBERG, A. (1953). The struggle between the desert and the sown. pp. 378–89 in *Desert Res. Proc.* Research Council of Israel, Jerusalem. 641 pp., illustr.

SHAPERA, I. (1936). The native inhabitants, pp. 20–50 in *Cambridge History of the British Empire.* Cambridge University Press, Cambridge, vol. VIII, xxviii + 1088 pp., illustr.

STEPHENS, C. G. & DONALD, C. M. (1958). Australian soils and their responses to fertilisers. *Adv. Agron.*, **10**, 167–256.

STEWARD, J. H. (1956). Cultural evolution. *Am. Sci.*, **44**(5), 69–83.

TALBOT, W. J. (1961). Land utilisation in the arid regions of southern Africa. *Arid Zone Res.*, **17**, 299–330.

WADHAM, S. (1961). The problem of arid Australia. *Arid Zone Res.*, **17**, 339–62.

WALLEN, C. C. (1966). Arid zone meteorology. pp. 31–50. in *Arid Lands: A Geographical Appraisal* (Ed. E. S. Halls) UNESCO, Paris, xviii + 461 pp., illustr.

WATSON, E. R. (1963). The influence of subterranean clover pastures on soil fertility. I. Short-term effects. *Aust. J. Agric. Res.*, **14**, 796–807.

WHYTE, R. O. (1961). Evolution of land use in south western Asia. *Arid Zone Res.*, **17**, 57–115.

WILLEY, G. R. (1960). New World prehistory. *Science*, **131**, 73–86.

WILLIAMS, C. H. & DONALD, C. M. (1957). Changes in organic matter and pH in a podzolic soil as influenced by subterranean clover and superphosphate. *Aust. J. Agric. Res.*, **8**, 179–89.

WOODBURY, R. B. (1963). Indian adaptation to arid environments. pp. 55–85 in *Aridity and Man* (Ed. C. Hodge) Am. Assoc. Adv. Sci., Washington D.C., xviii + 584 pp., illustr.

ZOHARY, M. (1954). Notes on ancient agriculture in the central Negev. *Israel Explor. J.*, **4**, 17–25.

VI

WATER RESOURCES AND THEIR DEVELOPMENT

World Water Needs on the Increase

The estimated increase in world population, currently at an average rate of about 2 per cent per annum, signifies that the world water-demand will double henceforth about every 35 years (U.N., 1964), at least for the time being. The increase in water requirements must be far more considerable in the dry regions than in the humid regions, for two main reasons: (1) that water is the limiting resource in the dry regions and therefore determines the extent to which other resources can be developed, and (2) that an increase in water requirements is concomitant with a rise in the standard of living. It is in the drier regions of the world that the standard of living tends to be the lowest, and therefore the rise should be the steepest.

Unfortunately, water resources are usually most abundant in regions in which they are least required, and scarce in those regions in which they form the key to agricultural development and, indirectly, to economic and industrial growth. Further, water-supply is generally most abundant during the seasons of lowest requirement, and relatively scarce during seasons of peak requirements.

Population pressures and development needs have generally created a situation whereby the development of a new water-resource is frequently possible only at the expense of previously available water-supplies. Upstream removal of water reduces the amounts available downstream, and may even reduce the quality of the remaining water to unsafe levels; the development of new wells affects the output of older wells that depend on the same underground reservoir; conveying water from one region to another may limit future development of the donor region.

The existing shortage of water-supply in the dry regions does not necessarily imply lack of water resources. In many of the dry regions there are potential water-resources which, with the necessary know-how and funds, could be developed. Much additional water can also be made available by improving water conservation, by increasing the efficiency of use of existing water-resources, and by a more rational allocation of water among competing demands.

Water Conservation Practices

INCREASING WATER STORAGE IN THE ROOT-ZONE

The Solum as a Moisture Reservoir

The soil, to the bottom of the root-zone, constitutes a moisture reservoir of vital importance to agriculture. Water that infiltrates into this reservoir can be stored with relatively little loss for fairly long periods—longer than is generally realized.

In humid climates, the amount of water that is stored in the soil depends on the water-holding capacity of the soil, and therefore heavy soils are more effective in this respect than sandy soils. In arid climates, it is the amount of precipitation that is the limiting factor; therefore, sandy soils are better water-reservoirs for rainfall than are heavy soils: percolation in the latter is slower, the depth wetted is less, and thus the proportion of precipitation lost by run-off and evaporation is greater.

In the Mediterranean region, crops such as sorghum, melons and sesame, can be grown during the hot summer months, without receiving a drop of rainfall from sowing to harvest; they can produce high yields on the moisture stored in the soil during the winter months. In the Great Plains of North America, numerous investigations have proven that the amount of moisture stored in the soil, prior to the sowing of winter wheat, largely determines the yield level that can be expected (Pengra, 1952). In irrigation farming, it is good practice to wet the soil to the full depth of the root system before sowing the crop, thereby forming a reserve of water on which the plant can draw throughout its growing period.

The amount of precipitation taken in by the soil depends on run-off and infiltration. By reducing run-off and increasing infiltration, the amount of water stored in the soil can be increased—with beneficial effects on crop production, provided the total amounts of rainfall are sufficient.

Run-off Control

It is somewhat paradoxical that the drier the climate is, the greater will be the havoc caused by rainfall. Precipitation in arid regions generally occurs as torrential rains, the sparse natural vegetation affords very little protection to the soil, and infiltration rates are rapidly reduced while run-off is increased. Most soil-conservation methods, such as strip cropping, contour ploughing, terracing, etc., aim at reducing run-off, and are therefore also effective in increasing the amount of water that is stored in the soil.

The protection provided by vegetation is usually a major factor in run-off control. Plants intercept part of the rainfall and reduce the velocity of raindrops; they also obstruct or slow down the movement of water on the soil surface. The indiscriminate destruction of the plant cover in watersheds increases soil erosion and causes excessively rapid run-off that cannot be controlled.

Increasing Infiltration

The maintenance of a high infiltration rate is an important objective of soil management; it reduces run-off and thereby increases the efficiency of rainfall and irrigation, at the same time minimizing soil losses by erosion.

The rate of infiltration of water into the soil depends on soil structure, soil cover, the degree of dryness of the soil, and the intensity and duration of the rainfall or irrigation. As rain falls on a bare soil, the surface aggregates are destroyed by the impact of the drops of water and the pores of the soil surface become clogged with fine particles; this rapidly reduces the rate of infiltration of the water.

The surface of many clay soils can be easily sealed, and infiltration capacity reduced drastically, by a few minutes of heavy rainfall (Jacks *et al.*, 1955). Mulches of straw or crop residues, by breaking the impact of the raindrops, markedly improve infiltration (Figure 21). Mulching was also found to reduce run-off by about half the amount that occurred without mulching (Zingg & Whitfield, 1957).

Cultivated row-crops, such as cotton, maize, and peanuts (groundnuts), usually afford less protection against the effects of rainfall on the soil than do densely-growing crops such as herbage and small grains, and therefore their soils tend to have a lower infiltration rate. The frequent cultivations which they require also reduce soil aggregation. The same is true for soils with a high stability of soil structure, such as results from a high organic-matter content. Hence, all factors that improve and maintain soil structure, also improve infiltration, and *vice versa*. However, if the subsoil is impervious, run-off will eventually occur even with a very permeable top-soil.

Whenever the rate of water supply is in excess of the rate of infiltration, ponding or run-off will occur.

The Water-table as a Source of Moisture Supply

The presence of a high water-table is of importance in determining the amount of water available to the crop. If the water-table is near to the soil surface, the root-zone of the crop will be confined to a shallow soil layer, with adverse effects on crop production, as ground-water rises to the soil surface by

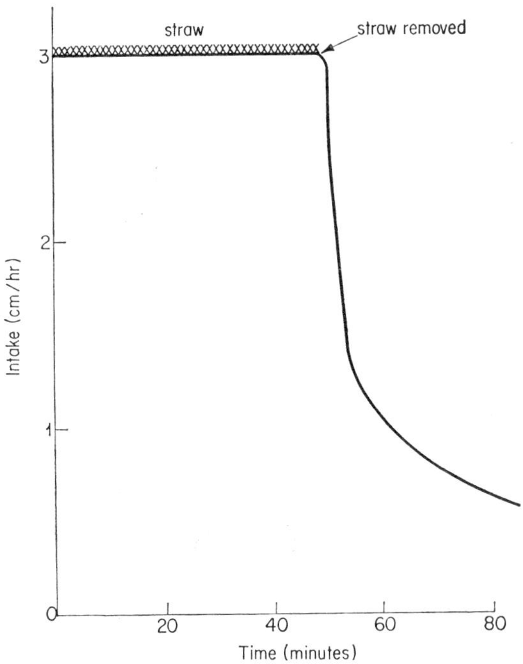

Figure 21. Effect of straw mulch on intake of water by soil. (From Duley (1939): by courtesy of the Soil Science Society of America.)

capillary flow, and may cause a dangerous accumulation of salts in the rootzone. If the water-table is sufficiently low not to interfere with normal development of the crop, a deep-rooted crop may obtain part of its water requirements from the capillary fringe of the water-table.

A fluctuating water-table is most dangerous, making it impossible for plants to adjust their rooting systems in the manner which they are able to do with a constantly shallow or a relatively deep water-table. When the water-table rises, roots are killed; when it recedes, the shallow soil-layer rapidly dries out and there are no roots to supply moisture from the lower depths. Lucerne, for example, which is normally deep-rooted, will make good growth on soils with a relatively high water-table, provided it is stable; if it fluctuates, the lucerne will be rapidly killed.

Appreciable quantities of water can move upwards, sufficiently to be used by plants, from depths of as much as 180 cm below the root-zone (Gardner, 1964). Upward movement from a not-too-shallow water-table can, therefore, under certain circumstances, make an important contribution to the water economy of plants.

Water Harvesting

Run-off Conditions in Arid Zones. In humid climates, run-off usually occurs when rain falls on a saturated soil. The more arid the region, the rarer are situations in which the soil is saturated by rainfall, and the more frequent is the occurrence of run-off from soil that is not saturated, following rains that exceed a certain intensity. It has been estimated that, in the central Sahara, a flood will occur when rainfall exceeds 5 mm with an intensity greater than 0·5 mm per minute (Dubief, 1953). Many soils of arid regions have a very low absorption capacity for rainfall. The sparse vegetation, the large proportion of bare rock, the low organic-matter content, and the hard impermeable crusts that form after the first minutes of rainfall, are all contributory factors.

Under very arid conditions, reducing run-off and increasing infiltration rates are ineffective, as the amounts of moisture that can be stored in the soil do not, in any case, suffice for crop production. Rather should efforts be made in the opposite direction: to increase run-off and reduce infiltration in certain areas which then serve as a source of water-supply for other areas.

Water Harvesting Methods. In principle, water harvesting consists in using water derived from an area that has been treated to increase run-off of precipitation in order to supplement soil moisture in an adjacent area, situated at a lower elevation (Plate 13). In many dry regions, in which contemporary precipitation alone is not sufficient to ensure a crop, harvested water plus that accumulated in the soil will suffice, if a proper ratio between donor area and recipient area is established.

Water harvesting was developed to a fine art about 2000 years ago by the Nabateans (*see* pp. 116–17); modern technology is attempting to develop more effective methods, aimed at reducing the ratio between donor and recipient areas. There are two main methods of water harvesting: conveying the water from barren hillsides to the adjacent relatively level land, and creating a micro-relief within a more or less level field.

Microwatersheds

Cropping systems have been developed in which run-off water from parts of a field are concentrated in strips in which crops are planted. The crop is sown in narrow strips between wide intervals that are ridged as artificial miniature watersheds. These latter are compacted to increase run-off of water to the crop rows. The relative widths of the water-shedding strips and of the crop-producing strips depend on the amount of annual precipitation that can be expected. The usual ratios are from 2:1 up to 4:1 (Figure 22). This system is considered to be more efficient than fallowing, in which water is conserved from one season to the other and the entire area is cropped in one year out of two (*see* Chapter XV) (Kemper, 1964).

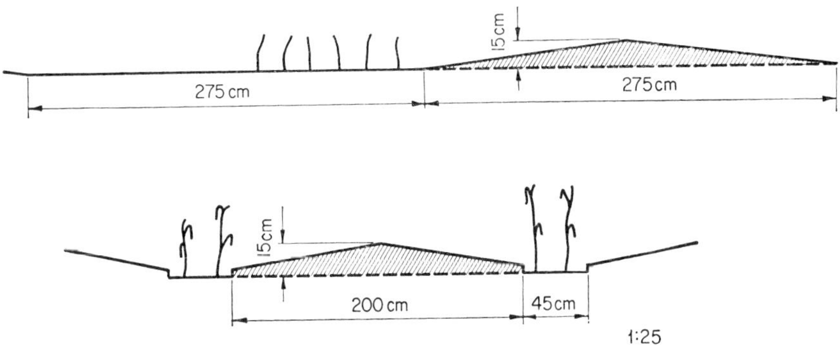

Figure 22. Microwatersheds. Above: microwatersheds for small grains. Below: microwatersheds for row crops. (From Kemper (1964): by courtesy of the U.S. Department of Agriculture.)

Various methods for increasing the run-off from the water-supplying strips are being investigated in the arid west of the United States, in Israel, and elsewhere. The two main approaches used are (1) ground covers of plastic films, rubber, or metal sheeting materials, and (2) waterproofing and stabilizing soil surfaces by spraying with low-cost materials (Myers, 1967).

Thin plastic, rubber, and metal, films have been used for building precipitation catchments. In one case, minor surface-ridges were formed which were almost completely covered (as to 90 per cent) with a plastic film, in order to concentrate rainfall penetration into the uncovered part. Results showed that the covering was effective in increasing the soil moisture-content, as well as the yields and water-use efficiency of maize, which served as an indicator plant (Willis *et al.*, 1963).

The main disadvantages of the various films are the high cost and susceptibility to damage by winds. Research is continuing on developing new, less expensive and more durable materials, as well as methods of reducing damage —such as bonding the films to special asphalt pavements (Myers, 1967). The most immediately promising approaches, however, appear to be the waterproofing and stabilizing of the soil surfaces.

The simplest of these methods consists of creating a cleared smooth surface. On a well-structured clay-loam soil, with a 4 per cent slope, this method produced 21 per cent run-off of a 195 mm rainfall (Hillel *et al.*, 1967). However, this method is applicable only when erosion hazards are minimal. Sodium salt can be used to reduce infiltration, by dispersing the soil colloids and sealing soil pores. Using sodium carbonate at the rate of 45 kg/ha, 70 per cent run-off was obtained from a cleared and smooth clay-loam soil (ibid.). This method, too, is dependent on effective erosion control, in the absence of which the treated soil may be removed within one season.

A number of water-repellent treatments with materials such as silicone, to make the soil hydrophobic, have given encouraging results. Finally, sprayed asphalt appears to offer promise of providing a relatively low-cost and enduring sealing and binding coating (Myers, 1967).

Conservation Bench Terraces

A new terrace system for semi-arid regions has been designed by Zingg and Hauser (1959) that not only reduces water-erosion, but also spreads run-off water over a levelled bench, where it is stored in the soil for use by the crop (*see* Figure 23). Experience in the field with these terraces shows that good

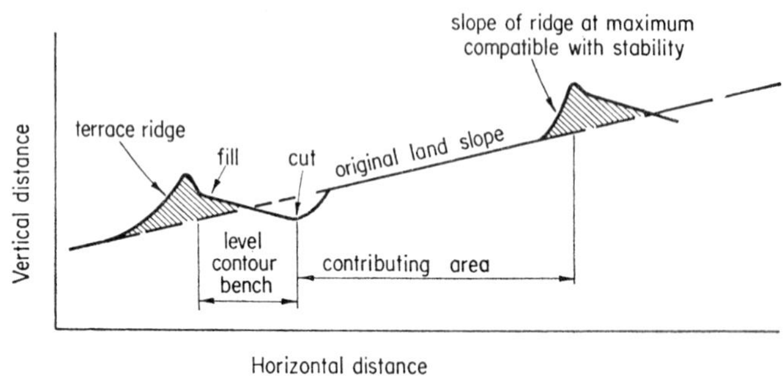

Figure 23. Cross-section of the slope-control practice of conservation benching. (From Zingg & Hauser (1959): by courtesy of the American Society of Agronomy.)

results can be achieved with this method, provided that (1) precision levelling is carried out, (2) there are adequate terrace outlets to handle excess water, and (3) fertilizers are applied so that the stored water is used efficiently. In a region with an average rainfall of 470 mm (mainly summer rainfall in the south-western Great Plains), it is recommended to use one-third of the field area for establishing the terraces, and the remaining two-thirds for watershed areas (Hauser & Cox, 1962).

REDUCING EVAPORATION LOSSES FROM THE SOIL

Soil Mulches

At the beginning of the century, King (1914) carried out his 'classical' experiment from which he claimed that a surface mulch of dry soil 5–8 cm deep, by obstructing the rise of water to the surface through capillary action, effectively reduced loss of water as compared with a soil having an undisturbed surface. Rarely has a mistaken conclusion drawn from an otherwise impeccable experiment given rise to such a wasteful practice as the frequent cultivations for moisture conservation adopted subsequently for many years in most dry regions.

In King's experiment, glass cylinders that were open at both ends, and each containing a column of soil, were placed with one end resting in a shallow water-container. A comparison was made between cylinders with an undisturbed soil surface and those with a dry surface mulch resulting from frequent stirring of the top-soil.

It was found after 100 days, that the surface mulch had reduced evaporation losses by 50 per cent. The only flaw in the conclusions and subsequent application of these results was that capillary rise of water to the soil surface is effective only when the water-table is sufficiently near to the surface; under these conditions, increased loss of water from the soil is probably more beneficial than detrimental!

Following on King's experiment and conclusions, a soil mulch for moisture conservation became standard procedure in dry regions, and dicta such as: 'two cultivations can replace one irrigation' became very popular. The too-frequent cultivations impaired soil structure, however, and increased soil losses by erosion. These effects were most pronounced on the row-crops, such as cotton, maize, and soybeans, which, for this reason, gained the reputation of being soil-destroying crops.

It has since been shown that after a rainfall or an irrigation, the top 7–10 cm of the soil dry out rapidly; in any case it would be impossible to prevent this loss by cultivation, because the drying out had already taken place by the time the soil surface was sufficiently dry for cultivation to be possible. Below the dry layer, moisture loss by evaporation from the soil becomes minimal, and further water losses are mainly due to plant transpiration—hence the favourable effect of weed control on water losses which are thereby reduced.

Straw Mulches

As opposed to the effectiveness of a 'dust mulch' to reduce evaporation, straw and the other crop residues left on the soil surface have proved to be very effective. It has been estimated that a straw mulch saved the equivalent of 50 to 75 mm of rainfall in dry weather (Stephenson & Schuster, 1946). The use and importance of stubble mulches will be discussed in detail in Chapter XIII.

Chemical Treatments

Experiments are being carried out with the object of reducing evaporation from the soil by using suitable chemicals. It has been found that Hexadonal mixed with the surface $\frac{1}{4}$ in. of the soil, reduced evaporation by 43 per cent (Olson *et al.*, 1964). This material, which is resistant to microbial activity, remained effective for more than a year. The surface layer of treated soil dried out more rapidly than that of untreated soil, creating a diffusional barrier to evaporation. Fewer cracks developed in the treated than in untreated soil, and the Hexadecanol also increased aggregate stability.

It has been shown that chemical weed control resulted in moisture storage that was equal to that achieved under stubble mulch and subsequent tillage, when the initial moisture conditions were equal (Wiese & Army, 1960). The loss of moisture from the soil, due to weeds, will be discussed in detail in Chapter XVI.

Weeds are one of the major maintenance problems on canal banks and irrigation channels. They reduce the carrying capacity and decrease the velocity of flow, thereby increasing sedimentation. By raising the water-level in the canals, they increase seepage losses—with a resulting rise in the water-table of the adjoining land, and increased salinity problems.

PREVENTING WATER POLLUTION

In humid regions, water pollution results mainly from the addition of organic wastes, which can be removed by biological sewage-treatment. In arid regions, the main and additional form of water pollution is salinity.

In nearly every stream that is used for irrigation, the concentration of salts increases downstream, as a result of pollution by the return flow of drainage waters from irrigated lands. For example, the dissolved solids content of the Rio Grande was found to increase progressively in a downstream direction, being nearly 16 times as great at Fort Quitman in Texas, as at an upstream point on the Colorado–New Mexico state line (Wilcox & Durum, 1967).

The removal of excess salts is possible only by expensive processes, which are still in the stage of development—hence the importance of measures aimed at preventing the build-up, beyond a tolerable level, of salts in the water supplies.

SURFACE WATERS

STREAMS AND RIVERS

The streams and rivers in arid regions are of three types:

(a) ephemeral streams that originate within the region;
(b) streams that originate in a humid watershed, in close proximity to the arid lands, and usually from higher altitudes within the region; and
(c) perennial rivers that originate in distant humid climates and flow through the arid region.

Ephemeral Streams

In arid areas approximately 5 per cent of the storms cause over 50 per cent of the stream-flow, and approximately 15 per cent of the floods produce nearly 90 per cent of the stream-flow (Burgy *et al.*, 1967). Small rains, and even large rains of low intensity, do not generally yield any water for stream-flow. Floods in arid regions, when they do occur, are usually destructive. The barren, and frequently steep, land surfaces cause rapid run-off, and the water soon flows with torrential speed along the previously dry beds of the numerous channels. Even a rainfall of 10 mm on a watershed of 50 km^2 may cause a flash-flood.

In arid regions, springs with a copious and continuous flow of water are relatively rare. Stream-flows that originate within the region are usually intermittent in nature: torrential flows may occur for a few days or even hours at a time, to be followed by long periods during which the stream-bed is completely dry. Certain streams flow fairly continuously for a few months during the rainy season, and then dry up progressively.

Flood-water Spreading

In most desert and semi-desert regions, flash-floods occur during which enormous amounts of water sometimes stream through the previously dry channels for short periods of time, varying from a few hours to a few days. These waters can be diverted, partially or entirely—depending on their volume and velocity—to inundate fields especially prepared for this purpose.

Essentially, techniques for such flood-water spreading have not changed markedly in comparison with those used by primitive people, as described in Chapter IX.

A diversion dam is erected at a suitable location in the stream channel, to direct the water into a ditch—the dimensions of which are calculated in accordance with the velocity and size of the stream to be diverted. The slope of the ditch has to be so adjusted as to ensure that the velocity of the stream is reduced to manageable proportions, or else the ditch will be rapidly destroyed. The velocity must be sufficient to reduce sedimentation to a minimum, as otherwise it would soon put the ditch out of service.

The dams themselves can be relatively simple earthen structures on small waterways; on larger stream-channels, however, more elaborate structures are required which may make the whole method uneconomical.

Sedimentation may, in the course of time, raise the soil level of the watered fields to such a degree that water spreading is no longer possible, unless the dam is relocated. In order to avoid this contingency, the water may first be released into sedimentation basins. These have the additional function of enabling some degree of regulation of the flow of water to the fields to be irrigated, as they can accumulate a larger quantity of water in a short time than could be effectively spread in the field. The field itself should be prepared as for ordinary irrigation. Basin irrigation is the most suitable method for water spreading. The ditches will usually be much larger than conventional irrigation ditches, in order to handle relatively large streams of water. The main difference of water spreading from conventional irrigation is that, in the former, water is not applied in accordance with crop requirements but whenever it suddenly becomes available.

The water can be applied before sowing in order to wet the root-zone to an appropriate depth. It can also be applied to a growing crop. The most suitable crop for this purpose is a perennial grass or a legume such as lucerne, that is capable of withstanding long periods of drought, but reacts quickly to favourable soil moisture conditions.

In view of the unreliable supply of water it is doubtful whether stable crop production, providing an acceptable standard of living to modern farmers, can be based exclusively on water harvesting or water spreading methods. These may, however, play a useful role in producing forage reserves for supplementing range grazing, as an adjunct to irrigation farming, or as a supplemental moisture supply in semi-arid regions where rain-fed crops are a possibility.

Lasting Rivers and Streams

Many rivers originate in relatively close proximity to arid land—usually at higher altitudes, where they are fed by melting snows and the water serves to irrigate the adjacent arid plains. Examples are the Syr Daria and Amu Daria of Soviet Asia, the Tigris and Euphrates of Iraq, the Sacramento and San Joaquin of California, and the Murray of Australia.

Others rivers and streams originate in humid climates that are very distant from arid regions. If, like the Congo River, they flow at constant tropical latitudes, they remain within the humid region, and the enormous amounts of water which they collect are discharged into the ocean. If, however, they flow meridionally like the Nile, their waters eventually reach an arid region and can serve for irrigation.

Interregional Transfer of Waters

The development of many arid and semi-arid regions is apt to be dependent on the transfer, by canals or other means, of waters from streams flowing in more or less distant humid regions. A few examples are: the transfer of the waters of the Indus and its tributaries by canal to the arid Sind and the semi-arid Punjab; the waters of the Colorado River also transferred by canal, to the Imperial Valley in California; and transporting the water from the sources of the Jordan in Upper Galilee in a system of open channels and closed large-diameter pipes to the Negev Desert in southern Israel.

In many cases, waters of one river system are transferred through canals and tunnels to a different river basin, to supplement the water-supply of the latter.

WATERSHEDS

'A watershed is an area of land and water bounded by a drainage divide within which the surface run-off collects and flows out of the watershed through a single outlet into a larger river, lake, or other body of water' (Brown, 1960).

Small watersheds are drainage areas of a few thousands to a few tens of thousands of hectares. Larger drainage areas are usually called river basins. In general the tendency in dry regions is for closed drainage basins to occur, in which sediments accumulate. Saline lakes are also of frequent occurrence (Dixey, 1966).

An ideal watershed management project would integrate plans for soil conservation, soil improvement, irrigation, drainage, flood prevention, and development of water-supplies for all purposes. It requires community cooperation, organized administration, taxation, and political action (Brown, 1960).

The relative proportion of local precipitation that remains as surface run-off or that sinks into the ground to join the underground aquifers, depends largely on the types of rock and overlying soil. Fine-textured clays and shales are relatively impermeable and, especially when torrential rains occur, a large proportion of the precipitation contributes to run-off. By contrast, limestone rock and its derived overlying soil readily absorb rainfall; most of the precipitation falling on it that is not used for evapotranspiration, eventually joins the underground water reservoir (Dixey, 1966).

The kind of vegetation that is established on the watersheds will determine to a large extent the amount of water that percolates deeply and replenishes the water-table and consequently the water yield from springs and streams. In California, the control of large areas of brush and its replacement by a cover of grasses increased the amounts of water available to the underground water-reserves (Biswell & Schultz, 1957). Water yields from forest land can also be increased by altering the type of vegetation. In North Carolina, cutting the woody vegetation on an entire drainage area increased the water yield by 65 per cent during the first year after cutting (Storey, 1960).

Strip-cutting of pine increased water yields by about 30 per cent (Love, 1960). Trees and brush species usually have a dense canopy, causing a high ratio of interception to loss; they grow throughout the dry season and, having a deep root-system, use all the water available to a considerable depth.

In the drier regions, grasses usually become dormant during the dry season. They intercept less of the precipitation, and use less of the available water in the soil, than do most shrubs, so that with them more water becomes available for the recharge of aquifers.

In general, changing or removing the vegetation in order to increase deep percolation is effective only in regions with sufficient rainfall. Where precipitation is low, virtually all the moisture will either be lost from the soil by evaporation in the absence of a plant cover, or, if much vegetation is present, will be used for evapotranspiration. This is regardless of the *nature* of the plant cover: changes in vegetation are effective in increasing water-supplies only where precipitation occurs as snow or is sufficient to saturate the soil frequently (Price, 1956).

CONTROL OF SURFACE WATER

Disastrous floods are of regular occurrence in many arid regions, affecting small to large areas. Their occurrence and size depend on the rate, duration, and extent, of rainfall, and on the condition of the land on which it falls (Ford *et al.*, 1955). When the resulting run-off exceeds the capacity of the stream channels, floods result. Contributing factors include permanent or accidental barriers in the stream, siltation, and the breakdown of dams or dykes. Crops and pasture are damaged, floodplains are eroded and gullied or covered with infertile sediments, water control systems are destroyed, etc.

Flood control includes the following measures (Ford *et al.*, 1955):

(a) *Land treatment:* reducing run-off and increasing infiltration on the lands of the watershed, by mechanical and vegetative methods, will lessen the amounts of water and sediment that reach the farm drainage-ways and the small streams.
(b) *Upstream water-flow retardation:* is based on appropriate measures in the small streams and tributaries of the watersheds, such as stream channel improvements to increase water-carrying capacity, gully stabilization works, flood-water retarding structures, sediment detention basins, etc.
(c) *Downstream flood-control measures:* consist of large reservoirs and levees constructed on the major waterways to control the flow of floods in the major river valleys.

Reservoirs

The more arid the region, the greater the seasonal fluctuations in the flow of surface water. Reservoirs are meant to achieve a more regular water-supply by correcting the great variation in stream-flow; but the larger the reservoir, the greater the evaporation losses, so that complete stabilization of surface flow is not possible in an arid climate (Langbein, 1962). However, in the absence of a reservoir, it is the periods of minimum flow which usually coincide with periods of peak requirements, that determine the quantities of water which can be relied on for agriculture.

Reservoirs are basically of two kinds (Riesbol *et al.*, 1967):

Onstream: the reservoir is fed directly by the stream and therefore receives the entire sediment load. It must store the maximum floods, or pass off the excess water. The Aswan Dam is a typical example.

Offstream: the reservoir is fed by diversions from the stream; it receives only part of the sediment load and must store only part of the floods.

In many arid regions of the world, spectacular works are undertaken to store river water on a large scale. Enormous masonry dams are built, behind which artificial lakes are formed. Evidence is increasing that in many cases these projects are ill-conceived; they are also always expensive and moreover extremely wasteful of water. Evaporation is very high in the arid regions which they are to serve and in which they are located, amounting to as much as 125 cm annually (Stamp, 1961).

As much as three-quarters of the water stored may be lost by evaporation (Dixey, 1966), and the remaining water becomes increasingly saline. As a result of the extremely high rate of evaporation, the total yield of a watershed in an arid region is actually decreased when large reservoirs are built. There is even a point at which the building of additional reservoirs causes a net deficit in usable water—when aggregate evaporation losses exceed the amounts of usable water gained (Burgy *et al.*, 1967).

'Reservoirs are also mortal', according to Addison (1961). Experience in South Africa has shown that reservoirs have filled up with sediment within a dozen years. In the Levant and in Sinai, relatively small reservoirs have filled up in a still shorter period. It is true that certain reservoirs have a longer life-expectancy: in the U.S.A., the great reservoirs are expected to remain effective for one or two centuries (ibid.). However, even in the U.S.A. there are examples of reservoirs that have silted up completely within a few years (Stamp, 1961).

Reduction of Evaporation Losses. Evaporation is usually the principal cause of water-loss from reservoirs and may assume considerable proportions. Many investigators have studied the possibility of reducing evaporation from a free water-surface by the use of surface films that are multi- or mono-molecular in thickness. The multi-molecular films, comprising mixtures of oils and spreaders, were found to be more effective in reducing evaporation than the mono-molecular films; but they were thermodynamically unstable, and lose very rapidly their ability to reduce evaporation (Mansfield, 1967).

Liquid mono-molecular layers, though somewhat less effective in reducing evaporation, were found to be more stable than the multi-molecular layers. To be effective, chemicals that can serve for forming mono-molecular layers should be inexpensive, non-poisonous, and stable.

A number of compounds are known that form such a 'monolayer' film on the surface of water in reservoirs, and thereby impede the evaporation of stored water. The molecules of the compounds that form monolayer films have a common characteristic structure: part of the molecule is hydrophilic and part is hydrophobic, repulsing water. The most promising of these substances in field trials up to the present have been fatty alcohols such as hexadecanol and octadecanol, or mixtures of the two. The reduction in evaporation is accomplished without markedly reducing the transfer of oxygen from air to water, or impeding the escape of carbon dioxide from the water. No toxic action was found on micro-organisms, aquatic plants, fish, or other animals (Quintela, 1967). An automatic system for applying the hexadecanol consists of a number of containers spaced around the perimeter of the reservoir, releasing the material through valves. Only those placed upwind operate, and the rate of application is regulated according to wind-speed (Langbein, 1962). Although practically insoluble in water because of the extremely small mass involved, a breakdown in the film usually occurs as a result of wave action, dust, and wind.

The best results with monolayers have been obtained on average-sized reservoirs. Experiments in Australia over several years on reservoirs with surfaces of 160 hectares or more, showed overall reductions in evaporation of 15 per cent at a cost of about 3 cents per thousand gallons of water saved (Mansfield, 1967). With appropriate techniques, effective results were obtained on storage areas down to approximately one hectare. For still smaller surfaces, costs were prohibitive.

At the other extreme, no techniques have yet been developed for reducing evaporation losses from very large reservoirs. For the enlarged Aswan reservoir, the loss by evaporation has been estimated as equivalent to 10 per cent of the annual flow of the Nile (Addison, 1961).

Salinity of Water Stored in Reservoirs. The salinity of the water flowing into the reservoir may vary considerably in the arid regions, depending on rainfall distribution, the salinity of the watershed, and the changing rates of evaporation.

All water evaporating from reservoirs and lakes must leave a residue of salts which increases the salt-content of the stored waters. When the rate of evaporation exceeds the rate of water inflow, the situation may deteriorate at a fairly fast rate, the loss of water being concomitant with increasing salinity.

The salt-content of the water in the reservoir will be at its lowest after the main flood-flow during the rainy season, and will reach its highest concentration towards the end of the dry season. The higher the surface:volume ratio is, the greater will be the proportionate loss by evaporation and, as a result, the higher the salt-concentration.

Small-scale Water Storage Methods

Water storage is an age-old tradition in arid lands. In many of the arid and semi-arid regions, the only source of water is rainfall. The more sparse and infrequent the rainfall is, the greater have to be the ingenuity and effort to store part of the precipitation for use throughout the year. Cisterns, tanks or ponds, and small reservoirs, have all served for this purpose.

CISTERNS

It was only after the invention of a mortar impervious to water, which made it possible to build cisterns for storing rainwater (Plate. 14), that the mountainous regions of the Middle East could be settled. This occurred during the Early Middle Iron Age (Reifenberg, 1953).

Within inhabitated areas, the roofs of buildings, paved roads, and courtyards, serve as catchment areas, and the water is led to the narrow opening of a funnel-shaped cistern. As these cisterns are either closed or have a very narrow opening, and their walls are lined with mortar, losses by evaporation and seepage are practically nil.

Outside the built-up areas, cisterns are designed to catch flood-water and are widely dispersed throughout the deserts. Small catchment areas are carefully prepared for this purpose, and channels lead the water to the cistern excavated into the side of the slope. In order to reduce the amount of sediment introduced into the cisterns, in many cases the water first passes through a settling basin. Frequently, cisterns are excavated alongside wadi-beds, and are filled during flash-floods.

Some of these cisterns may have quite a considerable capacity, and when a number of them are grouped together, the aggregate quantity of water stored may be quite large (Amiran, 1966).

The Nabatean culture maintained itself in the desert by depending to a large extent on cisterns, cut into the ground and lined with cement. Carefully planned diversion systems led the sporadic run-off from relatively large areas to the cisterns.

TANKS OR PONDS

Tanks or ponds are square or rectangular holes, excavated in the ground; the earth that is removed is piled in the form of an embankment on three sides. The tanks store run-off water or water from ephemeral streams, filling to ground-level. The size of the storage is therefore equal to the size of the excavation, making the tank a very costly way to store water (Geddes, 1964).

Tanks or ponds are of great importance for watering stock. They enable the grazing of potentially valuable range in regions in which ground-water is not available economically. Whilst they rarely provide an all-year-round water supply, they usually dry out when most of the grazing has already been effectively used. Their main disadvantage is the great loss by evaporation: the drier the season, the greater this loss.

Seepage losses are sometimes controlled by lining the tank or pond with plastic material.

SMALL RESERVOIRS

An alternative solution to large dams is a number of micro-dams, starting from the higher parts of small valleys and going down to the bottom of the valley. These create a number of terraces which absorb a large proportion of the water and prevent erosion.

Small reservoirs have been in use in the dry regions of central Asia for ages past. The contents of one of these reservoirs may sometimes suffice for irrigating from a few to several hundreds of hectares.

UNDERGROUND WATER RESERVOIRS

CHARACTERISTICS OF GROUND-WATER

Definitions

The water that accumulates in a porous zone, situated at a moderate depth below the soil surface, is known as ground-water. The upper surface of the saturated zone is called the water-table. Water that is held above the water-table by capillary force is called the capillary fringe. The thickness of the capillary fringe, and the amount of water held, depend on soil texture. The lower limit of the saturated zone is the level at which the underlying rock is practically impermeable to water.

During wet periods, or following continuous heavy irrigations, the water-table rises, whereas during dry periods, heavy withdrawals will cause the water-table to fall. If there is no impervious layer between the ground-water and the soil surface, the former is known as 'phreatic' or 'free' ground-water.

The ground-water reservoir or aquifer consists of saturated porous rock materials that are sufficiently permeable for the water to move through them by gravity and yield water freely to wells (Thomas, 1951).

Importance of Ground-water

The ground-water reservoirs are the source of water in wells and springs and the flow of streams during rainless periods (Figure 24). Ground-water is, indeed, the main source of water-supply in dry regions.

In general, ground-water reservoirs are in dynamic balance with precipitation, evaporation, and drainage to the sea.

The volume of ground-water stored at depths of less than 800 m is generally estimated to be 3 000 times as great as that contained in all rivers of the world.

In contrast to surface supplies, the development of ground-water resources is relatively fast and inexpensive, and is therefore a major factor in the economic development of arid regions.

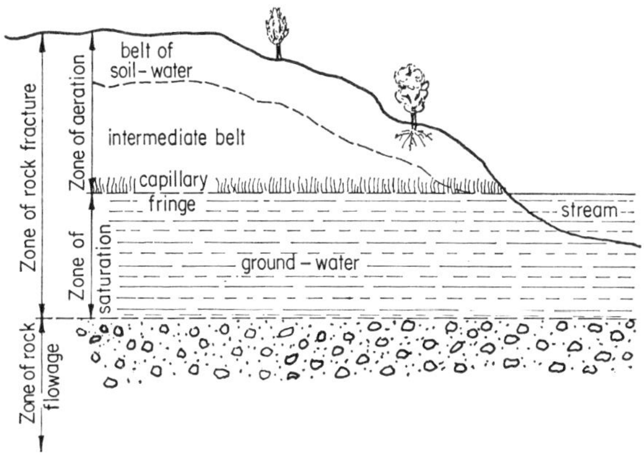

Figure 24. Ground-water zones and belts. (By courtesy of *World Farming*.)

Depth of Ground-water

Ground-water that is relatively near the surface can be used by plants. As a large concentration of roots develops in the capillary fringe above the water-table, even a temporary rise in the water-table may cause considerable damage by asphyxiating the roots (Bernstein, 1962). Phreatophytes (cf. p. 74) flourish where the depth of the water-table is less than 3 metres; the growth of some can, however, still be vigorous when the water-table is at a depth of up to 10 metres (Robinson, 1967).

However any potential benefits of ground-water near the surface are insignificant in relation to the dangers of high water-tables, of which salinization is the most important. This aspect is discussed in detail in Chapter IX.

Natural Recharge of Underground Water Reservoirs

Underground water supplies are replenished either by water that reaches the aquifer from a watershed or from precipitation that falls directly on the soil surface over the aquifer. Other sources are excess irrigation water, subterranean streams flowing into the aquifer, seepage from surface streams, and artificial recharge.

The rate at which water derived from a watershed percolates to the underground water reservoir or, alternatively, is lost to the ocean, depends on the nature of the watershed and its slope.

If the watershed is covered by vegetation, water-loss will be considerably increased. The nature of the vegetation will influence annual evapotranspiration, and hence the amounts of water which remain available for percolation. If the surface of the watershed is bare rock, the nature of the rock and the degree to which it is fissured will determine the relative amounts of run-off and percolation.

Precipitation in excess of that needed for wetting the soil to field-capacity may percolate deeply enough into the soil to be no longer subject to evaporation or accessible to plant roots. Eventually it may become part of the ground-water reservoir.

Normally, the drier a region is, the smaller will be the proportion of natural precipitation that will reach the ground-water reservoir. In regions with less than 200 mm annual rainfall, rain usually falls on non-saturated ground. Practically no precipitation will infiltrate to a water-table—either because the moisture deficiency in the soil is too great, or because the amount of precipitation is too small.

Where the underground waters are replenished by waters that percolate laterally from a humid region, the situation is far more favourable, and there is less danger of excessive withdrawal. Deep boreholes can then tap artesian supplies that have originated at high elevations on the border of the desert.

Quality of Ground-waters

The ground-waters in arid regions almost invariably have a high mineral content, but the degree of salinity depends on the nature and structure of the rocks holding the water (Dixey, 1956). Saline waters are frequently associated with a high clay fraction in the water-holding beds, whilst sandstone and grit usually yield water of low salinity (Bosazza, 1954).

Usable and unusable water-supplies are frequently found within a short distance of each other. Whilst the water may frequently be acceptable for drinking by livestock and even for human consumption, it is usually far more difficult to find water that is suitable for crop irrigation, for which the quality requirements are far more stringent (*see* Chapter IX).

Movement of Ground-water

The movement of ground-water is downward and lateral. Its flow is extremely slow—sometimes not more than a few metres per year, and generally not exceeding 3 metres per day—except in large underground channels. Underground flow is therefore not a very efficient method of transporting water from the fount of precipitation to the location where it is required (Thomas & Peterson, 1967).

The yield of wells may be more dependent on the rate of movement of the water through an aquifer than on the replenishment of the aquifer. If water is pumped from a well or group of wells more rapidly than it can move through the aquifer towards the wells, the depth of the water-table in their immediate vicinity will drop, the cost of pumping will increase, and in extreme cases the wells may be pumped dry even though there is ample water in the aquifer as a whole (Thomas, 1955). The pumping itself may increase the rate of movement in the aquifer by increasing the hydraulic gradient, but this may still be insufficient to balance the rate of withdrawal.

However, by and large, aquifers serve as transmission systems which partially replace, over relatively large distances, surface distribution systems of canals or pipelines, and therefore make possible economies in investment.

UTILIZATION OF GROUND-WATERS FOR IRRIGATION

There are two basic approaches to ground-water utilization (Hall, 1957): (a) the *concept of 'safe yield'* is based on the rule that annual ground-water withdrawals should not have undesirable effects, such as sea-water intrusion, land subsidence, increased pumping requirements, etc. (b) the *concept of 'mining' ground-water*: in this case, withdrawal exceeds annual recharge, sometimes considerably.

Planned overdraft of ground-water may be justified in the initial stages of a project in order to sustain agricultural development, or during the period of construction of dams, reservoirs, canals, etc.

Ground-waters used According to the Concept of 'Safe Yield'

When an underground water reservoir exists in an arid region, it provides a more reliable and regular water-supply than do surface waters of local origin. It can be drawn on throughout the year, and may even be overdrawn for a period of one or two years of drought. Where underground water-supplies in arid regions are dependent on local precipitation, the stored water has accumulated over many generations. It is not, however, inexhaustible. Withdrawal of water by pumping may exceed the annual restoration capacity of the aquifer to such a degree that the level of the aquifer falls and the wells become less productive. The less the precipitation occuring in a given season, the greater will be the effect of withdrawal of water in order to maintain an adequate irrigation supply to the crops.

The 'point of no return' may be reached, when the water-level becomes so low that pumping costs become prohibitive, or when sea water intrudes into the aquifer as frequently occurs in coastal areas. Even when overwithdrawal does not lead to exhaustion, it almost inevitably does lead to an increase in salinity.* All such withdrawal exceeding annual recharge is known as 'mining' of ground-water (*see* below).

Effects of Excessive Ground-water Withdrawal

Salt-water Encroachment. When a coastal aquifer is open to the sea, sea water forms a wedge underlying the fresh water flowing seaward. Excessive pumping will cause a lowering of the water-table. When the level of the water-table

* The opposite has occasionally been known to occur, when the extracted saline water is replaced by a new infiltration of sweet water (Dixey, 1956).

becomes lower than that of the saline water, the latter is drawn into the aquifer. This is a frequent occurrence in coastal areas. For example, sea-water invasion of aquifers has been observed in about 20 ground-water basins along the coast and inland bays of California (Nace, 1960).

Drawing in salt water is even more detrimental than simply depleting an aquifer. An emptied aquifer may eventually be recharged, but in a polluted aquifer, withdrawal of water has to cease even when plenty of water is still available. Reclamation is also more costly and difficult.

Methods for Controlling Intrusion of Sea Water. The following methods have been proposed for controlling the intrusion of sea water into an aquifer (Todd, 1961):

(a) Modifying the pumping distribution pattern so as to locate the most concentrated pumping areas farthest away from the coastline. This minimizes the landward gradients resulting from the pumping, and slows down sea-water intrusion.
(b) The artificial recharge of the aquifers, which creates a fresh-water mound or ridge, inshore from the sea.
(c) A line of wells, located parallel to the coast, makes it possible to create a pumping trough: the wells pump a mixture of fresh water and of salt water which is discharged to the sea. The salt water does not advance because of the converging flow on both sides of the line.
(d) Constructing a sub-surface cut-off wall. This is, of course, feasible only under very special circumstances, as in a narrow valley of shallow depth.
(e) Creating a pressure ridge by a line of wells, parallel to the coast, into which fresh water is injected.

Subsidence of the Overlying Land Surface. Excessive pumping may also lead to a subsidence of the land surface, resulting in compaction and the reduction of pore-space in the aquifer. The loss of underground pore-space is a permanent reduction of the underground storage capacity and is therefore an irreparable loss (Thomas & Peterson, 1967). An example of land subsidence as a result of the lowering of the artesian head in pumped aquifers is found in the San Joaquin Valley of California, where the rate of subsidence has been approximately 26 cm annually (Poland, 1960). Wetting of non-compacted soil and subsoil by irrigation may also cause subsidence (Nace, 1960).

Peatlands under irrigated cropping are particularly subject to subsidence as a result of the oxidation of the peat. The surface of the peatlands in the Sacramento-San Joaquin delta of California sank approximately 2 metres between 1962 and 1966 (Penman *et al.*, 1967).

'Mining' Ground-water

Some ground-water reservoirs are replenished only occasionally: this is the case in many deserts, after a heavy rainstorm which may occur once in several years (Thomas, 1951).

Some aquifers are completely sealed off by layers of impervious material, such as clay, so that replenishment is impossible.

In some cases the basins are so great, and have such huge reserves, that a very long period of time is necessary before any assessment can be made of what constitutes a safe rate of withdrawal. The rate of flow of the springs of Djerid in Tunisia has remained constant over a period of 50 years, notwithstanding the fluctuations in the water-supply to the aquifer from which they are derived—indicating an enormous reserve which can absorb annual fluctuations (Tixeront, 1956). A similar situation occurs in deserts where waters have accumulated underground, during a more humid period of the recent geological past; these waters are no longer replenished and are therefore called 'fossil' water. Typical examples are the ground-water reservoirs of the Libyan and Arabian deserts (Butzer, 1961). Withdrawing water from such reservoirs is indeed a 'mining' operation!

A justification for 'mining' water is when it is first used in order to develop and strengthen the local economy, so that at a later stage it will be possible to bring in water from other regions or to develop additional water resources in the region. This really constitutes a 'planned overdraft'.

WITHDRAWAL OF GROUND-WATER

Wells and Bore-holes

Normally, the term 'well' is reserved for structures that are more than one metre in diameter. Boreholes, or tube-wells, are bored by a drill, and consist of perforated tubes, of relatively small diameter, which reach into the lower aquifers.

Wells. Since time immemorial, wells dug by hand to reach the zone of saturation have supplied water to the inhabitants of arid regions. In many deserts, localities carry the names of the well or wells found on the spot. Most of the wells in arid regions were developed in the earliest times of civilization and continue to be used, without much improvement, by the present inhabitants of the area.

A characteristic feature of many wells in desert regions is the presence of a thin layer of good water (usually not more than 1–2 metres deep) overlying water with a far greater degree of salinity. This occurs when newly percolated rain reaches the water-table and does not mix with the body of saline water (Shotton, 1954). If the inflow of fresh water were greater, it would not remain unmixed with the salt water (Dixey, 1956); such wells, however, have only a very small yield, and are of little or no significance for the irrigation of crops. Their importance lies in making possible the grazing of desert areas, and therefore cannot be overestimated.

Wells have, however, also served as a source of irrigation water since prehistoric times. The depth of the water-table in these wells generally ranged from 1 to 30 metres. Primitive devices, activated by hand or by animal power, served to lift the water.

Boreholes or Tube-wells. The development of large-scale well irrigation was made possible by a succession of technical innovations: these included the wind-driven pump, the gasoline-engine pump, mechanical deep-well drilling equipment, and the submerged centrifugal pump. However, it was the deep-well turbine pump which made it possible to lift water economically from quite considerable depths (Gregor, 1959).

Artesian Wells

When an aquifer that originates at a high elevation is confined between two impervious layers, hydrostatic pressure is engendered which may be sufficient to cause the water to rise in wells located at a lower elevation. Such water is known as artesian water (Figure 25).

Artesian wells are an important source of water-supply in many arid regions. Important examples are the great artesian basins of Australia and the deep artesian wells of Tunis and Tripolitania. The depressions of Kharga and Dakhla in Egypt receive good water from a geological structure of Nubian

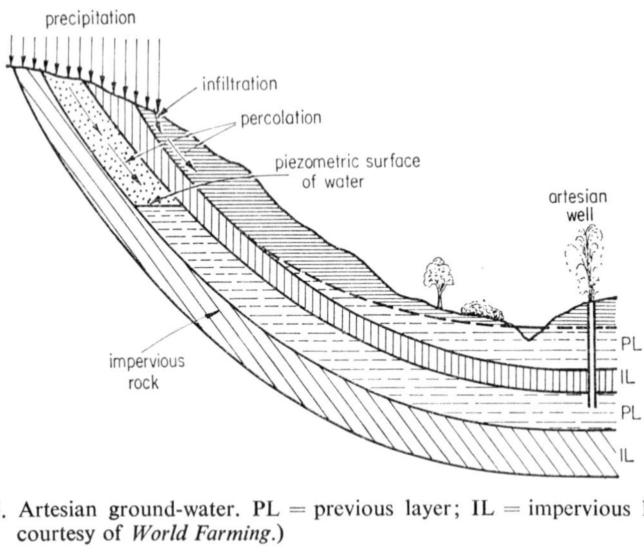

Figure 25. Artesian ground-water. PL = previous layer; IL = impervious layer. (By courtesy of *World Farming*.)

sandstone, which outcrops in the Sudan, over 1 000 kilometres to their south (Shotton, 1954).

In many regions, in particular in several North African countries, the potentialities of artesian waters have not yet been fully exploited (Dixey, 1966).

Shotton (1954) lists the following conditions for artesian supplies in the desert to be available in large and useful amounts:

1. The aquifer has to outcrop in a region of adequate rainfall;
2. The aquifer should have high permeability beneath the desert, so that water movement is sufficient to make high water-withdrawal rates possible;
3. The water should not be contaminated by saline water on its way from the intake area; and
4. Water withdrawal should be balanced by intake.

In the aforementioned depressions of Kharga and Dakhla, artesian water of good quality was abundant, but the increase in the number of boreholes, and their depth, were out of proportion to the renewal rate, so that water pressure has dropped continuously.

In the Great Artesian basin of Australia, much of the water becomes salinized on the way from the intake area to the basin, and cannot be used for irrigation. There appears to be no suitable geological structure for artesian supply in the Great Australian Desert (Shotton, 1954).

In general, the development of artesian wells requires costly drilling, frequently to depths of 1 000 to 2 000 m or even more. The limitation may therefore be not lack of water, but its high cost (Dixey, 1964).

*Quanats**

When an aquifer in the plainland at the foot of a mountain is wide and shallow, the conventional circular well is not efficient for tapping this supply. An underground system of infiltration tunnels for taking water from these aquifers to the land to be irrigated was developed thousands of years ago. Whilst their origin is obscure, these underground aqueducts were already used by the Persians in prehistoric times, and occur in many countries of the Near East.

A quanat is established by digging a number of vertical shafts to the water-table, and then excavating between them a gallery with a gently sloping flow (Figure 26). Water collects in the gallery and flows by gravity until it emerges into the open air. The distance which the water may be transported by the

* Called 'karez' in Persia, 'foggaras' in North Africa, 'puquio' or 'socavones' in South America; it signifies a subterranean canal in Arabic.

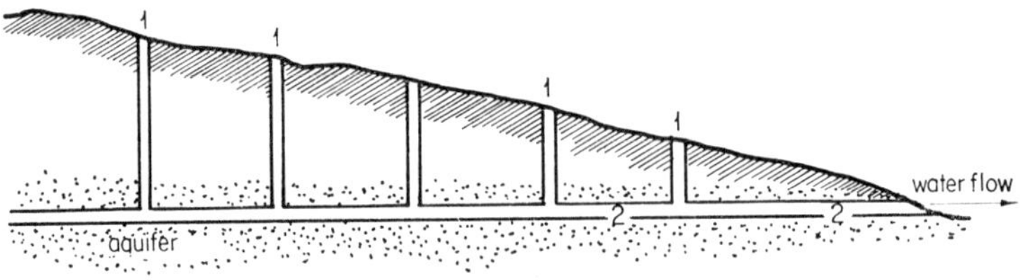

Figure 26. Schematic representation of quanat system. 1 = Access shafts. 2 = Water-collecting gallery with slight slope.

quanats may vary from several hundred metres to a number of kilometres (Dixey, 1966). In Iran there exist thousands of quanats, which are the sole water-supply of numerous villages and supply irrigation water for extensive areas (Cressey, 1958).

According to Chinese records, a system of quanats was already operating in the Shansi Province during the Han era (about 200 B.C.). A vast network of quanats is still in operation in the region of the Siang–Kiang in China, where it forms the most important irrigation system, with a total of 1500 underground canals irrigating in all some 20 000 ha (Kovda, 1961). Extensive quanat systems exist also in North Africa.

A system similar to the quanats of Iran, called 'puquio', is found in the central Andes (Kobori, 1964). Opinions are divided as to whether the puquio were invented by the Indians, independently of the Old World (C. Troll, as cited by Koboti, 1964), or were introduced by the Spaniards (Amiran, 1966).

Quanats have the advantage of flowing both day and night, without requiring a source of energy; since the water is derived from high up in the alluvial fan, it is fresh and cool, and pollution is avoided. Losses by evaporation are prevented, but much water may be lost by seepage.

Water yields of quanats vary widely, depending on ground-water characteristics, the porosity of the soil, and the season. Quanats that derive their water from a stable ground-water source may provide an almost constant flow throughout the year, whilst the supply from others may be intermittent and dry up completely in the dry season (Cressey, 1958).

Construction of quanats is a specialized trade, and has frequently served as a family occupation for generations. Enormous amounts of hand labour are involved in their construction and maintenance (Plate 15).

The average life of a quanat system is between 100 and 150 years, the underground channels becoming gradually silted up. When the flow of water becomes too sluggish for effective use, a new system has to be dug alongside the old one, and frequently five or more parallel lines of old wells may be discerned.

Chotts and Sabhas

In some desert regions run-off water concentrates in depressions which form closed underground basins, usually filled with a very fine silt. These depressions, called chotts or sabhas, are usually covered by patches of typical and abundant halophytic vegetation. The continuous evapotranspiration results in a high salt-content of the water, frequently shown by the salt incrustations that form on the surface of the soil. Tixeront (1956) states that it is not uncommon to find layers of less salty waters below the salt-saturated ones near the soil surface.

Drainage Lines

The sparse vegetation of desert regions, the surface dryness, and the occasional torrential rainfall, cause a high rate of run-off, which concentrates between the catchment areas into the dry river-beds that are the typical drainage lines of desert countries. Along these drainage lines, water may accumulate in shallow underground pools that can easily be tapped by water holes or wells. Typical are the meandering, narrow strips of vegetation which mark the location of the drainage lines in the desert and are in contrast with the barren countryside.

Subsurface Flow in Dry River-beds

In all arid regions there are river-beds* which carry occasional torrents, but which are dry during most months of the year. In some of these, water continues to flow under the surface even when the surface flow has ceased. Where special conditions at the site make it possible to construct a subsurface dam, underflow may be held up to form a subsurface reservoir that may supply water throughout the year. The sudden appearance of a considerable flow of water from an apparently dry stream-bed in the middle of the desert is indeed a striking sight! Unfortunately, it is usually difficult and even rare to find a section of the stream-bed that is suitable for building a dam, the requirement being a site where the sides and floor of the stream-bed are sufficiently impervious (Dixey, 1956).

RECHARGE OF GROUND-WATER—UNDERGROUND STORAGE

In the arid regions, the proportion of rainfall reaching the ground-water is very small, and usually does not exceed 5 per cent (Dixey, 1956). Therefore, the artificial recharge of ground-water is assuming increasing importance. As has been mentioned previously, the excessive withdrawal of water from the underground reservoir is a common feature in irrigated regions; not only do increased pumping costs and increased salinity result, but the very future of

* Called 'wadis' in the Near East, 'oueds' in North Africa, and 'arroyos' in America.

the water-supply may be endangered by the intrusion of salt water into the aquifer. On the other hand, experience has shown that, in arid regions, ground-water storage has many advantages over surface reservoirs: it is not subject to losses by evaporation, and annual fluctuations in precipitation do not have such a marked and immediate effect as is the case in surface reservoirs.

In the semi-arid regions, in which a period with adequate rainfall alternates with an absolutely dry period, underground water-supplies can help to offset the variability of the rainfall and to prolong the growing-season.

Preconditions for successful recharge are an effective absorption surface and a suitable aquifer (Dixey, 1956). Surplus water from springs or streams, at a time when they are not required for irrigation, can be used for replenishing ground-water. Flood-waters can also be used for recharge, although the sediments which they carry may cause difficulties by reducing the infiltration capacity of the soil. The silt may have to be removed from the water before it is led to the infiltration basins overlying the ground-water to be recharged.

The building of dams and weirs in streams increases the infiltration of water along the streams' channels, replenishing water-supplies to the wells and boreholes in the vicinity (Dixey, 1956).

When flow channels reach a valley floor they usually spread out, depositing the coarsest material first—thereby forming, in time, an extensive fan of coarse sediments which have a high percolation rate. By establishing a system of relatively simple checks and diversion dams, a proportion of the stream-flow is forced underground, whence it reaches the aquifer.

The main disadvantage of this method is that the water is at a great distance from the withdrawal region, and therefore recharge is not very effective at the site where the water is needed (Hall, 1957).

Water Spreading. This consists in diverting excess water, from whatever source it may be available, and spreading it over porous ground overlying the aquifer (Mitchelson & Muckel, 1937).

In principle, the methods used for water spreading approximate those used for surface irrigation, the main difference being in the objectives: in the former, the objective is to encourage deep percolation into the underground basin; in the latter it is desired to confine the applied water to the root-zone. One of the difficulties experienced in recharging an aquifer is surface clogging, due to the suspended solids in the recharge water and to microbial growth. The former is the most important factor when flood-waters are used for recharge, the latter when municipal wastes are used for this purpose.

When the basins remain dry for a substantial time, the surface soil again becomes permeable, and, after a period of rest, infiltration rates are considerably improved.

The main principles to be followed in water spreading are (Schiff & Muckel, 1967):

(a) To free the water from silt as far as possible;
(b) To pond the water to relatively high levels in order to increase hydraulic conductivity;
(c) Rotational flooding, so as to enable the land to dry out periodically, in order to recover the original infiltration rate; and
(d) To maintain and improve, as far as possible, the structure of the surface soil. Maintaining the original plant cover, or, alternatively, scraping or hoeing after surface drying, improves infiltration.

There are three methods of water spreading: using basins, furrows, or flooding. Combinations of these methods can be used, depending on local conditions such as the topography of the soil surface, the slope, the amount of silt carried by the water, and the stream-flow characteristics.

Basins. This method is used in land with irregular ground-surface: the area is divided into basins surrounded by banks of suitable height. During flooding, the water is impounded in the basins, excess water from one basin passing into the next, lower basin; most of the silt is thereby deposited in the higher basins. The ground-surface tends to puddle, gradually reducing the percolation rate. This can be overcome by harrowing the surface between floodings. The basin method has the great advantage of helping to control flood-peaks, by providing temporary surface storage of excess water. Its main disadvantage is that it is not suitable for water that carries a large load of silt.

Furrows. Shallow, flat-bottomed, and closely spaced furrows, are used to flow the water over the recharge area. The gradient of the furrows is adjusted to maintain sufficient velocity of the water to keep silt in suspension.

The furrows follow the contour of the land's surface. When a furrow reaches the border of the spreading area, a switch back is made, so that the water is moved in zig-zag fashion over the land, until it reaches the lower limit of the spreading area. Other distribution arrangements can be used. The furrow system is suited mainly to conditions in which flood hazards prevent the establishment of permanent works.

Flooding Between Borders. This method consists in spreading the water in a thin sheet, at a low velocity, over the spreading area. This is the ideal method for land that has a gentle slope and a fairly even surface. The best results are obtained when the native vegetation is left undisturbed, losses by evapotranspiration being insignificant in relation to the improved percolation rate achieved.

USE OF BORES OR WELLS

In areas in which an unsuitable soil surface, or an impervious layer under the soil surface, makes the methods used for surface spreading impracticable, special bores can be drilled, or old wells used, in order to replenish the aquifer. Percolation rates are usually high, but the limited area precludes use of large amounts of water. Active wells are sometimes used for recharge purposes, by reversing the flow of the water. This is not a method to be recommended, because even small amounts of silt may seal the walls and bottom of the well and put it out of action. It is practically impossible to clear the sealing material from the walls of the well. Bacterial contamination of the domestic water-supply is also a possibility. When clear or filtered and chlorinated water can be used, this recharge method can occasionally be used on a small scale (Hall, 1957).

Recharge wells have been successful in some areas and have failed in others, and much has still to be learned in improving the techniques of artificial recharge (Thomas, 1951).

A new approach to ground-water recharge is the intentional use of excessive amounts of water during irrigation of crops in the off-irrigation period, so that the surplus will be added to the ground-water.

Some of the methods of ground-water recharge described above are already being used on a large scale, for example in the U.S.A. and in Israel. For some regions, ground-water recharge is the only possible method of water storage; for others, where surface storage is also possible, underground storage may have obvious advantages—less costly works being required, losses from evaporation being reduced, etc.

It is generally accepted that most of the water stored underground is recoverable (Mitchelson & Muckel, 1937); such water is recovered by pumping from wells.

Unconventional Water Resources

WATER RE-USE

The increasing demand for water in the regions makes re-use, sometimes repeated re-use, of return flows imperative. Before re-use is possible, treatment for quality improvement may be essential. The techniques of improvement vary according to the nature of the contamination and the purpose for which the water is to be used.

Recovery of Excess Surface Irrigation Water

By constructing an irrigation return-water system at the low end of irrigated fields, it is possible to re-use water of a quality similar to that of the original water, and at relatively low cost. For this purpose, a reservoir or sump is constructed which collects all surplus irrigation water, which is then re-pumped to the head of the field for re-use (Hagan *et al.*, 1967). However, the return of drainage waters to the main stream causes an increase in the total concentration of soluble salts, and in the proportion and quantity of sodium and chloride whilst calcium, magnesium, sulphate, and bicarbonate, are reduced by precipitation.

For example, the salt-content of the Colorado River, as it flows into Mexico, has increased alarmingly, as a consequence of the pumping of subsoil saline waters from irrigated areas, which are then diverted into the river (Anaya, 1967).

Sewage Treatment

The processes in use at present for sewage treatment have not changed basically since their large-scale application at the beginning of the century (Weinberger & Stephan, 1967). Biological treatment processes remove a substantial portion of the organic matter, and, in their course, many pathogenic organisms may be destroyed—though the water may still remain dangerous from the point of view of human health. In general, even a well-run biological sewage treatment plant will not remove more than 90 per cent of the pathogenic organisms present in the raw sewage (Shuval, 1967). Some chemicals and toxic materials may be oxidized into harmless substances. However, certain salts, such as the chlorides, remain unaffected by the treatments, and the percentages of nitrates may be increased. The concentration of detergents remains unaffected.

Certain kinds and sources of pollution are becoming of increasing importance, including refractory organic materials, radioactive wastes, and inorganic salts.

A large number of processes are being investigated as to their technical and economic feasibility for the purification of sewage. A few of these techniques are: adsorption on activated carbon for the removal of colloidal and suspended solids (including the soluble, refractory organics), electrodialysis for removing the added inorganic salts, a foam separation process for removing detergents, and chemical oxidation for the destruction of organic molecules in dilute solution (Weinberger & Stephan, 1967).

Use of Municipal Waste-water for Irrigation

In view of the health hazards involved, in many countries regulations limit the use of reclaimed sewage waters to certain crops not used for human consumption, or to those consumed only after processing and/or cooking. Methods of irrigation which prevent direct contact between the marketable product and the reclaimed sewage, such as furrow irrigation, present less risk of contamination than do other methods, such as overhead irrigation (Shuval, 1967).

Municipal waste-water can be used for irrigation only during the dry season. During the rainy season, it can serve for ground-water recharge. This has the considerable advantage of improving the sanitary condition of the water, by the removal of most of the micro-organisms if the water is withdrawn from the aquifer at a reasonable distance from the location of the recharge area; it may even become suitable for drinking purposes. Any undesirable materials contained in the effluent will be diluted by the water already present in the aquifer (Amramy, 1967).

After adequate treatment, *municipal waste-water* can be an important source of water for irrigation. The main characteristics of municipal waste-waters are the presence of disease-causing organisms, and of contaminants from home and industry—which include toxic materials, salts, organic compounds, detergents, etc.

INCREASING PRECIPITATION

The presence of clouds alone does not ensure precipitation. The droplets of moisture which constitute the cloud are frequently too small to reach the weight required for them to fall earthwards as precipitation. For this reason only a small fraction, usually less than 10 per cent of the moisture in the atmosphere, reaches the ground as precipitation (Schleusener & Grant, 1967).

The general explanation of the formation of raindrops in the cloud is, in the most simple terms, that with decreasing cloud temperatures, ice crystals are formed, which attract additional moisture, increase in size, and attain the necessary weight to fall as snow-flakes. When the temperature is appropriate, these melt during their fall earthwards and change into raindrops.

On the basis of this theory, a number of methods have been developed for artificially stimulating the process described above. They are all based on 'seeding' the cloud with nuclei, on which the moisture will condense. Seeding suitable clouds with dry ice (solid carbon dioxide) reduces their temperature and triggers the process of formation of ice crystals. Conditions are particularly favourable when the temperature of the seeded cloud is as near as possible to the critical temperature for ice-crystal formation (–15° to –25°C).

Another approach—artificial nucleation—is based on the fact that dust or other particles may serve as nuclei of condensation, enabling rain droplets to form at temperatures that are somewhat above those normally required for ice-crystal formation. In particular, crystals of silver iodide are known to be effective as an ice nucleating agent at a temperature of –5°C (Neumann, 1967).

Seeding of clouds can therefore be effective in causing rainfall only when conditions are not too different from those under which the precipitation would occur naturally, at which time seeding may 'tip the balance': hence the difficulty of deciding unequivocally whether cloud-seeding has been instrumental in causing rainfall or not. In the absence of direct proof, the effect of seeding is assessed by statistical comparisons between the amounts of precipitation obtained with and without seeding in adjacent regions—a process that of necessity is non-sensitive and inaccurate. Even differences of 20 to 25 per cent in precipitation are not statistically significant (Bowen, 1958).

It is now generally agreed that positive results can be achieved from seeding with silver iodide when the conditions are favourable for natural precipitation. Even when precipitation is released from a cloud, whether or not it reaches the ground is also dependent on a number of factors: the quantity released, the height of the cloud-base from the ground, and the dryness of the air between cloud and ground (Bowen, 1958).

In the western U.S.A., it was found that seeding of winter storm-clouds in the mountainous areas produced an average increase in precipitation of 10 to 15 per cent, whilst in the non-mountainous area no statistically significant increase could be demonstrated (Advisory Committee on Weather Control, 1957). In Israel, an average increase in rainfall of 18–19 per cent over five seasons was achieved by cloud-seeding. The gain in precipitation due to seeding generally occurred as the result of large increases in a few days of seeding, whilst on most days of seeding only minor or no increases were recorded (Neumann, 1967).

Rain cannot fall from a relatively dry atmosphere, and therefore even the most successful rain-making can, at best, only contribute to an increase in rainfall, but cannot consistently provide rain to arid regions.

DESALINATION

Practically untapped major sources of water for arid lands are sea water and brackish lake and ground-water. Methods of desalination which would provide fresh water from these sources at an economical cost, would solve the problem of water for many areas in which no other solution is in sight.

Desalination Methods

A number of methods have been developed to produce fresh water from saline water. These conversion processes are based on distillation, passage through membranes, freezing, humidification, and chemical processes.

The methods that are most advanced technologically, for converting water in the quantities required for irrigation, are based on the first two methods, namely distillation for conversion of water such as sea water, with very high salt-content (35 000 to 38 000 p.p.m. total soluble salts (TSS)), and on passage through membranes for converting brackish water (3 000 to 4 000 p.p.m. TSS) into water that can be used for irrigation (500 p.p.m. TSS) (U.N., 1967).

Distillation. The concept of distillation for producing fresh water has been known, and applied on a small scale, for centuries. In principle, such distillation consists in applying heat to salt water to vaporize the water, and then removing the heat in order to condense the steam produced. Increased efficiency in the process is achieved by transferring the heat from the vapour that is being condensed, to the water that is being heated—thereby recovering part of the free energy for recycling. As the energy requirements for this method are almost independent of the amount of salts in solution in the 'raw' water, distillation is most suitable for water with a very high salt-content, such as sea water.

From the basic principle of distillation have been developed a number of processes, of which the most promising at present is multistage flash distillation (Figure 27). This process is based on the fact that, the lower the pressure,

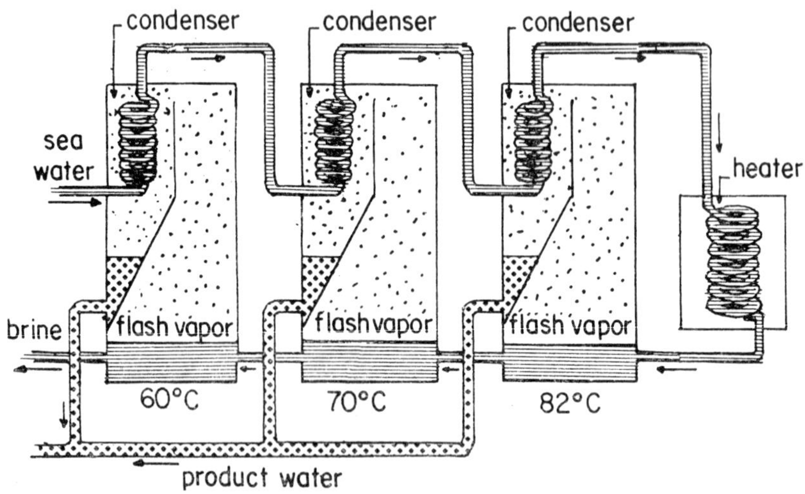

Figure 27. Multistage flash distillation process. (By courtesy of the U.S. Department of the Interior.)

the lower the temperature at which water starts boiling. As the name indicates, the water is converted in a number of stages.

Heated sea water is introduced into a chamber in which the pressure is sufficiently low to cause part of the water to boil immediately, and 'flash' into steam. As a result, the temperature of the remaining water is lowered. It flows into the next chamber, where pressure is lower than in the first chamber, so that part of the water is again 'flashed' into vapour, even though its temperature has been reduced. This process is repeated a number of times, in chambers with progressively lower pressures.

The vapour is condensed by coming in contact, in a heat exchanger, with the incoming salt water, so that part of the heat is transferred from the steam to the sea water. The remaining brine, which is returned to the sea, is only slightly warmer than the incoming sea water, as approximately 90 per cent of the heat-energy has been recovered (U.S. Department of the Interior, 1966).

Membrane Processes. The membrane processes are based on osmosis through semipermeable membranes. Of the three principal methods that have evolved—reverse osmosis, electrodialysis, and transport depletion—the most advanced, technologically, is electrodialysis.

The basic unit of a conversion plant employing electrodialysis is an electrolytic cell containing two different ion-selective membranes—one allowing the passage of cations and the other that of the anions (Figure 28). When electricity is applied, the positively-charged cations migrate towards the cathode, passing through one membrane, and the negative anions migrate towards the anode, passing through the other membrane. Fresh water remains between the two membranes. Actually, an electrodialysis cell is made up of many altern-

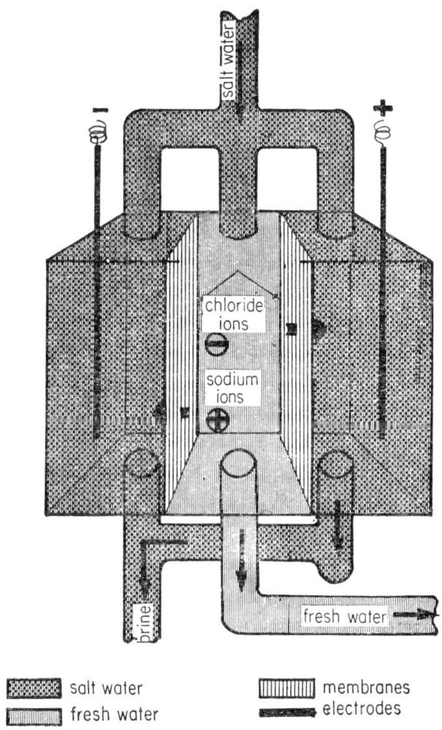

Figure 28. An electrodialysis cell. (By courtesy of the U.S. Department of the Interior.)

ating cation- and anion-permeable membranes which form channels for the flow of the water. The salt-content of the water is depleted and enriched in alternating channels when electricity is turned on in the end electrodes.

The amount of electric current that is required depends on the quantities of salt to be removed. Therefore, the cost of the energy required for the process is proportional to the concentration of salts in the water to be converted. For this reason, this method has promise mainly for removing part of the salts of brackish water, to a level safe for irrigation, but is too expensive for the conversion of sea water into pure water.

A special application of electrodialysis is the demineralization of brackish water that is being pumped in order to prevent its intrusion into a fresh-water zone, thereby achieving two objectives at the same time (U.N., 1964).

In reverse osmosis, pressure is extended on the salt water, forcing fresh water to diffuse through the semipermeable membrane in the opposite direction to normal osmotic flow. The principle of reverse osmosis is illustrated in Figure 29(a) and a diagram of the actual process is shown in Figure 29(b). The salt water is first pumped through a filter to remove solid particles and is then raised to operating pressure by a second pump before being introduced into

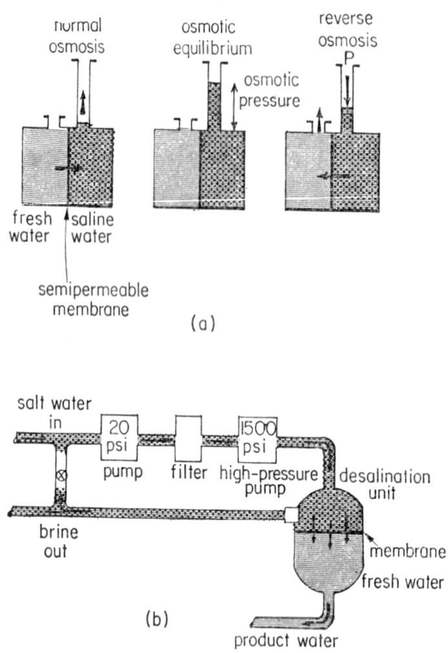

Figure 29. Schematic representation of the reverse osmosic process.
(a) Principles of reverse osmosis.
(b) Schematic diagram of the reverse osmosis process.
(By courtesy of the U.S. Department of the Interior.)

the desalination unit. Part of the water permeates the membrane and is collected at the bottom of the unit, whilst the remaining brine is discharged at the top of the unit.

Multipurpose Plants

When generating electric power, the steam that has activated a steam turbine which is driving an electric generator, still contains a considerable amount of heat-energy. By using the residual heat of the exhaust steam as a source of energy for a distillation conversion process, the cost of energy for desalination is reduced considerably.

At present, a conventional steam turbine power-plant, combined with a multistage flash evaporator distillation plant, gives the greatest promise for reducing the production costs of desalinated water (Sporn, 1966). Tripurpose plants are also envisaged. In Kuwait, for example, in addition to desalination and the production of power, caustic soda, salt, chlorine, and hydrochloric acid, are all extracted from the residual brine (U.N., 1967). The brine could also be a source of magnesium and potash (U.S. Department of the Interior, 1966).

Further Expected Developments

Further developments, which are expected to contribute to a lowering of the cost of converted sea water, are the use of nuclear power as a heat source to furnish steam, and increasingly larger power and distillation units to effect economies of scale (Sporn, 1966).

Economic Aspects of Water Use

In many regions, the development of agriculture is dependent on irrigation; in others it is a prerequisite to higher productivity of limited land resources, to production of high-value crops, and to the diversification and intensification of agriculture.

As a result, irrigation development has a considerable impact on overall economic development: increased employment and generally increased living standards provide an internal market for industrial products; moreover, raw materials such as textiles, sugar, vegetable oils, etc., are produced for industrial processing.

Frequently, the development of water resources is undertaken by governments even when the cost of the water that is eventually made available to agriculture is more than the producer can afford to pay; water is therefore frequently subsidized in one form or another. The vital role of water in food production and as an essential starting-point for overall economic development justifies, in these cases, the development of water resources as a national asset, while indirect benefits are sometimes sufficient to pay for the maintenance, operation, and interest charges, of the project.

ALTERNATE USES OF WATER

In developing countries in the arid regions, the bulk of the available water is used for irrigation. Unlike the situation in humid areas, where land has its own intrinsic value, in arid regions land is almost valueless without water; its value is determined by the amount of water available to the land (Koenig, 1956).

An economic evaluation of alternate uses of water shows that agriculture is far less productive in its use of water than are other users (Ackerman, 1960). The productivity value of water for industry is frequently 100 times as great as for agriculture (Olivier, 1967). In the San Juan basin in Colorado and New Mexico, it was estimated that the productivity of an acre-foot (=1 233·5m^3) of water in irrigation was $14 to $18—as compared with $1 200 to $3 300 when it was used for industrial or municipal purposes (Wollman *et al.*, 1959, as cited by G. F. White, 1962). In a survey by the United Nations of developing countries in dry regions, it was found that—in addition to industry—mining, ports, tourist centres, and all other economic activities, could afford to pay far higher costs for water than could agriculture (U.N., 1964). In the United States, the water that will support one worker in arid-land agriculture will support about 60 workers in industry (Koenig, 1956).

However, factors other than the economic return per unit of water have to be considered—such as the need to provide for a minimum local production of food, the displacement of the agricultural population, and even, in certain cases, security needs which may be overriding, All these considerations, which are certainly legitimate, have to be weighed against the economic evaluation of alternate uses of water.

As capital is one of the scarcest resources in a developing economy, errors in the economic evaluation of irrigation projects can have far-reaching consequences in retarding economic development.

In the long run, economic considerations will be overriding in the determination of priorities for the use of water, which is a basic resource for which demand increases far more rapidly than does supply.

VALUE, COST AND PRICE

The price of water is what the user pays for it; it may be less than cost if the price is subsidized. Value may range from zero, for a farmer paying a flat rate per unit of area farmed, to infinity, for a wanderer dying of thirst in the desert. A country will generally first develop its lowest-cost water resources, which are usually in proximity to the areas to be irrigated. As needs increase, more distant water sources have to be drawn on; water is pumped from greater depths, and more and more expensive water resources are developed.

At present, the price paid for irrigation water by the farmer in most countries is usually below total costs. In many cases, he does not even pay on the basis of the actual amount of water used, but according to the area irrigated or the crop grown. There is little doubt that this approach is conducive to the waste of water, to inefficiency in its use, and to the production of crops that are not economically justified. If the objective is to subsidize agriculture, more appropriate means can be adopted.

Yaron (1966) proposes a system whereby a relatively low price is fixed for the bulk of the farmer's water allocation, and a far higher price, in the range of the value of the marginal output of water on efficient farms, for the remaining. The result of such a policy would be that only efficient farms would purchase and use the full quantity of water allocated to them, and that therefore even the marginal segment of the water-supply would be efficiently used.

Marginal Costs of Water

In countries with a high level of agricultural production, linear programming makes it possible to calculate the amounts of water that it is economical to apply, at various price-levels for water. In California, for example, it was found that at the current farm prices, the critical price for water is 1·3 U.S.c/m^3 for small farms, and 1·6 U.S.c/m^3 for large farms (Clark, 1967). In the Negev of Israel, a range of 5 to 7·2 U.S.c/m^3 is considered to be the marginal value product of water in field-crops at the present time (Yaron, 1966).

In primitive agriculture, such as is exemplified by a village in India, the marginal value product of a cubic metre of water is about 1·9 U.S.c/m^3 (at world prices), which is equivalent to 0·25 kg of wheat (Clark, 1967). In Madras State, the use of water for rice production yields a gross agricultural income of 5c per 1000 gallons, which has to pay for the cost of land, labour, etc., in addition to that of water (Barnea, 1965). It is interesting to note that the raising of water from wells by primitive methods, by man and animal power, is barely remunerative—even at a relatively high value for grain and a very low cost of labour (Clark, 1967).

The cost limit of the marginal value product of water can rise, as a result of improved technology involving more efficient use of water, reduced production costs, higher yields, or the growing of higher-value crops.

Where water is the limiting factor, it is usually preferable to aim for the highest yield per unit of water applied than per unit of land area. The total production from the water may be much higher when the supply is spread over a fairly large area in amounts calculated to give less than maximum yields than when it is concentrated over an area on which it will produce maximum yields. In Tunisia, it was found that 80 per cent of the maximum possible yield of maize was obtained with 60 per cent of the water needed for potential evapotranspiration (Brichambaut & Wallen, 1963).

Such 'under-irrigation' is, however, possible only in those areas in which, during at least part of the year, precipitation is sufficient to leach the salts beyond the root-zone.

Cost of Water from Different Sources

Costs of water for irrigation are naturally dependent on the source of supply, being generally lowest when it is drawn by gravity from flowing streams or springs, higher when it is pumped from wells, and generally highest with water from reservoirs created by constructing dams. Thus, for example, in the United States, the costs of water drawn from stream-flow are around 0·1c/m^3, whereas water pumped from streams costs an average of 0·33 c/m^3, and water pumped from wells costs about 0·68 c/m^3 (Clark, 1967).

Ground-water costs naturally depend on the depth of pumping. In a particular location in California, the costs of well-water are as follows (Barnea, 1965):

TABLE VI

Costs of Ground-water in California, in Relation to Depth of Pumping

Depth (m)	*Capacity* (gallons/min)	*Cost per 1000 gallons* (cents)
50	300	2·7
50	1200	1·7
120	300	5·9
120	1200	3·5

In Madras, wells 50 m deep and with a capacity of 800 gallons/minute can provide water at 2·4c/1000 gallons (U.N., 1964).

The advantage of developing ground-water supplies is in the relatively low capital requirement.

The cost of water from reservoirs, based on the cost of constructing the dams, varies from 6–9 c/m^3 for small dams, to 0·7 for the largest (Clark, 1967). However, the price of the water on the farm, from behind a large dam, may be considerably increased by the cost of the necessary distribution system. On the other hand, the cost of water from favourably situated reservoirs may still be low: from 1·1 to 3·0 c/1000 gallons at certain sites in Ecuador (Barnea, 1965). However, in most of the developing countries, low-cost water resources are already fully exploited and further development requires heavier investment and higher unit costs.

Where the dams also serve to generate electricity, the net annual cost of water from large dams in India and Pakistan is estimated at 0·13 to 1·35 c/m^3.

Desalinated Water

The key to the use of desalinated water is, of course, substantial reduction in the costs of conversion. Although considerable progress has been achieved, the final goal—that is, a cost of conversion compatible with using the desalinated water for irrigation—has not yet been achieved.

The major costs involved in desalination are: capital investment, the cost of the energy used for the process, and operation and maintenance costs. To the cost of desalination must be added the expense of conveying the converted water to where it will be used. As the source of supply is always at sea-level, costs of lifting may be quite considerable. As a point of reference, the cost of water that can be economically justified for most agricultural crops is 5–10 cents per 1000 gallons (Sporn, 1966), and for high-value crops, 13·2 c (Hammond, 1967).

The cost of water produced by existing desalting plants exceeds $1 per thousand gallons (U.S. Department of the Interior, 1966). Yet it is estimated that a dual-purpose plant, generating 125 to 200 MW of electricity, and distilling 50 to 75 million gallons of water per day, would produce water at a cost of 50–60 c per 1000 gallons (Sporn, 1966). If the dual-purpose plant were upscaled to a capacity of 150 million gallons a day, fresh water could be produced at a cost of 22 c/1000 gallons (U.S. Department of the Interior, 1966). This is still double the economical level for crop production.

When considering the cost of desalinated water, its effect on the quality of the overall water-supply should also be considered. By blending it with waters which would otherwise be too saline for use, supplies of water available to irrigation can be considerably increased. Thus in Tunisia, a blend of desalinated sea-water with brackish water was achieved at a cost of 40 c/1000 gallons—a cost unachievable through desalination or any other method alone (U.N., 1964).

It is, however, estimated that the cost of desalinated water, based on predictable desalting costs for the early 1970s, will still be too high for conventional agricultural production, and may even be higher than the value of the end-product (Wiener, 1965).

Conversion of brackish water by electrodialysis may already be competitive in areas where the water-supply is limited and water costs are marginal. It is estimated that conversion of water can be achieved at a cost of about 7 c/m^3 to 16 c/m^3, depending on the quality of the raw water treated (Matz, 1967).

Water Legislation

Historically, in certain countries, water was considered as a free commodity, like air, which could be used by anyone having the need and the opportunity to do so. The rational development of water cannot, however, be achieved without appropriate legislation, that defines the rights and duties of the water users.

In those countries in which water legislation already exists, the growth of the population, technological progress, and additional uses for water, frequently make the existing laws obsolete. New legislation is then essential if progress in water development and use is not to be hamstrung.

Cano (1967) lists the following desiderata for national water laws: water legislation should define the national water-policy principles covering standards for planning, definition of priorities, government authorization for private works, and also measures for integration of water resources for multiple use. Water laws must also cover relations among individuals and between individuals and government. Proceedings for registration and protection of rights and uses must be established in order to ensure security and stability of investments. Legal rules are required on technical methods to ensure conservation and prevent waste, as well as on those relating to prospecting, extraction, and use of underground water.

Two examples of the basic principles of water legislation are given below: one is in force in the United States, a country in which the water law is no longer in accordance with present-day requirements, while the other is in force in Israel, a developing country with scarce water resources on which development is entirely dependent.

WATER LEGISLATION IN THE UNITED STATES

In the United States, water legislation in the humid regions is derived from English law and based on riparian doctrine which accords water rights only to owners of land that is adjacent to a stream (Riesbol *et al.*, 1967). All riparian owners are entitled to fair shares of the water, which could be used only for riparian lands. Whether or not the owner uses his prerogatives, has no effect on his rights.

The arid-regions doctrine, called 'appropriation doctrine', was based on the customs of miners who diverted waters for their operations, as well as on the experience of the early Spanish settlers who conveyed water from the streams to their lands for use in irrigation. According to this doctrine, the first to appropriate the water has priority of rights, which are subject to his actually using the water; lack of use usually results in loss of rights. The water can be used for riparian lands, or at a distance from the stream-bed, at the discretion of the appropriator. The great advantages of the appropriation doctrine, are the requirements of beneficial use and the discretion to convey the water to where it is most needed. Its main disadvantage is that a farmer with a right to much water would tend to be wasteful of this resource.

Ground-water is usually subject to the English rule that gives to the owner of the land absolute control of the underlying ground-water. However, this law entirely neglects the basic fact that water under a certain piece of land belongs to a common underground reservoir, and that surface water and underground water cannot be treated separately (U.S. Geological Survey, 1951). After the development of efficient pumping systems, and their extensive use, the shortcomings of the common law in this respect have become increasingly evident.

The absolute right of the owner of a well to pump, without any restrictions, generally leads to over-use of ground-water resources. As a result, cones of depression have developed in wells in many areas (Thomas, 1955). Natural recharge often being very small, stored water continues to be depleted.

WATER LEGISLATION IN ISRAEL

In Israel, *all* water resources of the country—surface and underground waters, flood-waters, waste and drainage waters—are public property and are controlled by the State (Aloni, 1967). No water can be owned by a private individual, even if it is found on his own land; nor does this confer on him any preferential rights. As agriculture uses about 80 per cent of the total supply, the implementation of the water law is the responsibility of the Minister of Agriculture, under whom a Water Commissioner manages all the water affairs of the country.

One of the main provisions of the law is that water rights, allocated to an individual, are linked to a specified purpose. If the purpose ceases to exist, or is not adhered to, the water right lapses. Water allocated to a farmer, for example, can be used for growing any crop he wishes, but if he ceases to farm his land, his water rights lapse, and cannot be transferred to another location.

The Minister of Agriculture can prescribe norms and rules for the efficient and economic use of water in all parts of the country.

REFERENCES

ACKERMAN, E. A. (1960). Water resource planning and development in agriculture. pp. 3–14 in *Water and Agriculture* (Ed. R. D. Hockensmith) Amer. Assoc. Adv. Sci., Washington D.C., 198 pp., illustr.

ADDISON, H. (1961). *Land, Water and Food. A Topical Commentary on the Past Present and Future of Irrigation, Land Reclamation and the Food Supply they Yield.* Chapman & Hall, London, xii + 284 pp., illustr.

ADVISORY COMMITTEE ON WEATHER CONTROL (1957). *Final Report.* U.S. Government Printing Office, Washington, D.C., 422 pp.

ALONI, S. (1967). Modern water legislation and development. pp. 538–44 in *Water for Peace*, U.S. Government Printing Office, Washington D.C., Vol. 5. vii + 1061 pp., illustr.

AMIRAN, K.H.D. (1966). Man in arid lands: I—Endemic cultures. pp. 219–38 in *Arid Lands: A Geographical Appraisal* (Ed. E. S. Hills) UNESCO, Paris, xviii + 461 pp., illustr.

AMRAMY, A. (1967). Re-use of municipal waste water. pp. 421–34 in *Water for Peace*, U.S. Government Printing Office, Washington D.C., Vol. 2, vi + 970 pp., illustr.

ANAYA, M. (1967). Mexico and its water resources policy. pp. 682–91 in *Water for Peace*, U.S. Government Printing Office, Washington D.C., Vol. 8, v + 827 pp., illustr.

BARNEA, J. (1965). Water costs in developing countries. First international Symposium on *Water Desalination*, Washington D.C., Doc. **SWD/98**, 7 pp.

BERNSTEIN, L. (1962). Salt-affected soils and plants. *Arid Zone Res.*, **18**, 139–74.

BISWELL, H. H. & SCHULTZ, A. M. (1957). Flow affected by brush. *Calif. Agric.*, **11**(10), 2–4, 10.

BOSAZZA, V. L. (1954). Problems of water supply in the arid areas. *Geogr. J.*, **120**, 119–22.

BOWEN, E. G. (1958). The artificial stimulation of rainfall and its application to the arid regions. *Arid Zone Res.*, **11**, 269–71.

BRICHAMBAUT, G., PERRIN DE & WALLEN, C. (1963). A study of agroclimatology in semi-arid and arid zones of the Near East. *World Met. Org. Tech. Work*, Geneva, 56, xv + 64 pp., illustr.

BROWN, C. B. (1960). Significance of small watershed programs to American agriculture. pp. 29–42 in *Water and Agriculture* (Ed. R. D. Hockensmith) Amer. Assoc. Adv. Sci., Washington D.C., 198 pp., illustr.

BURGY, R. F., FLETCHER, J. E. & SHARP, A. L. (1967). Watershed management. pp. 1089–107 in *Irrigation of Agricultural Lands* (Ed. R. M. Hagan, R. H. Haise & T. W. Edminster) Agronomy series; Amer. Soc. Agron., Madison, Wisconsin, II, xxxi + 1180 pp., illustr.

BUTZER, K. W. (1961). Climatic change in arid regions since the Pliocene. *Arid Zone Res.*, **17**, 31–32.

CANO, G. J. (1967). National water laws and administration as elements of water development. pp. 3–16 in *Water for Peace*, U.S. Government Printing Office, Washington D.C., Vol. 5, vii + 1061 pp., illustr.

CLARK, C. (1967). *The Economics of Irrigation.* Pergamon Press, Oxford, 116 pp., illustr.

CRESSEY, G. B. (1958). Quanats, karez and foggaras. *Geog. Rev.*, **48**, 27–44.

DE BRICHAMBAUT, G., PERRIN. *See* BRICHAMBAUT, G. PERRIN DE.

DIXEY, F. (1956). Variability and predictability of water supply. pp. 121–40 in *The Future of Arid Lands* (Ed. G. F. White) Amer. Assoc. Adv. Sci., Washington D.C., ix + 453 pp., illustr.

—— (1964). The availability of water in semi-arid lands: possibilities and limitations. *Arid Zone Res.*, **26**, 37–46.

—— (1966). Water supply, use and management. pp. 77–102 in *Arid Lands: A geographical Appraisal* (Ed. E. S. Hills) UNESCO, Paris, xviii + 461 pp., illustr.

DUBIEF, J. (1953). *Essai sur l'hydrologie superficielle du Sahara.* Direction du Serv. de la Colonisation et de l'Hydrologie, Service des Etudes Scientifiques, Alger, 457 pp., illustr.

DULEY, F. L. (1939). Surface factors affecting the rate of intake of water by soils. *Proc. Soil Sci. Soc. Amer.*, **4**, 60–64.

FORD, E. C., COWAN, W. L. & HOLTAN, H. N. (1955). Pp. 171–6 in Floods and a program to alleviate them. *Yb. Agric. U.S. Dept. Agric.*, Washington D.C., xiii + 751 pp., illustr.

GARDNER, W. H. (1964). Research for more efficient water-use: soil physics. Pp. 85–94 in *Research on Water*. Soil Sci. Soc. Amer., Madison, Wisconsin, vi + 121 pp., illustr.

GEDDES, H. J. (1964). The University of Sydney water-harvesting project. *Outlook Agric.*, **4**, 182–9.

GREGOR, H. F. (1959). Push to the desert. *Science*, **129**, 1329–39.

HAGAN, R. M. HOUSTON, C. E. & BURGY, R. H. (1967). More crop per drop: approaches to increasing production from limited water resources. pp. 610–21 in *Water for Peace,* U.S. Government Printing Office, Washington D.C., Vol. 3, vi + 968 pp., illustr.

HALL, W. A. (1957). Recharge studies—refilling underground water reservoirs. *Calif. Agric.*, **11**(4), 4.

HAMMOND, R. P. (1967). Desalted water for agriculture. Pp. 184–97 in *Water for Peace*, U.S. Government Printing Office, Washington D.C., Vol. 2, vi + 970 pp., illustr.

HAUSER, V. L. & COX, M. B. (1962). Evaluation of Zingg conservation bench terraces. *Agric. Engng, London,* **43**, 462–4, 467.

HILLEL, D., RAWITZ, R. & STEINHERDT, R. (1967). *Runoff Inducement in Arid Land.* An. Res. Reptr., Volcani Inst. Agric. Res., Bet-Dagan, Israel, 142 pp., illustr. (mimeographed).

JACKS, G. V., BRIND, W. D. & SMITH, P. (1955). Mulching. *Tech. Commun. Commonw. Bur. Soil Sci.*, **49**, 87 pp., illustr.

KEMPER, W. D. (1964). Microwatersheds. *Agric. Res.,* **13**(2), 10–11.

KING, F. M. (Ed.) (1914). *Physics of Agriculture.* Publ. F. H. King, Madison, Wisconsin. xvi + 604 pp., illustr.

KOBORI, I. (1964). Human geography of methods of irrigation in the Central Andes. *Arid. Zone Res.*, **26**, 135–8.

KOENIG, L. (1956). The economics of water resources. pp. 320–30 in *The Future of Arid Lands* (Ed. G. F. White) Amer. Assoc. Adv. Sci., Washington D.C., ix + 453 pp., illustr.

KOVDA, V. A. (1961). Land-use development in the arid regions of the Russian plain, the Caucasus, and Central Asia. *Arid Zone Res.*, **17**, 175-215.

LANGBEIN, W. R. (1962). Surface water (including sedimentation). *Arid Zone Res.,* **18**, 3–22.

LOVE, L. D. (1960). Water yield from mountain areas. pp. 3–15 in *Symposium on Water Yield in Relation to Environment in the southwestern United States* (Ed. B. H. Warnock & J. L. Gardner). Amer. Assoc. Adv. Sci. & Sul. Ross. Coll., Alpine, Texas, 74 pp., illustr.

MANSFIELD, W. W. (1967). Evaporation control in Australia. pp. 557–61 in *Water for Peace,* U.S. Government Printing Office, Washington D.C., Vol 2, vi + 970 pp., illustr.

MATZ, R. (1967). Application of electrodialysis in Israel. Pp. 43–57 in *Water for Peace,* U.S. Government Printing Office, Washington D.C., Vol. 2, vi + 970 pp., illustr.

MITCHELSON, A. T. & MUCKEL, D.C. (1937). Spreading water for storage underground. *Tech Bull. U.S. Dept. Agric.*, **578**, 80 pp., illustr.

MYERS, L. E. (1967). New water supplies from precipitation harvesting. pp. 631–40 in *Water for Peace*, U.S. Government Printing Office, Washington D.C., Vol. 2, vi + 970 pp., illustr.

NACE, R. L. (1960). Water management, agriculture and groundwater supplies. pp. 43–61 in *Water and Agriculture* (Ed. D. Hockensmith) Amer. Assoc. Adv. Sci., Washington D.C., 198 pp., illustr.

NEUMANN, J. (1967). Cloud seeding and cloud physics—a review of activities in Israel. pp. 375–88 in *Water for Peace*, U.S. Government Printing Office, Washington D.C., Vol. 2, vi + 970 pp., illustr.

OLIVIER, H. (1967). Irrigation as a factor in promoting regional development. pp. 266–81 in *Water for Peace*, U.S. Government Printing Office, Washington D.C., Vol. 7, vi + 1004 pp., illustr.

OLSON, S. R., WATANABE, F. S., CLARK, F. E. & KEMPER, W. D. (1964). Effect of hexadecanol on evaporation of water from soil. *Soil Sci.*, **97**, 13–18.

PENMAN, F., MINASHINA, N. G., KONONOVA, M. M. & MALINKIN, N. P. (1967). Some effects of irrigation and drainage on soils. pp. 509–62 in *Irrigation and Drainage of Arid Lands*. F.A.O./UNESCO, Paris, 663 pp., illustr.

PENGRA, R. F. (1952). Estimating crop yields at seeding time in the Great Plains. *Agron. J.*, **44**, 271–4.

PERRIN DE BRICHAMBAUT G. *See* BRICHAMBAUT, G. PERRIN DE.

POLAND, J. T. (1960). Land subsidence in the San Joaquin Valley, California, and its effect on estimates of groundwater resources. *Int. Ass. Sci. Hydrol.*, **52**, 324–35.

PRICE, R. (1956). Possibilities of increasing and maintaining production from grass and forest without accelerating erosion. pp. 233–44 in *The Future of Arid Lands* (Ed. G. F. White) Amer. Assoc. Adv. Sci., Washington D.C., xi + 453 pp. illustr.

QUINTELA, R. M. (1967). Reduction of evaporation from free water surface. pp. 175–83 in *Water for Peace,* U.S. Government Printing Office, Washington D.C., Vol. 2, vi + 970 pp., illustr.

REIFENBERG, A. (1953). The struggle between the desert and the sown. Pp. 378–91 in *Desert Research Proc.* Research Council of Israel, Jerusalem, 641 pp., illustr.

RIESBOL, H. S., MILLIGAN, C. H., SHARP, A. L. & KELLY, L. L. (1967). Surface water supply and development. pp. 52–66 in *Irrigation of Agricultural Lands* (Ed. R. M. Hagan, R. H. Haise & T. W. Edminster) Agronomy series; Amer. Soc. Agron., Madison, Wisconsin, **11**, xxxi + 1180 pp., illustr.

ROBINSON, T. W. (1967). The effect of desert vegetation on the water supply of arid regions. pp. 622–29 in *Water for Peace,* U.S. Government Printing Office, Washington D.C., Vol. 3, vi + 968 pp., illustr.

SCHIFF, L. & MUCKEL, D. C. (1967). Ground-water recharge and storage. pp. 92–103 in *Irrigation of Agricultural Lands* (Ed. R. M. Hagan, R. H. Haise & T. W. Edminster) Agronomy series; Amer. Soc. Agron., Madison, Wisconsin, **11**, xxxi + 1180 pp., illustr.

SCHLEUSENER, R. A. & GRANT, L. O. (1967). Weather variation and modification. pp. 40–47 in *Irrigation of Agricultural Lands* (Ed. R. M. Hagan, R. H. Haise & T. W. Edminster) Agronomy series; Amer. Soc. Agron., Madison, Wisconsin, **11**, xxxi + 1180 pp., illustr.

SHOTTON, F. W. (1954). The availability of underground water in hot deserts. *Water & Water Engng*, **249**, 218–26.

SHUVAL, H. I. (1967). Public health implications of waste-water utilisation. pp. 690–701 in *Water for Peace*, U.S. Government Printing Office, Washington D.C., Vol. 3, vi + 968 pp., illustr.

SPORN, P. (1966). *Fresh Water from Saline Water.* Oxford University Press, London, x + 35 pp., illustr.

STAMP, L. D. (1961). Introduction. *Arid Zone Res.*, **17**, 17–24.

STEPHENSON, R. E. & SCHUSTER, C. E. (1946). Straw mulch for soil improvement. *Soil Sci.*, **61**, 219–30.

STOREY, H. C. (1960). Increasing water yields by forest management. pp. 107–18 in *Water and Agriculture* (Ed. R. D. Hockensmith) Amer. Assoc. Adv. Sci., Washington D.C., 198 pp., illustr.

THOMAS, H. E. (1951). *The Conservation of Ground Water.* McGraw-Hill, New York, 327 pp., illustr.

—— (1955). Underground sources of water. pp. 62–78 in *Water.* Ybk Agric. U.S. Dept. Agric., Washington D.C., xiii + 751 pp., illustr.

—— & PETERSON, D. F., JR. (1967). Groundwater supply and development. pp. 70-84 in *Irrigation of Agricultural Lands* (Ed. R. M. Hagan, R. H. Haise & T. W. Edminster) Agronomy series; Amer. Soc. Agron. Madison, Wisconsin, **11**, xxxi + 1180 pp., illustr.

TIXERONT, J. (1956). Water resources in arid regions. pp. 85–113 in *The Future of Arid Lands* (Ed. G. F. White) Amer. Assoc. Adv. Sci., Washington D.C., ix + 453 pp., illustr.

TODD, D. K. (1961). The salt water intrusion problem in coastal aquifers. *Symposium on Water Improvement* (Ed. J. A. Schufle & J. E. Fletcher) American Association for the Advancement of Science, Denver, Colorado, 73 pp., illustr.

U.N. (1964). *Water Desalination in Developing Countries.* United Nations, New York, viii + 325 pp., illustr.

U.N. (1967). *Studies on Selected Development Problems in various Middle East Countries.* United Nations, New York, iv + 87 pp.

U.S. DEPARTMENT OF THE INTERIOR (1966). *The A-B Seas of Desalting.* U.S. Dept. Int., Washington D.C., 24 pp., illustr.

U.S. GEOLOGICAL SURVEY (1951). Water law with special reference to ground water. *Circ. U.S. Geol. Surv.*, **117**.

WEINBERGER, L. W. & STEPHAN, D.C. (1967). Technology of advanced waste treatment. pp. 547–56 in *Water for Peace*, U.S Government Printing Office, Washington D.C. Vol. 2, vi + 970 pp., illustr.

WHITE, G. F. (1962). *The Changing Role of Water in Arid Lands.* Riecker Lecture 6, November 1960. Univ. of Arizona Press, Tuscon, 15 pp.

WIENER, A. (1966). The desalting plant as a component within a water-supply system. pp. 225–49 in U.N. Proc. International Seminar on the *Ecomomic Application of Water Desalination*, Athens, Greece, 367 pp. (mimeographed).

WIESE, A. F. & ARMY, T. J. (1960). Effect of chemical fallow on soil moisture storage. *Agron. J.*, **52**, 612–13.

WILCOX, L. V. & DURUM, W. H. (1967). Quality of irrigation water. pp. 104–19 in *Irrigation of Agricultural Lands* (Ed. R. M. Hagan. R. H. Haise & T. W. Edminster) Agronomy series; Amer. Soc. Agron. Madison, Wisconsin, **11**, xxxi + 1180 pp., illustr.,

WILLIS, W. O., HASS, H. J. & ROBINS, J. S. (1963). Moisture conservation by surface or subsurface barriers and soil configuration under semi-arid conditions. *Proc. Soil Sci. Soc. Amer.*, **27**, 577–80.

YARON, D. (1966). *The Demand for Water by Israeli Agriculture.* The Hebrew Univ. of Jerusalem, Faculty of Agriculture, Rehovot, 211 pp. (mimeographed).

ZINGG, A. W. & HAUSER, V. L. (1959). Terrace benching to save potential runoff for semi-arid land. *Agron. J.*, **51**, 289–92.

—— & WHITFIELD, C. J. (1957). Stubble-mulch farming in the western States *Tech. Bull. U.S. Dept. Agric.*, **1166**, 56 pp., illustr.

VII

WATER IN THE SOIL–PLANT–ATMOSPHERE SYSTEM

The availability of soil moisture, its uptake by the plant, its translocation through the plant, and its evaporation into the atmosphere, are various steps in the transfer of water through the soil–plant–atmosphere system. Hence plants, in their relationship to water, are influenced by the interactions of at least three components of the system, and not by the individual values of each component viewed separately (Lemon *et al.*, 1957).

Water Movement in the Soil

The movement of water in the soil is the result of differences in the forces with which it is held in different parts of the soil.* The water always moves along a gradient from 'regions of higher to regions of lower free energy'. This movement may be lateral, upwards, or downwards—depending on whether the suction forces are greater than the gravitional forces, or *vice versa*.

Water movement in the soil comprises three phases (Miller & Klute, 1967): infiltration, the phase during which water is being applied; redistribution, the phase during which it is achieving equilibrium; and withdrawal, consisting mainly of evaporation at the soil surface, and transpiration, which affects the entire root-zone.

INFILTRATION

After water enters a dry soil, it is subject to attractive forces and to gravity. At first it moves downwards, fairly rapidly filling all the pore spaces and

* In the past the terms gravitational, capillary, and hygroscopic, water were used regularly to describe the water in soils at different stages of wetting. These terms have been largely abandoned in recent works, as they imply sharp distinctions (which scarcely exist) between the different stages of wetting, or even qualitative differences between kinds of water.

saturating the soil to a depth which depends on the amount of water applied and on specific soil characteristics.

REDISTRIBUTION AND EQUILIBRIUM

Expressed in its most simple form, redistribution of water and equilibrium generally occur as follows:

After from one to three days,* the moisture films covering the soil particles become so thin that moisture movement becomes negligible, and drainage practically ceases. The moisture content at this stage is called 'field-capacity' (FC).† At field-capacity, there is a fairly distinct boundary between the wetted part of the soil and the dry soil underneath. It must be stressed that the above applies only to a *well-drained soil*. When an impermeable layer is present near the soil surface, or the ground-water table is high, the root-zone may remain waterlogged for long periods. Another factor affecting the precision of the field-capacity concept is extraction of water by the growing crop, during the period of time between the application of water and the attainment of field-capacity in the soil. This affects the rate of downward flow and losses beyond the root-zone, and cannot be considered as a negligible influence within reasonable times after irrigation (Miller, 1967).

WITHDRAWAL

Soil-water is lost from the soil by evaporation from the soil surface and through transpiration by the plants, until a moisture content is reached at which the leaves of a test plant growing in the soil will wilt and fail to regain turgidity in a saturated atmosphere. This moisture content is called *wilting point* (WP) or *permanent wilting percentage*, and has generally been found to correspond to a tension of approximately 15 atmospheres.

The amount of water held in the soil between field-capacity and wilting point generally represents the reservoir of water available for plant use.‡ The capacity of this reservoir will depend on the depth of the soil, its texture, and its structure. For each crop, in turn, it will depend on the depth and extent of the rooting system. This will depend not only on the genetic characteristics of the crop, but also on the presence of hard-pans, poor drainage, toxic substances, nutrient deficiencies, etc.

* There are soils in which it is difficult to determine the time when downward movement has ceased.

† Field-capacity is defined as the moisture content of a deep, permeable, well-drained soil several days after a deep wetting.

‡ A certain amount of water may be taken up by plants from soils with water contents above FC and also below WP (Vaadia *et al.*, 1961).

Wilting point is not reached at the same time throughout the root-zone. When plant roots remove water from the soil, moisture in the immediate vicinity of the roots may be depleted to wilting point very rapidly. The high soil-moisture tension that develops in the soil enveloping the roots creates a tension gradient, as a result of which there is a tendency for water to move towards the roots. This movement is, however, extremely slow, and limited to a few centimetres. The slow rate of movement of the soil-water towards the roots may frequently cause a lag in water uptake in relation to transpiration demands, and as a result plants may wilt temporarily—especially during the hot hours of the day.

In order to maintain a satisfactory rate of water intake that will be sufficient for sustained growth, the roots, especially those of young plants, must grow actively towards the moist regions. The soil may be at field-capacity at a distance of a few centimetres from a zone that has been depleted to wilting point by the roots. As the root system becomes more and more extensive and completely permeates the root-zone, the soil moisture content is reduced to the wilting point throughout progressively expanding root-zones.

VAPOUR MOVEMENT

Water also moves through unsaturated soils by vapour transfer and as a result of temperature differences, the gradient being from warm to cold areas. Soil air, under natural conditions, is always saturated with water vapour—except in the uppermost $\frac{1}{2}$–1 cm layer of soil. Below a depth of 5–10 cm, the vapour pressure is generally greater than that of the atmosphere (Lebedeff, 1928). During the day, much of the solar energy is absorbed by a shallow layer of soil, which is therefore warmer than the atmosphere above and the underlying soil layers. As a result, water vapour will move upwards from the surface layer into the atmosphere (evaporation), and downwards into the colder, dry layers, where the moisture condenses.

By night, the opposite movement occurs in the soil. Vapour moves from the lower layers and condenses near the still colder soil surface. When skies are clear, radiation by night may cause vapour movement from the soil to the atmosphere; the vapour then condenses on cold surfaces in the form of dew.

Condensation from the air on to the soil surface may also occur under appropriate conditions: the absolute humidity of the atmosphere is occasionally greater than the vapour pressure in the surface layer; atmospheric moisture then penetrates into the soil and condenses. In the Odessa region this may occur on 200 nights out of the year, with an average condensation of 0·26 mm in a single night, totalling 72 mm per year; part of this moisture is transmitted during the day to lower soil layers (Lebedeff, 1928).

There are conflicting opinions as to the importance of vapour movement as a source of water-supply to the plants. Whilst water in the liquid form may amount to 30 per cent of the weight of the soil, the weight of water vapour is rarely in excess of 5 p.p.m. of soil weight (Day *et al.*, 1967). However, the relative importance of vapour movement increases as soil moisture is depleted. In dry soils, it is probably the major form of water transfer (Slatyer, 1960). Therefore, according to Lebedeff (1928), vapour movement may make significant contributions to the water-supply of plants in semi-arid regions—especially at moisture contents near WP.

In the Egyptian desert, Migahid (1962) found that the moisture content of the surface soil increased every night by 0·1 to 1·0 per cent of the dry-weight of the soil (a quantity equivalent to 0·4–2·2 per cent of the saturation capacity). This slight increase barely raised the water content of the soil above WP, and only for a few hours nightly in the top 2 cm of the soil.

Generally, the main effect of vapour movement is to contribute to the loss of water through evaporation from the soil, as the water vapour pressure in the soil atmosphere is generally greater than that of the atmosphere above the soil. Losses by evaporation are usually confined to the surface few centimetres of the soil; vapour loss from layers deeper than 15–20 cm is usually negligible (Richards & Wadleigh, 1952).

Availability of Water to Plants

The availability of water to the crop depends on a combination of soil, crop, and climatic, factors.

SOIL FACTORS

Water is held in the soil in the same way as in a porous matrix. The pores between the soil particles, and between their aggregates, are linked by narrow passages having some similarity to capillary tubes. The water fills the interstices between the soil particles and is held as a film on the soil particle surfaces. The area of solid surface of the soil particles can range from 1000 cm^2/gm of soil in coarse sands to 1 000 000 cm^2/gm for clay soils (Day *et al.*, 1967). Amongst the different kinds of clay minerals, montmorillonite absorbs more water than illite, which in turn absorbs more than kaolinite (Holmes *et al.*, 1967). Moderate compaction of the soil increases its water-storage capacity; very loose soils have a very high air content but a relatively low water-storage capacity (Jamison, 1956).

The water is therefore held in the soil by capillary forces and by physical adsorption—the former are dominant in wet soils, whilst the latter is dominant after part of the water has been removed from the interstices and only the films remain on the soil particle surfaces. Work is required to remove the water from the soil to offset these forces of attraction. Various parameters are used to define the work needed for this purpose, of which the most recent is the *matric potential** and the most widespread is the *soil moisture tension.* †

A second factor, which increases the amount of work required from the plant to remove water from the soil, is the concentration of salts in the soil solution. This is not usually an important factor in humid climates; but in arid conditions, especially in saline soils, the increased osmotic concentration may play an important role in increasing the work needed to remove water from the soil. The concentration of soluble salts in the soil solution tends to increase in arid climates as a result of the high rates of evaporation and transpiration. The total amount of work required of the plant to remove water from the soil, against the combined effects of soil moisture tension and osmotic potential, is measured by a number of parameters such as *total water potential,*‡ *total soil moisture stress, diffusion-pressure deficit*§ (expressed in atmospheres or *soil pF*¶), and *free energy deficit.*

In experiments carried out by Wadleigh *et al.* (1965), it was shown that the effect of total soil-moisture stress on growth was the same, irrespective of whether this stress was due predominantly to soil moisture tension or to osmotic pressure. However, a high osmotic pressure of the soil solution, in addition to its effect on water availability, also has undesirable indirect effects on the plants—by reducing root growth and the permeability of roots to water (Kramer, 1959).

The amount of work required to remove water from the soil increases progressively as the soil dries. At 'field capacity' the water is held at a soil moisture tension of approximately 1/10 atmosphere (Russell, 1959). As plants take up water from the soil, the water is first removed from the wider pores and capillaries, in which it is replaced by air. Progressively finer capillaries and pores contribute their water-supply, until the remaining water is held so strongly by the adsorption forces of the soil particle surfaces that many plants are practically unable to remove the remaining water.

* *Matric potential* (previously called 'capillary potential') measures the attractive forces resulting from the interaction of soil particle surfaces with water (expressed in ergs/gram, using the potential of free water as a base of reference).

† *Soil moisture tension* is the negative pressure of soil-water, expressed in atmospheres of pressure which must be applied to remove water at a given soil moisture content.

‡ *Total water potential* is the difference in free energy between the water contained in the soil (or the plant system) and pure free water at atmospheric pressure. It has the advantage that it can serve as an index of water status of both soil and plants.

§ *Diffusion-pressure deficit* is the amount by which the diffusion pressure of water in the soil is less than that of pure water, expressed in atmospheres. It is a measure of the difference between the free energy of the soil moisture and that of free water.

¶ *Soil pF* is the logarithm of the metric potential or the free energy when the latter is expressed in centimetres of water.

A distinction has been made between the wilting point—which is the moisture content at which the plant ceases growth—and the *ultimate wilting point,* at which the plant can no longer maintain life. The differences between the two is called the *wilting range* (Furr & Reeve, 1945). In most soils, the wilting point is reached when soil moisture tension reaches 15 atmospheres.

Soil texture affects in two ways the amount of available water that can be stored in the soil: it determines the total water-holding capacity, and it influences the tenacity with which water is held in the soil. The water-holding capacity of coarse-textured soils is less than that of more finely-textured soil. Soil texture is important in determining how near to the WP plants can withdraw moisture without suffering setbacks in growth. In a light-textured soil, 80–90 per cent of the water is held at tensions of less than 1 atmosphere, whilst only 10–20 per cent is held at tensions of 10–15 atmospheres. By contrast, in a clay soil the tension increases markedly as moisture content decreases (Kelley, 1954). As a result, in sandy soil the moisture content can be reduced to near WP without affecting plant growth, whilst effects of soil moisture tension on plant growth will make themselves felt at moisture levels above WP in the heavy soil (Figure 30).

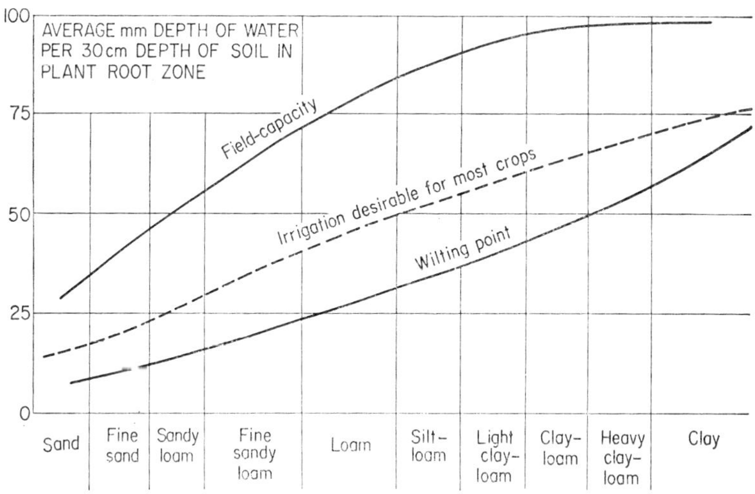

Figure 30. Typical water-holding characteristics of different textural soils. (By courtesy of the U.S. Department of Agriculture.)

AERATION

An insufficient air-supply to the roots may cause water absorption by them to decrease or to be stopped; hence the paradoxical situation whereby plants may suffer from water deficits in waterlogged soils. A high concentration of carbon dioxide has an even more drastic effect on water absorption than does a low concentration of oxygen.

CROP FACTORS

For many years, the important question of crop response to various levels of soil moisture has been debated by research workers; one view holds that soil moisture is equally available to plants over the whole range from FC to WP, whilst the opposing view contends that plant growth is adversely affected as soon as the soil moisture stress is increased beyond a certain level, long before the WP is reached. In a review of 80 research papers dealing with crop response to soil moisture régimes, the growth of crops was found to respond to differences in soil moisture levels within the available range in 66 cases; in most cases, yields declined as moisture stress increased. In the minority of studies, in which no significant effects of increased moisture stress up to wilting point were found, perennial crops and fruit crops were those generally involved (Stanhill, 1957). It is now widely agreed that soil moisture becomes progressively more limiting to plants as soil moisture-stress increases. However, the response of plants to soil-water régimes depends not only on soil moisture but also on plant factors and weather. It is *primarily* a function of the water status of the plant tissues, and only indirectly of the moisture régime of the soil.

The water status of the plant tissues depends on:

(a) Resistance to flow of water in the soil, which varies with water content;
(b) Resistance to flow from the stomata into the atmosphere, which varies with atmospheric conditions; and
(c) Resistance to flow within the roots and other tissues of the plant, which depends on physiological factors and is by no means constant. Hence it is usually impossible to predict the internal water status of a plant from conditions in the soil alone or the atmosphere alone, and it is possible for a plant to wilt with abundant water in the soil and to remain turgid when the soil is relatively dry (Gardner, 1964).

The principal plant factors which affect the availability of water are rooting characteristics and the ability to withstand an adverse water-balance.

ROOTING CHARACTERISTICS

Both root proliferation and depth of penetration are important characteristics. Roots may have an enormous surface area in intimate contact with soil particles.

When a major portion of the plant's root system is subject to water-stress at the WP, permanent injury to the plant may result; the degree of injury and the rate of recovery will depend mainly on the length of time that elapses before water is added to the soil, and on the proportion of the root system that is in soil with a moisture content above the WP. When the plant has an extensive and efficient root system, permeating a deep, well-structured soil, it can, under favourable climatic conditions, maintain an optimum supply of moisture to its aerial parts until part of the root-zone reaches WP, without growth being markedly affected.

This combination of circumstances is the exception rather than the rule, especially in arid climates. Young plants, with an underdeveloped rooting system, will be adversely affected at a level of soil moisture at which full-grown perennial plants, such as most grasses, remain unaffected. The latter have a wide and deep root system which thoroughly permeates the soil to a considerable depth. Therefore, when part of the root-zone is already at WP, other parts will continue to supply water from soil that is still moist. As water movement in an unsaturated soil is extremely slow, a continuous supply of water to the plant will depend largely on the growth of the roots towards the water. Crops whose root system develops rapidly will be able to explore the soil for moisture and will therefore be at an advantage over crops whose root systems grow slowly. The latter will suffer as soon as the water-supply in the soil which is in intimate contact with their roots is exhausted. Root elongation decreases as a soil becomes water-deficient (Gingrich & Russell, 1956), so that not only does it become progressively more difficult for the roots to extract water, but their ability to seek it is also affected adversely.

Deep penetration by their roots provides many crops with a continued supply of soil moisture, even when the upper layers, which usually dry out between rains or irrigations, have already reached WP. However, there are great differences between different crops, and even between different varieties of the same crop, in the characteristics of their rooting systems.

TABLE VII

Characteristics of Rooting Systems of Three Crops (Dittner, 1938)

Crop	Root surface area (in^2) per in^3 of soil	Number of root-hairs per in^3 of soil
Oats	15	150 000
Rye	30	300 000
Kentucky Bluegrass	65	1 000 000

In the examples presented in Table VII, Kentucky Bluegrass should be the most efficient, and oats the least efficient, in extracting moisture from the soil.

Plant growth may be retarded by the drying out of the upper soil layers, even when there is ample moisture available in lower layers of the root-zone, in cases in which some essential element is concentrated in the dried-out layer and therefore becomes unavailable to the plants.

There are indications that certain plants are capable of depleting water from the soil to much drier levels than the WP, indeed to an ultimate extent that is equivalent to several hundred atmospheres of stress (Slatyer, 1958).

Wheat plants with a well-developed root system were found to be capable of absorbing water at tensions greater than 26 atm. (Haise *et al.*, 1955). The moisture absorbed at levels below WP was not effective in maintaining vegetative growth but affected the yield and quality of the grain (Kozlowski, 1964).

Therefore, in considering the amount of water available to native plants in arid soils, the WP is of little significance as a lower limit of soil moisture. The lowest limit of availability will depend on the species involved and on the prevailing conditions (Figure 31).

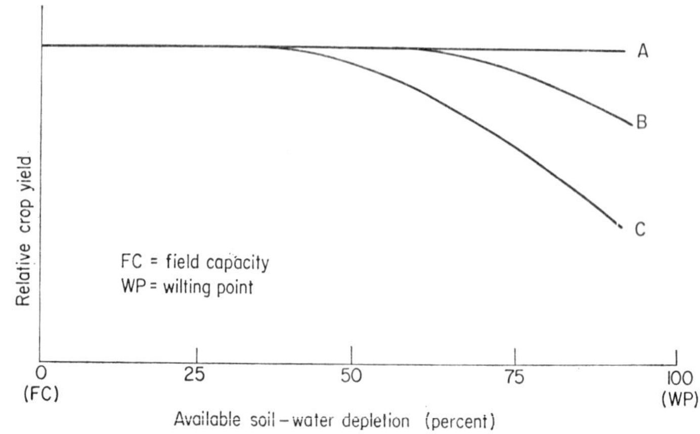

Figure 31. Various crop-yield responses to soil moisture depletion. The three curves represent respectively:
A: cases where moisture is equally available over the entire available moisture range during each irrigation cycle, so that no appreciable yield increases are obtained by irrigating before soil-moisture content falls to near wilting point.
B: cases where yields diminish as soil-water stress begins to increase rapidly.
C: cases in which high moisture stress values, even during brief time intervals, have considerable effect on crop response.
(From Hagan (1957): by courtesy of *California Agriculture*.)

A further possibility raised by Slatyer (1958) is that moisture may even be absorbed by plants from an unsaturated atmosphere, as long as there is a favourable diffusion-pressure deficit gradient. This is of special significance in semi-arid and arid regions, where high humidity levels and dewfall are frequent at night and may be considered of importance for survival, if not for growth.

Weather

The desiccating effect of the atmosphere is the main factor influencing the water needs of the plant. The plant is a water-conducting system between the soil and the atmosphere. The overall free-energy difference of the water moving into, through, and out of, the plants is from 0·1 to 15 atm in the soil, and about 1000 atm in the atmosphere, when relative humidity of the air is 47 per cent (Richards & Wadleigh, 1952).

The sensitivity of plants to increasing soil-moisture stress increases under conditions of high temperature, low relative humidity, high wind velocity, and increased light intensities—all conditions which increase the transpiration rate considerably.

Therefore, on a hot dry day, the rate of transpiration is so much in excess of water uptake that plants will wilt at a soil moisture content at which on a cooler or more humid day they would not have shown any signs of stress. For each day, there is therefore a stress-point, which is the moisture content of the soil that is required to keep plants from being subjected to moisture stress under the prevailing atmospheric conditions. Days on which a moisture content is reached that is lower than the stress-point are called 'stress days'.

In maize, there is a linear relation between dry-matter production and the number of stress days (Shaw & Burrows, 1966), indicating that the relative water availability fluctuates according to weather conditions as well as to the soil moisture content.

Summing up, it may be said that the growth and development of plants are influenced by their internal water balance, which in turn depends on both water intake and water expenditure by the plant. Therefore, water availability of the plant will not depend on soil moisture conditions alone but also on atmospheric conditions and plant characteristics. The work required to withdraw water from the soil increases gradually as soil moisture is progressively depleted—until near WP, when, in certain soil types, it increases steeply. There are plants that are able to reduce their transpiration rates when atmospheric conditions are conducive to high water losses or when soil moisture becomes limiting, and that have an extensive and efficient root system. For these, the gradual increase in the work required to remove water from the soil against the increasing tension as soil moisture is progressively depleted, will not have an adverse effect on yields until WP is reached. This is especially so if the marketable part of the crop is not particularly sensitive to moisture stress.

In most crop plants, however, which are generally mesophyllic in nature and are less able to maintain a favourable water balance under adverse atmospheric conditions, the considerable amount of work required to remove water from the soil, as it is gradually depleted, will definitely affect growth and development and, therefore, crop yields. For these crops, maximum yields are generally dependent on maintaining the soil moisture between the limits of FC and some critical range of soil moisture that is higher than the WP. This critical range will depend on the type of crop and its stage of development, on the nature of the marketable produce, on soil characteristics, and on the environmental conditions under which the crop is grown.

Properties of Water and Its Role in Plants

PROPERTIES

Water has a unique combination of characteristics which make it biologically most outstandingly important (Kramer, 1959). Its specific heat, latent heat of melting, and heat of vaporization, are very high. Therefore the evaporation of water has a strong cooling effect, and the condensation of water vapour has a warming effect.

Water has a high density, high surface-tension, and considerable tensile strength. It is adsorbed by surfaces, such as clay and cellulose, which have many oxygen atoms with which hydrogen bonds can be formed.

Water is also an excellent solvent for electrolytes and numerous other substances.

ROLE

Water is essential to the life and growth of plants: it is the main *constituent* of the protoplasm, and makes up 85–95 per cent of the fresh-weight of most of their green tissues. It is involved in essential *physiological processes* of the plant, such as photosynthesis and various metabolism reactions. It is essential for the maintenance of plant *turgor*, on which depend normal growth and the maintenance of the form and mechanical strength of various plant organs.

Water–Soil–Plant–Atmosphere Relationships

Water movement in relation to plant growth involves three phases: water uptake by the plant, water movement through the plant, and transpiration by the plant.

As water is generally free to move across the plant–soil, soil–atmosphere, and plant–atmosphere, interfaces, Philip (1966) stresses the importance of considering the water transfer system in the soil–plant–atmosphere continuum as a whole. However, in order to present a clear picture of the system, it is necessary to describe its individual components.

WATER UPTAKE

Mechanism of Water Uptake by Plants

Cultivated plants absorb water from the soil mainly through their roots and root-hairs. Some details on root proliferation and the number of root-hairs have already been given (pp. 182–3). The root system of most crop plants presents an enormous surface area that is active in water absorption from the soil. Water moves from points of high water potentials to points of low water potentials.

The plant's water potential has three components: turgor pressure, inhibitional pressure, and osmotic pressure. Increased turgor pressure increases the total water potential, whilst increased inhibitional and osmotic pressures have the opposite effect.

Roots absorb water both passively and actively.

Passive Absorption. This occurs when water is drawn into the roots by negative pressures in the conducting tissues, resulting from transpiration. It has long been known that the rate of loss of water by transpiration has a great influence on the rate of absorption through the roots (Kramer, 1939). As evaporation takes place from the leaves, there is a general lowering in the water potential of the leaves, and the resulting potential gradient is the driving force of the passive absorption of water from the soil. Water then moves from the soil surrounding the roots, along the water potential gradient through the plant, and so into the atmosphere.

Active Absorption. Under conditions during which there is little transpiration loss and yet ample soil moisture is available to the plants, the roots of many plants are known to absorb water actively (Slatyer, 1960). The water is then exuded into the stem under pressure. The main mechanism involved is osmosis: the higher the osmotic pressure of the fluid in the tissue, the greater the ability of the tissue to absorb water from its surroundings.

Under conditions of normal high transpiration, the contribution of active absorption to the water-supply of the crop is negligible, as compared with passive absorption, and is usually less than 10 per cent of the total water absorption of the plant (Kramer *et al.*, 1967).

Aerial Absorption of Water. It has been found that certain plants, growing in soils at wilting point, are able to absorb moisture from the atmosphere. It is therefore assumed that plants with tissues that are mechanically sufficiently strong to resist collapse, are capable of such so-called 'negative transpiration' (Stone *et al.*, 1950).

Direct absorption of water by leaves that are wetted by rain, dew, or overhead irrigation, can help to resaturate dehydrated leaf tissues; but proof of a real contribution to the amounts of water absorbed by the plants, causing greater growth or productivity, is lacking (Slatyer, 1967).

The main contribution of surface water on the leaves is apparently in reducing transpiration somewhat, thereby reducing the water deficit and prolonging survival (Slatyer, 1967).

Environmental Factors Affecting Water Uptake

Atmospheric Factors. All factors affecting transpiration (*see* p. 191) indirectly affect water uptake, as they affect the plant's need for water and consequently the demands which it makes on the local supply.

Soil Factors. The subject of the availability of soil-water has already been discussed on pp. 179–86.

The temperature of the soil has a major effect on the rate of water uptake by the roots; low soil temperatures reduce the absorption rate of water. Watering plants with cold water may actually cause wilting (F. E. Clements & Mashon, 1934).

The decrease in water uptake at low soil temperatures is probably due to the increased viscosity of the water, which impedes its movement through the soil, into the roots, and through the roots into the plants (Curtis & Clark, 1950). The permeability of cells to water also decreases, as a rule, at lower soil temperatures. These effects are compounded by the reduced growth of roots at low temperatures; not only is absorption then more difficult, but the absorbing root surface is also decreased (Kozlowski, 1964).

Plants differ in their sensitivity to soil temperatures. Water absorption is more reduced at low temperatures in warm-weather crops such as cotton or maize, than in typical cool-season crops, such as the small grains. The temperature below which water absorption by warm-weather crops becomes limiting is approximately 20°C (Kramer *et al.*, 1967).

As regards the osmotic condition in the root-zone it may be noted that, in arid regions, salt accumulation in various degrees is the rule. Even a small increase in the osmotic pressure of the soil solution may cause relatively large decreases in water uptake (Kozlowski, 1964).

WATER MOVEMENT THROUGH THE PLANT

Water movement through the plant is facilitated by the presence of specialized elongated cells (tracheids) or rows of cells with the ends broken down and forming a more or less continuous system of 'vessels'. Tracheids and/or vessels constitute the main functional components of the xylem, through which water can flow with a minimum of resistance, from the roots, through the stem, and into the leaves. This is the situation in all normal vascular and other 'higher' plants, including the seed plants or Spermaphyta to which virtually all crop plants belong.

The flat leaves of most crop plants are capable of absorbing considerable amounts of solar radiation and of exchanging gases with the atmosphere. Between the leaf-cells are numerous interconnected air spaces which lead to the atmosphere through small openings in the leaf—the stomata. When the water vapour content of the atmosphere in contact with the leaf is lower than that in the air spaces within the leaf, water diffuses from the leaf into the surrounding air; the rate of this transpiration will be proportional to the difference in concentration of water vapour between the two media. Most of the water that is lost to the atmosphere by transpiration escapes through the stomata, whilst relatively little passes directly through the cuticle of the leaf. However, under conditions of water-stress, when the stomata are closed, transpiration through the cuticle may become an important factor in water-loss (Milthorpe, 1960).

As the leaves become dehydrated by transpiration, a chain reaction occurs: the turgor of the leaf-cells, and consequently the cell volume, are reduced and the water potential in the leaves decreases. The resulting energy gradient causes water to be absorbed from the water-conducting tissues. The resulting reduction of pressure on the water extends right through the conducting system, to the roots and into the soil. Water is absorbed by the roots and moves upwards through the vessels and/or tracheids of the stem, and finally into the atmosphere.

The two processes of transpiration and water absorption are closely correlated, the transpiration by the leaves producing the tension which causes passive water absorption. However, because of the resistance to the movement of water through the roots, the xylem, and the leaf, absorption tends to lag behind transpiration (Kramer, 1959).

Such lags in absorption, due to the internal resistance to water transport within plants, are common even at high levels of soil moisture. The lag is aggravated as soil moisture-tension increases or soil temperature is reduced, or when environmental conditions favour a high rate of transpiration.

In order for the water to ascend to the top of the plant, it must have sufficient cohesion to withstand the tension needed to pull it upwards. When water is confined to small, completely wettable tubes, such as the xylem elements, it can be subjected to a pull of many atmospheres and still maintain its cohesion (Dixon, 1924).

TRANSPIRATION

Role

There are two schools of thought regarding the role of transpiration. The one considers transpiration as a *necessary* evil, the other as an *unavoidable* evil. Transpiration is to some extent an evil, because inevitably absorption of water lags behind transpiration, causing water deficits in the plant tissues—with adverse effects on photosynthesis and on growth, with disruption of metabolic processes, and in extreme cases with resultant death of the plants.

Transpiration is considered a necessary evil, because the foliage of a crop intercepts a quantity of radiation that is far in excess of its ability to utilize effectively. Leaves subjected a heavy radiation load must be able to dissipate the excess energy quickly, or else their temperature would rise to lethal levels within less than one minute (Idso *et al.*, 1966). This dissipation of excess heat is made possible by three mechanisms: thermal radiation of the leaf, removal of heat by convection currents, and transpiration. The literature is full of conflicting claims regarding the relative importance of these mechanisms in dissipating excess heat from leaves.

One group of authors maintains that, for most cultivated crops, the transpiration of large quantities of water is the main mechanism for this purpose (Thornthwaite & Mather, 1954), each gram of water evaporated removing the energy equivalent of 580 calories.

Another school of thought states that the role of transpiration in cooling plants has little practical significance. Many research workers have shown that there is little relationship between leaf temperature and the rate of transpiration. The temperature of wilted leaves was found to increase by only 2–5°C, when transpiration was slowed down. For them, the main factor in preventing an excessive rise in leaf temperature is the removal of heat by air movement (Ansari & Loomis, 1959).

There is no doubt that transpiration can be effective in dissipating excess heat; when leaf temperature approaches a lethal level, transpiration—by reducing leaf temperature by only a few degrees—may swing the difference between survival and death (Gates, 1964). When transpiration is reduced, convective cooling increases; only under extreme conditions of high radiation-incidence and very low velocity, would there be a danger of excessive leaf temperatures as a result of reduced transpiration (Gale & Hagan, 1966). Moreover in hot, arid regions, the stomata of plants close during the hottest hours of the day, when the cooling effect of transpiration would be most needed.

Transpiration is considered an unavoidable evil, because the entrance of carbon dioxide into the leaf for photosynthesis is essential; a plant structure that makes this possible will inevitably make possible water-loss through transpiration. It has been shown that plants which have very low rates of transpiration also have low rates of photosynthesis and grow slowly (Hygen, 1953). Maximum rates of photosynthesis and growth are related to a maximum rate of transpiration (Penman, 1952). Transpiration can also have considerable effects on respiration rates and various biochemical activities of the plant (Gates, 1964); it is considered as the probable energy-source for the translocation of organic and inorganic materials between adjacent cells and within cells.

All this does not imply that it might not be possible to reduce transpiration to some extent, and without adverse effects. For example, it has been found that small increases in carbon dioxide concentration can cause large decreases in stomatal apertures, and in transpiration, whilst net photosynthesis was increased (Moss *et al.*, 1961).

Environmental Factors Affecting Transpiration Rate

The factors which influence evaporation from a moist surface have a similar effect on transpiration. These factors are radiation, air humidity, temperature, and wind. In addition visible light, which has no direct effect on evaporation, affects the rate of transpiration by influencing the stomatal aperture (Kramer, 1959). Cloudy weather, especially during the cool season, causes decreased stomatal opening in most species that are native to sunny habitats (Wilson, 1948). The availability of soil moisture also modifies transpiration, with water deficits causing a rapid decrease in transpiration.

EVAPOTRANSPIRATION

Water is lost by evaporation from a bare, moist soil surface at about the same rate as from a free water surface having the same exposure and temperature.

Paradoxically, in arid regions, losses of water by evaporation from the soil are relatively low, the main component of water-loss being plant transpiration. The surface soil dries quickly, preventing water-loss from lower layers.

Capillarity and water vapour diffusion move rather little water to the surface from these moist layers. After the upper soil layer has become too dry to contribute to evaporation, plants continue to take up water from the lower moist layers. The high conductivity of the plants for water makes up for the reduced evaporation from the dry soil surface, and the plant cover will therefore maintain the level of water-loss for some time; the more dense the plant cover is, the greater will be the water-loss. However, as available water is exhausted, the decrease in availability counterbalances the effect of increased plant cover on water-loss. Therefore, after a while, the density of plant cover has no further effect on water-loss. By contrast, in irrigated soils, especially if light irrigations are given at frequent intervals, losses by evaporation from the soil may be considerable.

In field measurements it is difficult to distinguish between the two sources of losses; they are therefore usually estimated together and called evapotranspiration (ET), or consumptive use. When the water-supply is unlimited, evapotranspiration is equal to the evaporation from a free water surface, and therefore reaches the highest level possible under the prevailing conditions of radiation, wind velocity, temperature, air humidity, etc. It is then called *potential evapotranspiration* (PET).

Research on water requirements of crops has shown that, in a large number of cases, maximum yields can be achieved with rates of actual ET that are far lower than PET; in other words, maximum yields are possible even when the level of soil moisture is not constantly maintained at, or somewhat below, field-capacity.

Methods for Estimating Potential Evapotranspiration

Calculating the Energy Balance. This method of calculating potential evapotranspiration is based on the assumption that as transpiration is, in effect, the evaporation of water from the surface of a plant, transpiration rates are controlled by the same physical factors as those which control evaporation. Therefore, fundamentally, evapotranspiration can be considered as a physical phenomenon which requires energy to supply the heat of vaporization that is necessary for the process (Penman, 1948).

In addition, evapotranspiration can only take place when the atmosphere is not completely saturated; it is therefore dependent on a mechanism that will move water vapour from moist to drier levels in the atmosphere.

The principal factors affecting evapotranspiration are air temperature, atmospheric humidity, and cloudiness, whilst differences in reflection from the soil surface or plant cover are of minor importance. As a result, the rate of water-loss from an actively growing short green crop that completely shades the ground, when water is non-limiting, is about the same as that from a free water surface, being accordingly equal to PET (Penman, 1952).

Within certain limits, therefore, PET is independent of the type of crop or kind of soil on which the crop is growing, and is primarily determined by the prevailing weather. In consequence, according to Penman (1956) there should be no effective differences in water consumption of different irrigated crops at a given time and place, water consumption being dependent on the available energy.

By using the appropriate meteorological and aerodynamic data,* Penman has established a formula for calculating the energy balance, and deduces therefrom the values for PET (Penman, 1948).

Empirical Formulae based on Meteorological Data. A number of empirical formulae have been established for calculating PET, based on the following assumptions: (a) that there is a direct correlation between seasonal water requirement and temperature, day-length, and length of the growing period, and (b) that it is possible to disregard other climatological factors, such as wind velocity and direction, relative air humidity, light intensity, etc.

There is no doubt that the last-named factors influence the water requirement of crops. However, in attempting to establish a relatively simple formula, which would be practical and easy for determining water requirements, it was decided to use a minimum number of meteorological factors on which data are readily available.

In one method temperature and day-length were chosen and the other factors disregarded— because in any case the latter vary with temperature, and

* Daily mean sunshine hours, temperature, humidity, and wind velocity, at two-metres height.

transpiration is mainly active during the hours of daylight (Thornthwaite, 1948).

Blaney & Criddle (1962) were of the opinion that the water requirements of different crops are not identical, even under conditions of identical temperature and day-length. They proposed to determine experimentally an empirical coefficient K, which would be specific for each crop. The Blaney-Criddle formula for determining water requirement is:

$$U = K \times F$$

where U is the consumptive use of the crop (or ET) in inches for any period, K is the empirical coefficient, and F is the sum of monthly consumptive-use factors for the period (f): $f = t \times p$, where f = monthly consumptive-use factor, t = mean monthly temperature, and p = monthly percentage of daytime hours of the year. Computations of K for irrigated crops in arid and semi-arid areas have been found to be as follows: lucerne, 0·85; maize, 0·80; cotton, 0·65; pasture, 0·75; and rice, 1·20.

Recent work in Israel has shown that there is good agreement between the results of field experiments designed to determine optimum water requirements and the optima calculated according to the Blaney–Criddle formula, as shown in the examples in Table VIII.

TABLE VIII

Optimum Water Requirements (mm/growing-season) of Cotton at Two Locations in Israel.

Location	*As found experimentally*	*Acc. to Blaney–Criddle formula*
Negev	780	670–770
Esdraelon Plain	688	683–755

Equipment for the Estimation of PET. The most commonly-used apparatus for *estimating* PET is the so-called *Class A evaporation pan*, from which water losses are recorded daily.

Lysimeters are soil tanks especially designed for growing crops under what are essentially field conditions, the soil being kept at or near field-capacity, and in which water-loss can be measured directly by periodical weighing of the apparatus.

Measurements of ET from undisturbed plots of natural vegetation have been made by using a circular, frameless plastic tent through which air is blown. The difference in absolute humidity of the inflowing and outflowing air, multiplied by the ventilation rate, gives the ET rate (Decker *et al.*, 1962).

Comparative Value of Different Methods. An assessment of the comparative value of the different methods for measuring PET was made by Stanhill (1961).

He found that the most accurate results based on exact theory are obtained by calculating the energy balance according to Penman (1948). The estimates of energy balance are, however, based on the assumption that water is non-limiting. In arid regions, water is generally limiting: even irrigation is rarely planned so as to maintain non-limiting moisture conditions continuously.

When water is freely available, about 80 per cent of the net radiation is used for evapotranspiration; when water is not freely available, a greater proportion is used to heat the soil and the air (Shaw & Burrows, 1966).

In a humid region there is little vertical transfer of heat to the soil surface, so that when the soil is completely shaded with a well-watered crop, evapotranspiration is approximatively equal to net radiation. In arid regions, by contrast, net radiation does not give an accurate estimate of evapotranspiration, as considerable advective heat-transfer may occur (Tanner, 1960), especially when hot, dry winds are blowing. Thus in a large field of irrigated cotton in Texas, daily evapotranspiration was found to be 1·75 times greater than the net radiation (Lemon *et al.*, 1957).

The actual situation in arid regions it characterized by intermittent periods of low moisture supply and an almost continuous and high energy supply. Under these conditions the assumption that evapotranspiration is fundamentally a physical phenomenon, dictated exclusively by meteorological factors, must be qualified, and account must be taken of the response of plants to moisture stress and the restricted flow of water in unsaturated soil.

The generally most practical method, with satisfactory accuracy, is to use the Class A evaporating pan. Large weighing lysimeters are fairly accurate but expensive; yet in recent years their use for determining PET directly has become more widespread than formerly. Their main importance is to serve as an independent control on the suitability of other methods, and in particular in calculating empirical formulae used for estimating ET (Tanner, 1967).

The usefulness of the empirical meteorological methods (Thornthwaite, 1948; Blaney & Criddle, 1962), for determining seasonal water requirements or consumptive use, after they are adjusted to arid-region conditions, has been established beyond doubt. The information obtained can be used for planning large irrigation projects and individual farm layouts. The methods are also useful for planning the allocation of the water available on the individual farm to different crops, in order to ensure its most efficient and economical use. An important drawback of the methods for calculating the irrigation needs of a crop on the basis of weather data is that these methods do not take into account the differential response of certain plants to soil moisture conditions at different stages of growth. For example, plants may be most responsive to irrigation at certain critical periods, and tolerate very dry conditions at other stages.

The objective of irrigation is not to maintain maximum transpiration rates, which may be conducive to maximum yield of dry matter, but to produce the maximum yield of marketable produce per unit of water applied. This requires considerable flexibility in the moisture régime maintained during different stages of growth of the crop (Salter & Goode, 1967).

Therefore, the most efficient use of water for irrigation requires an integration of different approaches. Soil moisture should be maintained at a high level at sensitive stages; at other growth stages, irrigation can be postponed until the estimate of soil moisture deficits from weather data shows that a predetermined limiting deficit, specific for a given crop on a given soil, has been reached.

Consumptive Use or Real Water-requirements (Actual Evapotranspiration)

The real water-requirements of a crop are determined by: (a) potential evapotranspiration, which integrates climatic conditions; (b) the soil moisture régime; and (c) the nature of the crop and its physiological reactions to moisture stress.

The relationship between actual ET and the moisture régime of the soil is affected by the soil's structure, its water-holding capacity, the nature of the crop, and the rate of PET itself. Slatyer (1956) has shown that there is a progressive decline in actual ET, relative to PET, as the amount of available soil moisture decreases. At field-capacity, the initial values of ET are approximately equal to PET, for all crops; as soil moisture-stress increases, differences between crops, in the rate of decline of ET, become apparent (*see* Figure 32). Thus sorghum, with its extensive root system and greater ability to withstand atmospheric desiccation, maintains ET at a relatively high level for some time, whereas in cotton and peanuts, with their less extensive and efficient root systems, ET is reduced almost immediately soil-water depletion commences.

Plants cannot maintain a transpiration rate in excess of 7 mm/day (Slatyer, 1958). Actual transpiration therefore lags behind PET when a value of 7 mm/day is reached. Consequently, water deficits in plants may occur even when soil moisture is at FC.

When the water-supply to the plant is optimal, leaf stomata tend to be completely open during day-time; water-loss through transpiration from the leaves is then approximately equal to the evaporation from an open water surface. When soil moisture-tension increases, the stomata close proportionately, the rate of transpiration is reduced, and actual ET becomes less than PET. Stomatal closure is therefore effectively a mechanism for conserving water, and is of particular importance in arid regions.

WATER-USE EFFICIENCY

Water-use efficiency (WUE) is the yield of marketable crop produced per

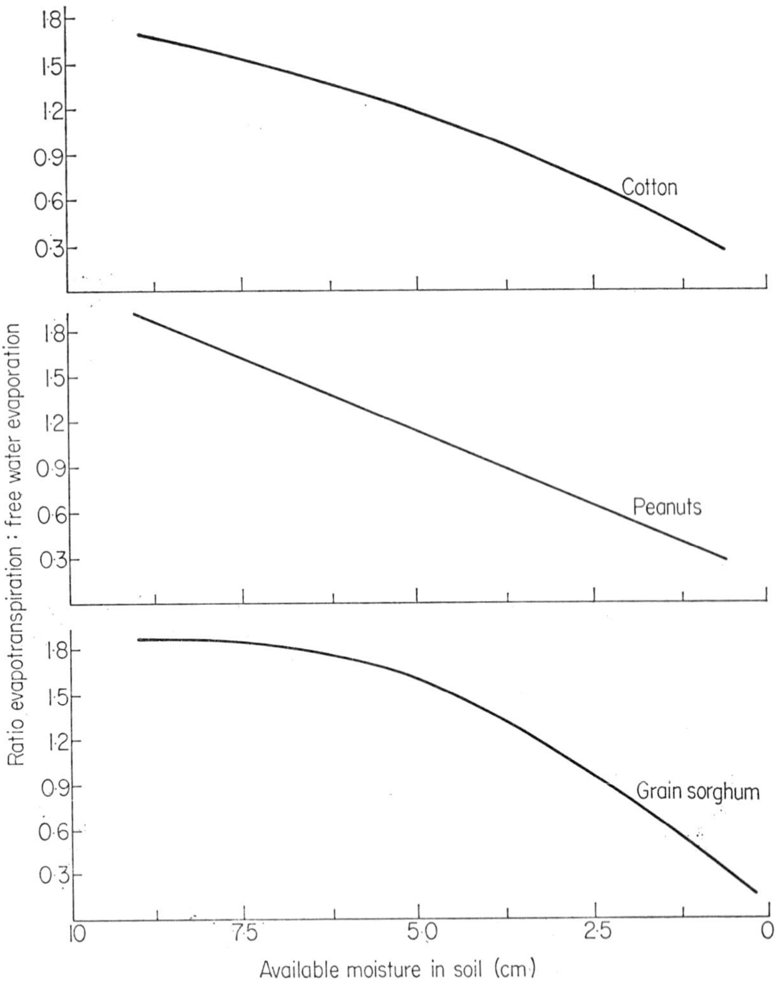

Figure 32. Actual evapotranspiration in relation to potential evapotranspiration as a function of soil moisture. (From Slatyer (1956): by courtesy of the *Netherlands J. Agricultural Science.*)

unit of water used in evapotranspiration. Therefore, $WUE = \frac{Y}{ET}$, where Y = yield of marketable crop and ET = evapotranspiration.

If yield were proportional to ET, water-use efficiency would be a constant. Actually, both the numerator and the denominator of the formula are influenced independently or differently by crop management and environment. The numerator Y is the more amenable to crop management practices, whilst

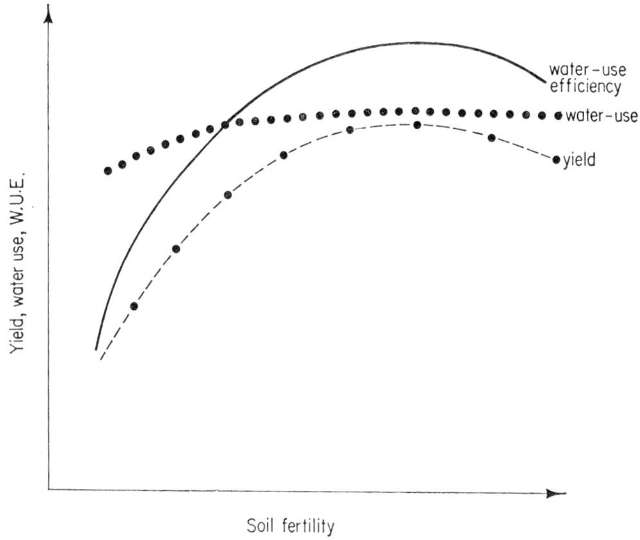

Figure 33. Generalized relationships between soil fertility and yields, water-use and water-use efficiency.

the denominator ET is mainly dependent on climate and on soil moisture supply. Fertilization and other cultural practices for top crop-yields usually increase water-use efficiency, because crop yields generally increase relatively more than crop water-use. Any increases in water-use accompanying fertilization are often negligible, as is shown in Figure 33. Once the crop covers the soil completely, crop management—with the exception of water-supply—has little or no effect on ET.

As a general rule it can be stated that considerable increases in crop yields can be obtained by improved management practices without causing a corresponding increase in ET. An example of the effect of fertilization on evapotranspiration is show in Figure 34. Nitrogen-fertilized maize used considerably more water during the period from 40 to 65 days after sowing than did the maize that received no nitrogen; but there was little difference between treatments in total evapotranspiration for the entire period of growth (Linscott *et al.*, 1962).

The premise that under an optimal water-supply ET is *not* dependent on the kind of the plant canopy, provided the soil is adequately covered, has already been mentioned (p. 192). Under these conditions, increasing the amount of plant canopy will have little or no effect on ET. Conditions of unlimited water-supply are the exception in arid-land agriculture, even under irrigation. But even when water deficiencies occur, whether they be transient or continuous, crop production may be increased by judicious irrigation without markedly

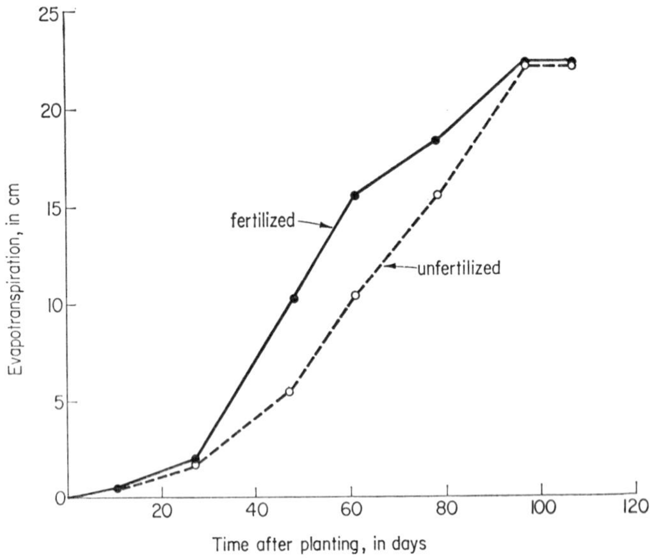

Figure 34. Evapotranspiration from fertilized and unfertilized maize throughout the growing period. The fertilized maize received 157 kg/ha of nitrogen. (From Linscott *et al.* (1962): by courtesy of the American Society of Agronomy.)

increasing ET. This results from the fact that water deficiencies have a greater effect on transpiration than on photosynthesis.

Factors Affecting Water-use Efficiency

The Nature of the Plant. There are considerable differences in the amounts of water which different plant species use to produce a unit of dry matter. If we compare two crops, such as lucerne and maize, grown for forage under identical conditions, the former requires from four to six times as much water to produce one kilogram of dry matter as does the latter. There may also be considerable differences between varieties of the same crop in this respect. Among plants that are most economical in their use of water are maize, sorghum, and fodder beets; barley, oats, wheat, and cowpeas, have medium efficiency; lucerne has, in relation to the others, a much lower WUE.

A high WUE is *not* an indication of drought-resistance. Lucerne, for example, is far more drought-resistant than maize, which is far more efficient in water-use.

Climatic Conditions. Weather affects both the numerator and the denominator of the WUE formula (*see* above). The amount of radiation determines the rate of photosynthesis and hence the potential yield, while other components of climate, such as temperature, day-length, rainfall, etc., influence vital physiological processes and thereby determine the actual yields. Evapotranspiration is still more drastically affected than are these physiological processes.

Transpiration rates increase almost in proportion to the intensity of solar radiation, whilst in certain crops, of which sugar-beet is an example, the rate of photosynthesis increases less rapidly. Certain species may therefore produce more dry matter per unit of water transpired in a cloudy, humid climate than in a sunny, arid climate (Monteith, 1966).

Relative Air Humidity. The lower the relative humidity is, the greater will be the ET. When comparing the WUE of lucerne in a number of locations in the Great Plains of North America, it was found that there was an almost constant inverse relationship between the WUE and evaporation rates. Thus the WUE in North Dakota, with an average daily evaporation rate of 3·98 mm, was almost double that obtained in Texas, where the evaporation was 7·65 mm (Briggs & Shantz, 1913).

Temperature. This factor also affects the WUE. For cool-climate plants, such as wheat and oats, the WUE decreases with increasing temperatures, whilst—within certain limits—the opposite is true for warm-climate crops, such as maize and sorghum.

In short, factors such as low relative air humidity, which increase transpiration without a corresponding increase in the production of dry matter, will reduce WUE. Factors such as light or temperature, that normally affect both transpiration and dry-matter production, will either increase or decrease WUE —according to which of the two effects predominates.

Soil Moisture Content. An inadequate supply of soil moisture has an adverse effect on plant growth and productivity; the same is true of excess moisture in the soil. Both conditions are therefore conducive to low WUE. There is therefore no basis for the assumption that plants grown under dryland conditions will generally be more efficient in water-use than the same crop grown under irrigation, or that water-stress improves WUE. For each crop and combination of environmental conditions there is a narrow range of soil moisture levels at which WUE will be higher than with a lesser or greater supply of water. Shimshi (1966), working on the interrelationships between fertilization and soil–water–plant relationships, showed that when soil fertility was not limiting, WUE increased when the supply of water to the crop was increased. By contrast, at low levels of fertilizer application, WUE decreased with increasing water-supply (Figure 35).

In general, yields of most crops increase when the soil-moisture level is maintained as near to field-capacity as is practicable. However, the efficiency of water-use is generally found to be improved at somewhat lower levels of soil moisture (Doss & Ensminger, 1958).

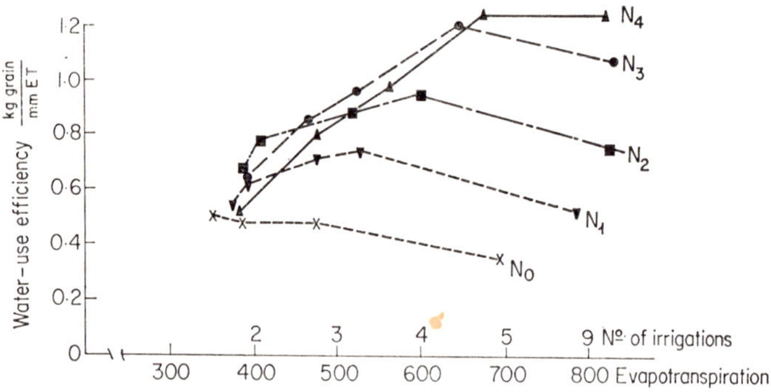

Figure 35. Relationship between soil moisture régime and water-use efficiency at five levels of fertilizer application.

N_0 = control
N_1 = N – 52·5 kg/ha
N_2 = N – 105·0 kg/ha
N_3 = N – 210·0 kg/ha
N_4 = N – 420·0 kg/ha

(Shimshi, 1966.)

Methods for Decreasing Transpiration

Stomatal Closure

Spray materials used for various purposes, such as certain herbicides, fungicides, metabolic inhibitors, and growth hormones, have been found to cause the closing of the guard-cells of the stomata and thereby reduce transpiration (Gale & Hagan, 1966). However, their concurrent effects on other plant processes were in most cases not recorded. Phenylmercuric acetate (PMA), a fungicide, was found to decrease transpiration to a greater degree than photosynthesis, in a number of plants (Zelitch & Waggoner, 1962). The main drawback of PMA is the relatively short period during which it remains effective (Gale & Hagan, 1966).

The chemicals that up to the present have been found to be specifically active in causing stomatal closure, such as sodium azide and potassium cyanide, are toxic, and so cannot be used for practical purposes (Waggoner, 1965). With the exception of PMA, they reduced plant growth to a greater degree than transpiration, so that water-use efficiency actually decreased (Slatyer & Bierhuizen, 1964).

Film-Type Antitranspirants

Plastic and waxy materials that cover the stomata with a film which has more resistance to the passage of water than to that of CO_2 and O_2, are being sought. The characteristics required, in addition to selective gas-permeability, are easy application, inexpensiveness, and non-phytotoxicity—also the ability to form a continuous film on the leaf surface, to resist degradation, and to retain elasticity (Gale & Hagan, 1966). The main objection to antitranspirants is that, by reducing transpiration, they might cause an increase in leaf temperatures to critical levels. It has, however, been estimated that under field conditions, transpiration cannot reduce the temperature of leaves by more than 5 per cent, and that under most conditions the higher temperature would not be critical (Gale & Polyakoff-Mayber, 1965). The best antitranspirants reduce transpiration by 30 to 40 per cent at the most.

Increased CO_2 Concentration in the Air

If a practical and economical method of increasing the normal carbon dioxide concentration in the air were developed, it would be possible to cause the closure of stomata as a result of increased CO_2 concentration (cf. p. 190), concurrently with increased photosynthesis, provided sunlight was not limiting. When CO_2 concentration of maize plants was increased from 0·0310 per cent to 0·0575 per cent, transpiration was reduced 23 per cent and net photosynthesis increased by 30 per cent (Moss *et al.*, 1961).

Growth Retardants

Certain growth retardants have been found to reduce the plant's susceptibility to water-stress, but this may be imputed to a lower top: root ratio and not to reduced transpiration (Plaut *et al.*, 1964).

Increasing Leaf Albedo

Reflective materials might reduce the amount of solar radiation absorbed by the leaves (Gale & Hagan, 1966). To the best of our knowledge, these possibilities have not yet been explored.

Weed Control

Since time immemorial controlling weeds has been known to be one of the most effective means of increasing the amount of water available to the crop, and therefore of increasing water-use efficiency. Weeds frequently transpire greater amounts of water per unit of dry matter produced than do the crop plants with which they grow in association.

Planting Patterns

One of the adaptations to limited rainfall in arid climates is the use of wide spacings between the rows of crop plants, so that, during a relatively long period of plant growth, more soil is exposed to radiant energy than is the case in humid climates. As a result, the rate of evaporation from the soil is greater. If precipitation occurs in the form of relatively frequent and light rains, the wide spacing may result in loss by evaporation of a very large proportion of the total precipitation.

Planting Date

In arid and semi-arid regions, the choice of an appropriate planting date may have a considerable effect on water-use efficiency (in relation to economic yield) by ensuring that the pattern of growth of the crop is adjusted to the pattern of precipitation.

Effect of Fertilizers

Under adequate irrigation, suitable fertilization generally increases yields considerably, with a relatively small increase in evapotranspiration, and therefore markedly improved water-use efficiency.

Under dryland conditions, nutrient deficiencies will reduce WUE, and therefore a moderate amount of suitable fertilizers, adjusted to the soil-moisture level, will increase WUE. If, however, the fertilizers increase water-use in the early stages of growth, so that severe water-stress occurs at the critical stages, the opposite effect will result.

Summing up we may say that the examples given above suffice to demonstrate the general rule: all factors which increase plant production do so without markedly increasing the amount of evapotranspiration, and will therefore improve water-use efficiency. Hence we conclude that a favourable environment and good crop management are preconditions for the most efficient use of water by the crops.

Physiological Effects of Water Deficits

Water-deficits occur in the plant whenever transpiration exceeds water absorption: this may be due to excessive water-losses, to reduced absorption, or to both.

According to the degree of internal water-deficit and its duration, one can distinguish between incipient, temporary, and permanent, wilting (Kramer, 1959). A small loss in turgor, causing incipient wilting, is an almost daily occurrence in warm, dry weather, even when the soil is moist. This does not produce visible symptoms of wilting. A more severe loss of turgor, causing leaf drooping and subsequently spreading through the plant, will reduce growth. If plants regain their turgidity when the water-supply is re-balanced, this is termed 'temporary wilting'. A longer period of dehydration causes 'permanent wilting', from which plants are no longer able to recover—even in a saturated atmosphere.

EFFECT OF WATER-STRESS ON PHOTOSYNTHESIS

As soil moisture-stress increases, photosynthesis drops to the compensation point (Pallas *et al.*, 1962). Water-stress can affect photosynthesis directly, by affecting various biochemical processes involved in photosynthesis, and indirectly, by reducing the intake of CO_2 through the stomata of the leaves, as a result of their closure in response to water-stress. Soil water-stress and atmospheric water-deficit act additively in reducing the rate of photosynthesis (Baker & Musgrave, 1964).

An adverse water-régime also reduces leaf area and hastens leaf senescence, thereby decreasing the productivity of the crop to a greater extent than the depressive effect resulting from a reduced net assimilation rate (Fisher & Hagan, 1965). For example, water-stress that reduced the net assimilation rate by only 50 per cent was sufficient to cause complete cessation of leaf growth, indicating that leaf area is more affected by water-stress than is the case for net assimilation rate (Ashton, 1956).

RESPIRATION

Respiration generally responds to loss of turgor in the inverse direction of photosynthesis: respiration rate first increases, but, as the deficit becomes more severe, it decreases (Evenari, 1962).

ANATOMICAL CHANGES

Periodical water-stress leads to many anatomical changes: these include decreases in the sizes of cells and of intercellular spaces, thicker cell-walls, and greater development of the mechanical tissues (May & Milthorpe, 1962). However, the number of stomata per unit of leaf surface tends to increase (ibid.).

METABOLIC REACTIONS

The proportion of water in the plant is far higher than in the soil, and very small fluctuations, within very narrow limits, can interfere with active metabolism (Slatyer, 1967).

Almost all metabolic reactions are affected by plant water-deficits. Hydrolytic reactions increase: starch is hydrolyzed into sugars, and proteins into amino-acids. It has been found in cotton that, as a result of water-stress, the sugar content of the leaf increases and starch content decreases. By contrast, in the stems, the concentration of both sugar and starch increases—indicating that the reduced water-supply not only affects photosynthesis, but also that the ability of the plant to utilize the products of photosynthesis is still more impaired (Eaton & Ergle, 1952). Similar results were obtained with wheat, in which the percentage of soluble carbohydrates in the leaves of plants under moisture stress increased steadily, reaching a level from three to four times as high as that in irrigated plants—after which death followed (Petinov, 1961).

The conversion of starch to sugars, and their accumulation, increases osmotic pressure. This causes a reduction in the permeability of the plant cell, so that internal resistance to water flow increases, thereby lessening water-loss (Shields, 1958).

Water-deficits impair the nucleic acid system as a result of increased R-nase activity; the degradation of RNA proceeds more rapidly than its synthesis, and this, in turn, affects enzyme production and growth (May & Milthorpe, 1962). Protein synthesis is intimately connected with the activities of the nucleic acid system, and is therefore also affected by drought (Kessler, 1961).

GROWTH

The relations between turgor and growth are not fully understood (Kramer, 1959). Growth is reduced by a decrease in relative turgidity* to below 90 per cent (May & Milthorpe, 1962). Water-deficits cause a dehydration of the protoplasm, associated with a loss of turgor. The expansion of cells and cell division are reduced, resulting in a decrease in the growth of stems, leaves and fruits. The final effect on growth will depend on the frequency and the duration of the periods of water-stress. It is therefore necessary to determine, for each crop, at what stages of development it can tolerate soil moisture-stress, so that an irrigation schedule can be planned that will produce the maximum economic yield per unit of water or per unit of area—whichever is most limiting.

Water-stress does not affect all organs of the plant equally; as a rule, there is a decrease in the ratio of leaves to stem. Older leaves and those that are shaded are usually the first to die under water-stress (Peters & Runkles, 1967).

REPRODUCTION AND GRAIN DEVELOPMENT

Moisture stress during the vegetative phase may stimulate the reproductive phase in some plants (Richards & Wadleigh, 1952). Aerial or storage parts of the plant may be adversely affected by a certain low level of soil moisture, whilst the economic yield may remain unaffected. For example, cotton was found to show marked differences in growth at different levels of moisture supply that had no significant effect on the yield of cottonseed (Adams *et al.*, 1942).

For many crop plants, moisture stress at the flowering period is critical. Even a short period of water-stress at anthesis will markedly reduce the number of flowers that set seed (May & Milthorpe, 1962). The critical period in cereals begins with the appearance of pollen mother-cells (pollen viability appears to be particularly susceptible) and ends after pollination (Henckel, 1964). This is especially marked in crops with determinate flowering, in which the period of pollination through seed-set is relatively short. Grain yields of

* Relative turgidity is the ratio of the water actually present in the tissues to that held at full turgidity.

maize were found to be reduced by 25 per cent when soil moisture was depleted to WP for two days during the tasselling period, and by 50 per cent when the period of stress was increased from six to eight days (Robins & Domingo, 1953). Water-stress duing maturation of the grain will usually result in small or shrivelled grains.

SEQUENTIAL EFFECTS OF WATER-DEFICITS ON PLANTS

Slatyer (1967) has described the sequence of effects on plants of water-deficits as they become progressively more severe.

A plant that has an adequate water and nutrient supply, and normal transpiration, will have accelerated breakdown of RNA and possibly DNA (without any cessation of RNA synthesis). Metabolism will, at first, be impaired only during the daily periods of maximum water-deficit; but these periods will become progressively longer. Stomatal closure during these periods will lead to reduced transpiration and also CO_2 absorption; photosynthesis is reduced and leaf temperatures are increased. Increased respiration will further reduce apparent photosynthesis.

As cell-turgor decreases, the rate of cell enlargement is reduced, causing a slowdown in the expansion of leaf area, resulting in reduced growth rates.

As the moisture stress in the soil approaches the PWP, turgor pressure in the plants approaches zero. Cell enlargement virtually ceases, and the rate of cell division is markedly reduced. The stomata remain closed for most of the day, and transpiration is limited to cuticular transpiration. Leaf temperature will increase markedly. Most metabolic processes, including respiration, slow down, apparent photosynthesis and production of dry matter becoming practically nil.

The disruption of normal cell-metabolism is accompanied by the breakdown of proteins and carbohydrates, causing an increase in the concentration of sugars; leaf phosphorus and nitrogen migrate from the older leaves to the stems.

Finally, as protoplasmic dehydration continues, individual cells and tissues die. Root-hairs die and the roots suberize. In some cases the tops of plants die before the roots, in others the opposite occurs.

EFFECTS OF RENEWED MOISTURE SUPPLY AFTER STRESS PERIODS

Whilst cases of crop desiccation and hence complete loss are not rare in arid and even semi-arid regions, what generally happens is that the crops suffer periods of stress of varying duration and severity, terminated each time by a renewed supply of moisture. This is true even under irrigation—either because the farmer does not have full control over timing of his water-supply, or because he does not have the full quantities of water required. Competing demands of other crops for water at the same time, the need to economize in the use of water, exceptional atmospheric conditions, mechanical breakdowns in water-supply, the normal cycle of wetting and drying between irrigations—all these and other factors ensure that practically every crop, even under irrigation, experiences periods of stress. It is consequently important to understand crop reactions to a renewed water-supply after periods of stress of various degrees of severity and duration, and at different stages of growth.

Recovery of photosynthetic efficiency occurs immediately on reduction of moisture stress (Baker & Musgrave, 1964). The initial decrease in the rate of photosynthesis as a result of water-deficit may even be followed by an increase to a point exceeding the original level (Eckardt, 1953). This phenomenon is not well understood (Evenari, 1962). Rapid drying has a more adverse effect on the rate of photosynthesis than does slow drying (Stocker, 1960).

The ability of stomata to recover after moisture stress may be impaired, so that the plant is unable to re-establish its normal functions even after regaining turgescence. There are, apparently, differences between various crops in the ability of their stomata to recover from the effects of water-stress, and this may be a factor in their apparent drought-resistance. Presumably, this is one of the reasons for the greater ability of sorghum to recover from the effects of drought as compared with maize (Glover, 1959).

If the moisture supply is renewed only shortly before death would have occurred, metabolic processes do not immediately return to normal (Denmead & Shaw, 1960). The death of the root-hairs and the suberization of the roots delay recovery—because of the reduced capacity for water uptake by the stressed plant, the complete dislocation of the metabolism of the plant, and the senescence of the previously active leaves (Slatyer, 1967).

The adverse effects of water-stress are less severe at some stages of growth than at others. At certain stages, there may be irreversible damage which will affect final yield; at other stages, a renewed water-supply may stimulate growth to such an extent that the plant may 'catch up'.

ROOT DEVELOPMENT

Root development is affected by soil-water potential in a number of ways, among which we should note the following:

1. The growth-rate of roots decreases with increasing water-stress; however, growth of the roots is less affected by water shortage than is that of the aerial parts, so that the overall shoot: root ratio is increased (Peters & Runkles, 1967).
2. Roots grow towards water in the soil, provided the distance from the water is small (Hunter & Kelley, 1946).
3. When only a small fraction of the potential root-zone is wetted, as happens

when irrigations are light and frequent, root penetration will be restricted and the crop will be particularly sensitive to drought periods at a later stage.

CROP YIELD

The effect of water-stress on yield will depend largely on what proportion of the total dry matter produced is considered as useful material to be harvested (Fisher & Hagan, 1965).

When the yield consists of most or all of the aerial part of the crop (forage crops, tobacco, etc.), the effects of stress on yield are much the same as those on total growth.

When the yield consists of storage organs other than seeds or fruits (stock beets, potatoes, etc.), it will generally be as sensitive to moisture stress as is the plant's total growth.

When the yield consists of the seeds or fruits, the situation is very different, and it has been shown for a number of crops that the dry matter stored in the seeds or grain is mainly the result of photosynthesis that occurred after flowering (Thorne, 1966). The effect of water-stress will therefore depend on the stage of growth at which it occurs. At an early stage, the number of primordia formed may be reduced; on the other hand, the drastic effect of stress at the time of flowering has already been indicated (p. 205). In general, water-stress, except at the most critical time, will have less effect on the yield of grain than on the plant's total growth (Figure 36).

For certain plants, such as lucerne, moderate water-stress during flowering may improve seed-set, and hence yield. This is ascribed to the increased sweetness of the nectar, which attracts a greater number of pollinating insects.

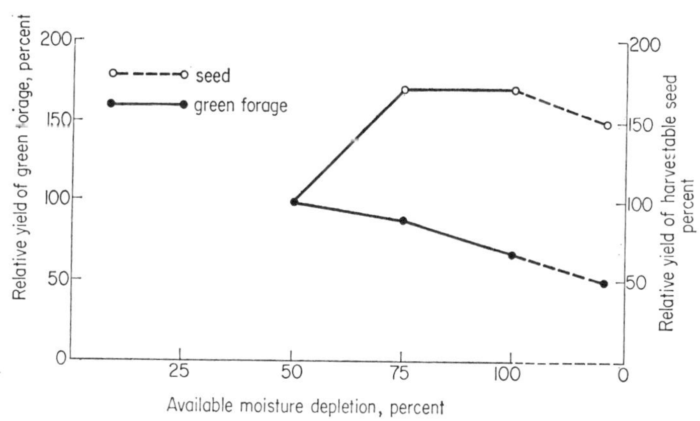

Figure 36. Response of seed yield and vegetative dry-weight of Ladino clover to soil moisture potential. (From Hagan (1957): by courtesy of *California Agriculture*.)

When the yield is a chemical constituent (sugar, pharmaceuticals, fibres, etc.), the economically valuable part of the crop is only a small fraction of the total dry matter produced, and moderate stress that affects growth may have no, or even a beneficial, effect on yield. For example, the content of essential oils in *Mentha* decreases sharply shortly after irrigation (Lvov, 1954), the assumption being that wilting favours the decomposition of starch and proteins, which stimulate the formation of the essential oils. Cool and humid growing conditions have been found to reduce the alkaloid content of many plants (Mothes & Romeike, 1958).

Effect of Moisture Supply on Crop Quality

CARBOHYDRATES

Carbohydrates are the cheapest and most abundant source of calories for human nutrition: they include sugars, starches, and, to a lesser degree, cellulose.

Below certain moisture levels, the rate of photosynthesis is progressively reduced (Eaton & Ergle, 1948), both as a result of the dehydration of protoplasm which affects its photosynthetic ability, and because the closure of stomata reduces the supply of CO_2 (Stålfelt, 1956). During periods of drought there may also be losses of carbohydrates, owing to increased respiration (Aslying, 1956). The combination of decreased photosynthesis and increased respiration implies a potential decrease in carbohydrate reserves in the plant, which may be partially compensated for by the reduction in vegetative growth and the concomitant decrease in utilization of carbohydrates in growth (Richards & Wadleigh, 1952). By reducing the amounts of food available for various biochemical processes in the plant, important changes in its chemical composition may be brought about. Under conditions of water-stress, the hydrolysis of starch is accelerated through increased amylotic activity (Woodhams & Kozlowski, 1954) and sugar concentration is increased, resulting in increased osmotic pressure (Mothes, 1956). As a consequence, plants that have been submitted to water-deficits usually show a decrease in their proportion of starch to total sugars, and, in most instances, a decrease in total carbohydrate reserves (Richards & Wadleigh, 1952). These effects may be noted even in cases when moisture tension is not sufficiently severe to cause wilting (Woodhams & Kozlowski, 1954).

The effects of moisture stress on sugars are less consistent than those on starch. Richards & Wadleigh (1952) state that this is undoubtedly related to the transitional nature of sugars in various plant tissues. Whilst H. F. Clements (1943) showed that sugar levels in cane are negatively correlated with growth and moisture supply, Hartt (1934) presented evidence that more sucrose is synthetized in leaves of sugar-cane that is well supplied with water than in plants under moisture stress. Woolley & Bennett (1962) showed that the percentage of sucrose in sugar-beets responds in a linear manner to increasing soil moisture.

Moisture supply affects sugar accumulation in plants in different ways:

1. An adequate moisture supply is essential for efficient photosynthetic activity and hence for sugar synthesis.
2. Favourable moisture régimes increase growth and hence increase the utilization and depletion of sugars.
3. Moisture stress causes a direct loss of water from the tissues and increases hydrolysis of starch, resulting in an increased concentration of sugars.

It follows that, under both moist and dry régimes, there are conflicting effects of moisture supply on sugar accumulation in plants.

Arnon (1962), working with twenty-two varieties of sweet sorghum, showed that sugar concentration in this plant is a result of soil moisture-variety interactions. Varieties that are generally low in sugar content show an increase in sugar concentration under conditions of moisture stress, whilst in varieties with a potentially high sugar content, the sugar concentration is adversely affected by moisture stress.

Janes (1948), working on snap beans, also reported a variety-moisture régime interaction in relation to sugar content. There was a significant difference in total sugar content expressed as per cent of dry-weight: the beans from the light irrigation plots had the lowest sugar content in the variety with an inherently low sugar content, whereas no difference was observed between plots in the variety with an inherently higher sugar content.

PROTEINS

Although favourable soil-moisture conditions increase the speed of nitrification and thereby increase the amounts of nitrogen available to plants, irrigation may carry readily soluble nitrogen beyond the root-zone. This would explain the contradictory results found in some cases, e.g. with tomatoes, in which nitrogen uptake was reduced by irrigation (Cannel *et al.*, 1962), and with lucerne, clover, and grasses, in which it was increased (Kilmer *et al.*, 1960).

Increasing moisture tension in itself does not appear to limit the ability of plants to extract nitrate from the soil, as long as moisture is above the wilting point. Whilst uptake of nitrate can be maintained under conditions of moisture stress, the reduced growth as well as the decrease in available phosphorus (*see* below) reduce nitrate utilization. This explains why the concentration of nitrogen in plants almost consistently increases with reduced water-supply (Bourget & Carson, 1962; Bennett, 1964). This holds true on both a fresh and a dry-weight basis. The stage of growth at which moisture stress occurs may affect this process (Kramer & Russell, 1959).

There is consistent evidence that moisture stress during maturation increases the protein content of grain. Yields of grain may be adversely affected, but most studies show that there is an inverse relationship between moisture availability during maturation and high protein content of wheat (Dubetz, 1961; Sosulki *et al.*, 1963) and maize (Greaves & Nelson, 1925).

MINERALS

Water affects the kind and amount of minerals supplied to the plant and hence may affect plant composition.

The general, overall effect of increasing soil-moisture supply is to increase the total uptake by plants of all the major elements concerned. Presumably, increased bacterial activity, made possible by favourable soil-moisture conditions, renders available greater quantities of mineral elements to the plants.

Where growth is markedly limited by increased moisture tension, the limitation in nutrient intake is not necessarily of the same magnitude as that of growth.

As a soil dries, the capillary tensions in the soil increase from 0·5 to around 15 atmospheres. At the same time the concentrations of the salts in the soil solutions tend to double (Eaton, 1959). Simultaneous and more or less parallel changes occur within the plant. With increasing moisture tension, growth is curtailed, movement of solutes into new tissues is reduced, and the overall result is an increase in concentration of solutes in the old tissues (ibid.). It is therefore logical to conclude that, under conditions of adequate nutrient supply, the mineral content of plants will be higher when growth is limited by moisture supply than when moisture is not a limiting factor.

Moisture stress has a variable effect on calcium concentration in plants (Richards & Wadleigh, 1952), though the general tendency is for there to be an increase in the calcium content of plants growing under conditions of limited moisture supply (Jenne *et al.*, 1958).

PHOSPHORUS

Phosphorus uptake is generally reduced under moisture stress, and plants generally contain less P when soil moisture is a limiting factor. This was found in sugar-beets (Haddock, 1952), forage crops (Kilmer *et al.*, 1960), and unfertilized oats (Bourget & Carson, 1962).

However, results for P-content are not always consistent (Jenne *et al.*, 1958), probably as a result of differences in the fixing power for P of different soils, as conditioned by soil-moisture content. This is borne out by the fact that when oats received P fertilizers, no reduction in P-content was experienced under conditions of a moisture stress which induced a decrease in P-content in the control plots (Bourget & Carson, 1962).

POTASSIUM

The potassium content of plants usually decreased in specimens grown under favourable soil-moisture conditions, probably as a result of dilution, (Bennett, 1964). Where potassium was lower, in plants, grown in drier soils, the possible reason was that the availability of this element at low soil-moisture content was decreased to a greater extent than was the decrease in uptake due to the limitation in growth (Richards & Wadleigh, 1952).

Whilst variations in K-content of the plant may have little direct relation to the quality of the product, the reduction in K below a certain level may have far-reaching results by affecting various physiological processes and thereby indirectly affecting quality. The best-known effects of this kind are those on the sugar and starch contents of plants.

MAGNESIUM

The magnesium content of plants generally increases under conditions of water-stress (Jenne *et al.*, 1958).

VITAMINS

Relatively little research has been carried out on the influence of soil moisture on the inherent capacity of varieties of plants to build up vitamins.

The effects of moisture tension on vitamins were generally found to be opposite to the effects on yields. Somers & Beenson (1948) found the highest moisture tension in plants with the highest carotene content (on a fresh-weight basis).

In general, it is possible to find relatively large differences in vitamin content when it is expressed on a fresh-weight basis, whilst differences may become small or non-existent when expressed on a dry-weight basis. Such differences as are observed seem to be merely ones of concentration and not actual differences in total amounts (Janes, 1948).

References

ADAMS, F., VEIHMEYER, F. J. & BROWN, N. (1942). Cotton irrigation investigations in San Joaquin Valley, California, 1926 to 1935. *Calif. Agric. Exp. Stn. Bull.*, **668**, 93 pp., illustr.

ANSARI, A. Q. & LOOMIS, W. E. (1959). Leaf temperatures. *Am. J. Bot.*, **46**, 713–17.

ARNON, I. (1962). Sweet sorghum as a secondary crop for sugar production. *Israel J. Agric. Res.*, **12**(1), 13–43.

ASHTON, F. M. (1956). Effects of a series of cycles of alternating low and high soil-water contents on the rate of apparent photosynthesis in sugar-cane. *Pl. Physiol.*, **31**, 226–74.

ASLYING, H. C. (1956). Water consumption in plant production. pp. 685–94 in *Handbuch der Pflanzenphysiologie* (Ed. W. Ruhland), Springer Verlag, Berlin, Band III, xxii + 1073 pp., illustr.

BAKER, D. N. & Musgrave, R. B. (1964). The effects of low-level moisture stresses on the rate of apparent photosynthesis in corn. *Crop Sci.*, **4**, 249–53.

BENNETT, O. L. (1964). Effect of soil moisture régime on yield, nutrient content, and evapotranspiration for three annual forage species. *Agron. J.*, **56**, 195–8.

BLANEY, H. F. & CRIDDLE, W. D. (1962). Determining consumption use and irrigation water requirements. *U.S. Dept Agric. Tech. Bull.*, **1275**, iv + 59 pp., illustr.

BOURGET, S. T. & CARSON, R. B. (1962). Effect of soil moisture stress on yield, water-use efficiency, and mineral composition, of oats and alfalfa grown at two fertility levels. *Can. J. Soil Sci.*, **42**, 7–12.

BRIGGS, L. J. & Shantz, H. L. (1913). The water requirement of plants. *U.S.D.A. Bur. Pl. Indus. Bull.*, **284**, 49 pp.

CANNEL, G. H., LINGLE, J. C. & GARBER, M. J. (1962). Temperature, moisture and phosphorus aid tomatoes. *Crops and Soils*, **14**(8), 26.

CLEMENTS, F. E. & MASHON, E. V. (1934). Effect of soil temperature on transpiration in *Helianthus annuus*. *Pl. Physiol.*, **9**, 619–30.

CLEMENTS, H. F. (1943). The primary Index, its meaning and application to crop management with special reference to sugar-cane. *Hawaii. Plrs' Rec.*, **47**, 257–98.

CURTIS, O. F., & CLARK, D. G. (1950). *An Introduction to Plant Physiology.* McGraw-Hill, New York, xiii + 752, pp., illustr.

DAY, P. R., BOLT, G. H. & ANDERSON, D. M. (1967). Nature of soil water. pp. 193–207 in *Irrigation of Agricultural Lands* (Ed. R. M. Hagan, R. H. Haise & T. W. Edminster) Agronomy series; Amer. Soc. Agron., Madison, Wisconsin, **11**, xxxi + 1180 pp., illustr.

DECKER, J. P., GAYLOR, W. G. & COLE, F. D. (1962). Measuring transpiration of undisturbed Tamarisk shrubs. *Pl. Physiol.*, **37**, 393–7.

DENMEAD, O. T. & SHAW, R. H. (1960). The effects of soil moisture stress at different stages of growth on the development and yield of corn. *Agron. J.*, **52**, 272–4.

DITTNER, H. J. (1938). A quantitative study of the subterranean members of three field grasses. *Am. J. Bot.*, **25**, 654–7.

DIXON, H. H. (1924). *The Transpiration Stream.* Univ. of London Press, London, 80 pp.

DOSS, B. D. & ENSMINGER, L. E. (1958). Effect of nitrogen fertilization on water use efficiency. *Amer. Soc. Agron.*, 50th Ann. Meet., Div. VI, Abstr., p. 36.

DUBETZ, S. (1961). Effect of soil type, soil moisture, and nitrogen fertilizer, on the growth of spring wheat. *Can. J. Soil Sci.*, **41**, 44–51.

EATON, F. M. (1959). Water uptake and salt accumulation. *Calif. Citrogr.*, **44**, 322, 336.

—— & ERGLE, D. R. (1948). Carbohydrate accumulation in the cotton plant at low moisture levels. *Pl. Physiol.*, **23**, 169–87.

—— & —— (1952). Fibre properties and carbohydrate and nitrogen levels as affected by moisture supply and fruitfulness. *Pl. Physiol.*, **27**, 541–62.

ECKARDT, F. (1953). Transpiration et photosynthèse chez un xérophyte mesomorphe. *Physiologia Pl.*, **6**, 253–61.

EVENARI, M. (1962). Plant physiology and arid zone research. *Arid Zone Res.*, **18**, 175–96.

FISHER, R. A. & HAGAN, R. M. (1965). Plant-water relations, irrigation management, and crop yield. *Expl. Agric.*, **1**(3), 161–77.

FURR, J. R. & REEVE, J. C. (1945). The range of soil moisture percentage through which plants undergo permanent wilting in some soils from semi-arid areas. *J. Agric. Res.*, **71**, 149–70.

GALE, J. & HAGAN, R. M. (1966). Plant antitranspirants. *Ann. Rev. Pl. Physiol.*, **17**, 269–82.

—— & POLYAKOFF-MAYBER, A. (1965). Effect of antitranspirant treatment on leaf temperatures. *Pl. Cell Physiol.*, **6**, 111–5.

GARDNER, W. H. (1964). Research for more efficient water use: soil physics. pp. 85–94 in *Research on Water.* Soil Sci. Soc. Amer., Madison, Wisconsin, vi + 121 pp., illustr.

GATES, D. M. (1964). Leaf temperature and transpiration. *Agron. J.*, **56**, 273–7.

GINGRICH, J. R. & RUSSELL, M. B. (1956). Effect of soil moisture tension and oxygen concentration on the growth of corn roots. *Agron. J.*, **48**, 517–20.

GLOVER, J. (1959). The apparent behaviour of maize and sorghum stomata during and after drought. *J. Agric. Sci., Cambridge*, **53**, 412–6.

GREAVES, T. E., & NELSON, D. H. (1925). The influence of irrigation water and manure on the composition of the corn kernel. *J. Agric. Res.*, **31**, 183–91.

HADDOCK, J. L. (1952). The influence of soil moisture condition on the uptake of phosphorus from calcareous soils by sugar-beets. *Proc. Soil Sci. Soc. Am.*, **16**, 235–8.

HAGAN, R. M. (1957). Water—Soil—Plant Relations. *Calif. Agric.*, **11**(4), 9–12.

HAISE, H. R., JENSEN, L. R. & ALLESSI, J. (1955). The effect of synthetic soil conditioners on soil structure and production of sugar beets. *Proc. Soil Sci. Soc. Am.*, **19**, 17–19.

HARTT, C. E. (1934). Water and cane ripening. *Hawaii. Plrs' Rec.*, **38**, 193–206.

HENCKEL, P. A. (1964). Physiology of plants under drought. *Ann. Rev. Pl. Physiol.*, **15**, 363–86.

HENDERSEN, D. W., HAGAN, R. M. & MIKKELSEN, D. S. (1963). Water-use efficiency in irrigation agriculture. *Bett. Crops*, **47**(1), 46–53.

HOLMES, J. W., TAYLOR, S. A. & RICHARDS, S. J. (1967). Measurements of soil water. pp. 175–98 in *Irrigation of Agricultural Lands* (Ed. R. M. Hagan, R. H. Haise & T. W. Edminster) Agronomy series; Amer. Soc. Agron., Madison, Wisconsin, **11**, xxxi + 1180 pp., illustr.

HUNTER, A. S. & KELLEY, O. J. (1946). Extension of plant roots into dry soil. *Pl. Physiol.*, **21**, 445–51.

HYGEN, G. (1953). Studies in plant transpiration. II. *Physiologia Pl.*, **6**, 106–33.

IDSO, S. B., BAKER, D. G. & GATES, D. M. (1966). The energy environment of plants. *Adv. Agron.*, **18**, 171–218.

JAMISON, V. C. (1956). Pertinent factors governing the availability of soil moisture to plants. *Soil Sci.*, **81**, 459–71.

JANES, B. E. (1948).The effect of varying amounts of irrigation on the composition of two varieties of snap beans. *Proc. Am. Soc. Hort. Sci.*, **51**, 457–62.

JENNE, E. A., RHOADES, H. F., TIEN, C. H. & HOWE, O. W. (1958). Change in nutrient element accumulation by corn with depletion of soil moisture. *Agron. J.*, **50**, 71–4.

KELLEY, O. J. (1954). Requirement and availability of soil water. *Adv. Agron.*, **6**, 67–94.

KESSLER, B. (1961). Nucleic acids as factors in drought resistance in higher plants. *Recent Adv. Bot.*, **2**, 1153–9.

KILMER, V. J., BENNETT, O. L., STAHLEY, V. F. & TIMMONS, R. D. (1960). Yield and mineral composition of eight forage species grown at four levels of soil moisture. *Agron. J.*, **52**, 282–5.
KOZLOWSKI, T. T. (1964). *Water Metabolism in Plants.* Harper & Row, New York, xx + 227 pp., illustr.
KRAMER, P. J. (1939). The forces concerned in the intake of water by transpiring plants. *Am. J. Bot.*, **26**, 784–91.
—— (1959). Transpiration and the water economy of plants. pp. 607–726 in *Plant Physiology* (Ed. F. C. Steward) Academic Press, New York, vol. II, xvii + 758 pp., illustr.
—— BIDDULPH, O. & NAKAYAMA, F. S. (1967). Water absorption, conduction and transpiration. pp. 320–9 in *Irrigation of Agricultural Lands* (Ed. R. M. Hagan, R. H. Haise & T. W. Edminster) Agronomy series; Amer. Soc. Agron., Madison, Wisconsin, **11**, xxxi + 1180 pp., illustr.
—— & RUSSELL, M. B. (1959). Plant-water relations. *Adv. Agron.*, **11**, 43–51.
LEBEDEFF, A. F. (1928). The movement of ground and soil waters. *Proc. 1st Int. Congr. on Soil Sci.*, **1**, 459–94.
LEMON, E. R., GLASER, A. H. & SATTERWHITE, L. E. (1957). Some aspects of the relationship of soil, plant, and meteorological factors to evapotranspiration. *Proc. Soil Sci. Soc. Am.*, **21**, 464–8.
LINSCOTT, D. L., FOX, R. L. & LIPPE, R. C. (1962). Corn root distribution and moisture extraction in relation to nitrogen fertilization and soil properties. *Agron. J.*, **54**, 185–9.
LVOV, S. D. (1954). Significance physiologique du processus de la formation des essences pour la plante. *Essais de Botanique, Moscou*, **2**, 563–74.
MAY, L. H. & MILTHORPE, F. L. (1962). Drought resistance of crop plants. *Fld Crops Abstr.*, **15**, 171–9.
MIGAHID, A. M. (1962). The drought resistance of Egyptian desert plants. *Arid Zone Res.*, **16**, 213–33.
MILLER, D. E. (1967). Available water in soil as influenced by extraction of soil water by plants. *Agron. J.*, **59**, 420–2.
—— & KLUTE, A. (1967). The dynamics of soil water—Part I: Mechanical forces. pp. 209–40 in *Irrigation of Agricultural Lands* (Ed. R. M. Hagan, R. H. Haise & T. W. Edminster) Agronomy series; Amer. Soc. Agron., Madison, Wisconsin, **11**, xxxi + 1180 pp., illustr.
MILTHORPE, F. L. (1960). The income and loss of water in arid and semi-arid zones. *Arid Zone Res.*, **15**, 9–36.
MONTEITH, J. L. (1966). The photosynthesis and transpiration of crops. *Expl. Agric.*, **2**, 1–14.
MOSS, D. N., MUSGRAVE, R. B. & LEMON, E. R. (1961). Photosynthesis under field conditions. III. Some effects of light, carbon dioxide, temperature, and soil moisture, on photosynthesis, respiration, and transpiration, of corn. *Crop Sci.*, **1**, 83–7.
MOTHES, K. (1956). Der Einfluss des Wasserzustandes auf Fermentprozesse und Stoffumsatz. pp. 656–63 in *Handbuch der Pflanzenphysiologie* (Ed. W. Ruhland) Springer Verlag, Berlin, Band III, xxii +1073 pp., illustr.
—— & ROMEIKE, A. (1958). Die Alkaloide. pp. 989–1059 in *Handbuch der Pflanzenphysiologie* (Ed. W. Ruhland) Springer Verlag, Berlin, Band VIII, xxi + 1310 pp., illustr.

PALLAS, J. E. Jr., Anson, R. B., Harris, D. G., Elkins, C. B. and Parks, C. L. (1962). Research in plant transpiration. *Prod. Res. Rep. U.S. Dep. Agric.*, 87.

PENMAN, H. L. (1948). Natural evaporation from open water, bare soil, and grass. *Proc. Roy. Soc., Ser. A.*, **193**, 120–45.

—— (1952). The physical bases of irrigation control. *13th Intern. Hort. Congr., London*, pp. 913–24.

—— (1956). Evaporation: introductory survey. *Neth. J. Agric. Sci.*, **4**, 9–29.

PETERS, D. B. & RUNKLES, J. R. (1967). Shoot and root growth as affected by water availability. pp. 373–86 in *Irrigation of Agricultural Lands* (Ed. R. M. Hagan, R. H. Haise & T. W. Edminster) Agronomy series; Amer. Soc. Agron., Madison, Wisconsin, **11**, xxxi + 1180 pp., illustr.

PETINOV, N. S. (1961). Physiological principles of raising plants under irrigated agriculture. *Arid Zone Res.*, **16**, 81–92.

PHILIP, I. R. (1966). Plant–water relations: some physical aspects. *Ann. Rev. Pl. Physiol.*, **17**, 245–68.

PLAUT, Z., HALEVY, A. H. & SHMUELI, E. (1964). The effect of growth-retarding chemicals on growth and transpiration of bean plants grown under various irrigation régimes. *Israel J. Agric. Res.*, **14**, 153–8.

RICHARDS, L. A. & WADLEIGH, C. H. (1952). Soil water and plant growth. pp. 73–225 in *Soil Physical Conditions and Plant Growth* (Ed. B. Shaw) Academic Press, New York, xv + 491 pp., illustr.

ROBINS, J. S. & DOMINGO, C. E. (1953). Some effects of severe soil deficits at specific growth stages of corn. *Agron. J.*, **45**, 618–21.

RUSSELL, M. B. (1959). Water and its relation to soils and crops. *Adv. Agron.*, **11**, 1–131.

SALTER, P. J. & Goode, J. E. (1967). Crop responses to water at different stages of growth. *Res. Rev. Commonw. Bur. Hort. Plantn Crops*, **2**, 246 pp.

SHAW, R. H. & BURROWS, W. C. (1966). Water supply, water use, and water requirement. pp. 122–42 in *Advauces in Corn Production* (Ed. W. H. Pierre, S. A. Aldrich & W. P. Martin) Iowa State Univ. Press, Ames, Iowa, vii + 476 pp., illustr.

SHIELDS, LORA. M. (1958). Morphology in relation to xerophytism. pp. 15–22 in *Bioecology of the Arid and Semi-Arid Lands of the Southwest* (Ed. L. M. Shields, & L. J. Gardner) N. Mex. Highlands Univ. Press, Las Vegas, 69 pp., illustr.

SHIMSHI, D. (1966). [The effect of nitrogen fertilization on some aspects of soil–water–plant relationships—in Hebrew]. Ph.D. Thesis, Hebrew University of Jerusalem, Israel, viii + 145 pp., illustr. (mimeographed).

SLATYER, R. O. (1956). Evapotranspiration in relation to soil moisture. *Neth. J. Agric. Sci.*, **4**, 73–76.

—— (1958). Availability of water to plants. *Arid Zone Res.*, **11**, 159–64.

—— (1960). Absorption of water by plants. *Bot. Rev.*, **26**, 331–92.

—— (1967). *Plant–Water Relationships*. Academic Press, New York, xii + 366 pp., illustr.

—— & BIERHUIZEN, J. F. (1964). The effect of seasonal foliar sprays on transpiration and water-use efficiency of cotton plants. *Agric. Met.*, **1**, 42–53.

SOMERS, G. F. & BEENSON, K. C. (1948). The influence of climate and fertilizer practices upon the vitamin and mineral content of vegetables. *Adv. Fd Res.*, **1**, 291–324.

SOSULKI, F. W., PAUL, E. I. & HUTCHEON, W. C. (1963). Influence of soil moisture, nitrogen fertilization, and temperature, on quality and amino-acid composition of Thatcher wheat, *Can. J. Soil Sci.*, **43**, 219–28.

STÅLFELT, M. G. (1956). Wasserzustand und Photosynthese. Pp. 655 in *Handbuch der Pflanzenphysiologie* (Ed. W. Ruhland) Springer Verlag, Berlin, Band III, xxii + 1073 pp., illustr.

STANHILL, G. (1957). The effect of differences in soil moisture status on plant growth. *Soil Sci.*, **61**, 205–14.

—— (1961). A comparison of methods of calculating evapotranspiration from climatic data. *Israel J. Agric. Res.*, **11**, 159–71.

STOCKER, O. (1960). Physiological and morphological changes in plants due to water deficiency. *Res. Rev.*, **15**, 63–104.

STONE. E. C., WENT, F. W. & YOUNG, C. L. (1950). Water absorption from the atmosphere by plants growing in dry soil. *Science*, **111**, 546–8.

TANNER, C. B. (1960). Energy balance approach to evapotranspiration from crops. pp. 173–95 in *Water and Agriculture* (Ed. R. D. Hockensmith) Amer. Assoc. Adv. Sci., Washington D.C., 198 pp., illustr.

—— (1967). Measurements of evapotranspiration. pp. 534–55 in *Irrigation of Agricultural Lands* (Ed. R. M. Hagan, R. H. Haise & T. W. Edminster) Agronomy series; Amer. Soc. Agron., Madison, Wisconsin, **11**, xxxi + 1180 pp., illustr.

THORNE, G. N. (1966). Physiological aspects of grain yield in cereals. pp. 88–105 in *The Growth of Cereals and Grasses* (Ed. F. L. Milthorpe & J. D. Ivins). Butterworth, London, xxii + 359 pp., illustr.

THORNTHWAITE, C. W. (1948). An approach toward a rational classification of climate. *Geogr. Rev.*, **38**(1), 55–94.

—— & MATHER, J. R. (1954). Climate in relation to crops. *Recent Studies in Bioclimatology. Am. Met. Soc. Meteorological Monographs*, **2**(8), 1–10.

VAADIA, Y., RANEY, F. C. & Hagan, R. M. (1961). Plant water-deficits and physiological processes. *Ann. Rev. Pl. Physiol.*, **12**, 265–92.

WADLEIGH, C. H., RANEY, W. A. & HERSCHFIELD, D. M. (1965). The moisture problem. pp. 1–17 in *Plant Environment and Efficient Water Use* (Ed. W. H. Pierre, D. Kirkham, J. Pesek, & R. Shaw) Amer. Soc. Agronomy and Soil Sci. Soc. Amer., Madison, Wisconsin, xii + 295 pp., illustr.

WAGGONER, P. E. (1965). Decreasing transpiration and the effect upon growth. pp. 49–68 in *Plant Environment and Efficient Water Use* (Ed. W. H. Pierre, D. Kirkham, J. Pesek & R. Shaw) Amer. Soc. Agronomy and Soil Sci. Soc. Amer., Madison, Wisconsin, xxii + 295 pp., illustr.

WILSON, C. C. (1948). The effect of some environmental factors on the movements of guard-cells. *Pl. Physiol.*, **23**, 5–37.

WOODHAMS, D. H. & KOZLOWSKI, T. T. (1954). Effect of soil moisture stress on carbohydrate development and growth in plants. *Am. J. Bot.*, **41**, 316–20.

WOOLLEY, D. G. & BENNETT, W. H. (1962). Effect of soil moisture, nitrogen fertilization, variety, and harvest date, on root yields and sucrose content of sugarbeets. *J. Am. Soc. Sug. Beet Technol.*, **12**, 233–7.

ZELITCH, I. & WAGGONER, P. E. (1962). Effect of chemical control of stomata on transpiration of intact plants. *Proc. U. S. Nat. Acad. Sci.*, **48**, 1297–9.

VIII

IRRIGATION: PRINCIPLES AND PRACTICE

The two central problems of efficient irrigation are:

(a) The choice of the most appropriate methods of irrigation, in relation to local conditions.
(b) Proper scheduling of irrigation and determining the correct amounts of water to apply at each irrigation.

Methods of Irrigation

Irrigation methods are generally designated according to the manner in which water is applied to the soil, as follows:

Surface methods: Irrigation by flooding
Furrow irrigation
Corrugation irrigation
Overhead irrigation
Sub-irrigation
Trickle irrigation

SURFACE METHODS

Irrigation by Flooding

In flood-irrigation, the objective is to spread a thin sheet of water over the surface of the land. The water should 'pond' for long enough to refill the storage capacity of the root-zone.

If flooding is to be reasonably efficient, a balance must be struck between the size of the stream, the infiltration rate of the soil, the amount of water to be applied, and the size of the units irrigated.

Wild Flooding. In this method, a field that was not initially prepared for irrigation is divided by irrigation ditches into strips from 20 to 25 metres wide. The water is diverted into the field from openings in the ditches, and allowed to spread more or less at random. This method is used for irrigating close-growing crops that are suffering as a result of a delay in rainfall, in areas in which irrigation water is available in plenty and at low cost. Wild flooding requires much manual labour, has a low water-application efficiency, and gives very uneven distribution.

Flooding from Contour Ditches. Irrigation ditches are established on the contour at relatively close spacing, the distance between the ditches depending on the slope of the terrain. Water is allowed to flood down the slope, between one ditch and the next. This is a laborious method—especially if more or less uniform wetting is to be achieved.

Border-strip Irrigation. Border-strip irrigation (Plate 16) is used on permeable soils for close-growing crops, on land with a uniform and gentle grade not exceeding 3 per cent—except on permanent pasture-land, for which a steeper slope of up to 6–7 per cent is permissible.

The field is divided into a number of strips between low parallel levees and so oriented that the water flows slowly in a thin sheet, until it reaches the end of the strip. The border strips should be level between the levees, and have a slight slope in the direction of the water flow.

In very primitive irrigation agriculture, the strips are very narrow and short—approximately 2 × 10 m. Only very small streams can be used, and the amount of labour involved is enormous.

In modern irrigation systems, the borders are much wider and longer. Where expert planning and execution of irrigation are available, the strips can be very large. The irrigator is able to handle large unit streams* of water of 500 to 800 m^3 or even more per hour. The width of the strips is adjusted to the size of the streams to be handled, being usually from 10 to 20 metres. The length of the strips is determined by the percolation rate into the soil, and varies from 100 to 400 metres. The higher the percolation rate is, the shorter the strip has to be.

In the border-strip method of irrigation, water distribution is generally not uniform down the slopes: there is a tendency for excessive water penetration towards the middle of the strip. As a result, salt usually accumulates in the middle region.

With border-strip irrigation, large streams of water can be handled safely, without causing erosion, when once the crop covers the soil. However, special

* A unit stream is the rate of water delivery to the irrigated unit, be it border, basin, furrow, etc.

care must be exercised until the crop is well established: a method frequently adopted is to use corrugations within the borders, for irrigating until a stand is established.

Basin Irrigation. Basin irrigation consists of applying irrigation water to level areas enclosed by levees. It is the simplest and oldest of all irrigation systems (Buringh, 1956), having been practised in ancient Egypt and Mesopotamia, when a river flooded the adjacent plain.

Fairly level fields are well graded and then divided by levees into rectangular basins, the dimensions of which are adjusted to the kind of soil and the head of stream available. A large stream of water—150 to 200 m^3 per hour—is used to fill these basins quickly to the required depth. The water then percolates slowly into the soil. The method is suitable both for highly permeable soils and for soils with a very low infiltration rate. As the basin is filled rapidly, uniform distribution is achieved even in permeable soils. As infiltration takes place after the basin is filled, even soils with low permeability will be irrigated evenly.

Furrow Irrigation

Furrow irrigation consists of running a stream of water in a small channel (Plate 17) with a gentle, nearly uniform slope. This method is especially suited to row crops, and can be applied on various soil types and slopes of up to 5 per cent. Water streams that are too small for efficient border-strip irrigation can be used for furrow irrigation. Large water streams in the field ditch are simply divided equally into the appropriate number of furrows; siphon tubes are frequently used for introducing into each furrow a stream of 6 to 10 m^3 per hour. In general, long runs are more efficient than short ones, but the optimum length of furrow varies from 50 to 300 m—depending on intake rate, soil texture and erodability, steepness of slopes, and size of the streams.

With furrow irrigation, it is difficult to prevent some erosion, as the soil of the furrow is not protected by vegetation. It is therefore desirable to select a gentle grade—0·2 to 0·6 per cent—depending particularly on the soil type and size of stream. On steeper slopes, the furrow should be laid out on the contour, within the grades mentioned above.

Corrugation Irrigation

In this method, water flows in very shallow (about 10 cm deep) and narrow furrows (corrugations). The water spreads laterally, saturating the area between the corrugations.

This is a system of irrigation that is adapted to close-growing crops, on sloping land and on soils with low permeability. It consists of applying water very slowly in small, shallow furrows, 50 to 150 cm apart, running down the slope from field ditches, or preferably from portable gated pipes made of aluminium or from hosepipes made of various plastic materials—in either case with outlet tubes. These movable pipes make the method considerably more efficient. The outlet tubes are spaced according to the width between the corrugations. The hosepipes can be rolled up and carried from field to field.

The length of time taken by the water to flow from one end of a corrugation to the other varies from 8 to 48 hours; the streams used are very small—1 to 2 m^3 per hour per corrugation. The length of the corrugations varies from 40 to 120 metres; the grade is usually 2 to 6 per cent, or occasionally even 10 per cent.

The advantages of the corrugation method are that it makes it possible to irrigate on relatively steep slopes, without causing erosion. No crust forms after irrigation, as water infiltrates sideways, and rises to the soil surface by capillary action. However, this method is very conducive to increasing salinity, because of the large amounts of water used, the capillary rise of the water, and the absence of any leaching effect. It is also generally conducive to excessive watering—especially when irrigation is continued until the entire soil surface is visibly moist.

OVERHEAD IRRIGATION—SPRINKLER IRRIGATION

Sprinkler irrigation has been practised on relatively small areas for several decades. The principal types of sprinklers are rotating sprinkler heads, fixed jets (Plate 18), and perforated pipes.

The introduction of light-weight aluminium tubing, and more recently of plastic tubing, and the development of quick-coupling systems, has made portable systems technically possible, and thereby economically competitive with the more traditional methods of irrigation. Rotating sprinkler heads, mounted on portable tubing, are now the most widely-used type of overhead irrigation.

Overhead irrigation has made possible, both technically and economically, the irrigation of terrain that is too steep or too uneven for surface irrigation, as well as of very sandy soils, and therefore makes available for development large tracts that were formerly excluded.

On land with a more regular topography, levelling is not as essential as for surface irrigation, and certainly does not need to be as elaborate. Not only is the expense involved in levelling often avoided when sprinklers are used, but also great damage to the land, due to the removal and/or covering of the topsoil, which sometimes requires many years of expensive reclamation with resulting yield losses.

Further Advantages of Sprinkler Irrigation

Water application can be far more uniform with sprinkler systems than with surface irrigation, except during times of high winds. The amount of water required can be calculated and then applied with great precision, so that overall water-use is more efficient with sprinkler irrigation than with surface irrigation, and soil erosion damage can be reduced. Sprinkling during the hot hours of the day may improve the microclimate, prevent transient wilting, increase stomatal opening, and thereby improve the photosynthetic effectiveness. On the other hand, the humid microclimate under the foliage may favour the spread of plant pathogens, though this danger has apparently been overestimated.

The elimination of field ditches increases the net area available for crop production and reduces water-losses to seepage and percolation.

The manual work involved in sprinkling is less arduous and requires less skill than that needed for surface irrigation. Agricultural Research Service engineers in the U.S.A. have even developed a radio-controlled automatic irrigation system which permits precise water application, requiring little labour (Renne, 1965).

In regions in which light frosts occur occasionally, sprinkler irrigation is an effective means of preventing frost damage.

Drawbacks of Sprinkler Irrigation

There are a number of drawbacks, however, to sprinkler irrigation. Thus the capital investment for equipment is relatively high, even with portable apparatus, and the water has to be applied under relatively high pressure, which increases water cost.

Water-loss due to evaporation and the interception of water by the foliage is greater with sprinklers than with the traditional methods, but this can be alleviated by night irrigation. The aluminium pipes are subject to corrosion when saline water is used.

Sprinkler irrigation is not well adapted to windy areas. Wind-speeds above 12 km per hour make the irrigation inefficient and cause lack of uniformity in water application. For this reason, sprinkler irrigation is usually confined to certain hours of the day and night. Giant sprinklers, which compensate for wind effects by variable rotation speeds, have, however, been developed.

Another disadvantage of overhead irrigation is the need for scheduling irrigation so as not to wash off chemicals used for plant protection before they have taken effect.

All in all, however, the drawbacks of sprinkler irrigation are more than offset by the advantages, and the areas under sprinkler irrigation have increased considerably in recent years—both absolutely and in relation to other irrigation methods.

SUB-IRRIGATION

Sub-surface irrigation is achieved by raising or maintaining the water-table at a predetermined depth, usually 30 to 75 cm, from which moisture rises by capillary action into the root-zone. This is possible when there is a high permanent water-table or a relatively impermeable soil-stratum not too far from the soil surface.

In one system, the water is held continuously in properly spaced field ditches; the water level in the ditches is regulated so as to maintain a constant supply of water to replace water lost by evapotranspiration. In another system, the ditches are filled periodically and the water is held in them until the surface soil is moist. The excess water is then drained off (Hagan *et al.*, 1967). The drainage system serves alternatively for irrigation and for drainage.

This method is commonly used for organic soils, in order to prevent excessive oxidation and subsidence. It tends to cause salt accumulation in the root layer and can therefore be used only where the soil is regularly leached by rainfall, the irrigation ditches then serving as a drainage system.

TRICKLE IRRIGATION

In this method, plastic 12–16 mm pipes are placed on the ground, along the row-crop to be irrigated. In the pipes, at distances of one metre apart, are calibrated orifices from which water trickles at the rate of 2 to 10 litres per hour per orifice. At this slow rate of application, water percolates immediately downwards and sideways into the soil.

This method was first applied for irrigation in greenhouses but is giving very promising results for row-crops grown in the desert—especially on sandy or gravelly soils on which the conventional surface methods are not practicable, and where the water-losses by evaporation from overhead irrigation would be excessive. The moisture régime and soil aeration are optimal under the tricklers. Irrigation is possible at all hours of the day and night, the supervision required being minimal. Plant nutrients can be supplied through the irrigation system.

All the usual water-losses experienced with other methods of irrigation are avoided in trickle irrigation. However, the method favours the accumulation of salts in the root-zone, and these must therefore be leached out periodically. If natural precipitation is too limited for this purpose, fairly large amounts of water must be applied, using one of the conventional irrigation methods; generally portable sprinklers are employed for this purpose.

CHOICE OF METHODS OF IRRIGATION

The following factors determine the choice of irrigation method: topography, soil characteristics, the crop(s), and the quantity, quality, and cost, of the water available.

Sprinkler irrigation can be used for all crops (even upland rice); furrows are suitable for row-crops; basins are the only method suited to paddy rice, but can also serve for other cereals and for forage crops; border strips are used mainly for fodder crops, pastures, and close-rowed cereals.

The steepness of the slope, its direction and regularity, are the principal topographical factors to be considered. Surface irrigation methods require land of fairly uniform and not excessive slope. Levelling is usually necessary to achieve this requirement; slopes should not exceed 4 per cent, or else excessive levelling operations will be required. Land with a very irregular topography should be irrigated by sprinklers, corrugations, contour furrows, or flooding from contour channels. Level land or land with a regular and gentle slope can be irrigated by furrows or flooding between borders.

The infiltration rate of the soil is of great importance in deciding on the irrigation method to be used. Generally, the rate of intake decreases rapidly from soon after irrigation is started, until it reaches a fairly constant level. Surface-flow methods are not very suitable for soils with a very high intake rate, while in ponding methods, very rapid filling of the basins is essential if uniform distribution is to be achieved. Soils with a very low infiltration rate require irrigation methods in which the water can be ponded on the soil surface for a relatively long period.

Soil texture and structure also affect the choice of irrigation method. On very light soils, only sprinkler irrigation is efficient. Soils with a very unstable structure may break down easily under irrigation, to form hard, impenetrable crusts. Such soils require very careful irrigation—especially during the period of germination and emergence.

The nature of the crop must also be considered. High-value and perennial crops can justify elaborate irrigation methods that would not be justified for low-value and annual crops. Furrow irrigation is not suitable for crops other than row-crops.

When the quantities of water available are restricted, or their cost is very high, the efficiency of irrigation assumes a still greater economic importance than under less stringent conditions, though maximum efficiency of water-use should always be the aim.

The size of the head of water available must also be considered. Small heads can be used for furrow irrigation or small borders. Large heads require appropriate irrigation methods if the water is to be controlled properly, and if damage to soil, crops, and the waterworks, is to be avoided.

When water with a high content of salts is used, irrigation methods must be designed so as to make it possible to leach the salts periodically from the soil. Certain irrigation methods, such as corrugation or sub-irrigation, have to be avoided under these circumstances.

The choice of an inappropriate method of irrigation may result in the squandering of large sums of money that have been invested in soil preparation and purchase of equipment; moreover operating costs may be unnecessarily high, with inefficient use of labour and water. Yields may be reduced by an excess of water in one part of the field and a shortage in another.

Efficient Use of Water for Irrigation

Efficient use of water for irrigation implies preventing moisture stress in the root-zone from reaching a level at which the yield of the marketable product is adversely affected by improper scheduling of irrigation; also applying, at each irrigation, the amount of water required to restore the whole root-zone to field-capacity.

PREVENTION OF EXCESSIVE MOISTURE STRESS

Numerous studies have shown that it is possible to deplete the soil-water to certain levels—which depend on the crop, the economic marketable product, and the root-zone depth—before the yield or the quality of the product is economically affected (Jensen *et al.*, 1967). Plate 19 shows maize plants suffering from a deficiency of moisture.

One can consider, as the optimum soil-moisture content of the root-zone, the range of moisture fluctuating between field-capacity (as the upper limit) and the soil-moisture tension level at which the yield of the marketable product is reduced (as the lower limit). This can be termed the 'critical soil-moisture tension' (Figure 37).

The critical soil-moisture tension will depend on the nature of the crop and, in particular, on the effect of soil-moisture stress on the marketable product.

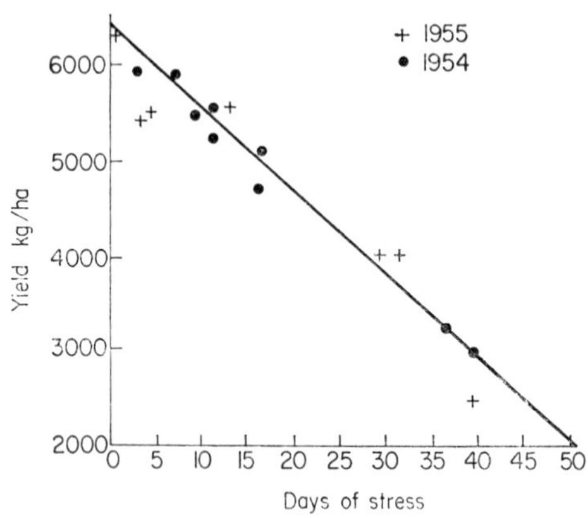

Figure 37. Relationship between grain yields of maize and the number of days of critical soil-moisture tension (soil moisture below 12 per cent on a loess soil). (From Rubin & Bielorai (1957).)

The differential response of crops to water-stress at different stages of growth has been discussed in Chapter VII; for the 'critical soil-moisture tension' is not a constant throughout the growing period even for a given crop. Adjusting irrigation schedules in accordance with the critical periods of growth, so as to prevent soil-moisture tension from exceeding the critical level, has a considerable bearing on yields as well as on total water requirements. There is a general tendency in farming practice to over-irrigate the crop in its vegetative growth stages, and to underestimate water requirements at periods of maximum susceptibility to water-deficits—namely, when the plant is flowering and fruiting. Over-irrigation during the later stages of growth will usually cause a delay in maturity, without necessarily improving yields.

Reliable information on optimum soil-moisture content for each crop can only be obtained on the basis of field experiments in which the reaction of each crop to different soil-moisture régimes is determined.

INTEGRATED MOISTURE STRESS

Taylor (1952) suggested that the soil-moisture régime cannot be described adequately by indicating the soil-moisture content at the time of irrigation of a specified soil layer. He proposes that the whole distribution of local and temporal moisture states, at all depths throughout the root-zone and throughout the growing period, should be taken into account. Taylor therefore proposes an index called 'integrated moisture stress' (IMS) to integrate all the local and instantaneous moisture tension values over the relevant soil profile and the growth period.

SCHEDULING OF IRRIGATIONS

When information is available on optimum soil-moisture content, it is necessary to be able to determine when soil-moisture content reaches the critical level at which it has been decided to irrigate. A variety of methods have been developed for this purpose, based on soil-moisture measurements, meteorological measurements, or plant characteristics.

Techniques of Soil-moisture Measurement

Gravimetric Sampling Methods. Soil samples are taken from the root-zone and dried at 105°C, after which the weight-loss is determined. If the specific weight of the soil is known, the percentage of water in relation to soil volume can be calculated.

The main disadvantages of this method are (a) the time-lag of at least 24 hours between sampling and the availability of the results, (b) the need for a large number of samples, and (c) the large amount of labour involved. However, the method is simple and straightforward, does not require expensive equipment, and gives very accurate results.

Electrical Resistance Units. This method is based on the linear relationship between the electrical resistance and moisture content of the soil.

Units that measure electrical resistance, consisting of two electrodes embedded in blocks of suitable porous material—such as gypsum, nylon, or fibreglass—are buried at appropriate depths in the root-zone. Equilibrium is established between the soil-moisture content and the moisture content within the blocks (which respond to fluctuations in soil moisture). If the blocks have been individually calibrated, for the soil type in which they are being used, the electrical resistance within each block can be measured by a Wheatstone bridge and from this the moisture content can be deduced.

The advantages of this method are the relatively low cost of the units, the speed with which readings can be made, and the suitability of the method for automatization in remote reading and recording (Stanhill *et al.*, 1967).

Tensiometers. A tensiometer consists of a porous ceramic cup and a pressure-gauge. The cup is buried in the soil at the appropriate depth, and equilibrium is established, through the porous material of the cup, between the water in the apparatus and that in the soil. Fluctuations in soil-moisture tension are accurately reflected by variations in the pressure-gauge.

Tensiometers are effective at moisture ranges from saturation to 0·8 atm soil-moisture tension, and will therefore encompass moisture fluctuations involving 90 per cent of the available water in sandy soils, but not more than 50 per cent in clay soils. When moisture tension increases beyond 0·8 atm, air penetrates into the porous cup and the whole system becomes ineffective.

In view of the high cost of the tensiometers, they are generally located in the zone of maximum soil-water withdrawal, and their use is commonly limited to high-value crops for which soil moisture has to be maintained at continuously high levels. The tensiometers can be used for automatic regulation of overhead sprinkler systems.

Neutron Scatterers. In recent years, equipment based on the use of radioactive materials has been developed for measuring soil moisture.

When fast neutrons collide with the nuclei of hydrogen atoms in the soil, they are transformed into slow neutrons. As water is the main source of hydrogen in the soil, the number of collisions will depend on the soil's moisture content. The number is determined by a slow-neutron counter, the neutron source being lowered into the soil to the required depth—into special access-tubes inserted in the soil. The number of slow neutrons is correlated with the moisture content of the soil.

This method has a number of advantages. The measurements determine moisture content directly in relation to soil volume, and can be expressed as mm of water in relation to soil depth. The moisture determinations are not affected by salt content of the soil at normal field salinity, or by compaction. The method is extremely accurate and rapid.

The main disadvantages are the high cost of the equipment, the radiation hazard, and the need for a high level of maintenance and of skilled operators.

Meteorological Methods. The so-called 'bank-account' or 'book-keeping' procedures are based on the use of daily soil-water balances, in which the amounts of water added to the soil by precipitation or irrigation are deduced from the losses due to evapotranspiration. The objective of this approach is to provide a method whereby reliable information on the scheduling of irrigation can be made available for a fairly large irrigated tract, without having to rely on direct soil-moisture measurements on the individual farms. For experience has shown that very few individual farmers are prepared to make the necessary investment in equipment and time required for this purpose, even in countries with an advanced agricultural technology.

A number of formulae have been developed for calculating evapotranspiration on the basis of physical and meteorological calculations. The most important of these have been described, and their relative values discussed, in Chapter VII. In short it may be said that, whilst the Penman formula gives a proper evaluation of the effect of solar radiation and other factors in computing evapotranspiration, there are only a few stations that are equipped to supply the data required. Thornthwaite's and Blaney & Criddle's formulae are empirical, but they are easy to use and generally give fair approximations.

In New Zealand, a relatively good agreement was found between soil moisture as measured by the gravimetric method, and the figure obtained by using the Thornthwaite formula for calculating the day-to-day values of evapotranspiration (Rickard, 1957).

In Washington, it was found that evaporation-tank rates gave a much closer estimate of actual evapotranspiration than either the Blaney & Criddle or the Thornthwaite procedures (Edminster & Schilfgaarde, 1955).

In general, it can be said that the reliability of the empirical methods for evaluating potential evapotranspiration increases when the methods are calibrated for each crop in each region (Tanner, 1967). This is necessary because the effective rooting depth and the permissible water-deficit for each crop, as well as soil-water retention characteristics, are factors that have to be taken into consideration. In addition, the considerable advection of energy from unirrigated surroundings (the so-called 'oasis effect') affects the energy balance of each location (Haise & Hagan, 1967).

The 'Book-keeping' Methods in Agricultural Practice. 'Book-keeping' procedures for scheduling irrigation application have been developed (Penman 1949; Bavel 1960) and successfully applied in a number of places

(Bavel, 1960). These methods are applied on a community service basis, with a central office serving a major irrigation project or district having uniform climatic conditions. Data on evapotranspiration, derived from evaporation installations located in a number of representative sites, are conveyed to the central office, where the calculations on moisture use are made, employing the appropriate crop coefficients. This information is then conveyed to the irrigators in the district by 'phone, radio, television, newspaper, or any other appropriate means of communication.

The general moisture content of the soil at the beginning of the irrigation period, as determined by random sampling for each district, serves as the basic starting-point for the irrigation season's moisture record. Evaporation values are converted into weekly water-use based on water-use curves, established experimentally for the crops grown in the area. These are subtracted from the base starting-point, whilst precipitation is added. Irrigation requirements as well as the anticipated time of irrigation are then forecast. These recommendations take into account the water-holding capacities and readily available moisture values as determined for the major soils of the district (Steed, 1967).

The book-keeping methods for scheduling irrigation have been most successful in humid regions, because occasional heavy rains cancel the cumulative errors that may have been made in estimating the water used, and permit 'a new start' to be made several times during the season. In arid regions the cumulative errors have, up to the present, made the method very problematical (Phelan, 1964).

Physiological Indicators

The possibility of using the actual plant which integrates all the factors involved in its own water potential—soil conditions, atmospheric conditions, and plant characteristics—as an indicator for the timing of irrigation, has been investigated and a number of methods have been proposed that are based on physiological indications of water needs.

The Appearance of the Plant. An adverse water-balance is soon reflected in the appearance of the plant. Transient wilting of the plants at certain hours of the day are the first signs of an incipient imbalance in water-supply. Delaying irrigation until wilting becomes more severe will usually affect crop yields adversely. Fortunately, water-stress does not usually occur at once in the whole field, but appears first in patches; this gives some advance warning of impending water need.

Another approach is to grow, in the field, 'indicator plants' which are more sensitive to moisture stress than is the main crop, so that indications of the impending need for irrigation are given before the crop suffers a setback due to lack of available moisture (Haise & Hagan, 1967).

Colour of the Foliage. Most farmers are aware of the fact that with increasing moisture-stress, the leaves of most crops change colour—from the bright green of well-watered plants to the dark green typical of water-stressed plants. In this case, too, isolated patches of dark-green plants may give advance warning of soil moisture-stress. Yet nitrogen application, which also affects leaf colour, may make it difficult to use this characteristic as an indicator of water-stress.

Rate of Growth. Growth depends first and foremost on turgor, which in turn is dependent on a favourable water-balance. Plants are so sensitive to the internal water-balance, that daily fluctuations in the moisture content of the plant tissues are reflected in parallel fluctuations in the growth-rate of expanding organs.

Stalk elongation is markedly reduced when soil moisture-stress approaches the critical level, but accelerates again after irrigation. Criteria can be developed for different crops. For example, in cotton the rate of elongation of the internodal space below the growing tip can serve as a physiological indicator of water-stress. In a moderately fertile soil, it was found that a rate of elongation of 0·5 to 1 cm daily of the internodal space indicated the approach of a critical soil-moisture tension and the need for irrigation (Aharoni, 1964).

Measuring the Stomatal Aperture. In many crop plants the degree of closure of the stomata reflects accurately the internal water-balance of the plant. Various methods have been developed for rapidly and accurately measuring stomatal opening under field conditions, as a diagnostic tool for predicting the timing of irrigation. The following are noteworthy:

(*a*) *Infiltration Methods.* These measure the rate of infiltration into the leaf tissues of liquids with low viscosity; the rate of infiltration reflects the size of the stomatal aperture which, in turn, depends on the water supply to the plant. A mixture of paraffin oil and turpentine has given good results in cotton (Shmueli, 1964). As can be seen in Figure 38, the rate of infiltration of the mixture used remains fairly constant until the plant becomes stressed, approximately after 60 per cent of the available moisture in the 0–90 cm soil layer has been exhausted. The method is reliable only if standardized procedures are adopted, such as using leaves of a similar age, and carrying out the tests on a predetermined region of the leaf, under certain conditions of light intensity, and at a given hour of the day. The method is simple, does not require expensive equipment, and gives an accurate estimate of soil-water tension, enabling timing of irrigation that ensures high yields. The main drawbacks of the method are the need for experienced operators in order to obtain objective results, and the fact that the method is suitable only during certain periods of plant development.

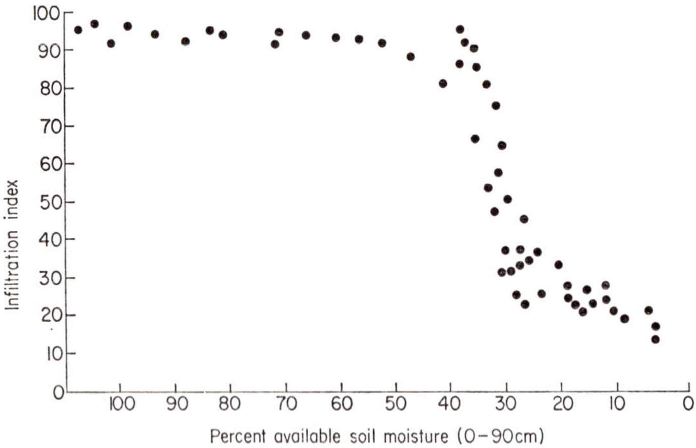

Figure 38. Relationship between infiltration index in cotton leaves and available soil moisture. The tests were made between 12.30 and 13.30 hours (Shmueli, 1964).

(*b*) *Field Porometers*. These measure the degree of resistance to air-flow through the leaf, on the assumption that the rate of resistance will be closely related to the degree of opening of the stomatal apertures. Special forms of equipment have been developed for this purpose, called porometers. These consist of a receptacle containing air under pressure, an apparatus for forcing the air through the leaf, and a gauge for measuring the resistance to the air-flow (Shimshi, 1965). The method is limited to crops which have stomata on both sides of the leaf; an increase in resistance to air-flow, during noon hours, is an indication that soil moisture-stress is approaching a critical level.

Plant Water Content. Soil moisture tension is reflected in the moisture content of the plant tissues and in their osmotic pressure. Many attempts have been made to develop simple and practical techniques, based on this relationship, that could serve for determining the timing of irrigations. Determining the concentration of solids in the plant sap by the use of a refractometer, and measuring the sap's electrical resistance with a Wheatstone bridge, are examples of the methods that have been tested.

Determining the Amount of Water to be Applied

The objective, at each irrigation, will be to replenish the whole root-zone with water to field-capacity. Plants do not withdraw water equally from the entire depth of the root-zone; the activity of the roots of most crops is far greater in the upper levels of the soil than at lower depths. For example, a deep-rooted crop, such as sugar-beet, removes 60 per cent of its water requirement from the uppermost 60 cm, whilst the remaining 40 per cent is contributed by the soil layers at a depth of 60–120 cm. The uppermost soil layer also has the most favourable air régime, the most intensive biological activities, and most of the plant nutrients. It is therefore easy to understand that, even if irrigation is applied before the critical soil-moisture content is reached throughout the entire root-zone, active growth of the plant will have been checked. Wetting only the upper soil layers is also not desirable: the roots will be concentrated in the upper layers, and as these dry out rapidly, irrigation will have to be frequent. As a result, there will be cycles of low and high soil-moisture tension, with little or no water reserves to draw on. Losses by evaporation from the soil will be relatively great in relation to losses by transpiration, and salts will tend to accumulate in the upper soil layer.

Excess water applications are not only a waste of water, but impair soil aeration for lengthy periods, increase leaching of plant nutrients, and cause a rise in the water-table level.

It is possible to calculate the amount of water required to remedy a water-deficit in the root-zone. This amount will depend on the depth of the root-zone, on the critical soil-moisture tension, and on the efficiency of irrigation.

The following formula is used for calculating the amount of water to be applied at each irrigation:

$$d = \frac{10\,(WF_e - W_o)\,SD}{E}$$

where d = mm of water to be applied,
WF_e = soil moisture percentage (by weight) at field-capacity,
W_o = soil moisture percentage (by weight) before irrigation,
S = specific weight of the soil,
E = irrigation efficiency, and
D = depth of the soil layer sampled.

Irrigation Efficiency

WATER REQUIREMENT AND IRRIGATION REQUIREMENT

The amount of water required to replace water lost from the soil by evapotranspiration is called 'water requirement', or 'consumptive use'. The amount of water actually applied in irrigation is usually in excess of consumptive use, and includes water-loss and waste during conveyance, distribution, and the actual application of water to the land, as well as percolation beyond the root-zone, etc. The total amount of water that has to be applied in order to replace effectively the soil moisture deficit in the root-zone is called the 'irrigation requirement'.

IRRIGATION EFFICIENCY IN AGRICULTURAL PRACTICE

The ratio between water requirement and irrigation requirement is a measure of irrigation efficiency.* With present irrigation practice, an efficiency of 100 per cent is never achieved. In addition to losses from such circumstances as interception by crop foliage, run-off, etc., some of which are unavoidable, there are others which depend on the efficiency of the irrigator and on the irrigation methods used. Lack of uniformity in the distribution of water in the field, deep percolation, losses of water by seepage from the irrigation ditches, and water leakage from taps, pipes, joints, etc., are all in excess of consumptive use by the crop and therefore reduce the irrigation efficiency.

In an area of the arid west of the United States, on an irrigated area of about half-a-million hectares, it was found that in a given irrigation season, the average amount of water used for irrigation was 2000 mm, with certain regions reaching 3050 mm. Calculated potential evapotranspiration for the period was 670 mm, of which precipitation supplied 60 to 120 mm. The efficiency of water use was therefore less than 30 per cent! In an adjacent area with the same potential evapotranspiration, farmers were getting good yields with 600 mm of water (Nace, 1960). Irrigation efficiency, on a world-wide average, is very low, being only around 30 per cent according to Olivier (1967). The main reasons for this are low technical competence, unrealistic pricing of water, and poor water distribution networks and water allocation methods.

WATER APPLICATION LOSSES

Water application losses result from uneven distribution of the water run-off from the field, and from excessive amounts of water being applied and resulting in percolation beyond the root-zone.

Losses due to evaporation will be relatively minor when intervals between irrigations are long, but will increase in importance with an increase in the frequency of irrigation. In overhead irrigation, evaporation losses may be of considerable magnitude.

FACTORS AFFECTING APPLICATION EFFICIENCY

Water application efficiency depends on many factors: soil topography, texture, and structure; the vegetative cover; the degree of care exercised in levelling and preparing the area to be irrigated; the experience and ability of the irrigator; the frequency of irrigation; the depth of the root system, etc.

Efficiencies of surface irrigation methods usually vary from 40 to 80 per

* Irrigation efficiency has been defined as 'the percentage of irrigation water that is stored in the soil and available for consumptive use by crops' (Grassi, 1967).

cent, though 20 per cent has been recorded and an efficiency of 60 per cent is considered reasonably good (Grassi, 1967). In well-planned overhead irrigation the range is usually narrower: 70 to 80 per cent.

It must be stressed that water percolation beyond the root-zone is not always to be considered a loss; as will be indicated in the next chapter, it is frequently essential and caused deliberately, in order to leach salts beyond the root-zone. It is the unplanned and unnecessary, excess amounts of water that are undesirable and even dangerous, as they usually cause water-logging and eventually a rise in the water-table.

Water Delivery

CONVEYANCE LOSSES

Conveyance losses are practically nil when the water is conveyed to the field through pressure pipes that are in perfect condition. However, leaks in the system, faulty valves, or other results of neglect, frequently cause considerable losses of water.

The conveyance losses in open ditches are usually considerable, and are due to seepage from the ditch, to transpiration by aquatic plants or those growing on the ditch banks, and, to a far lesser degree, to evaporation. The losses in an average unlined canal have been estimated as amounting to at least one-third of the water conveyed (Arthur & Thoroky, 1967).

It has been estimated that conveyance losses in the Punjab are 24 per cent lost in the canals and distributaries, and an additional 21·5 per cent lost in the water channels taking the water through the villages (ibid.). This is before even one drop has reached the field to be irrigated!

LININGS

A variety of linings can be used to reduce these losses, including

(i) *Hard-surface and exposed linings:* of asphalt, films of plastic, synthetic rubber, or concrete (Plate 20).
(ii) *Buried-membrane linings:* made of hot-applied asphalt or pre-fabricated plastic film.
(iii) *Earth linings:* Compacted earth-blanket linings made of impervious soils, as well as betonite-soil mixture blankets, are used.

Each type of lining has its advantages and disadvantages. A number of linings are still in the experimental stage, and no final evaluation of them is yet available.

Selection of Soils for Irrigation

REQUIREMENTS

When assessing land as to its suitability for irrigation, the emphasis should not be on the present condition and physical characteristics of the land, but on a prognosis of what changes are required and what may be expected to occur as a result of irrigation (Maletic, 1967). These last are usually changes for the worse, and are due to alterations in the water-table, to drainage and salinity problems, to the need for modification of the slope and microrelief by land-forming, or to the need for improving soil profile characteristics, etc. In view of the large investments involved, and as the land available for irrigation is usually in excess of the water resources, it is desirable to use only the most suitable soils for irrigation development.

The first requirement is that the soil should have sufficient *potential productivity* to justify the investment involved in irrigation. This evaluation must be based on what the soil will be like after being brought under irrigation.

Slope

The effect of slope on suitability for irrigation depends on a number of factors: the cropping system chosen, the intensity of precipitation, the water intake of the soil, etc. Methods of irrigation can be adapted to the dominant slope, or the decision may be made to modify the slope by levelling. These decisions have to be based on economic appraisals, in addition to the physical parameters mentioned above.

The maximum slopes considered as suitable for irrigation vary widely, from 3 to 35 per cent (Maletic & Hutchings, 1967). However, as a general rule, surface irrigation by gravity-flow begins to become difficult at slopes exceeding 2°, and very difficult on slopes exceeding 4° (Greene, 1966).

Economics of Preparing the Land for Irrigation

Extensive land development may be essential as a preliminary to irrigation: e.g. clearing vegetation (trees or brush), removing stones or rocks, terracing, levelling, deep ploughing, and shattering hard-pans. The costs of these operations will have to be considered in evaluating the suitability of land for irrigation.

Quality of Irrigation Water and Drainability of the Soil

The next requirement is that productivity can be permanently sustained under irrigation. The main threats to the permanency of irrigated agriculture are salinization and water-logging. These depend to a large extent on the quality of the irrigation water available—the evaluation of the suitability of a soil for irrigation cannot be divorced from the evaluation of the quality of the irrigation water—and the drainage conditions that will prevail when the land is irrigated. For 'irrigable land must be drainable' (Maletic & Hutchings, 1967).

The ability of the soil to allow percolation of the water needed for leaching, the drainability of the subsoil, the soluble salt levels, and the exchangeable sodium present in the soil profile, are important characteristics.

The salt problem can present itself in two ways: the land may already be affected by excess salts—the evaluation will then be concerned with whether the excess salts can be effectively and economically removed—or the soil may be still free from excess salts, in which case the evaluation will be based on the danger of salt accumulation and the possibility of sustaining productivity.

References

AHARONI, A. (1964). [Timing of irrigation in field crops.—in Hebrew.] *Bull. Extension Service Israel Ministry of Agriculture, Tel Aviv*, 59 pp., illustr.

ARTHUR, H. G. & THOROKY, G. N. (1967). Water conservation through transport loss reduction. pp. 562–72 in *Water for Peace*, U.S. Government Printing Office, Washington D.C., Vol. 3, vi + 968 pp., illustr.

BAVEL, C. H. M., VAN (1960) Drought and water supplies in agricultural soils of the lower Mississippi Valley area. *Tech, Bull.*, U.S. Dept. Agric., **1209**, 93 pp., illustr.

BURINGH, P. (1956). Soils of the oldest cultivated area of the world. pp. 89–91 in *Rapp. 6e Congr. Sci. Sol. E.*, 655 pp., illustr.

EDMINSTER, T. W. & SCHILFGAARDE, J. Van (1955). Technical problems and principles of drainage. Pp. 491–8 in *Water*. Ybk. Agric. U.S. Dept. Agric., Washington D.C., xiii + 751 pp., illustr.

GRASSI, C. J. (1967). Factors affecting irrigation efficiency and the contribution of agricultural engineering in the improvement of water use in Latin American irrigation projects. pp. 497–505 in *Water for Peace*, U.S. Government Printing Office, Washington D.C., Vol. 3, vi + 968 pp., illustr.

GREENE, H. (1966). Irrigation in arid lands. pp. 255–70 in *Arid Lands: A geographical Appraisal* (Ed. E. S. Hills) UNESCO, Paris, xviii + 461 pp., illustr.

GULHATI, N. D. (1967). Irrigation and drainage in the world. pp. 36–40 in *Irrrigation and Drainage of Arid Lands*. F.A.O./UNESCO, Rome & Paris, 663 pp., illustr.

HAGAN, R. M., HAISE, H. R. & EDMINSTER, T. W. (Ed.) (1967). *Irrigation of Agricultural Lands*. Agronomy Series, Amer. Soc. Agron., Madison, Wisconsin, **11**, xxi + 1180 pp., illustr.

HAISE, H. R. & HAGAN, R. M. (1967). Soil, plant and evaporative measurements as criteria for scheduling irrigation. pp. 577–604 in *Irrigation of Agricultural Lands* (Ed. R. M. Hagan, R. H. Haise & T. W. Edminster) Agronomy Series; Amer. Soc. Agron., Madison, Wisconsin, **11**, xxxi + 1180 pp., illustr.

JENSEN, M. E., LAWRENCE, R. & PHELAN, J. T. (1967). Improving irrigation efficiency. Pp. 1120–36 in *Irrigation of Agricultural Lands* (Ed. R. N. Hagan, R. H. Haise & T. W. Edminster) Agronomy Series; Amer. Soc. Agron., Madison, Wisconsin, II, xxxi + 1180 pp., illustr.

JEWITT, T. N. (1966). Soils of arid lands. pp. 103–24 in *Arid Lands: A Geopraphical Appraisal* (Ed. E. S. Hills) UNESCO, Paris, xviii + 461 pp., illustr.

MALETIC, J. T. (1967). Irrigation, a selective function—selection of project lands. pp 513–27 in *Water for Peace*, U.S. Government Printing Office, Washington D.C., Vol. 7, vi + 1004 pp., illustr.

—— & HUTCHINGS, T. B. (1967). Selection and classification of irrigable land. pp. 125–56 in *Irrigation of Agricultural Lands* (Ed. R. M. Hagan, R. H. Haise & T. W. Edminster) Agronomy Series; Amer. Soc. Agron., Madison, Wisconsin II, xxxi + 1180 pp., illustr.

NACE, R. L. (1960). Water management, agriculture, and government supplies. pp. 43–61 in *Water and Agriculture* (Ed. L. D. Hockensmith) Amer. Assoc. Adv. Sci., Washington D.C., 198 pp., illustr.

OLIVIER, H. (1967). Irrigation as a factor in promoting regional development. pp. 266–76 in *Water for Peace*, U.S. Government Printing Office, Washington D.C., Vol. 7, vi + 1004 pp. illustr.

PENMAN, H. L. (1949). Meteorology and agriculture. *Q. J. Roy. Meteorol. Soc.*, **75**, 198–324.

PHELAN, J. T. (1964). Research progress and future needs in farm use of irrigation water. pp. 42–47 in *Research on Water*. ASA Spec. Publ. 4, Soil Sci. Soc. Amer., Madison, Wisconsin, vi + 121 pp., illustr.

RENNE, R. R. (1965). A proper perspective of water in agriculture. pp. 42–47 in *Plant Environment and Efficient Water Use* (Ed. W. H. Pierre, D. Kirkham, J. Pesek & R. Shaw) Amer. Soc. Agronomy and Soil Sci. Soc. Amer., Madison, Wisconsin, xii + 295 pp., illustr.

RICKARD, D. S. (1957). A comparison between measured and calculated soil moisture deficit. *N. Z. Jl Sci. Technol. A*, **38**, 1081–90.

RUBIN, J. & BIELORAI, H. (1957) A study of the irrigation requirements and consumptive water use by sugar beet in the northern Negev. pp. 20–24 in *Final Report of the Ford Foundation Israel Project A5*. Spec. Bull. Volcani Inst. Agric. Res., No 6, 30 pp., illustr. (mimeographed).

SHIMSHI, D. (1965). The use of field porometers for the study of water-stress in plants. *Israel J. Agric. Res.*, **14**, 37–43.

SHMUELI, E. (1964).The use of physiological indicators for the timing of irrigation. *8th Congress of Soil Science, Bucharest*, Romania, II, pp. 423–31.

STANHILL, G. (1957), The effect of differences in soil moisture status on plant growth. *Soil Sci.*, **61**, 205–14.

——, RAZUMOVA, L. A., DOYLE, J. J., GANGOPADHYAY, M., BAIER, W. & WINTER E. J. (1967). *Practical Soil Moisture Problems in Agriculture* World Meteorological Organization, Geneva, 112 pp. (mimeographed).

STEED, G. L. (1967). Scheduling irrigation for increased efficiency of water and land use. pp. 573–84 in *Water for Peace*, U.S. Government Printing Office, Washington D.C., Vol. 3, vi + 968 pp., illustr.

TANNER, C. B. (1967). Measurement of evapotranspiration. pp. 504–55 in *Irrigation of Agricultural Lands* (Ed. R. M. Hagan, R. H. Haise & T. W. Edminster) Agronomy Series; Amer. Soc. Agron., Madison, Wisconsin, **11**, xxxi + 1180 pp., illustr.

TAYLOR, S. A. (1952). Estimating the integrated soil moisture tension in the root-zone of growing crops. *Soil Sci.*, 73, 331–40.

VAN BAVEL, C. H. M. see BAVEL, C. H. M. VAN.

IX

THE MAINTENANCE OF A PERMANENT IRRIGATION AGRICULTURE

The Problem

In view of the breakdown in the past of many civilizations that were based on irrigation agriculture, and the numerous cases of rapid soil-deterioration under irrigation in modern times—both in countries with a primitive agriculture and in those with the most advanced technologies—doubts are frequently expressed as to the possibility of maintaining irrigated agriculture permanently.*

Experience and research have shown unequivocally that the basic causes of the failure of crop production under irrigation are the combined and related effects of excessive salt-accumulation in the root-zone and the development of a high water-table. Because of the high evaporation-rates and low precipitation of the arid regions, many arid soils contain appreciable amounts of salts to start with, and natural drainage in these regions is rarely effective. As the salts generally accumulate in areas of low elevation, and as these are the lands that in their topography, are best adapted for irrigation, they are already predisposed to salinity hazards which are then aggravated by irrigation proper.

Irrigation water adds considerable amounts of salts to the soil, and frequently contributes to a rise in the water-table. When the water-table is within capillary range of the root-zone, the vicious circle is closed: additional salts are added to the root-zone by the capillary rise of saline water from the water-table, and evaporation of water and deposition of salt are increased. As a result, leaching of excess salts becomes impracticable and the inevitable restriction of the root-zone precludes normal development.

An understanding of the factors involved in the breakdown of irrigated agriculture indicates that this can be prevented by the maintenance of a satisfactory salt-balance in the root-zone. This is dependent on the quality of the

* The effect of irrigation on soil fertility will be discussed in detail in Chapter XI.

water used and on the effective removal of excess salts by leaching.

If the amount of precipitation or irrigation water is sufficient to leach the salts to a depth from which capillary rise to the root-zone cannot occur, a harmless zone of accumulation in the subsoil will result. Sooner or later, however, the excess salts have to be removed from the soil by an effective system of leaching and drainage. To make this possible, two conditions are essential: the maintenance of a satisfactory percolation rate—itself dependent on good structure—and an effective drainage system.

Quality of Irrigation Water

One of the first, essential steps before considering an irrigation scheme is to assess the quality of the irrigation water available. In general, the shortage of water resources in arid regions results in a tendency to use water of questionable quality. The quality of the water available will have a bearing on the type of soil that can be irrigated, the crops that can be grown, and the management techniques and irrigation methods to be adopted.

There are considerable differences in the salt-content, both quantitative and qualitative, of waters used for irrigation. A number of criteria have been proposed by various authors, for assessing water quality. Without going into details of the variously purposed approaches to the classification of waters according to their quality, certain criteria are generally accepted and can serve to predict the effect of irrigation water on soil and crops under specified conditions.

QUALITY CRITERIA OF IRRIGATION WATERS

Total Soluble Salt-content (*TSS*) *or Salinity Hazard*

The early assessments of irrigation-water quality were concerned mainly with the concentration of soluble salts in the water, irrespective of their composition. TSS-content was expressed in gm/l, meq/l, p.p.m., or as specific electrical conductivity.

The salt-content of most irrigation waters ranges from 70 to 3500 p.p.m. (Allison, 1964). Irrigation therefore adds appreciable amounts of salts to the soil, and the resultant salinization may, after a time, jeopardize the permanence of irrigation agriculture.

The principal cations of irrigation water are sodium, calcium, and magnesium, and these are generally present as chlorides, sulphates, and bicarbonates.

Though the total concentration of soluble salts is the most important single criterion of irrigation water, it is generally inadequate in itself to indicate quality, as it makes no distinction between different kinds of salts: some salts are more dangerous than others, and certain salts are even beneficial when not present in excess.

When considered in conjunction with additional criteria described below, waters are classified by the U.S. Salinity Laboratory (Richards, 1954) in four categories based on electrical conductivity values:

C_1 *Low-salinity water* has an electrical conductivity of less than 250 micromhos/cm and can be used safely for irrigation of most crops and on most soils.

C_2 *Medium-salinity water*, which has an electrical conductivity lying between 250 and 750 micromhos/cm, can be used safely, if a moderate amount of leaching is possible, for plants with moderate salt-tolerance.

C_3 *High-salinity waters* are in the range of 750 to 2250 micromhos/cm. They can be used only on soils with adequate drainage and for crops with fair salt-tolerance, and provided management practices for salinity control are applied.

C_4 *Very high-salinity waters* can be used only on salt-tolerant crops grown under special circumstances, and with adequate management practices on permeable soils. Their electrical conductivity is above 2250 micromhos/cm. Water classification based on salinity hazard, in relation to the permeability of the soil, is presented in Figure 39 (Fireman & Kraus, 1965).

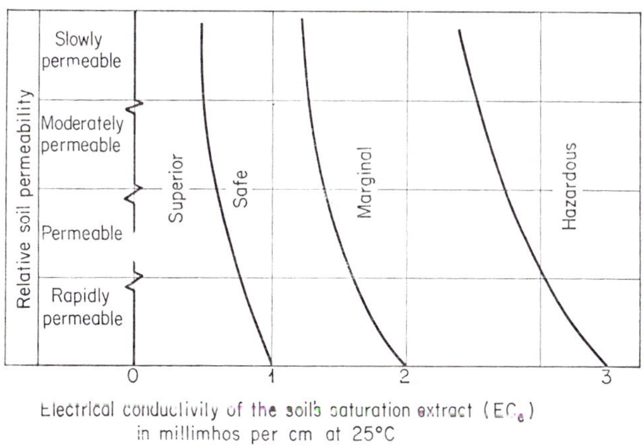

Figure 39. Water classification based on salinity hazard in relation to soil permeability. EC = electrical conductivity of the soils saturation extract. (From Fireman & Kraus (1965): by courtesy of 'Tahal-Water Planning for Israel'.)

In Table IX the water supply of four countries is classified according to their electrical conductivity values (Kovda *et al.*, 1967*b*).

Whilst the classification proposed by the U.S. Salinity Laboratory appears adequate for the classes of water used for irrigation in the United States and Israel, it is apparent from Table IX that 67 per cent of the water used for

TABLE IX

Percentage of Water-supply that falls within Indicated Salinity Ranges in Four Different Countries.

$EC \times 10^6$ at 25°C	U.S.A.[1]	Algeria[2]	India[3] (Rajasthan area)	Israel[4]
			No of samples	
	1018	79	576	1507
250–750	53	5	13	36
750–2250	37	28	35	60
2250–4000		28	22	3
2250–5000	10			
4000–6000		18	18	1
5000–20000		21		

[1] Richards (1954). [2] Durand (1958). [3] Darra *et al.* (1964).
[4] Shalhevet & Reininger (1964).

irrigation in Algeria, and 40 per cent of that used in India, should be considered as very high-salinity waters that can, at best, be used only under special circumstances. This has led to the conclusions that the classification of waters for irrigation based on TSS is of general descriptive value only, and that the limits should be modified for different circumstances. This has been done, for example, for Algeria by Durand (1958), and is based on soil type and plant tolerance, in addition to TSS (*see* Table X).

TABLE X

Upper Permissible Conductivity Limits for Irrigation Water for Two Groups of Plants and Five Soil Types (Durand, 1958)*

	Plant Tolerance Group, $EC_e \times 10^6$†			
	<4 000	4 000–10 000	>10 000	
Soil texture	EC_{iw}‡ $\times 10^6$	$EC_{iw} \times 10^6$	$EC_{iw} \times 10^6$ Forage crops	$EC_{iw} \times 10^6$ Field crops
Sandy	2500	6500	12000	10000
Loamy sand	1600	4000	7000	6000
Loamy	1000	3000	5000	4500
Loamy clay	800	2000	–	3500
Clay	400	1000	1800	1600

* Assuming ideal irrigation conditions (permeable soil and good drainage).
† EC_e = electrical conductivity of the soil's saturation extract.
‡EC_{iw} = electrical conductivity of the irrigation water.

If the TSS of the irrigation water is low, there is usually no need to determine its ionic content. If the concentration of TSS is high, it is necessary to determine the amount of each ion present, the ratios between certain ions, and total salt-content, as well as the ratios between groups of ions. Additional criteria, besides TSS, are therefore needed!

Sodium Hazard

In view of the major adverse effects of sodium on the soil and on the crop, its concentration in irrigation water is a major criterion of quality. However, classifying waters on the basis of sodium content alone is not a satisfactory measure of the sodium hazard, as it does not take into account a number of highly relevant factors. For example, the presence of Ca in the irrigation water can alleviate, to some extent, the negative effects of the Na cation—hence the considerable importance of considering the Na–Ca balance in the irrigation water.

Owing to the dynamic equilibrium between soluble and adsorbed ions, the amounts of calcium and magnesium displaced by sodium from the exchange complex of the soil depend on the relative proportions of these ions that are present in the soil solution. Very little calcium will be displaced from the soil exchange complex when sodium and calcium are present in the irrigation water in approximately equal concentrations. When sodium concentration is double or more that of the calcium, however, some of the calcium will be displaced, and with still higher concentrations of sodium, more and more of the calcium will be displaced from the exchange complex.

In the past, the ratio of sodium to calcium was usually expressed in the following formula, called the 'soluble sodium percentage' (SSP):

$$SSP = \frac{\text{Soluble } Na^+ \text{ concentration}}{\text{Total cation concentration}} \times 100$$

Sodium Adsorption Ratio

A somewhat more recent approach, using the sodium adsorption ratio (SAR), is based on the exchange reaction between monovalent and divalent cations, and therefore gives a more accurate appreciation of sodium hazard, because it relates more directly to the adsorption of sodium by the soil (Richards, 1954).

$$SAR = \frac{Na^+}{\sqrt{\frac{1}{2}(Ca^{++} + Mg^{++})}}$$

The concentration of the ions is expressed in milli-equivalents per litre (meq/l).

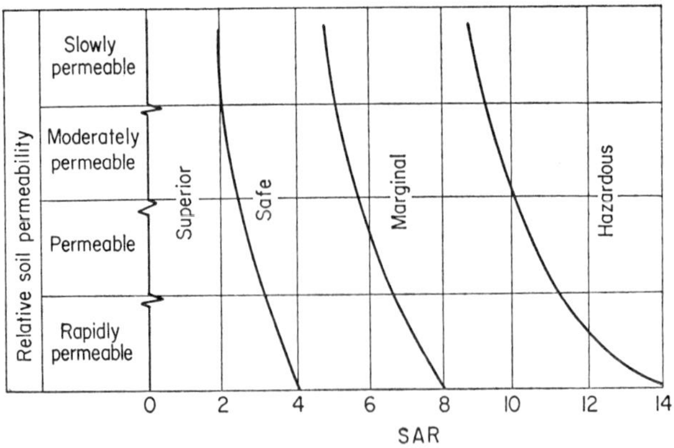

Figure 40. Water classification based on sodium hazard in relation to soil permeability. SAR = sodium absorption ratio. (From Fireman & Kraus (1965): by courtesy of 'Tahal-Water Planning for Israel'.)

Waters may be classified into four categories according to their sodium hazard (Figure 40; Wilcox, 1958):

S_1 *Low-sodium water* can be used for most soil types, with no danger of causing salinization.

S_2 *Medium-sodium water* will not cause difficulties when used on permeable soils with a high sand or organic-matter content. On fine-textured soils with a low calcium content, salinity problems may occur.

S_3 *High-sodium water* can cause salinity problems on most types of soil, and can only be used if certain precautions are taken and preventive methods adopted: these include regular leaching, adequate drainage, use of organic manures, etc. If the calcium content of the soil is high, there is little danger of rapid salinization. On soils that are low in calcium, however, the danger of salinization is always imminent, and regular applications of gypsum are indicated.

S_4 *Very high-sodium water* is generally not suitable for irrigation, and can be used only under exceptional circumstances for special crops.

The dividing points between the above categories depend on the total salinity hazard; with increasing salinity, the permissible SAR values decrease progressively. For example, at EC values of 100 micromhos/cm, the limits of the various groups are at SAR values of 10, 18 and 26, whereas at 2250 micromhos/cm, the corresponding SAR values are 4, 9 and 14 (Allison, 1964).

An empirical relation of high statistical significance* between the SAR of

* There are exceptions: e.g. in Algeria, considerable differences were found between calculated and determined ESP (Durand, 1958).

irrigation water and the exchangeable sodium percentage (ESP) of the soil, has been established (Wilcox & Durum, 1967). Therefore, after the SAR value of an irrigation water has been calculated, it is possible to estimate the ESP value of the soil when it is in equilibrium with the irrigation water.

Bicarbonate Hazard

Water in which the concentration of the bicarbonates is higher than that of calcium and magnesium together, is unsuitable for irrigation—even if the sodium content is within the permissible limits. After each irrigation the calcium and magnesium react with the bicarbonate, and as the soil dries out they become precipitated as insoluble carbonates. With less calcium in solution, the relative concentration of sodium in the soil solution, and hence the sodium hazard, increases. Following this process, infiltration rates are reduced, the pH increases, the organic matter in the soil is dissolved, and the soil takes on the typical black colour which is characteristic of soils in which this process is occurring (black alkali). For assessing water quality in relation to its carbonate and bicarbonate content, the residual sodium carbonate (RSC) is calculated (Eaton, 1950) according to the following equation:

$$RSC = (CO_3^{--} + HCO_3^{-}) - (Ca^{++} + Mg^{++})$$

(all expressed in milliequivalents, meq).

When the RSC is less that 1·25 meq/l, water quality is considered to be good. With an RSC between 1·25 and 2·50 meq/l, the water can still be used for irrigation, provided essential precautions are taken to avoid salinization. Water with more than 2·50 meq/l is not suitable for irrigation (Wilcox *et al.*, 1954). A diagram, combining the various criteria mentioned above for the classification of irrigation waters, is presented in Figure 41.

Phytotoxic Substances

Certain salts, such as boron, which at low concentrations are not harmful to plants, and may even be beneficial, are apt to accumulate in the soil following irrigation and may reach toxic concentrations. Water containing 1 p.p.m. of boron can still be considered as safe for most field-crops; semi-tolerant crops, such as lima beans, oats, sorghum, wheat, barley, field peas, and cotton, are sensitive to concentrations of 1 to 2 p.p.m., while broad beans, lucerne, and sugar-beets, can tolerate concentrations of 2 to 4 p.p.m., which is the upper permissible limit for all crops (Allison, 1964).

Boron is not uncommon in well-water, and a toxic level may be reached in the soil by irrigating with such waters. It may also be present in unusually high amounts in certain soils. Larger quantities of water are required for leaching boron out of the soil than are necessary for sodium salts, as the boron is held more strongly by the soil (Reeve *et al.*, 1955).

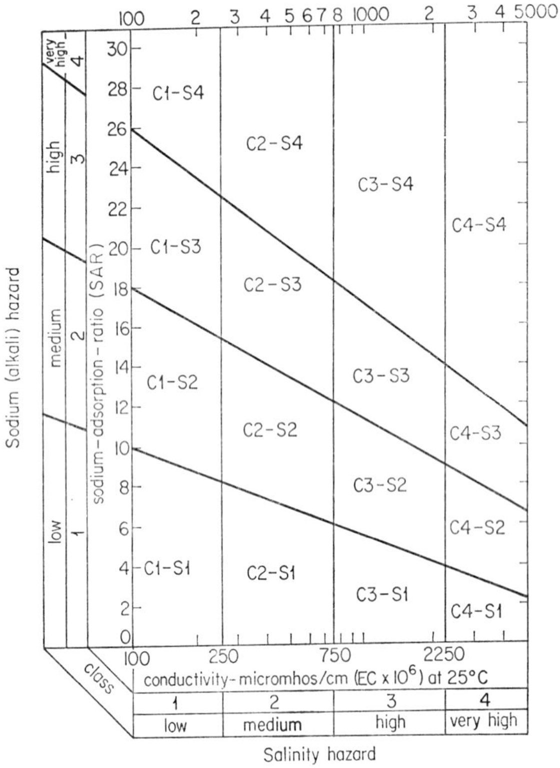

Figure 41. Diagram for the classification of irrigation waters. For explanation, *see* text. (From Richards (1954): by courtesy of the U.S. Department of Agriculture.)

Suspended Inorganic and Organic Materials

The silt content of some river waters that are used for irrigation is an important qualitative characteristic. The more clay aggregates and organic matter the silt contains, the higher is its value in adding fertility to the soil. However, sediment-laden water is more difficult than clear water to use for irrigation: water penetration is impaired by the sediment, and such water cannot be used for overhead irrigation without prior sedimentation.

GENERAL APPRECIATION OF WATER QUALITY

The suitability or otherwise of water for irrigation cannot be determined without reference to the soil to be irrigated, the crops to be grown, the agronomic practices adopted, and the environmental conditions prevailing in the region. Therefore, quality criteria for irrigation waters cannot be applied rigidly, and will frequently have to be adapted to regional conditions. Even good-quality water can cause damage when applied to an impermeable soil, while poor-quality water may be satisfactory when used for irrigating permeable, well-drained soils (Bernstein, 1962). Two aspects of the problem should be considered:

(*a*) *In Relation to Soil Types.* The effect of salts in irrigated soils on plant growth is not determined by the salt-content of the irrigation water *per se*, but depends on the maximum concentration of the salts in the soil solution, which is reached shortly before irrigation.

In very permeable soils, the differences between the concentration of salts in the soil solution, and the salt-content of the irrigation water, are usually small. In less permeable soils, the concentration of salts in the soil solution, shortly before irrigation, may be ten times as high as in the irrigation water. Therefore, the less permeable the soil, the greater will be the danger of using water with a high salt or a high sodium content. Furthermore, the situation usually deteriorates progressively, as percolation gradually becomes reduced following the replacement of calcium by sodium in the exchange complex of the soil.

In India it has also been found that the higher the pH of the soil is, the narrower will be the range of tolerance to salinity (Agarwal & Yadav, 1956).

Highly saline waters, with salt concentrations of 3 to 8 gm/l, are frequently used for irrigation during very long periods in Central Asia, the Near East, and North Africa. This is possible when salt-tolerant crops are grown, the soils are very sandy, the water is high in gypsum content, and is applied at high rates and under conditions of excellent natural drainage.

In southern Tunisia, water with over 5 gm/l of total salts, of which 1·2 gm/l is sodium, has been used under these conditions for long periods. Yields are low because of lack of fertilizers, inadequate pest-control, etc., so that salts are not a limiting factor under these circumstances (Kovda *et al.*, 1967*a*). Elsewhere in Tunisia, irrigation with water that had a total salt-content of 2·68 gm/l did not appreciably increase soil salinity after three years, though some increase in salt concentration in the root-zone was noted. During this period the yields of maize were reduced by 6 to 10 per cent (not significant), and those of lucerne by 11 to 12 per cent; by contrast, the yields of wheat were increased by 35 per cent, approximately, and those of cotton by 15 to 20 per cent (Cointepas, 1964). In Israel, water containing 2·3 gm/l of total salt, of which 1 gm/l is chloride, has been used on sandy soils without apparent damage (Shalhevet & Reininger, 1964). In Texas, water with more than 4 gm/l of total salts is occassionally used (Longenecker & Lyerly, 1959).

In certain regions of Italy, in particular in the province of Bari, irrigation with high-salt water has been practised for a long time (Bottini, 1961). The waters used contain up to 8–10 gm/l of salts. The soils are sandy to a depth of several metres, and are therefore very permeable, with excellent natural drainage; they are also rich in calcium and organic matter. During the growth period the soil is maintained at a high level of moisture by frequent irrigations.

Irrigation is practised in one summer out of two, whilst the salts are leached out by two successive seasons of winter rainfall amounting to 400–600 mm each. Even with these precautions, not all crops can be grown: maize—a moderately tolerant crop—gives very low yields under these conditions. The crop that is generally grown is tomatoes.

On the other hand, water with a relatively low salt-content may cause salinization of heavy soils with poor percolation and drainage in a relatively short time, as has occurred in the delta of the basins of the Tigris and Euphrates, in the Indus basin, and elsewhere. The process is actually accelerated if water applications are low, though less salts are in fact added.

(*b*) *In Relation to Crops*. The evaluation of irrigation water will depend on the tolerance of the crops, to be grown in the rotation, to total salinity and to specific ion concentrations.

Crops differ in their tolerance to soil salinity. The response to salinity of most economic crops has been determined on the basis of measurements of the electrical conductivity of a saturation extract of the soil. The degree of soil salinity at which a specific reduction in crop yield occurs, in comparison with the non-saline condition, serves as a criterion for establishing the relative tolerance of crops to salinity. Various levels of yield reduction have been used, such as 10 to 15 per cent by Bernstein (1964) and 50 per cent by Richards (1954).

The salt tolerance of various field and forage crops is shown in Table XI and Figures 42 and 43.

TABLE XI

*Salt Tolerance of Main Field-crops** (Simonneau *et al.*, 1967)

(EC_e)† Salts in dry soils‡(%):	16–10 1.0–0.5 (Highly tolerant)	10–4 0.5–0.2 (Moderately tolerant)	4–2 0.2–0.05 (Sensitive)
	Barley	Wheat	Red Clover
	Sugar-beet	Oats	Peas
	Cotton	Lucerne	Beans
		Rice	Sugar-cane
		Maize	
		Flax	

* Within each group, the crops are listed in the order of decreasing tolerance.

† $(EC)_e$ electrical conductivity of the saturation extract of the soil, in millimhos/cm.

‡ Soil of average texture.

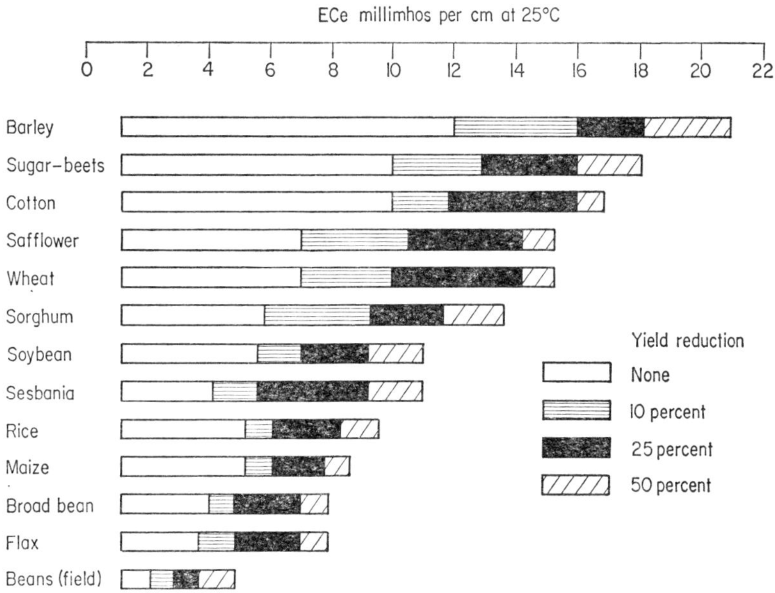

Figure 42. Salt tolerance of a number of field-crops. (From Bernstein (1964): by courtesy of the U.S. Department of Agriculture.)

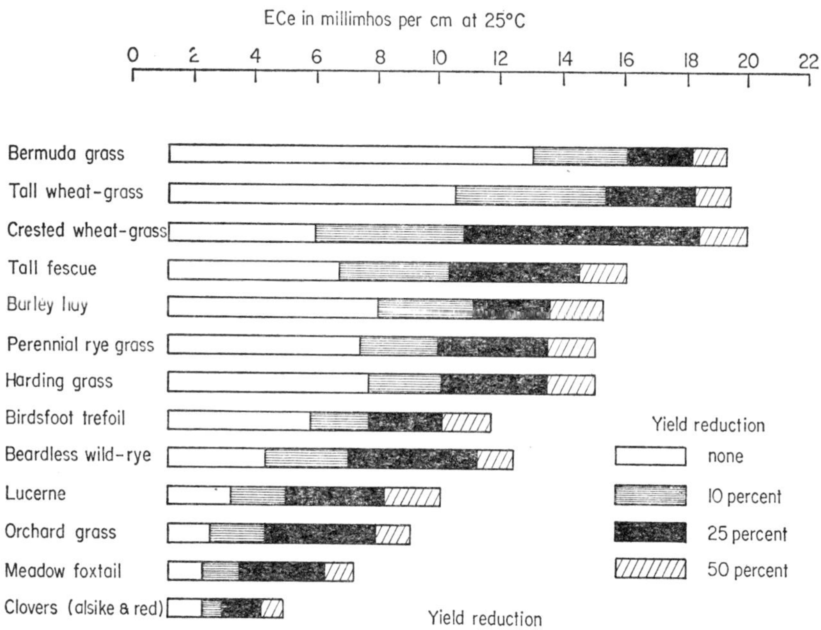

Figure 43. Salt tolerance of a number of forage crops. (From Bernstein (1964): by courtesy of the U.S. Department of Agriculture.)

Rainfall and Temperature

The level of salts that can be tolerated in irrigation water is directly related to the amount of precipitation. In low-rainfall regions or seasons, there is an increased demand for irrigation water, so that more salts are added to soil and at the same time leaching by rain is reduced. These effects are further enhanced by high temperatures which cause increased evapotranspiration. In regions with high seasonal rainfall, most of the salts that accumulate in the root-zone during the irrigation season can be leached out during the rainy season, providing that drainage—whether natural or artificial—is adequate. It is estimated that an annual rainfall of 400 to 500 mm completely removes any excess of salts (Tames, 1950). Consequently, the dangers of salinization will increase, the more arid the region is.

Seasonal Variations

The quality of irrigation water is not static, but shows seasonal variation. As a result of evaporation, the salt-content may rise in reservoirs as the summer advances. This rise is also a function of the size of the body of water, varying from 20 per cent for a large reservoir such as Lake Tiberias, to 100 per cent in small marshes (Kovda *et al.*, 1967*b*). The salt-content of flowing water may also vary within wide limits. During the flood stage, Nile water has a low salt-content and sodium percentage; the salt-content is more than doubled at the low-water stage (Bernstein, 1962). Deep ground-water tables generally do not show seasonal fluctuations in salt-content. By contrast, rainfall and irrigation may cause considerable changes in the salt-content of high ground-water.

Amounts of Water Applied, and Drainage

The soil solution usually has a salt concentration that is in excess of that in most waters used for irrigation. In most irrigated soils the salt concentration of the soil solution is generally from three to ten times as great as that of the irrigation water used (Tison, 1961). This relationship is affected by the actual concentration of salts in the irrigation water and in the soil, by the permeability of the soil, by the efficacy of drainage, by the climatic conditions, and by the amount of leaching.

The optimum concentration of soluble salts in soil solutions of irrigated soils is within the limits of 3–5 gm/1. Cultivated plants show slight stress at around 10–12 gm/1, and they die at concentrations of 20–25 gm/1 (Kovda, 1961). Therefore, irrigation with water having a relatively high salt-content (2–5 gm/1) will still serve to dilute the soil solutions. If the dilution is sufficient and the excess salts are leached out, no damage to the crops will result. It is only when salts are allowed to accumulate, because of insufficient leaching and/or inadequate drainage, that permanent damage results. Therefore, the allowable salt-content of irrigation water depends on the amounts of water applied and subsequently removed (Kovda *et al.*, 1967*a*).

Improving Water Quality

Frequently, the only alternatives open to the farmer whose water is of extremely poor quality are: not to irrigate at all or to use water for supplemental irrigation only. There are, however, cases in which it is possible to improve somewhat the quality of water with a relatively high sodium and low calcium content, by increasing the proportion of Ca^{++} ion in the irrigation water. Gypsum is an inexpensive source of relatively soluble calcium, which reduces the sodium: calcium ratio in the irrigation water. Added gypsum also increases the infiltration capacity of soils that have already been adversely affected by irrigating with sodic waters.

Management Practices

The methods of irrigation practised, the efficacy of drainage, the maintenance of a favourable soil structure and of a high level of soil organic matter, and other agronomic practices affecting soil fertility, all have a bearing on the quality of water that can be used safely for irrigation.

Effect of Irrigation Water on Soils

Irrigation water affects the soil by changing both the concentration of the soil solution and the composition of the constituents in solution as well as those adsorbed by the exchange complex (Bower, 1961).

The amount of soluble salts added to the root-zone will depend on the amount and quality of the water that passes through this zone during a period of time (Figure 44). The concentration of the soil solution also increases as the water-content of the soil decreases.

The magnitude of these effects will depend on the frequency of irrigation, the amounts of water removed by evapotranspiration, the depth of the ground-water table, the concentration of salts in the ground-water, and the permeability of the soil. Irrigation usually results in a gradual increase of salts in the root-zone, causing salinization of the soil. The abundance of soluble salts maintains the soil colloids in a flocculated condition, and therefore has a favourable effect on soil structure and permeability (Figure 45).

As soluble salts accumulate in the soil and the soil solution becomes more concentrated, the solubility limits of calcium sulphate, calcium carbonate, and magnesium carbonate, may be exceeded, causing precipitation of calcium and magnesium. This results in a corresponding increase in the relative proportion of sodium in the soil solution. There exists an equilibrium between the cations in the soil solution and those adsorbed on the exchange complex. A

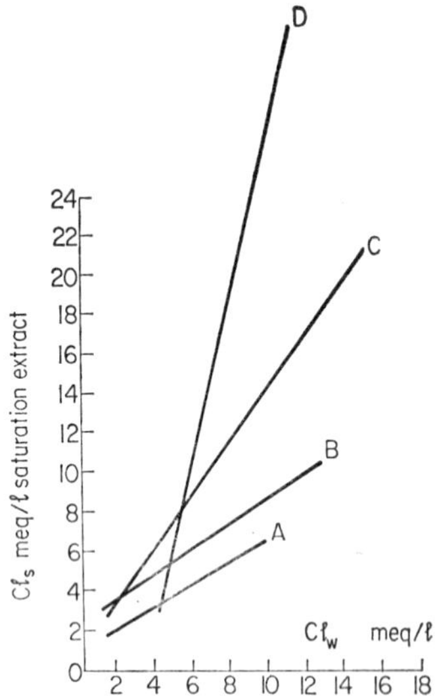

Figure 44. Accumulation of chloride in four different soils (Cl_s) as a function of chloride content of the irrigation water (Cl_w).
A = sandy soil. B = sandy clay-loam. C = clay-pan soil. D = grumosolic dark-brown.
(From Salinity Survey, Progress Report 1963, State of Israel.)

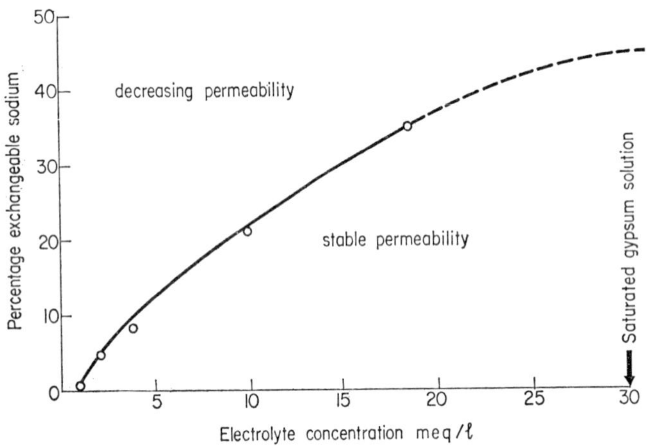

Figure 45. Effect of electrolyte concentration on soil permeability. An abundance of soluble salts maintains the soil colloids in a flocculated condition and therefore has a favourable effect on soil structure and permeability. (From Quirk & Schofield (1955): by courtesy of the British Society of Soil Science.)

the relative proportion of sodium in solution increases, more sodium is adsorbed on the exchange complex. This process will in due course cause the soil to become both saline and sodic.

The increase in exchangeable sodium has a marked influence on the physical properties of the soil, the main effect being the tendency to cause the dispersal of the soil colloids. As long as sufficient non-sodic salts are present, even sodium-saturated clays remain flocculated. However, if soluble salts are removed from the soil by leaching, changing the saline sodic into a non-saline sodic soil, the soil colloids are dispersed and the soil structure then deteriorates. Water applied then no longer percolates easily, the soil does not drain well, and oxygen levels become low. As the soil solution becomes more dilute, new equilibria are established between the cations of the soil solution and the exchange complex.

Effects of Salinity on Crops

Salts affect plants directly in two ways: by increasing osmotic pressure and by the specific toxicity of their ions. Which of these two is the major factor, is still subject to discussion; as the two kinds of effect occur simultaneously, it is very difficult to elucidate their mechanisms. Indirect effects may also result from nutritional disturbances, and from the effects of salts on micro-organisms and on soil structure—the last-mentioned often resulting in adverse physical conditions for plant growth. All these factors naturally affect crop yields in various degrees, depending on the susceptibility of the crop (Figure 46).

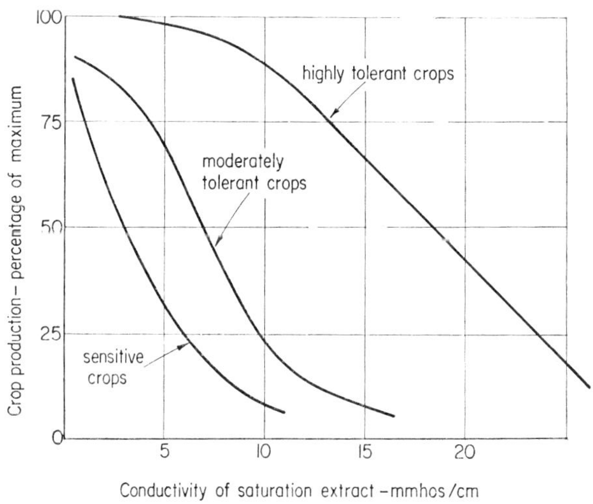

Figure 46. General relationship between crop production and soil salinity. (From Fireman & Kraus (1965): by courtesy of Tahal-Water Planning for Israel.)

OSMOTIC EFFECTS

The increase in osmotic pressure of the soil solution due to high concentrations of salts, irrespective of the kinds of salts, increases the amount of energy which the plant must expend to take up water from the soil. As a result, respiration is increased, and the growth and yield of most plants decline progressively as the osmotic pressure increases (Grillot, 1956). With increasing osmotic concentration of the soil, growth reduction of many crops is generally linear within a wide range. For most crops, relative yields are more depressed at a given osmotic concentration in warm climates than in cool climates (Magistad *et al.*, 1943).

As the soil dries out between two irrigations, the salt concentration of the soil solution increases progressively. Immediately after irrigation, the concentration is suddenly reduced. Plants are therefore submitted alternatively to an excess of salts, followed by a sudden excessively rapid uptake of water which may occasionally cause the rupture and death of the cells (Grillot, 1956).

TOXIC EFFECTS

Many of the ions which are harmless or even useful at relatively low concentrations may become toxic to plants at high concentrations. The ions that are most likely to be toxic to plants are chloride, sodium, bicarbonate, sulphate, and boron (Hayward, 1956). Soviet scientists generally emphasize the adverse effects of the anions (chlorides and sulphates) (Anon., 1962), whilst American scientists are generally more concerned with the cationic components of the salts (Allison, 1964).

Salt tolerance appears to be related either to an ability to control intake of Na^+ and Cl^- ions, or to tolerance to a high level of chloride content before leaf injury occurs (Vose, 1963). Susceptible trefoil varieties absorbed more chloride than did resistant varieties (Hayward & Wadleigh, 1949), whilst susceptible cotton varieties contained more sodium than did tolerant varieties (Richards, 1954). Similarly, salt-susceptible varieties of barley had higher Na and chloride and lower K contents than did resistant varieties. The grain yield and 1000 grain weight of the susceptible varieties were reduced at a relatively low level of NaCl, at which the resistant varieties remained unaffected (Greenway, 1962).

Most field-crops are, however, not harmed by the absorption of quite high concentrations of sodium or chloride. An excess of bicarbonate may be toxic to sensitive crops such as Dallis Grass and field beans (Allison, 1964); boron may be extremely toxic at certain levels, as already mentioned (p. 243).

INDIRECT EFFECTS

Nutritional Effects

The excess of certain ions frequently has a disturbing effect on the absorption of other ions. For example, high concentrations of sulphate generally decrease the uptake of calcium whilst promoting the uptake of sodium, thereby causing sodium toxicity in susceptible crops. High concentrations of calcium may interfere with the uptake of potassium (Bernstein & Hayward, 1958). Saline soil containing a high proportion of magnesium to calcium, may cause Ca deficiency in plants.

The increased pH, due to the increment of sodium carbonate (*see* p. 243), reduces the availability of P, Fe, Zn and Mn (Magistad, 1945).

Soil Micro-organisms

Barren alkali soils contain a rich microflora, capable of surviving under air-dry conditions for very long periods. The sensitivity of different micro-organisms varies widely: ammonia-producing organisms and *Azotobacter* are relatively resistant, whereas nitrate- and nitrite-producing organisms are far more sensitive. Alkali salts become toxic to nitrifying micro-organisms before they harm many higher plants, and it is possible that, at such lower concentrations, the harm done by salts is indirect, through their effect on nitrification (Greaves & Jones, 1941).

Effects of Impaired Structure of Sodic Soils

The impaired soil structure of sodic soils may adversely affect plant growth as a result of poor aeration, nutritional disorders, and poor permeability to water (Magistad, 1945).

TOLERANCE OF CROPS TO EXCHANGEABLE SODIUM PERCENTAGE (ESP)

Whilst most fruit trees are extremely sensitive to ESP, showing toxicity symptoms at ESP values of 2–10, the tolerance of field-crops ranges from moderate to high. Moderately tolerant crops are clover, oats, Tall Fescue, rice, and Dallis Grass, whose growth becomes stunted at ESP values of between 20 and 40; tolerant crops are wheat, cotton, lucerne, barley, and beets, whose growth is adversely affected only at ESP values of 40 to 60. The growth of Crested and Tall Wheat-grass, as well as of Rhodes Grass, is affected only when the ESP exceeds 60.

FACTORS INFLUENCING THE TOLERANCE OF CROPS TO SALINITY

Species and Varieties

There are, frequently, marked differences in the susceptibility to salinity of different species and varieties of the same crop. Thus the long-staple Egyptian cotton varieties are less sensitive than are the medium-staple American varieties, while marked differences in salt tolerance can exist between closely related varieties of soybean (Plate 21).

Tolerance to salinity is genetically controlled, but inheritance is relatively complex, and—like yielding ability—it varies with climate and environmental conditions (Peterson, 1961).

Age of the Plant

Plants are frequently found to be salt-tolerant either in the germination and seedling stage, or in a later stage of growth, or in both. Therefore, the degree of tolerance of a crop to salinity must be evaluated at three stages of its growth, namely those of germination and seedling, vegetative growth, and maturation.

Germination and Seedling Growth

At relatively low concentrations, increasing salinity causes a progressive retardation of germination; at higher concentrations the *ability* to germinate is reduced. Certain ions may also have a toxic effect that reduces germination or causes abnormalities in the embryo or seedling.

The higher the temperature is above optimum levels, the more will the sensitivity to salinity increase (Hayward & Wadleigh, 1949). The effects of salinity on germination increase as soil moisture-tension increases, and *vice versa*.

These effects are aggravated by the fact that salts tend to concentrate in the upper layer of soil, where the seeds are placed (Bernstein, 1962).

There is, however, not always a correlation between the relative resistance of crops to salinity at germination and that at later stages of growth. Thus most crops are more sensitive to salinity in the seedling stage than at later stages: sugar-beets, for example, are very sensitive to salts in the seedling stage, being affected by a conductivity of 4 to 5 mmhos/cm, but they become one of the more resistant crops as they develop, tolerating up to 12–15 mmhos/cm in their later stages of growth (Reeve & Fireman, 1967). There are, however, exceptions:

Maize, which is less tolerant during later stages of growth than in early ones, gives satisfactory germination at salinity levels which adversely affect the germination of sugar-beet and lucerne seeds (Hayward & Wadleigh, 1949). Rice germination is normal at salinities up to 10 to 15 mmhos/cm, but the seedlings usually die at salinities in excess of 5–6 mmhos/cm; at a later stage of growth, the plants are less sensitive to salinity (Pearson & Ayers, 1958). For this reason, it may be necessary to drain the paddy field during a sensitive stage, refilling it with fresh water in order to lower salinity (Allison, 1964).

Vegetative Growth

As already pointed out, vegetative growth is depressed as the osmotic pressure of the soil solution increases. At first, the rate of growth may be reduced without any external signs of injury, and it is not always easy to recognize salt injury in the field. However, with increasing salinity, visible modifications in leaf structure may appear: the cuticle becomes thicker, with a tendency towards succulence in some crops; the leaves develop a waxy bloom and a deep blue-green colour. At high salt levels, certain crops, such as barley, sorghum, and field beans, may show leaf-burn (Eaton, 1942).

Economic Yield

On passing from the vegetative to the reproductive stage, there is generally a marked increase in salt tolerance (Pantanelli, 1937).

High osmotic pressure of the soil solution usually causes reduced yields, but not in all cases are vegetative growth and yielding ability affected equally. For example, the vegetative growth of barley and cotton is more affected by salinity than are the yields of the economic product, grain and fibre, respectively, so that relatively stunted plants can still give fair yields (Reeve & Fireman, 1967).

Among the more salt-tolerant forage crops, yields of seed are less affected by salinity than are forage yields (Ayers *et al.*, 1952). By contrast, rice may produce normal straw yields at a level of salinity that adversely affects grain production (Kapp, 1947).

Climatic Conditions

Climatic conditions generally influence the sensitivity of most crops to salinity: e.g. sensitivity generally increases during dry, hot weather (Figure 47). However, not all crops behave alike in their reaction to the combined effect of salinity and climate. The yields of some crops are reduced equally by a given concentration of salts in the soil, irrespective of climate (Magistad *et al.*, 1943).

Limiting Factors

The criteria for salt tolerance of crops have been determined under optimum environmental conditions; very little is known on how crop tolerance is affected by other limiting factors such as soil fertility, moisture régime, etc. (Reeve & Fireman, 1967).

MANAGEMENT PRACTICES FOR MAINTAINING THE PRODUCTIVITY OF IRRIGATED SOILS

The salt balance in the soil depends on maintaining a proper water-balance. Therefore, the effective control of salinity depends on all management practices that are related to the amount of water used, the frequency with which it is applied, and its movement in the soil. Among these practices are:

1. The preparation of the land to be irrigated, and soil management;
2. The methods of irrigation adopted;
3. The amounts of water applied and the frequency of irrigation, according to crop needs;
4. The maintenance of a 'salt balance' by leaching; and
5. The removal of excess water and salts by drainage.

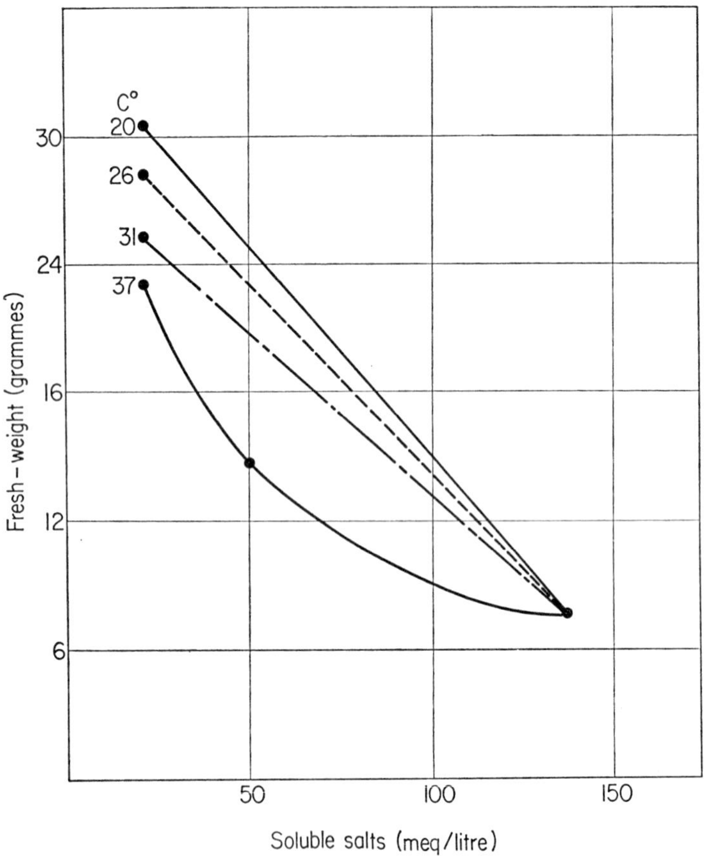

Figure 47. Effect of temperature on the response of dwarf beans to different levels of salinity. (From Lunt *et al.* (1960): by courtesy of Elsevier Publishing Company, Amsterdam.)

LEVELLING

A careful levelling of the areas to be irrigated as well as careful planning of the irrigation layout, are essential in order to ensure a uniform distribution of water in surface methods of irrigation. With overhead irrigation, this is less critical (cf. p. 223). When irrigation water is not uniformly distributed throughout the irrigated area or water infiltration is not uniform, leaching is not uniformly effective: there are then tendencies for excess leaching to occur in the low areas and for salts to accumulate in the high ones.

SOIL MANAGEMENT

All soil management practices that improve the infiltration rate and maintain a favourable soil structure will contribute to salinity control.

Mulching has been found helpful in keeping down salinity in localized saline-soil areas by reducing evaporation from the soil surface, so that less salt accumulates, and by improving percolation rates, so that excess salts are more easily removed by leaching.

On a saline-sodic soil with high salt-concentration, it was found that a cotton-bur mulch facilitated leaching of soluble salts by rainfall. Salt concentration was thus reduced to a safe level for field-crops in the surface 75 cm of soil (Fanning & Carter, 1963).

METHODS OF SOWING

Following irrigation, there are sometimes considerable differences in the concentration of salts in the soil, depending on micro-relief. Even in soils that are non-saline, enough salt may accumulate in seed-beds on flat-crowned ridges to prevent germination (Bernstein & Ayers, 1955). With furrow irrigation, for example, the salts will tend to concentrate mainly in the centre of the ridge between furrows and in a thin layer along the top of the ridge, decreasing on the slopes of the furrow, to reach a minimum at the bottom of the furrow (Bernstein, 1962). Therefore, the most dangerous location for the seed-row is in the centre of the ridge, and the safest is at the bottom of the furrow. At this location the emerging seedling may, however, suffer from crusting of the soil surface. Methods proposed for minimizing salt accumulation in the vicinity of the seeds include: additional irrigation under low evaporative conditions (e.g. at night), modifications in planting practice (such as maintaining furrows between flat-topped ridges as full of water as is practicable), using low ridges, planting in V-depressions on otherwise flat ridges, cf. Figure 48 (Bernstein & Fireman, 1957).

From research carried out by Bernstein & Ayers (1955) the most successful seed-bed would appear to be on a sloping ridge, with the seed planted on the slope of the ridge, several inches below the crown of the ridge.

The salt is carried beyond the line of sowing and therefore leaves a nonsaline zone for the germinating seed. Germination under these conditions has been found to be satisfactory, even when salinity in the ploughed layer as a whole was as high as 30 to 40 mmhos EC_e at the time of ridging.

CROP ROTATION

Crop rotation can be very effective in controlling salinity. The inclusion in the rotation of such crops as rice, that require higher-than-usual amounts of water, makes possible the leaching of excess salts without damage to the current crop. The inclusion of salt-tolerant crops in the rotation, during periods of leaching and reclamation, is also desirable.

IRRIGATION METHODS

Irrigation methods have considerable influence on salinity levels, and in

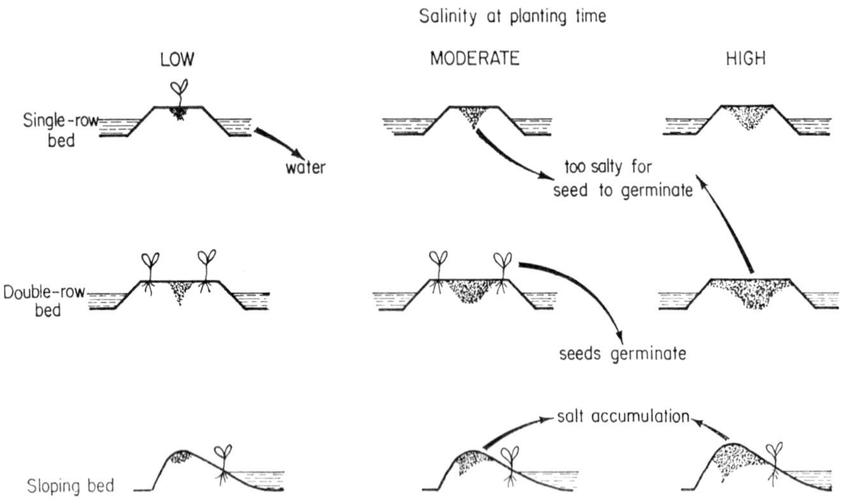

Figure 48. Effect of soil salinity and bed type on salt accumulation in a seeded area. Germination is prevented or delayed where salt accumulation is excessive. The stippling indicates the positions and relative amounts of salt accumulation. (From Lyerly & Longenecker (1964): by courtesy of *World Farming*.)

particular on the spatial distribution of the salts in the root-zone. Salts will be most uniformly distributed under overhead irrigation. Whilst the uniformity of application of water that is possible with sprinkler irrigation is a great advantage in the prevention of saline conditions, the general tendency is to apply less water than with surface irrigation. This may result in increased salt concentrations in sprinkler-irrigated soils as compared with surface-irrigated fields. Consequently, it is essential to apply sufficient water periodically for leaching, when overhead irrigation is used (Hobbs & Russell, 1967).

Furrow irrigation causes the maximum concentration of salts to occur in the location in which damage to plants is greatest (*see* Figure 49). Localized accumulation of salts is also typical for raised beds, ridges, or borders. Under conditions that are conducive to high salinity, the border method of flooding is more suitable than furrow irrigation, whilst corrugations are definitely unsuitable. The frequency of irrigation may also affect the degree of salt damage experienced by the crops. It is the concentration of the soil solution that normally causes injury to the crops and not the irrigation water itself. By irrigating at relatively short intervals, it is possible to prevent the soil solution from becoming excessively concentrated. The more saline the growing conditions, the more frequently the crops must be irrigated in order to reduce adverse effects on plant growth. Such more frequent irrigations are, however, associated with greater proportional losses of water by evaporation, and with reduced

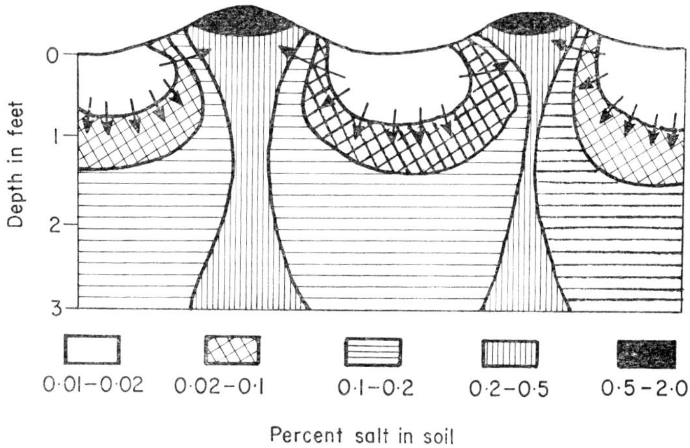

Figure 49. Effect of furrow irrigation on salt-distribution in the soil profile. The arrows show the direction of the flow of water and salt during and some time after irrigation. Salt leaches out of the soil under the furrows and builds up in the ridges. (From Lyerly & Longenecker (1964): by courtesy of *World Farming*) based on data from Wadleigh & Fireman (1949).

leaching effects—so that, in the long run, they are conducive to greater increments of salt.

SALT BALANCE AND LEACHING

In order to prevent salinization, a balance must be maintained between the quantity of dissolved salts introduced into the root-zone, and that removed from the root-zone.

As the irrigation water is the main source of added salts, it would appear at first sight that reducing the amounts of water applied, to the bare minimum requirement, is the first essential step in reducing salinity hazards. However, experience has amply proven that only by using water in excess of consumptive use, can salt accumulation in the root-zone be prevented unless, of course rainfall is adequate to do the leaching during the rainy season.

The irrigation farmer therefore faces two contradictory requirements: on the one hand he is required to apply water not in excess of crop requirements in order to avoid waste and also a rise in the water-table with all its attendant ills; on the other hand, soil salinity will increase unless he applies water in excess of what is actually needed by the crop. Small wonder that primitive agriculture systems were not able to solve this dilemma!

Leaching Requirement

The quantity of salts that accumulates in the root-zone during an irrigation season depends on the amount of irrigation water applied, and on its salt-content. The amount of additional water which must pass through the root zone in order to reduce soil salinity to a predetermined level, called the 'leaching requirement', is also dependent on the same two factors.

Bernstein (1962) has established a formula that gives an estimate of the leaching requirement (LR), based on maintaining a balance between the amount of salts that enter the root-zone and the amount that is removed. The equation gives an approximation only, as it does not take into account the precipitation of certain salts, their uptake by plants, fertilizer applications, etc.

The amount of salt that enters the root-zone with irrigation is:

$$EC_{iw} \times D_{iw} \times K \quad [1]$$

and that leaving the root-zone is:

$$EC_{dw} \times D_{dw} \times K \quad [2]$$

where EC = electrical conductivity,
D = depth of water applied per unit area,
K = a constant,
iw refers to irrigation water, and
dw refers to drainage water.

In order to prevent accumulation of salts in the root-zone, the amount of salts removed must be at least equal to the amount added, so that:

$$EC_{iw} \times D_{iw} \times K = EC_{dw} \times D_{dw} \times K,$$

from which is derived the equation:

$$LR = \frac{D_{dw}}{D_{iw}} = \frac{EC_{iw}}{EC_{dw}} \quad [3]$$

EC_{iw} is known from the analysis of the irrigation water and EC_{dw} is established according to the maximum permissible salt concentration of the crop involved.

The leaching requirement may vary from 15 per cent above the water requirement, when high-quality irrigation water is used, up to 70 per cent above the water requirement, when relatively saline water is used and the crops grown are particularly susceptible to salts (Figure 50). It is frequently possible to apply a heavy (leaching) irrigation at the beginning of the season, to be followed by irrigations at the normal rates (based on evapotranspiration and irrigation efficiency) throughout the growing period of the crop. Amounts of rainfall of less than 250 mm annually are insufficient to affect the leaching requirement appreciably (Eaton, 1954), whilst with 450 mm annually, irrigation leaching may become unnecessary (Kelley, 1951). The effectiveness of rainfall in leaching is considerably increased if the root-zone is left wet to field-

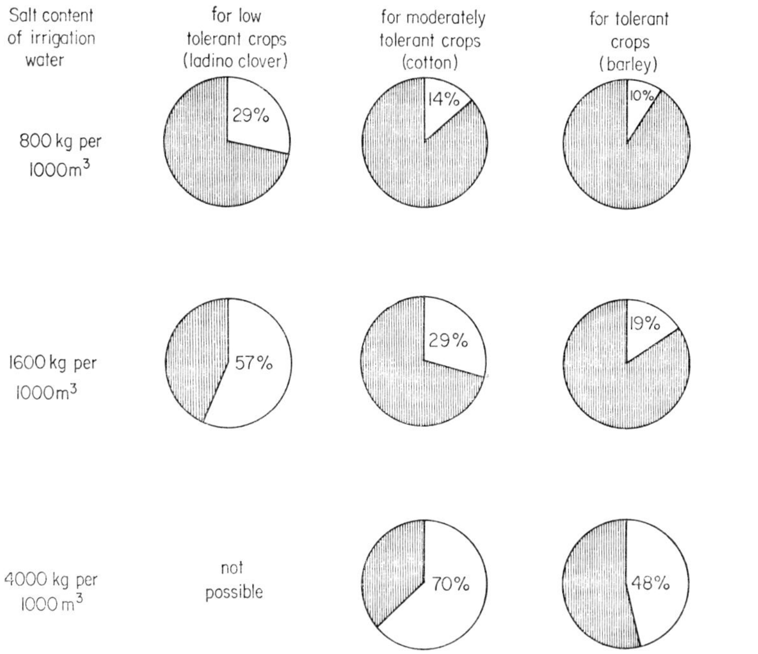

Figure 50. Leaching requirements in relation to water quality and the salt tolerance of crops. The leaching requirement expressed as percentage of the irrigation water (white area) that must pass below the root-zone to prevent an appreciable reduction in crop yields. (From Lyerly & Longenecker (1964): by courtesy of *World Farming*.)

capacity at the end of the irrigation season and the onset of the rains. Effective leaching is of course possible only if the water-table is very low or where drainage is satisfactory.

Leaching Methods

Basically, there are three methods used for leaching:

1. Continuous ponding (leaching prior to growing the crop).
2. Ponding with a rice crop (continuous leaching).
3. Applying water in excess of irrigation requirements during the growing of a crop (intermittent leaching).

Leaching is applied in the U.S.S.R. not only to remove excess salts below the root-zone, but also to replace saline ground-water with good-quality irrigation water. By leaching with a total of approximately 5 metres of water, changes in the quality of the ground-water up to a depth of 10 metres have been reported at the Mugan Experimental Amelioration Station in Azerbaijan (Anon., 1962).

FERTILIZATION

Fertilization is not effective in a very saline soil, but is doubly important after salinity has been reduced by leaching, as this also removes much of the plant nutrients, and in particular nitrates, from the root-zone.

Excessive fertilization may contribute to salinization. Certain fertilizer salts are more dangerous in this respect than others: the application of sodium nitrate, for 27 years, increased the exchangeable sodium content of a soil at Riverside, California, from an initial value of 2 per cent right up to 20–30 per cent (Bernstein, 1962).

DRAINAGE

As a general rule, irrigation results in a significant rise in the level of the ground-water, which, in arid regions, is usually much more saline than in humid regions (Kelley, 1951). Water in excess of consumptive use has to be applied in order to flush away the salts that have accumulated in the root-zone. In addition, the general tendency is to apply water in excess of crop requirements. Seepage from irrigation channels and other losses of water through faulty management, as well as precipitation during the rainy season, also contribute to a rise of the water-table. When the water-table has reached the root-zone, capillary rise of the salt-laden water and the removal of water by evapotranspiration, leave behind a greatly increased concentration of salts in the root-zone.

There are many cases in which the water-table rises to within one metre of the soil surface, creating conditions under which crop plants can no longer be grown profitably and leaching of excess salts is of course no longer possible. Indeed the introduction of irrigation in arid regions will generally result in drainage problems, unless adequate natural drainage exists.

Conditions under which satisfactory natural drainage can occur at all widely in arid regions are not common. The classic example is the area bordering on the course of the Nile, which has natural drainage towards the river after it has subsided. This drainage was adequate as long as basin irrigation was practised, but was no longer effective under the conditions of perennial irrigation, which has practically eliminated the low-water stage. Natural drainage by deep percolation is also possible when there is a very permeable substratum with a considerable ground-water gradient (Jewitt, 1966).

Drainage has the dual function of removing excess salts and excess water. *Surface drainage* is necessary to remove excess surface water that is due to irrigation and rainfall, preventing its percolation to ground-water and the formation of a water-table near or in the root-zone. *Subsurface drainage* is necessary to control the height of the water-table and to remove excess salts. This can be achieved only if the hydraulic conductivity is sufficient and if there are adequate outlets for the drained ground-water.

Magnitude of the Problem

In regions in which the main source of water-supply is from pumped wells, there is generally little danger of raising the water-table, and drainage problems are usually minor. Well-irrigation may even improve the situation in locations where the soil is waterlogged to begin with. In the San Joaquin Valley of California, for example, wells which supply irrigation water also serve successfully for water-table control (Donnan & Hudston, 1967). The water that is extracted from the aquifer is replaced by percolation from the waterlogged root-zone.

The problem generally becomes acute where water is 'imported' into the irrigation area. In West Pakistan, since 1870, when modern irrigation was started with water from the Indus and its tributaries, it is estimated that 40000 hectares *yearly* have become unfit for agriculture (Hamid, 1961)! In Iraq, about 60 per cent of the irrigation area is affected by salinity and a high water-table. In Mexico, the development of irrigation resources in recent times has already necessitated the 'rehabilitation' of 600000 hectares, of the total irrigated area of 1232000 hectares (Anaya, 1967)!

The 'irrigation district' of the Imperial Valley in California can serve as an example of the progressive deterioration of irrigated lands before attempts were made to overcome the problem. In this region an area of 200000 ha is irrigated with an annual quantity of water of approximately 3600 million m^3. The irrigation water contains approximately one metric tonne of salts per 1000 m^3 of water. Irrigation was started in the valley in 1901, and by 1919, 25 per cent of the irrigated land had already been affected by salt accumulation and a high water-table (Thomas, 1956). Drainage was initiated subsequently but was not sufficiently extensive to prevent continued damage in most of the area. By 1940 the water-table was at less than two metres' depth and salt was accumulating at the rate of two-and-a-half tonnes per hectare per year! By 1950, methods for adequate drainage had been developed and applied and, in the course of less than a decade, the entire area had been successfully reclaimed.

It has been estimated that drainage problems exist on at least one-half to two-thirds of the world acreage of irrigated land (Donnan & Hudston, 1967). Approximately 80 million hectares of irrigated land-area are at present protected by drainage (Gulhati & Smith, 1967), indicating that the importance of drainage as a complement to irrigation is now becoming more generally realized.

In the U.S.S.R., drainage is usually considered as the first step in the reclamation of saline soils, and is subsequently maintained both for salinity control and to control ground-water levels (Anon., 1962).

In general, farmers delay attacking the problem of drainage until the situation has deteriorated to the extent that economic crop production is no longer possible without reclamation. The carrying out of drainage works as a preventive measure is unfortunately still the exception. Yet, in many cases, artificial drainage would be unnecessary if conveyance losses to the field were eliminated and excessive quantities of water application to the crop were avoided.

Drainage is generally a regional problem, because even the most careful farmer may be penalized by waterlogging of his fields resulting from seepage from near-by irrigation canals over which he has no control. Frequently, excess water from high-level lands seeps through lateral aquifers, to reappear in lower-lying lands—sometimes many kilometres downstream (Luthin, 1957). In many cases, bringing upstream areas under irrigation has caused the loss of better areas downstream, as a result of the rise of water-tables in the latter.

For drainage to be effective, it is necessary to lower the water-table over a fairly large area. This cannot be achieved by the individual farmer, but requires a community effort (Kelley, 1951). In general, the costs of major drainage schemes are borne by governments which also administer the drainage projects of districts. The costs of drainage of individual land-tracts are generally borne by the owners—although government subsidies and loans on easy terms are available for this purpose to landowners in most countries in which drainage problems are acute.

Desirable Depth of Water-table

In the U.S.S.R., the accepted desirable depths of water-tables are based on the salt-content of the ground-water; this criterion is used in determining the depth and spacing of the drains (Anon., 1962). For example, in Tadzhik S.S.R. the norms indicated in Table XII are adopted:

TABLE XII

Norms for Determining Permissible Depth of Water-table

Salt content of ground-water (gm/l)	*Permissible water-table (metres)*
1–2	1·5
3–5	2·5 or deeper

The general tendency in the U.S.S.R. is not to lower the water-table more than is essential for salinity control. When the water-table is lowered by deep drainage, the quantities of water required for irrigation are increased considerably, 2–3 more irrigations per season being required with a low water-table as compared with a more shallow one which contributes to the water requirements of the crop (Anon., 1962).

Reclamation of Salt-affected Soils

DIAGNOSIS OF SALINE AND SODIC SOILS

It has been customary in the past to assess soil salinity in terms of the total salt-content of the soil. Soils were considered to be slightly saline when the salt-content was 0·3 per cent, moderately saline at 0·7 per cent, and strongly saline at 1 per cent. As field-crops generally respond to the osmotic pressure of the soil solution, the determination of the salinity hazard, based on the electrical conductivity of the saturation extract (EC_e) expressed in millimhos per centimetre at 25°C, gives a better appraisal of the salinity hazard in relation to plant growth than does the total salt-content of the soil (Richards, 1954) (Table XIII).

There is a definite relationship between the EC and the OP of the soil solution. For plants that are sensitive to a specific ion, the concentration of this ion has also to be determined.

The salt concentration in the saturation extract is about half of the concentration of the soil solution at field-capacity, and about one-quarter of the concentration at wilting point. This makes it possible to relate the salinity measurements at saturation to field moisture conditions (Allison, 1964), which is not possible when salinity is expressed in terms of percentages of total salts in the soil.

The most reliable index of *sodium status* of salt-affected soils is the sodium adsorption ratio (SAR) (Richards, 1954) (*see* p. 241). This index takes into consideration changes in the concentration and composition of the salts that are present in the soil solution and are in equilibrium with the soil (Allison, 1964).

TABLE XIII

Classification of Salt-affected Soils According to their Chemical Properties
(Richards, 1954)

Type of soil	Specific conductivity of saturation extract at 25°C in millimhos/cm		Exchangeable sodium percentage		pH	
	<4	>4	<15	>15	<8.5	>8.5
Normal	x		x		x	
Saline		x	x		x	
Saline-sodic		x		x	x	
Non-saline sodic	x			x		x

RECLAMATION PROCEDURES

An important characteristic of salt-affected soils is that the situation can be changed within a relatively short time, provided the necessary conditions are created.

Reclamation of salt-affected and sodic soils involves:

(a) the removal of excess soluble salts by leaching with adequate amounts of water;
(b) the displacement of exchangeable sodium by calcium, by adding either soluble calcium or acidifying agents that solubilize lime present in the soil;
(c) improving the physical condition of the soil; and
(d) providing adequate drainage.

Reclamation of Saline Soils

When the soil structure is favourable and drainage is possible, the reclamation of these soils is relatively simple, and can be achieved by leaching with a sufficient quantity of water. The amount of water required will depend on its quality and on the amount of salts that have accumulated in the soil. As a rule-of-thumb, in order to reduce salt-content of a soil to about 20 per cent of the initial high value, the depth of leaching water required is equal to the depth of soil to be leached (Allison, 1964).

Reclamation of Sodic and Saline-sodic Soils

Reclamation of these soil types is more difficult and expensive than that of saline soils. The saline-sodic soils have a good structure and relatively good infiltration, which are maintained as long as they contain a high percentage of soluble salts; when these are removed by the leaching processes, in the absence of an abundance of calcium, the saline-sodic soil is transformed into a non-saline sodic soil, which is characterized by a low permeability which makes leaching difficult.

In both soil types, calcium salts in suitable form, or substances that promote the solution of $CaCO_3$, have to be added prior to leaching or along with the irrigation water, in order to promote removal of exchangeable sodium. In the case of saline-sodic soils, the addition of soluble calcium prevents the soil from becoming non-saline sodic in the first place, and therefore the deterioration of the soil structure is prevented.

In the case of a soil that is already non-saline sodic, the calcium displaces the sodium from the exchange complex, converting the Na clay into a Ca clay, and thereby improving the soil structure.

Two types of soil amendments can be used to achieve this aim: materials such as gypsum (calcium sulphate) which supplies soluble calcium directly and is suitable for all types, or acidifying agents such as sulphuric acid, iron sulphate, or sulphur, for soils that are rich in calcium carbonate. The sulphur is inert until it is oxidized by soil micro-organisms to sulphuric acid, which may take time. The acid reacts with the $CaCO_3$ to form a suitable calcium salt ($CaSO_4$).

The solubility of limestone is too low to permit it to be an effective amendment.

The quantities of amendments required also depend on the quality of the irrigation water used and the amount of exchangeable sodium in the soil to be reclaimed. A usual yardstick is to apply sufficient amendment to replace the adsorbed sodium in the top 15 to 30 cm of soil (Reeve et al., 1953). Soluble amendments can be applied along with the irrigation water.

Sodic soils may be extremely impermeable to rainwater, and yet be relatively permeable to saline water. Such soils can therefore be reclaimed in a relatively short time by first saturating the soil with high-salt water, which serves as a flocculent and as a source of divalent cations. This method is applicable only when the SAR of the leaching water is appreciably lower than the SAR of the saturation extract of the soil to be reclaimed—otherwise more sodium will be adsorbed, rather than released from the soil exchange-complex. The saline leaching water transforms the sodic soil into a saline-sodic soil, thereby flocculating the soil and improving its permeability. After the initial leaching with high-salt water, there follow leachings with successively lower dilutions of the high-salt water with good irrigation water (Reeve & Bower, 1960). The progressive dilution reduces the SAR of the leaching water, thereby improving the sodium ratio in the soil.

Treating sodic soils with synthetic polyelectrolytes (e.g. CRD–186) was found to give a marked increase in permeability, in proportion to the 'rate' of treatment. No such effect was observed in normal soils (Allison, 1952).

SOIL MANAGEMENT PRACTICES AFTER RECLAMATION

Fertilizing and Manuring

Leaching not only removes undesirable salts, but also plant nutrients—particularly nitrates, whence the need for relatively high applications of nitrogen fertilizers. After leaching, it is desirable to apply large quantities of organic manure and/or to grow a green-manure crop. Such crops aid in rapidly improving soil structure and raising the level of soil fertility—without requiring the application of considerable amounts of fertilizers, which increase osmotic pressure.

Tillage

Deep ploughing (70 cm and deeper) may be helpful as a first step in reclamation by improving tilth, breaking hard-pans, occasionally turning up soil layers rich in gypsum, and overall improvement of percolation.

Crop Rotation

A transitional crop rotation, using salt-tolerant or salt-resistant plants, can be used during and after reclamation. Sunflowers, sugar-beets, barley, sweet sorghum, lucerne, Sweet Clover (*Melilotus alba*), and several perennial grasses such as Rhodes Grass (*Chloris gayana*), can serve for this purpose. The salt-tolerant crops are very effective in accelerating the reclamation process, by improving infiltration and by the effect of the CO_2 given off by their roots, which reacts with the $CaCO_3$ of the soil to form the more soluble calcium bicarbonate (Kelley, 1951).

Rice is an excellent crop for growing during the period of reclaiming saline soils, as it can support the abundant irrigation required for leaching the salts. However, salinity must be reduced to at most 0·5 per cent before the rice can be grown (Hamdan, 1961). After rice, clover and then cotton or sugar-beets can follow.

Under exceptional circumstances, wild plants that are highly salt-resistant and at the same time remove large amounts of salts, such as *Atriplex* and *Kochia* sp., have been recommended (Simonneau *et al.*, 1967).

DRAINAGE

After reclamation, an adequate system of drainage must be permanently maintained, otherwise a new cycle of salinization and alkalization will be started, and the whole process of reclamation will have to be repeated. With proper management practices and adequate drainage, the results of reclamation can be permanent (Kelley, 1951).

Summary

1. The *quality* of the water available for irrigation will determine the choice of crops to be grown, the soils that can be used, and the management and irrigation techniques that should be adopted.
2. Criteria for assessing the quality of irrigation water are based on the hazards to be expected following their use: *salinity hazard*, dependent on the total soluble salt-content (TSS); *sodium hazard*, based on the sodium adsorption ratio (SAR); *bicarbonate hazard*, based on residual sodium carbonate (RSC); and *toxicity hazard*, usually confined to excessive boron. These criteria cannot be applied rigidly, but must be adapted to regional conditions.
3. Irrigation usually causes the salinization of soils, and may frequently culminate, in the long run, in alkalization (accumulation of exchangeable sodium), which causes the soil to become both saline and sodic.
4. The salinization and alkalization of the soil can be prevented by careful management and irrigation practices and by maintaining the salt balance of the soil. Whenever necessary, water must be applied in excess of irrigation requirements, in order to leach excess salts beyond the root-zone.

5. Irrigation also frequently results in a significant rise in the water-table, which may create waterlogged conditions under which crops can no longer be grown profitably.
6. The twin requirements of removal of excess salts and of excess water frequently make *drainage* an essential complement to irrigation. Drainage is usually a regional problem to be solved by communal or governmental action.
7. Salt-affected soils can be reclaimed by leaching in the presence of an abundance of soluble calcium and by providing adequate drainage.
8. By preventing the accumulation of salts and waterlogging, and by maintaining a satisfactory level of fertility (*see* Chapter XI), the permanence of productivity of irrigated lands can be assured.

References

AGARWAL, R. R. & YADAV, J. S. P. (1956). Salinity and alkali scale to evaluate saline alkali soils for crop responses. *J. Indian Soc. Soil Sci.*, **4**, 141–5.

ALLISON, E. L. (1952). Effect of synthetic polyelectrolytes on the structure of saline and alkali soils. *Soil Sci.*, **73**(6), 443–54.

—— (1964). Salinity in relation to irrigation. *Adv. Agron.*, **16**, 139–78.

ANAYA, M. (1967). Mexico and its water resources. pp. 682–91 in *Water for Peace*, U.S. Government Printing Office, Washington D.C., Vol. 8, v + 827 pp., illustr.

ANON. (1962). *Soil Salinity and Irrigation in the Soviet Union.* Agric. Res. Serv. U.S. Dept. Agric., Washington D.C., 41 pp., illustr.

AYERS, A. D., BROWN, J. W. & WADLEIGH, C. H. (1952). Salt tolerance of barley and wheat in soil plots receiving several salinization régimes. *Agron. J.*, **44**, 307–10.

—— (1964) Salt tolerance of plants. *Inf. Bull. U.S. Dep. Agric.* **283**, 23 pp.

BERNSTEIN, L. (1962). Salt—affected soils and plants. *Arid Zone Res.*, **18**, 139–74.

——, & AYERS, R. S. (1955). Sloping seedbeds. *Calif. Agric.*, **9**(11), 8.

——, & FIREMAN, M. (1957). Laboratory studies on salt distribution in furrow-irrigated soil with special reference to the pre-emergence period. *Soil Sci.*, **83**, 249–63.

——, & Hayward, H. E. (1958). Physiology of plant tolerance. *Ann. Rev. Pl. Physiol*, **9**, 25–46.

BOTTINI, O. (1961). Tradition et recherche en Italie dans l'emploi des eaux saumâtres pour l'irrigation. *Arid Zone Res.*, **14**, 251–8.

BOWER, C. A. (1961). Prediction of the effects of irrigation waters on soils. *Arid Zone Res.*, **14**, 215–22.

COINTEPAS, J. P. (1964). Résultats d'un essai d'irrigation à l'eau salée en Tunisie. *Trans. 8th Int. Congr. on Soil Science*, Bucharest, Vol II, pp. 793–802.

DARRA, L. B., METHOK, N. & PARECK, B. L. (1964). Quality of irrigation water in Rajahstan. *J. Indian Soc. Soil Sci.*, **12**, 121–31.

DONNAN, W. W. & HUDSTON, C. E. (1967). Drainage related to irrigation management. pp. 974–84 in *Irrigation of Agricultural Lands* (Ed. R. M. Hagan, R. H. Haise & T. W. Edminster) Agronomy Series; Amer. Soc. Agron., Madison, Wisconsin, II, xxxi + 1180 pp., illustr.

DURAND, J. H. (1958). *Les Sols Irrigables.* Imprim. Imbert, Algérie.

EATON, F. M. (1942). Toxicity and accumulation of chloride and sulfate salts in plants. *J. Agric. Res.*, **64**, 357–99.

—— (1950). Significance of carbonates in irrigation waters. *Soil Sci.*, **69**, 123–33.

—— (1954). Formulas for estimating leaching and gypsum requirements of irrigation waters. *Misc. Publ. Texas Agric. Exp. Stn.*, **11**. 18 pp.

FANNING, C. D. & CARTER, D. L. (1963). The effectiveness of a cotton bur mulch and a ridge-furrow system in reclaiming sodic soils by rainfall. *Proc. Amer. Soc. Soil Sci.*, **27**, 703–6.

FIREMAN, M. & KRAUS, Y. (1965). *Salinity Control in Irrigated Agriculture.* Tahal-Water Planning for Israel Ltd, Tel-Aviv, 46 pp., illustr. (mimeographed).

GREAVES, J. E. & JONES, L. W. (1941). The survival of micro-organisms in alkali soils. *Soil Sci.*, **52**, 359–64.

GREENWAY, H. (1962). Plant response to saline substrates. *Austr. J. Biol. Sci.*, **51**, 16–38.

GRILLOT, C. (1956). The biological and agricultural problems presented by plants tolerant of saline or brackish water and the employment of such water for irrigation. *Arid Zone Res.*, **4**, 9–35.

GULHATI, N. D. & SMITH, W. C. (1967). Irrigated agriculture: an historical review. Pp. 3–9 in *Irrigation of Agricultural Lands* (Ed. R. M. Hagan, R. H. Haise, & T. W. Edminster) Agronomy Series; Amer. Soc. Agron., Madison, Wisconsin, **11**, xxxi + 1180 pp., illustr.

HAMDAN, G. (1961). Evolution of irrigated agriculture in Egypt. *Arid Zone Res.*, **17**, 119–39.

HAMID, S. (1961). *Programs of Waterlogging and Salinity Control in the Irrigated Areas of West Pakistan.* W. Pakistan Water Power Development Authority.

HAYWARD, H. E. (1956). Plant growth under saline conditions. *Arid Zone Res.*, **4**, 37–71.

—— & WADLEIGH, C. H. (1949). Plant growth in saline and alkali soils. *Adv. Agron.*, **1**, 1–38.

HOBBS, E. H. & RUSSELL, G. C. (1967). Salt accumulation as influenced by irrigation method. *Can. J. Soil Sci.*, **47**, 73.

JEWITT, T. N. (1966). Soils of arid lands. pp. 103–24 in *Arid Lands: A Geographical Appraisal* (Ed. E. S. Hills). UNESCO, Paris, xviii + 461 pp., illustr.

KAPP, L. C. (1947). The effect of common salt on rice production. *Bull. Ark. Agric. Exp. Stn.* **465**, 7, pp.

KELLEY, W. P. (1951). *Alkali Soils.* Reinhold, New York, vi + 176 pp., illustr.

KOVDA, V. A. (1961). Principles of the theory and practices of reclamation and utilization of saline soils in the arid zones. *Arid Zone Res.*, **14**, 201–14.

——, EGOROV, V. V. & HILLS, N. S. (1967*a*). Landscape in relation to irrigation, drainage, and salinity. Pp. 218–45 in *Irrigation and Drainage of Arid Lands.* F.A.O./UNESCO, Paris, 663 pp., illustr.

——, YARON, B. & SHALHEVET, J. (1967*b*). Quality of irrigation water. pp. 246–80 in *Irrigation and Drainage of Arid Lands,* F.A.O./UNESCO, Paris, 663 pp., illustr.

LONGENECKER, D. E. & LYERLY, P. J. (1959). Chemical characteristics of soils of West Texas as affected by irrigation water quality. *Soil Sci.*, **87**, 207–16.

LUNT, O. R., OERTLI, J. J. & Kohl, H. C. (1960). Influence of environmental conditions on the salinity tolerance of several plant species. *Trans. 7th. Int. Congr. on Soil Science*, vol. IV, pp. 560–9.

LUTHIN, J. N. (1957). *Drainage of Agricultural Lands*. Amer. Soc. Agron., Madison, Wisconsin, xii + 620 pp., illustr.

LYERLY, P. L. & LONGENECKER, D. E. (1964). Salinity control in irrigation in agriculture. *Wld. Fmg.*, **6**(10), 4–7, 34.

MAGISTAD, O. C. (1945). Plant growth relations on saline and alkali soils. *Bot. Rev.*, **11**, 181–230.

——, AYERS, A. D., WADLEIGH, C. H. & GAUCH, H. G. (1943). Effect of salt concentration, kind of salt and climate on plant growth in sand culture. *Pl. Physiol.*, **18**, 151–66.

PANTANELLI, E. (1937). Irrigasione con aqua salmastre. *Bul. Staz. Sper. Bari*, **26**.

PEARSON, G. A. & AYERS, A. D. (1958). Unpublished material, cited by Allison (1964) (q.v.).

PETERSON, H. B. (1961). Some effects on plants of salt and sodium from saline and sodic soils. *Arid Zone Res.*, **14**, 163–8.

QUIRK, V. P. & SCHOFIELD, R. K., (1955). 'The effect of electrolyte concentration on soil permeability', *J. Soil Sci.*, **6**, 163–78.

REEVE, R. C. & BOWER, C. A. (1960). Use of high-salt waters as a flocculant and a source of divalent cations for reclaiming sodic soils. *Soil Sci.*, **90**, 139–44.

—— & FIREMAN, M. (1967). Salt problems in relation to irrigation. pp. 998–1003 in *Irrigation of Agricultural Lands* (Ed. R. M. Hagan, R. H. Haise, & T. W. Edminster) Agronomy Series. Amer. Soc. Agron., Madison, Wisconsin, **11**, xxix +1180 pp., illustr.

——, PILLSBURY, A. F. & WILCOX, L. V. (1955). Reclamation of a saline and high boron soil in the Coachella Valley of California. *Hilgardia*, **24**, 61–91.

RICHARDS, L. A. (Ed.) (1954). Diagnosis and improvement of saline and alkaline soils. *Handbk U.S. Dept. Agric.*, **60**, U.S. Govt Printing Office, Washington D.C., vii + 160 pp., illustr.

SHALHEVET, J. & REININGER, P. (1964). The development of salinity profiles following irrigation of field crops with saline water. *Israel J. Agric. Res.*, **14**(4), 179–87.

SIMONNEAU, P., BERG, C. VAN DEN, STRAGONEV, V. V., KABANOV, V., BEYDEMAN, I. N. & CHAUDRI, I. I. (1967). Plants in relation to waterlogging and salinity. pp. 343–88 in *Irrigation and Drainage of Arid Lands*. F.A.O./UNESCO, Paris, 663 pp., illustr.

TAMES, A. C. (1950). *Calculo del Aqua Necessaria Para el Riego y Empleco de Salinas*. Min. de Agric., Madrid.

THOMAS, H. E. (1956). Changes in quantities and qualities of ground and surface waters. pp. 542–67 in *Man's Role in Changing the Face of the Earth* (Ed. C. O. Thomas, C. D. Sauer, M. Bates & L. Mansford) Univ. of Chicago Press, Chicago, xxxviii + 1193 pp., illustr.

TISON, L. T. (1961). Hydrologie et salinité des eaux souterraines. *Arid Zone Res.*, **14**, 25–36.

VOSE, P. B. (1963). Varietal differences in plant nutrition. *Herbage Abstr.*, **38**, 1–13.

WADLEIGH, C. H. & FIREMAN, M. (1949). Salt distribution under furrow and basin irrigated cotton and its effect on water removal. *Soil Sci. Soc. Amer. Proc.*, **13**, 527–30.

WILCOX, L. V. (1958). Determining the quality of irrigation water. *Inf. Bull. U.S. Dept of Agric.*, Washington D.C., **197**. 6 pp., illustr.

——, BALIR, G. Y. & BOWER, C. A. (1954). Effect of bicarbonate on suitability of water for irrigation. *Soil Sci.*, **77**, 259–66.

—— & DURUM, W. H. (1967). Quality of irrigation water. pp. 104–19 in *Irrigation of Agricultural Lands* (Ed. R. M. Hagan, R. H. Haise, & T. W. Edminster) Agronomy series; Amer. Soc. Agron., Madison, Wisconsin, II, xxxi + 1180 pp., illustr.

X

CROP INTRODUCTION AND IMPROVEMENT

Plant Introduction

INTRODUCTION SERVICE

Thousands of new, improved varieties of the principal crops are developed and released yearly by plant breeders all over the world. Introducing 'ready-made' varieties is the cheapest and frequently the most effective way of replacing inefficient varieties that are used in traditional agriculture. Moreover crop species, introduced into a new region free from their natural enemies, may be far more successful than in their country of origin.

Even for countries with an advanced agriculture, the introduction of improved varieties, as well as of varieties with special characteristics required for breeding work, is an essential part of any breeding programme. For example, not a single agronomic species or major crop plant grown in the U.S.A. actually originated within the borders of the country (Hodge & Erlanson, 1955). Even the hard red wheats, of which over 10 million hectares are grown annually in the United States, are derived from introductions. The same can be said of rice, oats, rye, and sorghum, varieties. Outstanding among forage grasses is a wheat-grass introduced from Siberia, while soybeans, providing crops having a value of a thousand million dollars a year in the United States alone, were introduced from Asia. Acala cotton, which has achieved almost world-wide distribution in warm lands, was introduced from Mexico.

One cannot, however, simply 'go shopping' for improved crop varieties. An *introduction service* must be established that is capable of testing introductions for adaptability, disease-resistance, and technological suitability, according to scientific procedures. In addition, the introduction of new varieties of crops, if carried out without the necessary safeguards and supervision, may cause untold damage by introducing at the same time pests, diseases, or weeds, that were previously non-existent in the country.

SOURCES FROM WHICH INTRODUCTIONS SHOULD BE MADE

For Improved Varieties

Preconceived ideas, based on climatological analogies, can be very misleading. Pragmatic experience in many countries indicates that it would be an unnecessary limitation to confine crop introduction to the relatively narrow limits of homoclimes. For some of the most conspicuous successes of plant introduction in many countries have been with varieties introduced from areas of very dissimilar climate. Conversely, failures have frequently been recorded with introductions from areas of similar climate.

In Israel, for example, excellent results were obtained with potato varieties from Ireland, peas and tomatoes from the United Kingdom, sugar-beets and flax from Holland, grasses from tropical Africa, sunflowers from Canada, barley from various temperate European countries, and many others. There is no certain guideline for deciding *a priori* from which countries successful introductions can be expected. The scientific level of the agricultural research of a country and its success in breeding work on a certain crop, are frequently more important criteria than the climate of the country!

In a programme of introduction, there is therefore no alternative but to establish as wide a collection as finances and man-power allow, without bias or preconceived ideas. To keep the scope of the work and expenses within reasonable limits, the first screening for adaptability, disease-resistance, etc., should be carried out as rapidly as possible, retaining for more detailed study a limited number of introductions that show the most promise.

For Breeding Material

In addition to 'ready-made' varieties, an introduction service must supply plant breeders with 'germ-plasm' for the improvement of the varieties on which they are working. Sometimes primitive or unadapted varieties, or even wild relatives of the crop to be improved, may be the carriers of a character which it is desired to incorporate in the accepted varieties. In former times, material of this kind could best be obtained by setting up an expedition to one of the 'centres of origin' of the crop under consideration. It is in the centre of origin of a crop that, according to Vavilov (1951), one finds the greatest variability and the best prospect of discovering ecotypes with the sought-after characteristics. However, plant breeders in the developing countries of the arid zones do not have to mount expeditions to find the germ-plasm they require. Large collections of varieties and 'banks' of germ-plasm have been established in a number of agricultural research institutes.

Papers appearing in the appropriate journals, and reports and lists from research institutes, will indicate sources of required germ-plasm, and in most cases it will be made available on request without any strings attached.

'Adapted' Varieties or Varieties with Wide Adaptation

A question that is relevant both in introduction work and in breeding work proper, is the type of variety which it is desirable to introduce or develop.

One of the traditional tenets of plant breeding is the importance of 'adapting varieties to the specific conditions of each region'. As a result, enormous efforts have been expended the world over in breeding varieties that are adapted to narrow ecological niches. This approach is, however, basically unsound—at least so far as arid regions are concerned. A variety with very specific adaptation is limited in its ability to adjust to changing conditions. But arid climates are characterized by their great variability—variability within a region and variability from one season to another.

It appears almost self-evident that the objective of the breeder, at least for arid regions, should therefore be to obtain varieties with the maximum ability to adjust to a wide range of ecological conditions. Kiesselbach (1951) pointed out that, of the approximately 100000 inbred lines of maize that had been developed in the U.S.A. up to 1950, only 60 were in use for the production of commercial hybrids. Eight of these inbreds, used in various combinations, were involved in the make-up of the most outstanding and widely adapted commercial hybrids. These inbreds had been isolated in a still more limited number of open-pollinated varieties.

In uniform varietal trials carried out in 18 localities in the Near East under the auspices of the International Wheat Programme, the variety Pitic 62 bred in Mexico, yielded close to 2275 kg/ha—about 650 kg/ha more than the average of the 25 varieties tested, and 1300 kg/ha more than the average national yields in that area. Three varieties: Pitic 62, Penjamo 62, and Nainari 60, gave the highest yields in 14 out of the 18 locations. Pitic 62 gave the highest yield in 9 locations (Stakman *et al.*, 1967).

The trials were located between 35° and 17°N latitude and at elevations from about 170 m below sea-level to 1700 m above sea-level. Soil fertility ranged from low to high; temperature, length of day, and incidence of pests and diseases, varied considerably; at some locations the wheat was rain-fed and at others it was irrigated. Obviously, the three outstanding varieties, and in particular Pitic 62, had a very wide adaptability.

An important characteristic of varieties with wide adaptation is their stability. In contrast to most varieties with specific disease-resistance, which commonly breaks down after ten to fifteen years of commercial production because new strains of the disease organism have evolved, a variety such as Florence × Aurore wheat, that is outstandingly successful in all the North African and Middle Eastern countries, has maintained its superiority unimpaired for over thirty years.

Fortunately, in most of the important crops, one or more varieties that perform extremely well over a wide area have 'happened' in the course of breeding work in different parts of the world. These include varieties such as Florence × Aurore wheat and more recently some of the Mexican semi-dwarf wheats, Rs 610 hybrid sorghum, Acala 4–42 cotton, Up-to-date potatoes, and US 13 hybrid maize (and its close parents). Perfection peas are further examples of varieties that are very successful over wide areas in arid and semi-arid lands. Of course, the term 'universal' is a misnomer; but it is accepted for denoting varieties with a very wide range of adaptability.

An example of the fallacy of the requirement to breed varieties with very specific adaptation is the early introduction and breeding work on wheat which was carried out in Palestine by the present author (Arnon, 1961). By tradition, almost every district in Palestine grew a standard local variety of wheat. These 'land' varieties were remarkably uniform in their respective types, probably as a result of mass selection and adaptation over a very long period. In view of this centuries-old tradition, it was originally thought that the breeding programme should aim at creating improved varieties adapted to the specific conditions prevailing in various districts. In consequence, a large number of selections and crosses were made among native varieties, and a comprehensive introduction programme was initiated. However, after a few years, it became apparent that the views on which this programme were based were not well founded.

In the course of the first six years of work on this crop in Palestine, 51 experiments were carried out at 11 locations, from Dan in the north to Negba in the south, each location being typical of a certain district. The annual rainfall recorded during these trials ranged from 300 to 940 mm; the varieties sown experienced drought, waterlogging and dry, hot winds, the so-called 'khamsins'. The experimental sites were located at altitudes varying from 100 m below to 500 m above sea-level, and the soil types on which the experiments were carried out ranged from loess to heavy soil. These trials were therefore representative of most conditions that were likely to be encountered in the country by wheat, and the generalizations drawn from the data could be considered well founded.

The surprise result of these trials was to find that it was not the 'adapted' varieties that gave the best overall results, but that one or two varieties were outstanding in almost all trials, notwithstanding the great range of environmental conditions encountered.

In the light of this work, the objectives of the breeding programme were revised. The development of special varieties for specific local conditions at district level was abandoned as an objective, and efforts were directed towards the development of varieties with as wide a range of adaptation as possible. This approach was justified subsequently by the results obtained from introduction and breeding programmes with such other field crops as barley, maize, sorghum, cotton and sugar-beets. In all these crops it was found that breeding designed to adapt varieties to suit each of the numerous districts in the country was a waste of effort; in practically every case it was found that the outstanding varieties were those which excelled in most districts and most seasons. In fact, outstanding varieties of all the crops mentioned showed not only country-wide adaptation but also regional adaptation, and proved successful in most countries in the Mediterranean region. Apparently these outstanding varieties combine high yielding ability with a great degree of adaptability.

The basic reasons for the superiority of varieties with a wide range of adaptability are not known. These varieties appear to be equipped with physiological and/or structural buffers that enable them to adjust to changes in environment (Grafius, 1956).

Crop Improvement for Efficient Water-use

Specific plant-breeding objectives for crop production in the arid regions are, of necessity, directed towards two diametrically opposite tendencies:

(1) breeding plants which are to be cultivated under near-arid conditions and which have drought resistance or tolerance and low water requirements;
(2) breeding crops to be grown under irrigation, and which are capable of making the most efficient use of the favourable environmental conditions provided.

In the final analysis, the aim will be to breed crops for efficient use of water —under conditions of a limited moisture-supply for rain-fed crops, and of a controlled water-supply for irrigated crops.

BREEDING FOR ADAPTATION TO DROUGHT

The amount of precipitation is the principal factor influencing crop yields in the semi-arid regions.

The principal wheat-producing areas of Canada, the United States, Argentina, Australia and the US.S.R., are in semi-arid regions, which periodically undergo severe drought conditions (Aamodt & Johnston, 1936). Little systematic work for the specific objective of breeding crop varieties for drought-resistance has been carried out. Drought-resistance is the result of many, frequently independent, morphological and physiological characters, whose interaction has not yet been sufficiently elucidated (Fuchs & Rosenstiel, 1958). Most physiological components of resistance are polygenic, and genetical analysis of the heredity of this complex has not yet been achieved.

Drought itself is not a uniform concept; the reaction of the plant is different to atmospheric drought and to soil-moisture deficiency, and will also depend on the stage of development at which drought occurs.

ATMOSPHERIC DROUGHT

This is caused by low air humidity, frequently accompanied by hot, dry winds. It may occur even under conditions of relatively high soil-moisture (Aamodt & Johnston, 1936). For example, hot, dry winds occurring at the time of earing of wheat may completely kill the young ears, irrespective of the amount of moisture available in the soil. Actually, plants that have grown continuously under favourable soil-moisture conditions, are the most susceptible to atmospheric drought.

SOIL DROUGHT

This occurs when soil-moisture supply lags behind evapotranspiration: it is usually gradual and progressive. Plants can therefore adjust, at least partially, to the increased moisture-stress.

In general, no satisfactory physiological, anatomical, or morphological, characteristics have yet been defined which could serve as reliable criteria in selecting for drought-resistance.

DROUGHT-RESISTANCE DEFINED

Drought-resistance can be defined in a number of ways, particularly in terms of the ability of plants to (1) survive under drought conditions, (2) endure drought without injury, and (3) be efficient in their use of water.

The Ability to Survive

The ability to survive under drought conditions is an essential characteristic of the xerophytic natural vegetation of the arid regions, and the characters involved have been described in Chapter IV.

Typically, these plants have adapted themselves to arid conditions by a reduced transpiring-surface resulting in dwarfed plants with very limited total leaf-surfaces. However, the farmer is not concerned with the *survival* of the species and varieties he grows; in case of crop failure he can always buy a new supply of seeds from a commercial source. The farmer is concerned rather with the *productivity* of his crops, whilst most of the mechanisms for ensuring survival of xerophytes under drought conditions usually achieve this latter aim by severe restriction of the assimilating surface and therefore at the *expense of productivity*. Typical xerophytic characteristics are therefore not, *a priori*, necessarily desirable in cultivars to be grown under conditions of limited water-supply.

Nevertheless, many attempts have been made to use xerophytic characters as indices of drought-resistance for breeding purposes. The osmotic pressure of germinating seeds (Buchinger, 1930), the bound-water content of wheat varieties and grass species (Newton & Martin, 1930), viscosity (Schmidt *et al.*, 1940), resistance to artificial atmospheric drought and heat in drought chambers (Aamodt & Johnston, 1936; Hunter *et al.*, 1936; McAlister, 1944), etc.; all these tests, at best, could serve for screening strains that are more or less capable of enduring drought, but without any direct relation to their productive ability.

Ability to Endure Drought Without Injury

A more modern school of thought believes that the only true criterion of drought-resistance is the ability to endure drought without injury. In this sense, drought-resistance is defined as the 'ability of plants to adapt to the effect of drought and to grow, develop, and reproduce normally under drought conditions, because of a number of properties acquired in the process of evolution under the influence of environmental conditions and natural selection' (Henckel, 1964).

The 'ratio of the yield under dry conditions to the yield under optimal conditions of water-supply' is considered as a valuable criterion of this concept of drought-resistance (Levitt, 1951). However, a few examples will suffice to show that this criterion of drought can be misleading in practice.

Harrington (1935), in varietal trials with wheat in semi-arid Saskatchewan, found that the variety Marquis gave more constant yields over a period of ten years and was less affected by drought than the variety Reliance. According to the above criterion, Marquis could be considered the less drought-susceptible variety of the two (Ashton, 1948). Yet during five of the seasons, which had a favourable moisture supply, Reliance out-yielded Marquis by an average of 645 kg/ha; during dry seasons, the difference in favour of Reliance was reduced to an average 41 kg/ha. The advantage for the whole period was clearly with Reliance; indeed this variety which is apparently more susceptible to drought, but which has a higher yielding potential, produced for the ten-year period 3430 kg/ha more than Marquis (Harrington, 1935). Even in drought years, the yield of Reliance was not lower than that of Marquis.

In variety trials of sweet sorghum, under dryland conditions as compared with irrigation, the following results were obtained (Arnon, 1962):

TABLE XIV

Yield of Dry Matter of Two Varieties of Sweet Sorghum (tonnes/ha)

	Dryland	*Irrigated*	*Dryland yield as per cent of irrigated*
Bet-Dagan 30	18600	63300	29·4
Norkan	15300	30200	50·6

If we assume that the smaller the reduction in yield due to adverse moisture conditions is, the more drought-resistant will be the variety, it follows that Norkan, whose yield was reduced by almost one-half under dryland conditions, is far more drought-resistant than Bet-Dagan 30, whose yield was reduced to less than one-third. Yet Bet-Dagan 30 out-yielded Norkan under dry conditions.

In these examples, if the farmer had chosen the 'more drought-resistant variety', he would have been the loser.

A further point that has to be considered is the great variability in rainfall from season to season that characterizes all dry regions—indeed, the greater the aridity is, the greater will be this variability.

A farmer will in all probability be better off if he grows varieties that have the yielding potential to give good returns in the good seasons, even if they are relatively disappointing in drought years. In general, the varieties with xeromorphic characters will also give poor yields in drought seasons, without being able to make up for this deficiency in seasons with favourable moisture conditions.

Efficient Use of Water

From the foregoing it is clear that the emphasis in selecting or breeding varieties for growing in semi-arid regions, in which moisture deficiencies occur in some seasons, or during part of the growing period, should be on achieving a high efficiency of water-use, as expressed in the ratio of yield produced per unit of water.

Basically, this will first and foremost be a question of selecting for high yielding potential under semi-arid conditions. This aspect will be discussed at length in the paragraphs concerned with breeding for yield (pp. 282 et seq.). Special consideration should be given to breeding lines with high specific net assimilation rates, which could compensate for the reduction in leaf area that is an almost inevitable consequence of drought (Fuchs & Rosenstiel, 1958).

Breeding for Dry Conditions

A number of additional breeding objectives can be useful in improving yielding ability under conditions of limited water-supply, and thereby improving the yield/water ratio.

(1) An important agronomic practice, enabling adaptation to limited soil-moisture supply, is the adjustment of plant population to the expected moisture régime. This is practically impossible to achieve with crop varieties that have a strong tendency to tiller. Tillering may be profuse in the early stages of growth, when soil moisture is still adequate, producing an excessive vegetative cover and a number of ears which cannot be brought to maturity because of

the early depletion of the limited moisture-supply. Breeding for limited tillering can therefore be very useful, in crops such as wheat, barley, sorghum, etc., for growing under dry conditions.

(2) In most crops, the adverse effects of drought vary in intensity according to the stage of growth. It is therefore important that the critical stage of development of the crop, when it is most susceptible to drought, should not coincide with the period when drought usually occurs. In cereals, for example, adverse moisture conditions during the period of earing, pollination, and seed formation, have disastrous effects on grain yield. Adjusting the growth rhythm of the crop to fit the precipitation pattern that is normal for the region is therefore an important breeding objective. Varieties with the desirable phenological characters can be selected or bred. Usually, though not always, this will entail breeding for early maturity, as the rainy season is commonly short in the dry regions. In Nebraska, early maturity was found to increase wheat yields by 122 kg/ha/day (Reitz & Salmon, 1959). However, there is generally an inverse relationship between early maturity and yielding potential, so that care must be taken not to shorten the growing period more than is absolutely essential. There are cases, as in certain regions of the Soviet Union, in which late maturity is an advantage, as dry periods occur relatively early in the growing-season (Fuchs & Rosenstiel, 1958).

(3) Probably the most promising and effective breeding measure for crops to be grown under conditions of limited moisture is by increasing the proportion of the total dry matter produced that is diverted to those parts of the plant which are of economic importance. In grain, for example, this would signify increasing the grain/straw ratio; outstanding successes of this approach are the dwarf and semi-dwarf wheat varieties, originally bred in Japan, and used in Mexican, Israeli, and other breeding programmes. In wheat, the importance of the assimilating surface of the ear for grain production has been demonstrated repeatedly (Buttrose, 1962). This makes it possible to reduce the leaf area considerably, without affecting the ability of the plant to produce photosynthates required for grain production.

(4) There may be circumstances under which breeding for specific characters is essential, if a crop is to be grown successfully in a dry region. An interesting example is that of Hopi maize. In order to reach moist soil at the time of planting, the Hopi Indians dig holes with a stick to a depth of 35 cm! Normally, maize could not reach the surface from such depths, but in Hopi maize the mesocotyl is greatly elongated. Hopi maize also develops a single root from the grain, which strikes deep into the soil to reach moisture, as opposed to the three seminal roots that are characteristic of normal maize varieties (Shantz, 1956).

(5) Components of the xerophytic vegetation of the arid regions are also amenable to improvement by breeding. In addition to xerophytic characters

which are essential for survival, breeding could aim at improving the grazing value of the more useful perennial plants by such means as lengthening the period during which they remain green and palatable, increasing their capacity to withstand the effects of grazing, and improving their nutritional value (Whaley, 1952).

FIELD TESTS

The limitations in our knowledge of the physiological factors involved in achieving maximum possible productivity under adverse conditions make field tests, under the appropriate conditions, the most practical method for final screening of the available breeding material. The field test, with all its limitations, integrates all the interactions of genetic and environmental factors to be expected in a given habitat. Field tests, when carried out in a number of locations for a number of seasons, also give a fairly accurate estimate of the range of adaptability of a variety.

Breeding for High Yields

The principal objective of plant breeders is to create crop varieties that will be capable of producing maximum yields under the conditions under which they are to be grown. An enormous effort has been expended on this objective practically the world over; it is, however, doubtful whether much of this effort was specifically directed toward high productivity *per se*. Varieties have been selected that are disease-resistant, early- or late-maturing, or possessed of other important characters influencing yield. Efforts have also been made to isolate the most productive individuals from within big populations, but without direct knowledge of the genetic and physiological factors responsible.

Impressive results have been achieved by these methods, but further major breakthroughs will require new approaches and a more basic understanding of the factors involved.

The ultimate yield of a crop is determined by the interaction of its *genetic characteristics* with the *environment* in which it is grown and the *management practices* to which it is subjected.

In the warm arid regions, the combination of favourable temperatures and the large amount of sunshine make possible very high levels of crop production. If the crop is grown under irrigation, with adequate management practices, it can be assumed that growth is not limited by light, temperature, water, or nutrients. The remaining limiting factor will therefore be the genetic characteristics of the plant, which should enable it to make full and efficient use of the growth-factors available to it.

Breeding for high yields can follow either of two pathways: eliminating characters that limit productivity or cause it to fluctuate from season to season, or combining characters that increase the potential productivity of the plant. The two pathways are not in conflict with each other—on the contrary, they are mutually complementary. Examples of the first approach are breeding for drought-resistance and for resistance to diseases and pests. These are discussed elsewhere (pp. 277–82 and 518–19; 544–7).

In the following paragraphs, we are concerned with breeding aimed directly at high yielding ability.

THE GENETIC APPROACH

There are no genes for yield *per se*, and therefore there can be no heritability of yield. There are only genes determining the components of yield (Grafius, 1959). However, a very large number of genes are involved, making the genetic analysis of the sum total of effects of yield components, as a basis for scientific breeding, extremely complicated. It is also always very difficult to distinguish between the effects of the various genotypes, as these are frequently masked or modified by the environment.

The planned maximum accumulation of genes for productivity into a single genotype, capable of transmitting this complex intact to its progenies, is not feasible at the present stage of our knowledge.

HETEROSIS

The great success of commercial hybrids in increasing the productivity of maize and, hence, the efficiency of water-use by this crop, is well known.

One of the most important developments in plant breeding of recent times has been the extension of the use of heterosis (hybrid vigour) to many other crops, by the use of male-sterility characteristics. In Texas, the best hybrid sorghum, in dryland tests, yielded on an average 39·5 per cent more than the best-adapted open-pollinated strain; under irrigation, the yield advantage of the hybrid was 27 per cent (King & Collier, 1961).

In Israel, the best hybrids out-yielded the best-adapted open-pollinated variety on an average by 20 to 28 per cent under dryland conditions, by 67 per cent with supplemental irrigation, and by 103 per cent with full irrigation. Their advantages were, therefore, considerable under conditions of both limited moisture supply and optimal soil moisture. Their ability to utilize the additional moisture-mainly for grain production was far greater than that of the best self-pollinated varieties (Arnon & Blum, 1964).

THE PHYSIOLOGICAL APPROACH

One of the principal tools of the plant breeder has always been the choice of appropriate parents for crossing programmes, in which the objective is to breed new genotypes, combining the favourable characters of the parents. A serious limitation of this method has been the lack of knowledge of the physiological processes on which crop productivity depends, making the choice of the parents a haphazard process (Boonstraa, 1934).

In recent years, great efforts have been made to clarify the physiological basis of yield and to draw practical conclusions that would be applicable to defining breeding objectives.

Maximum Dry-matter Production

The first prerequiste for high yields is a high production of total dry matter per unit area.

Carbon compounds account for 80–90 per cent of the total dry matter produced by plants. Photosynthesis is the basic process for the building of organic substances by the plant, whereby sunlight provides the energy required for reducing carbon dioxide, with sugar as the end-product of the process. This sugar serves as 'building material' for all the other organic components of the plant.

The amounts of dry matter produced will therefore depend on the effectiveness of photosynthesis of the crop and, furthermore, on plants whose vital activities are functioning efficiently.

The effectiveness of photosynthesis is dependent on (1) a large and efficient assimilating area, (2) an adequate supply of solar energy and carbon dioxide, and (3) favourable environmental conditions. If we assume that, under irrigation, in an arid climate, (2) and (3) will not be limiting, the total products of photosynthesis throughout the lifetime of the crop will depend on the size of the assimilating area, the efficiency with which it functions, and the length of the period during which it is active.

There is accordingly full justification in assuming that differences in productivity between cultivars will depend to a large extent on inheritable characteristics of the structure and function of the organs involved in photosynthesis (Stoy, 1963).

A proportion of the organic substances produced by photosynthesis will be needed for the vital processes of the plant, and consumed—by respiration—by the plant itself. The respiration intensity depends on hereditary characteristics and on growing conditions. The total yield of dry matter will therefore be the total amount of dry matter produced, less the photosynthates used for respiration.

Finally, the manner in which the net dry matter produced is distributed among the different parts of the plant will determine the magnitude of the economic yield.

Assimilating Area

The leaves of a plant are normally its main organs of photosynthesis, and the total area of leaves per unit area of land surface, called leaf area index

(LAI), has therefore been proposed by Watson (1947) as the best measure of the capacity of a crop for producing dry matter; this he calls its productive capital.*

Variation in total leaf area (LA) of a plant may result from changes in leaf number or in leaf size. The number of leaves present at any particular time equals the total number of leaves produced minus the number of leaves that have been lost by abscission. Leaf number depends on the number of growing-points, the length of time during which leaves are produced, the rate of leaf production during this period, and the length of life of the leaves. Leaf size is determined by the number and size of the cells of which the leaf is built, and is influenced by light, moisture régime, and the supply of nutrients.

Growth-curve of Leaf Area Index—Optimum LAI

After germination, LAI increases very slowly at first, over a fairly long period; then follows a period of rapid expansion (cf. Figure 51). As LAI in-

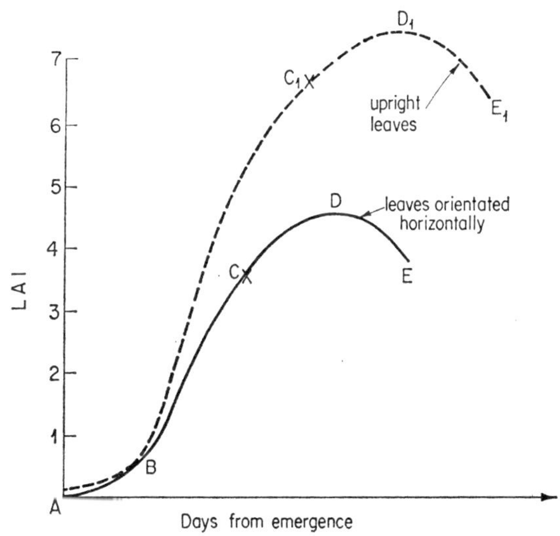

Figure 51. Schematic presentation of growth-curve of leaf area index (LAI). AB = period after emergence (LAIs increase slowly): BC and BC_1 = period of rapid growth (steep increase in LAIs); C and C_1 = optimum LAI, when lower leaves are at compensation point; D and D_1 = maximum leaf area.

creases, light absorption and the rate of dry-matter production increases, until the foliage becomes sufficiently dense to cause mutual shading. Then less light penetrates to the lower leaves, whose photosynthetic activity is therefore

* It would appear logical to include also the area of other green parts, such as leaf-sheaths and the ears of cereals. In these last, the assimilating area of the ears provides about half the assimilates used in grain formation (Asana & Mani, 1949).

reduced. As shaded leaves respire about as actively as leaves receiving full sunlight, their contribution to the total assimilation effort of the crop becomes less or often negative. There is therefore an optimum LAI for maximum dry-matter production, which is reached when the lowest leaves receive just sufficient light for photosynthesis to balance respiration—e.g. when the lower leaves are at the compensation point, and the canopy as a whole has reached maximum net assimilation. Below optimum LAI, light-energy is not being fully intercepted; above optimum LAI, the leaf area is not being utilized at maximum efficiency.

Differences in optimum LAI depend on the light intensity (being higher under high light intensity) and on the manner in which light is intercepted by the crop canopy. This latter factor varies with crop species, and even between varieties of the same species. The light received by a crop with a large optimal LAI is spread over more leaf surface than is the case with a crop having a low value for optimal LAI; this results in a lower intensity of illumination for most leaves, at which photosynthetic efficiency is higher than with more intense illumination (Watson, 1967). For crops with horizontally oriented leaves, such as Kale, optimum LAI is between 3 and 4. For crops with upright leaves, such as cereals and sugar-beets, optimum LAI may be as high as 6 to 9 (Figure 51, cf. Stoy, 1961). When LAI increases beyond the optimum, the lower leaves lose carbohydrates and will eventually die—usually long before the leaf canopy as a whole reaches the compensation point (Donald, 1961).

The rate of loss of the lower leaves increases as LAI increases, until a point is reached at which it equals the rate of production of new leaves; at this point LAI has become static, and leaf area is at its maximum. At this stage dry-matter production is still increasing, though usually at a somewhat slower rate than at 'optimum LAI'. When LAI exceeds its optimum, many of the leaves of the lower levels will actually be 'parasitic' even under conditions of full light intensity. At the same time the formation of new top leaves is stimulated, as is stem elongation; in many crops this is conducive to excessive vegetative development at the expense, for example, of grain formation (Nichiporovic, 1966).

Effect of Canopy Structure on Light Interception

The proportion of incident light that is intercepted by the canopy does not depend on LAI alone, but also on the so-called 'architecture of the plant community' (Loomis *et al.*, 1967).

In a crop, the leaves are not isolated but are arranged in a canopy. The way they are arranged will affect the proportion of incident light that is intercepted by the crop canopy as a whole. The extinction coefficient (K) is a measure of the light-intercepting efficiency of the leaf area (Figure 52; cf. Wilson, 1969).

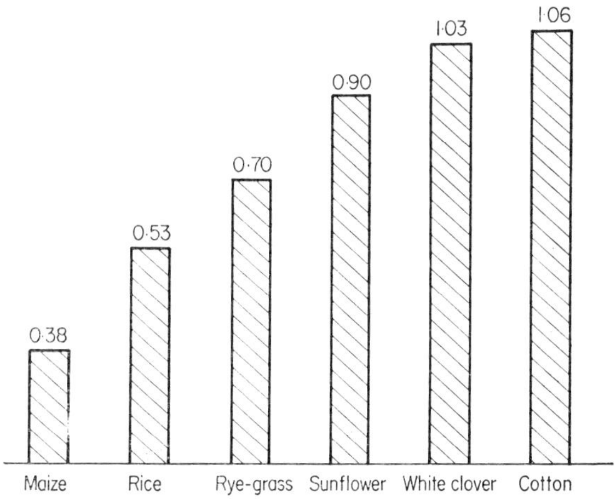

Figure 52. Extinction coefficient (K) for a number of crops. (After Wilson, 1969.)

Leaf characters of importance in this respect are: leaf angle, leaf area, continuity of leaf layers, and randomness of dispersion of leaves (Donald, 1961).

Interrelationship Between LAI and K

Under conditions of limited moisture-supply which can support only a relatively low plant population per unit area, the LAI optimum is low and a high K is therefore advantageous in maximizing interception (Figure 53). By contrast, when moisture is not limiting, a low K becomes beneficial because it enables optimum LAI to be high. Under these conditions, photosynthetic efficiency of light utilization is optimal: at high LAI, a lower K ensures that the total light available is dispersed over a greater area of leaves, but the light is proportionally weaker. Photosynthetic efficiency of light utilization being greatest in weak light, this is a manifest advantage in improving dry-matter production.

Figure 54 indicates that to achieve the maximum rate of net photosynthesis of the canopy per unit ground-area, light flux density should be as high as possible, K should be low, while the optimum value of LAI depends both on light intensity and on K (Wilson, 1969).

All these characters influencing K are controlled genetically, and could serve as a basis for breeding with the aim of increasing the optimum level of LAI. It has been stated that the high productivity of modern varieties of sugar-beet, as compared with the ancestral wild beet, is due to a different disposition of the leaves. Instead of the rosette of the ancestral type, the upright leaves of the cultivated varieties enable a more efficient use of light (Watson, 1956).

In a study of factors that are characteristic of high-yielding rice varieties, it

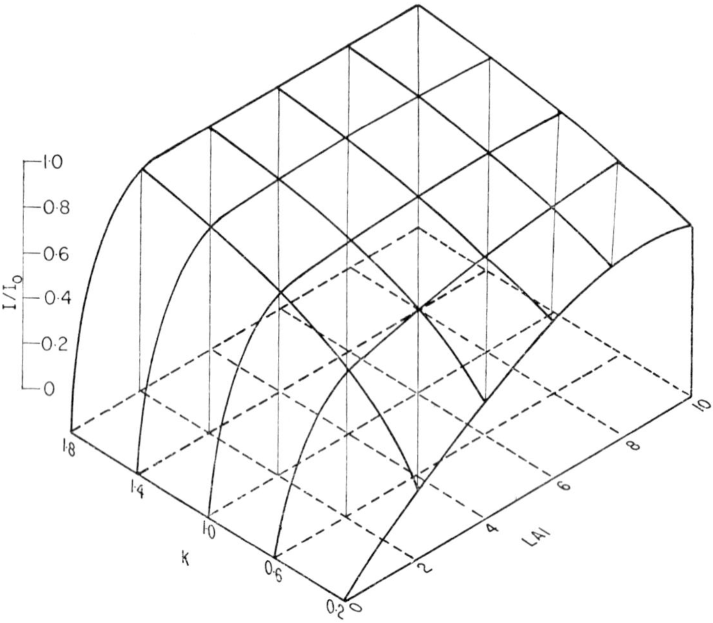

Figure 53. Effect of leaf area index (LAI) and extinction coefficient (K) on the proportion of incident light intercepted by a canopy (I/I_o). (After Wilson, 1969.)

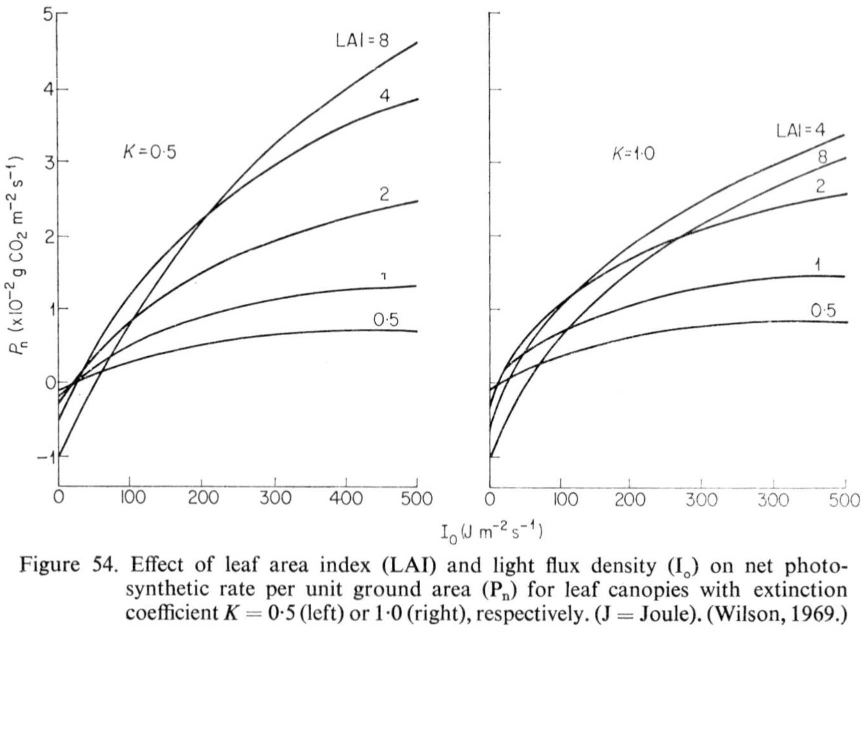

Figure 54. Effect of leaf area index (LAI) and light flux density (I_o) on net photosynthetic rate per unit ground area (P_n) for leaf canopies with extinction coefficient $K = 0{\cdot}5$ (left) or $1{\cdot}0$ (right), respectively. (J = Joule). (Wilson, 1969.)

was found that they have erect, short, narrow, thick, dark-green leaves and thick, sturdy stems. The thick, dark-green leaves lose less light through reflection than do others. The narrow leaves and erect positioning permit uniform light distribution throughout the canopy. However, as leaf area is small, these varieties produce heavy yields only if they are supplied with adequate amounts of nitrogen, and if the plant population density is high (Jennings, 1964).

In barley varietal trials it was found that the three highest-yielding varieties had narrow upright leaves which permitted light to penetrate deep into the leaf canopy, while three low-yielding varieties had wide, drooping leaves. Approximately 300 lines of wheat, oats, and barley, were evaluated visually on the basis of the angle of incidence and width of the leaves. With few exceptions, it was possible to select the high-yielding varieties on this basis alone (Tanner & Gardner, 1965).

An understanding of plant characters which are related to efficient utilization of light can therefore serve as a guideline for defining model plants for selection and breeding work (Figure 55 and Plate 22).

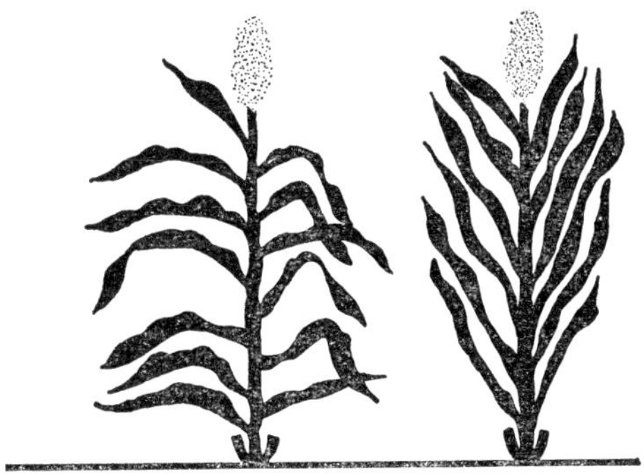

Figure 55. Contrasting plant types.
Left = plant with drooping leaves.
Right = plant with same number of relatively erect leaves.

Importance of High Initial Rate of Growth

For most annual crops, LAI is lower than optimum during most of the growing period. So long as LAI$<$1, some of the incident solar energy is not intercepted by leaves, but reaches the soil.

In the cereals of cool-temperate regions, LAI increases rapidly to a peak fairly early during growth and then falls rapidly; in maize, a high LAI is maintained for a much longer time. A certain hybrid maize, when grown with an ample supply of nitrogen, maintained a LAI of over 4 for 61 days from an age of approximately 2 months, and reached a peak of 5·6 five days before flowering which occurred at 82 days from sowing (Bunting & Drennan, 1966). If the crop could achieve an optimum LAI (3–4) within a very short period, and then maintain it during most of the growing-season, enormous yields could be achieved (Nichiporovic, 1960).

Watson (1947) has calculated that if LAI could be maintained at 4 for six months, yields of dry matter of 35 tonnes/ha for wheat and 60 tonnes/ha for sugar-beets could be obtained, instead of the present maximum yields of 10–12 tonnes/ha. It must, however, be stressed that the rapid increase of LAI, and its maintenance over a long period, create at the same time a transpiration mechanism that favours the loss of huge amounts of water which are needed to dissipate the enormous amounts of energy absorbed. This is of particular importance in arid-zone agriculture, and is a reminder of the close relationship between efficient photosynthesis and water régime.

In a crop that is grown from seed, it might be possible to shorten the period of very low LAI by breeding cultivars with a high initial rate of leaf growth, especially under irrigation, which provides favourable conditions for rapid growth (Stoy, 1963). Another breeding objective would be to create cultivars which have a large photosynthetically active area at the time when assimilates are being stored in the economic product. In cereals, for example, the grain receives most of its assimilates from the ear, the stem, and the upper leaves. The size and activity of these parts are therefore likely to be of paramount importance in determining yield.

Selection for leaf habits that would be conducive to a high optimum LAI cannot, of course, be effective in increasing yields if management practices remain primitive. Improvements resulting in increased light interception will result in higher productivity only if the supplies of nutrients and water are not limiting, plant-protection measures are adequate, and the plant population is adjusted to the new plant-form developed. Foliage canopies that are capable of utilizing high LAIs will be most effective in regions with high insolation and high solar elevation (Loomis *et al.*, 1967).

Efficiency of Photosynthesis

Maximum Potential Dry-matter Production. Penman (1968) calculates a theoretical efficiency of conversion of total radiant energy as between 8 and 10 per cent (*see* Figure 56). This figure should be compared with those obtained in actual farming practice: 0·03 per cent in India and 0·36 per cent in Europe.

Based on certain theoretical and empirical considerations, the maximum efficiency of production has been calculated by Loomis *et al.* (1967) to be a conversion of 5·3 per cent of the total radiant energy to chemical energy, or 15 μgm net of dry matter per calory of solar radiation. In an experiment in the Central Valley of California, carried out during the summer, with high daily levels of solar radiation, a maximum dry-matter production of 52 gm/m^2 per

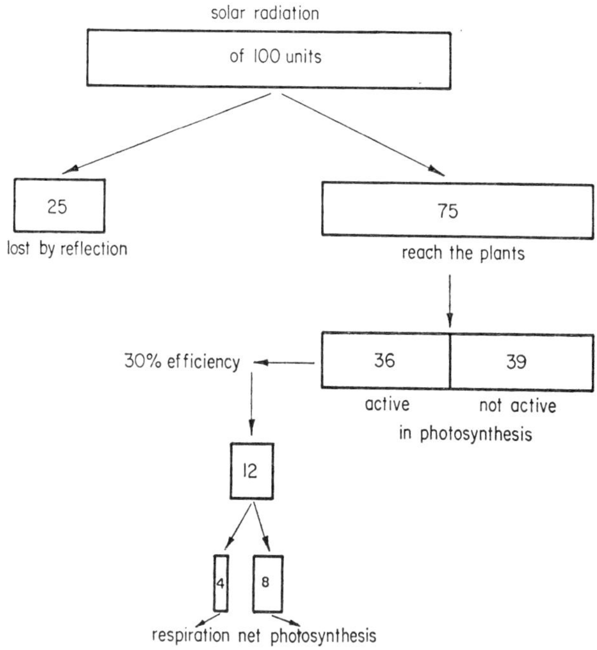

Figure 56. Theoretical efficiency of conversion of total radiant energy. (Based on data from Penman, 1968.)

day was obtained with a population of 71 maize plants/m² during a pretasselling period of 12 days. During this period, dry-matter production was a direct function of the proportion of solar radiation intercepted by the foliage canopy. The net energy conversion of this maximum accounted for 2·9 per cent of the solar radiation received, amounting to 47 per cent of the calculated maximum efficiency (Williams *et al.*, 1965).

Other high growth-rates that have been recorded for terrestial communities include: sugar-cane in Hawaii (shoots only), 37 gm/m²/day over a three-month period (Borden, 1942), and for maize 30 gm/m²/day during 65 days when the maize was near full-leaf development (LAI = 4·0–4·2) (Lemon, 1963). Bulrush Millet in the Northern Territory of Australia had a peak growth-rate of 54 gm/m²/day when insolation was 510 cal/cm²/day, representing a net energy conversion of 4·2 per cent of the total incoming radiation (Begg, 1965). However, peak production rates during very short periods, or with plant populations that are in excess of what would be practicable under field conditions, have relatively little relevance to economic crop productivity.

Wilson (1969) has calculated maximum annual yield production of dry matter as 8200 gm/m²—a yield level of total dry-matter production that is approached by certain perennial tropical crops (8000 gm/m²) but is far higher than that of annual crops (3000 gm/m^2 for tropical crops and 2000 gm/m^2 for temperate crops).

Growth Analysis. The gasometric methods which provide data on the uptake of CO_2, can only be used for measurements over very short periods. To estimate photosynthetic production over longer periods, measurements of CO_2 uptake in the field are not practical, and dry-matter gain is a more appropriate measure. A method called 'growth analysis' estimates the efficiency of the photosynthetic system in the field by determining the rate of increase of dry matter per unit leaf area at short intervals throughout the growth period (Watson, 1956). This method does not give a pure measure of photosynthesis per unit leaf area, but only of the *excess* of dry matter over loss by respiration. This value has been called the 'net assimilation rate' (NAR), and can be calculated by the following formula:

$$\text{NAR} = \frac{\text{Increment in dry weight}}{\text{Mean leaf area} \times \text{days}} \text{ (in gm/m}^2\text{/day).}$$

The Significance of NAR. NAR cannot be considered a measure of the intensity of photosynthesis, as it is also affected by changes in respiration rates. NAR varies between and within species, and is influenced by mineral nutrition, water supply, and seasonal climatic conditions. The variation between years for a given species grown at the same site, and between varieties in the same year, is, however, much greater for leaf area (LA) than for NAR.

Similarly, agronomic practices, such as the rate of supply of nutrients, affect dry-matter yields mainly by their influence on leaf growth, whilst variations of NAR are usually of minor importance and show no consistent trend (Figure 57; cf. Watson, 1956). In general, it can therefore be stated that increases of dry-matter yield depend much more on higher LA than on NAR (Heath & Gregory, 1938).

From these observations it has been concluded that there is little prospect of increasing yields through increasing NAR by breeding or by special treatments. Yet not all authors are convinced that differences in NAR between varieties have little selection value. Even small increases in the rate of photosynthesis can result in fairly big increases in dry-matter production, because the latter accumulates according to the principle of compound interest. Wallace & Munger (1966) found that the high relative growth-rate and large biological yield of one of their bean varieties was due, at least in part, to a 25 per cent advantage in NAR.

Sugar-beet (*Beta vulgaris*) and spinach (*Spinacia oleracea*) are closely related and have very similar LA values and leaf arrangements. Yet the NAR of sugar-beets is always higher than that of spinach, by an average of 56 per cent

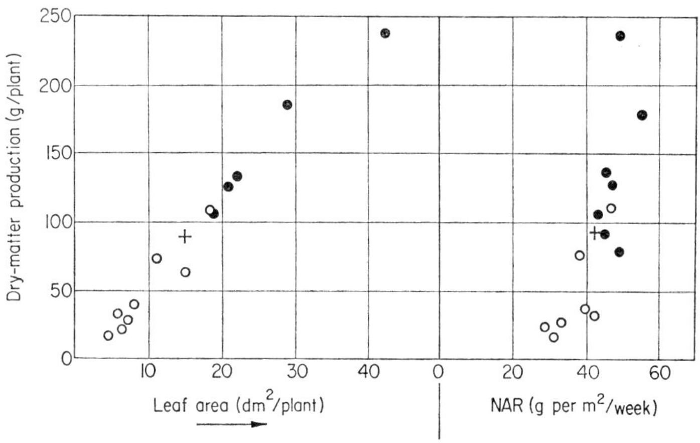

Figure 57. Relations between dry-weight increase, mean leaf area, and net assimilation rate in mangolds.

O = without N-fertilization
● = with N-fertilization

(From Watson (1947): by courtesy of *Annals of Botany* and the author.)

(Thorne, 1966), and the dry-matter accumulation of sugar-beets is much greater than that of spinach—presumably because the root of the sugar-beet provides a more efficient 'sink' for carbohydrate than does that of spinach.

Some of the factors that influence growth affect LA and NAR in the same direction; these include increases in N and water-supply. Others, such as light, may have opposite effects, and the resultant variation in dry-matter production will then tend to be small (Watson, 1952).

Water-supply (irrigation), in particular, has a considerable effect on NAR. A severe moisture-deficit normally reduces the rate of photosynthesis considerably—directly through changes in plasmatic structure, and indirectly through closure of the stomata (Stoy, 1963). It is therefore probable that part of the increase in yield that is due to irrigation is attributable to an increase in NAR.

If NAR cannot be markedly increased by most cultural treatments, efficiency in dry-matter production can be improved by making maximum use of the period during which climate favours high NAR, and this may be an important reason for differences in efficiency between species, cultivars, sowing dates, etc. (Watson, 1956). It might also be possible, by breeding, to improve the absorption of CO_2 by the plant and also its transport to the sites of photosynthetic activity (Stoy, 1963).

Respiration Rates

As has already been indicated, NAR does not merely reflect photosynthetic rates but is also influenced by respiration rates. The proportion of the photosynthates that is used for respiration varies widely with the crop and the environment but is generally quite substantial, sometimes amounting to as much as 40 per cent (Müller, 1960). However, not all of the dry matter used for respiration can be considered as wasted. Wilson (1969) distinguishes between three components of total respiration:

(*a*) *Constructive Respiration.* Respiratory energy is required for the synthetic processes on which structural growth depends. A close relationship has been shown to exist between relative growth-rate and respiration rate in a number of crop plants, indicating that, for each gram of tissue synthesized, about half a gram of substrate is respired to provide energy.

(*b*) *Maintenance Respiration.* In addition to providing energy for growth, respiration is necessary to maintain the existing organization, such as maintaining the concentration of mineral salts in the tissues.

(*c*) *Substrate-induced Respiration.* Respiration also occurs that is not related to productive metabolic processes. For example, respiration is known to increase when the level of sugars in the tissues exceeds a certain level.

(*d*) *Photorespiration.* Most higher plants release a considerable amount of CO_2 when they are photosynthesizing, a phenomenon known as photorespiration. The rate of oxidation of photosynthetic products by photorespiration is about five times the rate of respiration in the dark. Photorespiration appears to confer little or no benefit on the plant, and the reverse is even probably true (Goldsworthy, 1969).

A decrease in the rate of unproductive respiration of the latter kinds could therefore contribute to higher net production, provided sufficient energy remained available for the vital processes of the plant. It has been shown that there are great differences in the respiration rates of different species and even varieties, and it is speculated that high-yielding varieties have a relatively low consumption of respiration energy (Stoy, 1963). For example, *japonica* varieties of rice have lower rates of respiration than the far-less-productive *indica* varieties (Sato, 1949, cited by Murata, 1965). Respiration rates also depend on temperature: as plants are known to adapt to the temperature conditions of the region in which they are grown, it might be possible to select for varietal differences in degree of adaptation (Stoy, 1963).

Relationship Between NAR and LAI

Various experiments suggest that NAR is almost independent of LAI when the latter does not exceed 3, but decreases when LAI increases beyond this point.

If NAR decreases with increases of LAI beyond a certain range, crop growth-rate cannot be linearly related to LAI, but must increase more and more slowly with increasing LAI, eventually reaching a maximum and subsequently falling (Figure 57; cf. Watson 1947).

Leaf Area Duration

Yield of dry matter can be derived from the following formula:

$$Y = LA \times NAR \times t,$$

in which Y = total production of dry matter,

LA = assimilating area,
NAR = net assimilation rate, and
t = time

The length of the vegetative period (t) is obviously of great importance in determining the effect of LAI on yields. Crop improvement aimed at the most efficient use of the available growing-season may contribute considerably to increased productivity. An integral of LAI over the growth period is a measure of the combined action of LAI and time, and has been called 'leaf area duration' (LAD) by the English workers and 'photosynthetic potential' by the Russians. LAD is a measure of the opportunity for assimilation throughout the life of the plant.

The mean yields of dry matter for wheat, barley, potatoes, and sugar-beets, grown in several seasons at Rothamsted, England, were found to be approximately proportional to mean LAD, so that $\frac{\textit{dry-matter yield}}{\textit{LAD}}$ was practically the same for all these crops. This ratio is an estimate of dry-matter increase per unit time—that is, of mean NAR over the whole growth period. It confirms that variation in dry-matter production between different species grown on the same location, between varieties, between years, and between different fertilizer treatments, depends mainly on variation in LA, and that variations in NAR are generally of relatively little importance (Watson, 1947). That this view is not generally accepted has been pointed out above (p. 292).

Genetic Response to Light Intensities

There are marked differences between different species in their response to light intensities. Single leaves of some plants, such as tobacco, are saturated at near one-quarter of full sunlight, whilst other crops, such as sugar-cane, maize, and sunflower, respond to differences near to full sunlight (*see* Figure 58).

Productivity of plants is therefore correlated with their ability to benefit by high light intensities: for plants with continued response to high light intensities, there is practically no limit to the light requirement for maximum photosynthesis and their potential productivity is high. On the other hand,

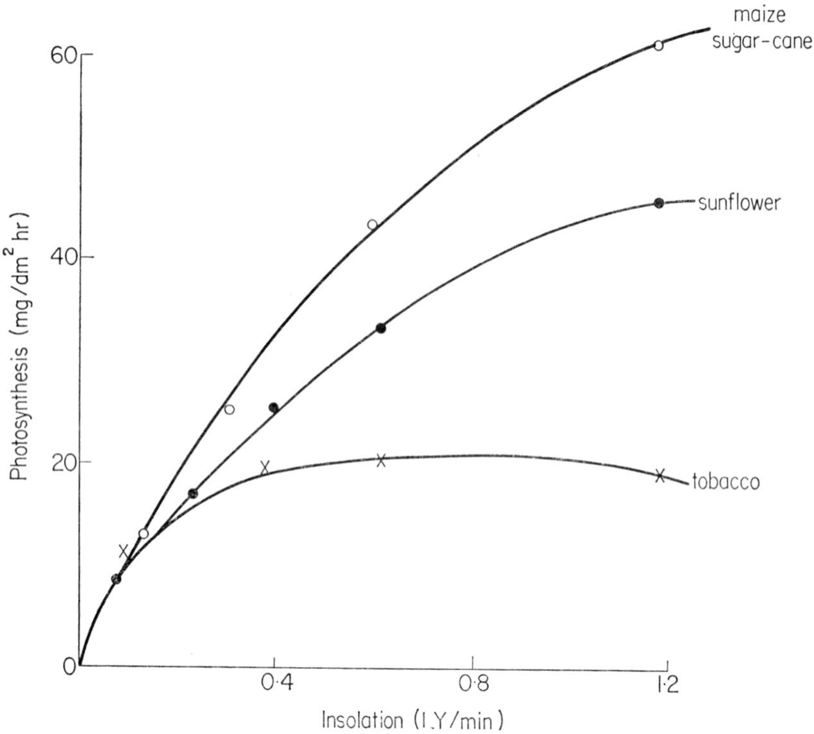

Figure 58. Effect of light intensity on photosynthesis in some crop plants. (After Waggoner *et al.* (1963): by courtesy of the American Society of Agronomy.)

plants that fail to respond beyond one-half of full sunlight have a low productive potential (Waggoner *et al.*, 1963).

Photosynthetic capacity is therefore a variable genetically controlled characteristic which can probably be increased by plant breeding.

The leaves of most crop plants are known to be reasonably efficient at low light intensities: the less light that is received by an individual leaf, the more efficiently will it be used. A shaded leaf may achieve a conversion efficiency of up to 20 per cent, whilst a light-saturated leaf will usually convert into assimilates no more than 1 per cent of the absorbed energy (Gaastra, 1959). A breeding objective could be to increase light-conversion efficiency at higher light intensities, either by raising the level of light saturation or by reducing light absorption above the saturation point (Loomis & Williams, 1963).

Summing up it may be said that improvement in yields must be sought mainly, though not exclusively, through increasing the LAI. However, the LAI cannot be increased indefinitely, for at a certain stage the NAR will decrease through the reduction in average light intensity at the leaf surface due to mutual shading of leaves (and possibly through decrease in CO_2 concentration resulting from increased uptake). For unirrigated crops in arid regions, there is the further limitation of an insufficient moisture-supply. Too large a leaf surface may threaten the very existence of the plant. Nichiporovic (1966) suggests that, under these conditions, the greatest promise is in increasing net assimilation rate. Unfortunately, no clear-cut method is yet available for achieving this aim.

BIOLOGICAL AND ECONOMIC YIELDS

We have discussed at length the ability of the plant to produce a high yield of carbohydrates. In certain crops, such as those produced for forage, the total dry-matter production or biomass has economic significance; however, this is not the case with most cultivated crops. Generally, only part of the crop is important to the farmer, the rest being either of inferior value (such as in the case of the straw of cereals), or a nuisance to be disposed of (as in the case of the stalks of cotton or haulms of potatoes). It is the part of the biomass that is converted into the economic product that has been called the 'economic yield' (Nichiporovic, 1960). The relationship of the economic yield to the total, or 'biological yield', is expressed as the 'coefficient of effectiveness' or 'harvest index' (K), according to the equation:

$$Y_{biol} \times K = Y_{econ}.$$

There exist, therefore, two different, though not necessarily antagonistic, pathways that can be followed in breeding for high yields. These are:

(a) increasing the total production of dry matter (Y_{biol}), which has already been dealt with, and
(b) increasing the proportion of the total dry matter produced that is accumulated in the parts of the plant which are of economic importance (Y_{econ}), e.g. by increasing the coefficient of effectiveness (K).

The causes of variations in the distribution of dry matter between different parts of the plant are not sufficiently understood. Information is still required on the following aspects (Thorne, 1966):

(1) Which parts of the plant produce the photosynthates that will be devoted to producing the dry matter of the economic part of the plant?
(2) When, during the life of the plant, are these sources active?
(3) What internal factors influence the distribution of the dry matter between different parts of the plant?
(4) What proportion of total photosynthates is used for respiration?
(5) What is the effect of environmental factors on the whole system?

A balance between the productive parts of the plant and the reserves which form the economic yield is essential. Experience has shown that a considerable increase in the yield of the economic product is usually dependent on an increase in the total dry matter produced. However, this is not an absolute relationship; indeed, in extreme cases, the opposite may be true. In maize, high plant populations, which produce the greatest biomass (and are therefore the most efficient in utilizing radiant energy), may produce little or no grain because of excessive competition between the plants for light.

In many cereals, most of the dry matter in the grain is produced by photosynthesis after the ears emerge. Grain yield therefore depends to a large extent on the photosynthetic activity of the parts of the plant that are still green after anthesis. In wheat, barley, and similar cereals, the photosynthetic activity of the ear, which is situated at the top of the stalk, makes a considerable contribution to grain formation. Practically all the dry matter of the grain is produced by the part of the shoot above the flag-leaf node; of this dry matter the ear contributes about 50 per cent in wheat (Asana & Mani, 1949) and up to 70 per cent in barley (Frey-Wyssling & Buttrose, 1959). However, the size and longevity of these organs, as well as the number and potential size of the grains, are determined during the vegetative phases. The importance of the influence of the vegetative phases is demonstrated by the effect of yellow rust of wheat, which, when it attacks the early leaves of wheat, causes a considerable decrease in the yields of total dry-matter and of grain (Bunting & Drennan, 1966).

Another possible way to increase the effective yield would be to extend the duration of the period of formation of the economic product. For example, the formation of grain in maize takes, on the average, about 35 days of the total growing period. It is estimated that with the varieties that are grown at present, every additional day of grain formation would increase yields by approximately 3 per cent (Army & Greer, 1967). Selecting for varieties or hybrids with a period of 50 to 75 days between pollination and grain maturity, would make possible a yield increase of approximately fifty per cent. This is of course possible only in regions with a long frost-free growing-season—another manifest potential advantage of the arid and semi-arid subtropical lands.

Plant-breeding efforts aimed at increasing the coefficient of effectiveness (K) have shown that it is definitely possible to increase the ratio of economic yield to biological yield. The most striking success achieved in this respect in recent years is probably the development of the semi-dwarf wheat varieties, which are characterized by very short straw, few leaves, and a very large ear. This is illustrated by a typical example from a trial carried out in Israel, comparing the standard wheat variety, Florence x Aurore, with a semi-dwarf variety which produced the same biological yield (Ephrat *et al.*, 1964):

TABLE XV

Biological and Economic Yields of Two Wheat Varieties (kg/ha)

	Y	*K*	Y_{econ}
Florence × Aurore	9500	0·43	4130
N 67 (dwarf variety)	9350	0·65	6220

Another example is that of the recently introduced sorghum hybrids. Thus in trials with sorghum in Israel, the grain-weight of outstanding hybrids equalled that of the vegetative material, whilst in the best open-pollinated variety—Martin—grain production constituted only one-third of the total above-ground production of the plant (Arnon & Blum, 1964).

When the biological yield is limited by external conditions, this does not always imply that the economic yield will be reduced, as surplus carbohydrate may move to the grain (Thorne, 1966). There may also be conditions in which an increase in K, accompanied by an overall decrease in potential biological yield, may be desirable—e.g. under conditions of drought, when a crop with less vegetative development than others will be less affected by drought. Thus cereal types with very short straw and large ears may be more efficient under relatively arid conditions than types having the opposite characteristics (Dobben, 1962).

The possibility has also been raised that an increase in size of the storage organs (the economic yield) results from a stimulation in the productivity of photosynthesis or from a greater transfer of photosynthates from other organs of the plant to the storage organs and that, therefore, an increase in K is possible without a corresponding increase in Y_{biol} (Stoy, 1963). The latter possibility is illustrated by the following example (Thorne, 1962):

TABLE XVI

Ear/Shoot Ratios of Two Varieties of Barley

	Plumage-Archer	*Kenia*
Emergence	13·8	1·20
Anthesis	22·5	20·0
Maturity	128	150

The relative ear/shoot ratios of the two varieties of barley changed completely between emergence and maturity. As the absolute dry-matter production values of the two varieties were similar, the assumption is justified that there exists a mechanism whereby the growing kernels are able to withdraw varying amounts of photosynthates from other parts of the plant. It is not known whether this is an active or a passive process, but it appears from work by Mothes (1956) that certain substances, such as kinetin and benzimidazole, have a profound influence on the movement of nitrogen to different parts of the plant. There might therefore be an activating principle, produced by the rapidly growing parts of the plant, in amounts that are genetically controlled, which affects the above-mentioned process.

In summary, it is important that after the harvestable part has been initiated, all dry matter in excess of what is required for the maintenance of an efficient photosynthetic structure should be diverted to the economic part of the plant. In addition, the maximum possible of dry matter should be translocated from the non-harvestable parts of the plant to the economic product during the later stages of maturation.

The Components of Economic Yield

Attempts have been made to achieve better understanding of the economic yield of many crops, so as to provide criteria for breeding purposes, by breaking down yield into yield components.

A familiar formula of yield is that for grain cereals:

$$Y = a \times b \times c \times d,$$

in which Y = grain-yield per unit area,

a = the number of plants per unit area,
b = the number of fertile tillers per plant,
c = the number of grains per ear, and
d = the weight of the individual grain.

At first sight, an increase in one of the components on the right-hand side of the equation should lead to an increase in yield. This expectation has led, in the past, to a considerable amount of attention being paid by plant breeders to these factors in the hope that yields could be increased by combining, in a single genotype, a combination of components such as increased tillering capacity, large ears, heavier grain, etc. This requires an understanding of the influence of the various components on yield, and the effect of crop management on each of them.

Tillering

The first genetically controlled component of yield is the number of ears or of fertile tillers per plant. The optimum number is, of course, closely dependent on the soil moisture régime during the growing period of the crop.

Most cereals have a considerable capacity to increase the number of tillers per plant, when sufficient space is available to the individual plant. An optimum number of ears per unit area can therefore be achieved in two different ways: by a low plant density, compensated for by an increased number of fertile tillers per plant, or by a high plant density, which will keep the number of tillers per plant low. At first glance, the former approach appears to be the more rational: savings in seed can be achieved by it, and the plant will auto-regulate the number of ears per unit area according to whether environmental conditions are favourable or not.

In practice, however, this approach has a number of drawbacks. Thus a sparse stand is more susceptible than a dense one to weed infestation, and unfavourable conditions may reduce plant population below the point at which increased tillering can compensate for missing plants. The principal limitation, however, is that the tiller is not an independent unit: it is considerably influenced by other tillers produced on the same plant. There is, therefore, a great difference in principle, in whether the number of ears per unit area is derived from a large number of plants with a few fertile tillers each or from relatively few plants with abundant tillers. The following considerations seem most pertinent in these connections:

(a) The ear-bearing capacity of an individual tiller depends on the number of tillers per plant. A negative correlation is known to exist between tiller production and tiller survival, indicating the existence of a regulatory mechanism which keeps the rate of ear production per plant down to a determined level (Frankel, 1935). It has been shown that only tillers formed before a certain period in the life of a plant will produce ears (ibid.). Therefore, on the same plant, tillers appearing at a relatively late date will have less chance of producing ears. The final number of ears per unit area can therefore be achieved by increasing either tillering or the proportion of those tillers that survive and produce ears. There appears to be better justification to produce fewer tillers and increase the survival rate than *vice versa*.

(b) The primary tiller usually produces the largest ear, each successive ear-bearing tiller producing ears of decreasing size. The second and third ears are generally nearer in size to the first ear than are the subsequent ones to be formed. The tendency in recent years has therefore been to breed for varieties with sparse tillering capacity but higher ear-producing tiller survival, and to plan plant density so as to decrease dependency on tillering as a means of achieving optimum density. However, retaining a limited ability to tiller (2–3 tillers per plant) enables auto-adjustment of the plant according to environmental conditions and compensation for missing plants, without adversely affecting the ultimate yield. It has also been shown that one-ear plants may produce smaller ears than two-ear plants—mainly because the former are frequently weak or diseased (Frankel, 1935).

Number of Grains per Ear, and Weight of the Individual Grain

Increasing the productivity of the plant, in terms of increased carbohydrate production, will contribute to increased yields only if the total storage-capacity of the plant is adequate—e.g. in the case of the cereals, if a sufficient number of grains are produced and their development is adequate; in other words, an adequate 'sink' is required, that is capable of receiving the carbohydrates in excess of the metabolitic needs of the plant.

It has been shown, in non-cereals, that there is a feed-back effect, in the sense that the size of the 'sink' affects the production and the movement of carbohydrates in the plant. It is therefore possible that in cereals, also, too few grains, or unsatisfactory development, may restrict the photosynthesis of the shoot (Thorne, 1966). Consequently, the number of grains per ear, and the weight of the individual grain, are important.

Generally speaking, factors acting early in the season mainly influence the grain number,* whilst the size of grain is affected by the factors acting after anthesis. In breeding wheat for drought-resistance, it was found that when soil moisture was adequate, the yield component which had the greatest effect was number of ears per unit area; under conditions of moisture stress the grain number per ear and, occasionally, average grain-weight, had as much effect as the number of ears. Asana (1962) therefore suggested grain-number per ear and 1000-kernel weight as criteria for selecting varieties for dryland production.

Attempts to obtain high-yielding varieties by combining high tillering capacity, large ears, and large kernels, into a single genotype, have generally not been successful.

The maximum yield that can be achieved under given environmental conditions has a ceiling which cannot be surpassed. Any increase in one of the components is therefore usually accompanied by a parallel decrease in another component when conditions are optimal. Experience has shown that an inverse correlation exists among the yield components, and that the highest-yielding cereal varieties are not those with extremely large heads or heavy kernels, but usually those in which these components attain only average size.

From the lack of achievements in breeding work resulting from the use of yield components as criteria for selection of desirable prototypes, it should not be assumed that no worthwhile significance can be attached to the traditional yield components. It is true that yields cannot be increased beyond an upper limit by manipulating individual yield components; but much can be done to approach this limit by breeding for yield components that are well-balanced and adjusted to the environmental conditions in which the crop is to be grown.

* Though a proportion of grains may degenerate at a later stage, when environmental conditions are unfavourable.

References

AAMODT, O. S. & JOHNSTON, W. A. (1936). Studies on drought resistance in spring wheat. *Can. J. Res.*, **14**(C), 122–52.

ARMY, T. J. & GREER, F. A. (1967). Photosynthesis and crop production system. pp. 321–32 in *Harvesting the Sun—Photosynthesis in Plant Life* (Ed. A. San Pietro, F. A. Greer & T. J. Army) Academic Press, New York. ix +342 pp., illustr.

ARNON, I. (1961). *Some Aspects of Research on Field Crops in Israel.* Div. of Publ., Nat. & Univ. Inst. of Agric., Rehovot, Israel. Abstract of Publication 372–E, 58 pp., illustr.

—— (1962). [Sweet sorghum as a secondary crop for sugar production—in Hebrew.] *Ktavim*, **12**, 13–44.

—— & BLUM, A. (1964). Response of hybrid and self-pollinated sorghum varieties to moisture régime and intra-row competition. *Israel J. Agric. Res.*, **14**, 45–53.

ASANA, R. D. (1962). Analysis of drought resistance in wheat. *Arid Zone Res.*, **16**, 183–90.

—— & MANI, V. S. (1949). Photosynthesis in the ears of five varieties of wheat. *Nature, London*, **163**, 450–1.

ASHTON, T. (1948). *Technique of Breeding for Drought Resistance in Crops.* Commonwealth Bureau of Plant Breeding and Genetics, Technical Communication No. 14, 34 pp.

BEGG, J. E. (1965). High photosynthetic efficiency in a low-latitude environment. *Nature, London*, **205**, 1025–6.

BOONSTRA, A. E. H. R. (1934). Physiologisch onderzoek ten dienste der plantenveredeling. *Med. Landb. Hoogesch*, **38(1)**, 1–99.

BORDEN, R. J. A. (1942). A search for guidance in the nitrogen fertilization of the sugar-cane crop. 1. The plant crop. *Hawaii Plrs' Rec.*, **46**, 191–238.

BUCHINGER, A. (1930). Die Bedeutung der Selektion nach der Saugkraft für die Pflanzenzüchtung. *Z. Zücht.*, **15**(A), 101–14.

BUNTING, A. H. & DRENNAN, D. S. H. (1966). Some aspects of the morphology and physiology of cereals in the vegetative phase. pp. 20–38 in *The Growth of Cereals and Grasses* (Ed. F. L. Milthorpe & J. D. Ivins) Butterworth, London, xii + 359 pp., illustr.

BUTTROSE, M. S. (1962). Physiology of cereal grain. III. Photosynthesis in the wheat ear during development. *Aust. J. Exp. Biol. Med. Sci.*, **15**, 611–18.

DOBBEN, W. H. van (1962). [The distribution of dry matter in cereals and grasses in relation to nitrogen nutrition—in Dutch, with English summary.] pp. 77–89 in *Jaarboek 1962, Inst. voor Biologisch en Scheikundig Onderzoek vah Landbouwgewassen*, Wageningen, 175 pp., illustr.

DONALD, C. M. (1961). *Competition for Light in Crops and Pastures.* pp. 282–313 in Symposia of the Society for Experimental Biology no. XV. Mechanisms in Biological Competition, University Press, Cambridge, vi + 365 pp., illustr.

EPHRAT, Y., KARCHI, Z., GRAMA, A. & GERECHTER, Z. (1964). [*Experiments with Dwarf and Semi-Dwarf Wheat Varieties in the Seasons 1961/62 and 1962/63*—in Hebrew, with English summary.] Prelim. Rep. Nat. Univ. Inst. Agric., Rehovot, No. 453, iii + 30 pp., illustr. (mimeographed).

FRANKEL, O. H. (1935). Analytical yield investigation in New Zealand. 2. Five Years' Analytical Variety Trials. *J. Agric. Sci., Camb.*, **25**, 466–509.

FREY-WYSSLING, A. & BUTTROSE, M. S. (1959). Plant physiology, photosynthesis in the ear of barley. *Nature, London*, **184**, 2031.

FUCHS, H. & ROSENSTIEL, K. (1958). Ertragsicherheit. pp. 365–442 in *Handbuch der Pflanzenzüchtung* (Eds. H. Kappert & W. Rudorf) Paul Parey Verlag, Berlin, Band I, xxiii + 848 pp., illustr.

GAASTRA, P. (1959). Photosynthesis of crop plants as influenced by light, carbon dioxide, temperature, and stomatal diffusion resistance. *Med. Landb. Hoogesch.* **59**, 1–68.

GOLDSWORTHY, A. (1969). Riddle of photorespiration. *Nature*, London, **224**, 501–2.

GRAFIUS, J. E. (1956). Components of yield oats; a geometrical interpretation. *Agron. J.*, **48**, 419–23.

—— (1959). Heterosis in barley. *Agron. J.*, **51**, 551–5.

HARRINGTON, J. B. (1935). Cereal crop improvement for dry farming conditions. *Sci. Agric.*, **16**, 113–20.

HEATH, O. V. S. & GREGORY, F. G. (1938). The constancy of the mean assimilation rate and its ecological importance. *Ann. Bot., N. S.*, **2**, 811–18.

HENCKEL, P. A. (1964). Physiology of plants under drought. *A. Rev. Pl. Physiol.*, **15**, 363–86.

HODGE, W. H. & ERLANSON, C. O. (1955). Plant introduction as a federal service to agriculture. *Adv. Agron.*, **7**, 189–211.

HUNTER, J. W., LAUDE, H. H. & BRUNSON, A. M. (1936). A method for studying resistance to drought injury in inbred lines of maize. *J. Am. Soc. Agron.*, **28**, 694–8.

JENNINGS, P. R. (1964). Plant type as a rice breeding objective. *Crop Sci.*, **4**, 13–15.

KIESSELBACH, T. A. (1951). A half-century of corn research. *Am. Sci.*, **29**, 629–55.

KING, J. G. & COLLIER, J. W. (1961). *Performance of Grain Sorghum Hybrids and Varieties*, 1957–60. Progr. Rep. Tex. Agric. Exp. Stn. No. 2182, 5 pp. (mimeographed).

LEMON, E. (1963). Energy and water balance of plant communities. pp. 55–78 in *Environmental Control of Plant Growth* (Ed. L. T. Evans) Academic Press, New York, viii + 449 pp., illustr.

LEVITT, J. (1951). Frost, drought and heat resistance. *Am. Rev. Pl. Physiol.*, **2**, 24–5 68.

LOOMIS, R. S. & WILLIAMS, W. A. (1963). Maximum crop productivity. *Crop Sci.*, **3**, 67–72.

——WILLIAMS, W. A. & DUNCAN, W. G. (1967). Community architecture and the productivity of terrestrial plant communities. Pp. 291–308 in *Harvesting the Sun —Photosynthesis in Plant Life* (Ed. A. San Pietro, F. A. Greer & T. J. Army) Academic Press, New York, ix + 342 pp., illustr.

MCALISTER, D. F. (1944). Determination of soil drought resistance in grass seedlings. *J. Am. Soc. Agron.*, **36**, 324–36.

MOTHES, K. (1956). Der Einfluss des Wasserzustandes auf Fermentprozesse und Stoffumsatz. Pp. 656–63 in *Handbuch der Pflanzenphysiologie* (Ed. W. Ruhland) Springer Verlag, Berlin, Band III, xxii + 1073 pp., illustr.

MÜLLER, D. (1960). Ökologische Energetik der Photosynthese. pp. 285–68 in *Handbuch der Pflanzenphysiologie* (Ed. W. Ruhland) Springer Verlag, Berlin, Band V(2), xvi + 868 pp., illustr.

MURATA, Y. (1965). Photosynthesis, respiration, and nitrogen response. pp. 385–400 in *The Mineral Nutrition of the Rice Plant*, Johns Hopkins Press, Baltimore, Maryland, ix + 494 pp., illustr.

NEWTON, R. & MARTIN, W. M. (1930). Physico-chemical studies on the nature of drought resistance in crop plants. *Can. J. Res.*, **3**, 336–427.

NICHIPOROVIC, A. A. (1960). Photosynthesis and the theory of obtaining high crop yields. *Fld Crop Abstr.*, **13**, 169–75.

—— (1966). *Photosynthesis of Productive Systems.* Nauka Publ. House, Moscow. (Translated from Russian by Israel Programme for Scientific Translations, iv + 182 pp., illustr.)

PENMAN, H. L. (1968). Inefficiency of world farming. *J. Austr. Agric. Sci.*, **34**, 235.

REITZ, L. P. & SALMON, S. C. (1959). Hard red winter wheat improvement in the plains. *U.S. Dept. Agric. Tech. Bull.*, **1192**, 117 pp., illustr.

SATO, K. (1949). [Some experiments on the respiration of rice plants.—In Japanese.] *Proc. Kyushu Nojishiken Kenkyn Happ yokai*, **5**, 7–10.

SCHMIDT, H., DIWALD, K. & STOCKER, O. (1940). Plasmatische Untersuchungen an Dürreempfindlichen und Dürreresistenten Sorten Landwirtschaflicher Kulturpflanzen. *Planta*, **31**, 559–96.

SHANTZ, H. L. (1956). History and problems of arid lands development. pp. 3–25 in *The Future of Arid Lands* (Ed. G. F. White) Amer. Assoc. Adv. Sci., Washington D.C., ix + 453 pp., illustr.

STAKMAN, E. C., BRADFIELD, R. & MANGELSDORF, P. C. (1967). *Campaigns Against Hunger.* Harvard University Press, Cambridge, Massachusetts, xvi + 328 pp., illustr.

STOY, V. (1961). Dry matter production as an objective in plant breeding. pp. 31–47 in *Eucarpia*, Third Congress of the European Association for Research on Plant Breeding, Paris, 336 pp., illustr.

—— (1963). Some plant physiological aspects of the breeding of high-yielding varieties. pp. 264–75 in *Recent Plant Breeding Research.* Wiley, New York, 346 pp., illustr.

TANNER, J. W. & Gardner, C. J. (1965). Leaf position important in barley varieties. *Crops and Soils*, **18**(3), 17.

THORNE, G. N. (1962). Survival of tillers and distribution of dry matter between ear and shoot of barley varieties. *Ann. Bot., N. S.*, **26**, 37–54.

—— (1966). Physiological aspects of grain yield in cereals. pp. 88–105 in *The Growth of Cereals and Grasses* (Ed. F. L. Milthorpe & J. D. Ivins) Butterworth, London, xii + 359 pp., illustr.

VAN DOBBEN, W. H. *See* DOBBEN, W. H., VAN.

VAVILOV, N. I. (1951). The origin, variation, immunity and breeding of cultivated plants. *Chron. Bot.*, **13**, 1–6, xviii + 366 pp. illustr.

WAGGONER, P. E., MOSS, D. N. & HESKETH, J. D. (1963). Radiation in the plant environment and photosynthesis. *Agron. J.*, **55**, 36–39.

WALLACE, D. H. & MUNGER, H. M. (1966). Studies of the physiological basis for yield differences. *Crop Sci.*, **6**(6), 503–7.

WATSON, D. J. (1947). Comparative physiological studies on the growth of field crops. I. Variations in net assimilation rate and leaf area between species and varieties and within and between years. *Ann. Bot., N. S.*, **11**, 41–76.

—— (1952). The physiological basis of variation in yield. *Adv. Agron.*, **4**, 101–45.

—— (1956). Leaf growth in relation to crop yield. pp. 178–90 in *The Growth of Leaves* (Ed. F. L. Milthorpe) Butterworth, London, x + 223 pp., illustr.

—— (1967). Physiological characteristics of the growth of sugar-beet crops on different soils. *J. Intern. Sugar Beet. Res.*, **2**, 225–9.

Whaley, W. G. (1952). Arid lands and plant research. *Scien. Monthly*, **75**, 228–33.
Williams, W. A., Loomis, R. S. & Lepley, C. R. (1965). Vegetative growth of corn as affected by population density. *Crop Sci.*, **5**, 211–19.
Wilson, J. Warren (1969). *Maximum Yield Potential.* Paper presented at Colloquium on the Transition from Extensive to Intensive Agriculture, Israel, 33 pp., illustr. (mimeographed).

XI

SOIL FERTILITY AND ITS MAINTENANCE

There is no generally accepted concept of the term fertility as applied to soils. For our purpose, soil fertility can be defined as the ability of the soil to produce high yields consistently, provided environmental factors such as light, moisture, temperature, etc., are not limiting and crop management practices are efficient.

The nature of soil fertility is, however, more complex than this definition implies; it depends not only on chemical and physical characteristics of the soil, but also on the biological processes resulting from the many and varied activities of the soil organisms. The soil is a very complex medium, in which chemical, physical, and biological, activities are proceeding continuously, with many and varied interactions taking place between them.

Under arid conditions the physical, chemical, and biological, processes of soil formation are slowed down considerably. The plant cover is at best diffuse and of low productivity. As a result, rock minerals break down slowly, profile characteristics are not well developed, and soil organic matter and soil nitrogen are always at a low level. Leaching, however, is also at a low level, and therefore there is frequently an excessive accumulation of soluble salts. The 'fertility cycle' under desert conditions is practically a closed system, with nutrients circulating at a very slow rate.

The breakdown and decomposition of plant residues is extremely slow under arid conditions, and nitrification is inhibited. Nutrients may accumulate in the soil during a cycle of more-than-usually dry years, permitting a fairly lush growth when rains do finally occur; this may give a mistaken impression of fertility, as evidenced by the poor growth that is usually obtained when there are two or more good rainfall seasons in succession.

The farmer has a considerable influence on the level of fertility of his soils: practically all agricultural operations affect soil fertility for better or for worse, according to the way in which they are executed. Tillage can improve soil structure or just as effectively destroy it; chemical fertilizers are one of the most efficient means of increasing the productivity of the soil, but their injudicious use can cause much damage to soil fertility. Irrigation can 'make the desert bloom'; it can, however, also transform fertile areas into deserts, in which even desert plants will find it difficult to establish themselves.

No single factor is capable of raising the fertility of the soil to its highest potential level if other relevant factors are neglected. A discussion of the main factors affecting soil fertility therefore involves a review of most soil characteristics and farming practices.

Physical Factors Affecting Soil Fertility

The physical properties of a soil largely determine the way in which it can be used most effectively. The soil system consists of three, physically separable phases—solid, liquid and gaseous. The solid phase consists of a porous matrix; the pore spaces within the matrix are occupied by the liquid phase (soil 'solution') and the gaseous phase (soil atmosphere).

The matrix consists of mineral and organic particles of various sizes; the relative proportions of mineral and organic components vary considerably in different soils. The mineral component is derived mainly from the weathering of the parent rock and, to a smaller extent, from the breakdown of animal skeletons—providing calcium carbonate and calcium phosphate, both of which are of considerable importance to soil fertility.

The organic component consists of plant and animal materials in various stages of decomposition.

THE SOLID PHASE

Soil Texture

All gradations in the size of particles normally occur in a soil. For purposes of classification, it has been found necessary to divide the particles into more or less arbitrary groups known as soil 'fractions' or 'separates'. These are determined by mechanical analysis.

The most generally adopted classification of soil separates is the so-called Alterberg system, which is also known as the 'international' system:

Stones, gravel	— larger than 2·0 mm in diameter
Coarse sand	— 2·0 to 0·2 mm
Fine sand	— 0·2 to 0·02 mm
Silt	— 0·02 to 0·002 mm
Clay	— smaller than 0·002 mm.

These differences in size of soil particles profoundly affect such soil properties as the movement and retention of water, circulation of soil 'air', adsorption of nutrients, and microbial activity; they also largely dictate management practices such as tillage methods, irrigation methods, choice of crops, etc.

As each fraction has different physical and chemical properties, the physical and chemical properties of a soil will be dominated by the properties of the fraction that is present in the largest amount.

Soils in which large particles (gravel and sand) predominate, do not swell on wetting or contract on drying; they have a very low water-holding capacity, very little capillarity, rapid water infiltration, low retention of nutrients, little or no cohesiveness and plasticity, high specific temperature, and excellent aeration. Soils with predominantly very small particles will swell when wetted, and shrink considerably on drying; they have a very high water-holding capacity, high capillarity, slow infiltration rate, high retention-ability for nutrients and gases, considerable cohesiveness and plasticity, low specific temperature, and poor aeration.

A predominantly sandy soil will have good aeration and will be easy to till; but it will have little inherent fertility and will not be able to retain water or plant nutrients to any large extent.

By contrast, clay soils may have a high potential fertility and considerable ability to retain nutrients and water, but they are apt to become stone-hard when dry, and sticky and poorly aerated when wet. The ideal soils are, therefore, in general, those in which particles of the various size-groups occur together in desirable proportions.

Predominantly sandy soils do have certain inherent advantages in arid regions—especially when irrigation is to be practised. Their greater permeability to water and lower ability to retain cations, reduce the dangers of their becoming saline. Though their water-holding capacity is lower, the water is far more easily available to plants than in heavy loams and clays—a point that is of importance for rain-grown crops when rainfall is sparse. For these reasons, alluvial sand areas in arid regions always carry more abundant vegetation than do alluvial clay soils, which are also more frequently saline (Kovda, 1961).

Briefly, it can be stated that the texture of a soil is its most permanent characteristic, in contrast with other physical characteristics such as soil structure, which can be rapidly modified—for better or worse—by management.

Standard Textural Names of Soils. Over the years, colloquial terms such as loam, sandy loam, silt-loam, etc., have been used to describe combinations of sand, silt, and clay, in various proportions. An attempt has been made to standardize these names and their meanings, based on a triangular diagram called a 'textural triangle' such as is depicted in most textbooks on soil science.

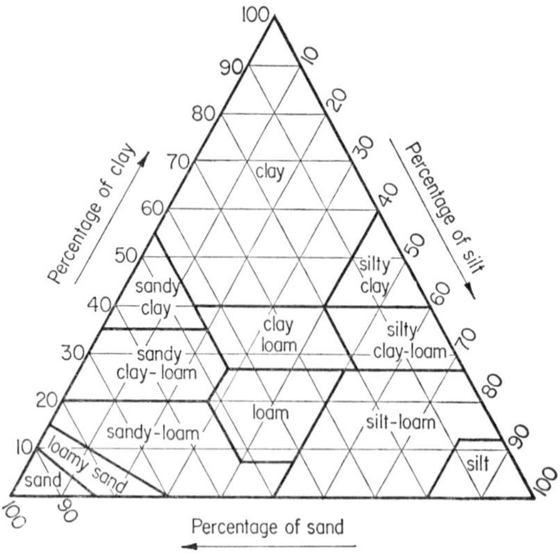

Figure 59. Basic soil-texture classes. (By courtesy of *World Farming* and U.S. Department of Agriculture.)

This triangle is broken up into twelve areas, with different proportions of sand, silt, and clay—each with its standard, conventional name (Figure 59).

Determination of Textural Grade. In order to place a soil in the proper class, the relative proportions of sand, silt, and clay, must be known. Most of the laboratory methods in use at present for making a mechanical analysis of soil are based on the principle that the rate of settling of the particles of a soil dispersed in water depends on the size of the particles. The larger the particle is, the more rapid will be the rate of settling—so that the various fractions can be separated and their relative amounts determined quantitatively.

In soil surveys in the field, the determination of soil texture is made by feeling the wetted soil between thumb and fingers. This procedure is of course far from precise; but it is surprising how close the results may be to those obtained by laboratory analysis, if the field scientist has skill and experience based on frequent checks against laboratory results.

Porosity or Pore Space

Porosity is the percentage (by volume) of a dry soil that is occupied by 'air'. Normally the pore space is occupied by gases and water in varying proportions. The *total porosity* is of importance from the standpoint of aeration and moisture storage. The aeration of soils (the exchange between soils and atmospheric gases) occurs mainly by diffusion. The rate of diffusion is proportional to the amount of free pore space—that is, the pore space unoccupied by water.

The size of the pores has a marked influence on the rate of water movement through soils. Obviously, the determination of both the total porosity and the size of the pores yields important information. The total porosity may be calculated easily provided the specific gravity and volume weight are known. The following formula relates these properties to porosity:

$$\text{Percentage porosity} = 1 - \frac{\text{Volume weight}}{\text{Specific gravity*}} \times 100$$

Crop plants, for their normal development, need a constant and balanced supply of water and air to their roots. These two requirements may be conflicting, since with increasing water-supply a deficiency in air-supply may arise. This will impair nutrient uptake by the roots, interfere with the proper functioning of many essential physiological processes of the plant, and have an adverse effect on the activities of many groups of beneficial soil micro-organisms. Total porosity and pore-size distributions are therefore important parameters of a tilled soil, being relevant to the availability of moisture and air, as well as to the ability of the soil to absorb water rapidly during a heavy rain.

Total pore space alone is a poor measure of aeration. A soil may have a high percentage of total pore space, but if the percentage of clay is excessively high, the pores will be extremely small and all pore space may, under certain circumstances, be filled with water, with a resultant lack of aeration. At the other extreme are the sandy soils, with a large number of very large pores, incapable of retaining water. Air-supply in this case is very favourable but plants easily suffer from a shortage of water. The ideal soil will therefore have sufficient pore space, more or less equally divided between large and small pores, the large pores ensuring a free movement of air, and the small pores a good water-holding capacity. This can be achieved only if the primary soil particles are aggregated. The size distribution and the interconnections of the pores forming the pore space also have a considerable effect on root growth. The more favourable the pore-space distribution is, in particular of the interconnected pores in the 0·1–0·5 mm size-range, the greater will be the depths to which the roots can penetrate during dry periods. This is of considerable importance in regions with erratic rainfall (E. W. Russell, 1966).

The proportion of solid to pore space varies greatly among the different types of soils. In general, the greater the content of soil colloids is, the higher will be the percentage of pore space. Total soil porosity in mineral soils ranges from 25 to 60 per cent, but most usually is within the range of 40 to 50 per cent (Gardner, 1966).

A compact soil, with poor structure and small total volume of pore space,

* The average value for specific gravity of all inorganic soils is found to range between 2·5 and 2·7.

has a high *bulk density*. Bulk density can actually serve as a measure of pore space in the soil.

Soil structure

The development of soil structure involves two processes: (a) the cementing of soil particles to form aggregates, and (b) the shaping and orientation of the aggregates to form a macrostructure characteristic of a soil or of its individual horizons. The first is mainly a physicochemical process and the second is a biotic process (Jacks, 1963).

Importance of Soil Structure. The basic requirement for effective crop production is that the solid phase of the soil should not constitute a compact mass, but be so structured as to facilitate water infiltration, retain moisture, allow free movement of 'air', and promote easy penetration and proliferation of the root system.

Soils that have broken down into a powdery dust by excessive tillage when dry—the classic example being the soils of the Great Plains in the U.S.A.—or soils that form a compact mass on drying, such as the alkaline soils resulting from faulty irrigation methods, cannot be a convenient medium for crop production under arid conditions.

The soil of virgin lands in semi-arid regions, usually covered by grass, generally have a very favourable soil structure. A clod taken from such a soil will break down easily into small aggregates about the size of wheat kernels. The aggregates themselves are fairly resistant to further breakdown.These granules are the foundation of 'crumb' structure, which is considered the ideal soil structure, as the crumbs leave ample pore space between them and provide pores of the right size for a balanced supply of 'air' and water to the roots of the crop.

Cultivated soils rarely achieve the ideal soil structure of the virgin soils that were covered with permanent vegetation; clods do not break down as easily into smaller aggregates, whilst the aggregates themselves are less resistant to breakdown than those of the virgin soil. The size of the aggregates is not uniform, but ranges mostly from the size of a pin-head to that of a pea—with a varying proportion of dust-sized particles. It has been shown that the most favourable size of aggregates needed for a good soil structure in semi-arid soils ranges from one to five millimetres in diameter (Kvasnikov, 1928). The smaller-sized aggregates are more desirable in semi-arid agriculture for rain-fed crops because their better water-holding capacity prevents excessive vegetative growth and too early depletion of soil moisture; the larger-sized aggregates are favoured for irrigation agriculture, as they drain more easily and ensure better aeration of the soil. They also provide the best soils under arid conditions in which precipitation is not sufficient to drain beyond the root-zone, so that high availability is not offset by losses due to percolation.

The soils aggregates should be small enough to provide adequate aeration and good contact between the moisture films on one hand and the seeds and roots on the other. Aggregates smaller than 1 mm in diameter tend to fill the pore spaces between the larger aggregates and so restrict air movement. Excessive breakdown of aggregates will also cause severe soil-crusting after each rain or irrigation. Aggregates larger than 5–6 mm in diameter leave excessively large pore spaces between them, thereby reducing water-holding capacity and preventing intimate contact between the roots and the soil particles. Relatively large aggregates and a rough micro-relief of the soil surface improve water infiltration.

Even a soil with good structure will, on drying, form hard, compact clods. However, these clods break down easily under tillage, and the favourable soil structure is easily restored. By contrast, soils with poor structure produce clods which are resistant to normal tillage operations. When wetted, they break down into a plastic mass; upon redrying, the soil is covered with a hard crust. These soils can be tilled only within very narrow limits of moisture content, and are rarely productive.

Experience has shown that there is a very close relationship between the fertility and stability of soils and their structure, as expressed by the percentage of stable soil aggregates of desirable size. All farm operations that improve soil structure can therefore be considered as fertility-improving, and *vice versa*.

Classification of Soil Structure. The structural type includes the morphology of the structural units as well as their arrangement.

Although there are no precise expressions for soil structure in the field, there are a number of fairly well-established morphological types, which vary in hardness and consistency. Some of the more important types of structure are:

1. *Structureless*
   (a) *Single grain*—Each 'grain' by itself, as in sand.
   (b) *Massive*—Large uniform masses of cohesive soil, sometimes with irregular cleavage, common to sandy loams, loamy sands, and silt-loam.

2. *With structure*
   (a) *Platy*—Thin, horizontal plates. Appears as aggregates with horizontal dimensions greater than vertical dimensions. Not ordinarily developed under cultivation.

(b) *Prism-like*

(1) *Prismatic*—Appears as vertically elongated aggregates in the shape of a prism.

(2) *Columnar*—Prism-like structure with rounded tops.

(c) *Block-like*—Aggregates developed equally along the three axes. Common in subsoils, particularly in forest soils or prairies.

(d) *Spheroidal*—Soil aggregates rounded.

Granular—Hard or soft; but firm, small aggregates, rounded.

Crumb—Generally soft, small, very porous rounded aggregates.

3. *Structure destroyed*

Puddled—If soils containing clay are ploughed when wet they become puddled or 'run together'. Puddled soils provide poor seed-beds, and poor stands of planted crops.

The structure of arid soils is usually ill-defined. When it is well defined, any of the morphological types of group 2 may occur (Aubert, 1962).

Formation of Soil Structure. Soil aggregates can be formed by the breakdown of clods or by the aggregation of discrete soil particles. The first process can be achieved mechanically, by tillage under appropriate conditions; the second is mainly the result of biological activity in the soil.

The formation of more stable aggregates from the primary soil particles depends on the presence of cementing materials in the soil. This purpose is fulfilled by the soil colloids. The organic colloids are the most effective in this respect, producing a stable structure that is capable of resisting a number of alternate wettings and dryings in the field. An example of the degree of stability that can be achieved is provided by aggregates formed from a loess soil, rich in organic matter, that can resist breakdown after having been immersed in water for several consecutive days. Most of the inorganic colloid cements are subject to alternate dehydration and hydration with each cycle of drying and wetting, thereby reducing the stability of the aggregates. Such soils may crust excessively (Yoder, 1937).

In a productive soil, improving structure is a gradual process, resulting from the activities of living organisms. In infertile soils, such as alkali soils, whatever structure exists is mainly due to abiotic processes (Jacks, 1963).

The Stability of Soil Structure. Soils differ in their ability to retain a favourable soil structure under the impact of rainfall, irrigation, farming equipment, and the solvent effects of soil moisture. When the structure is lacking in stability, rains or irrigation may cause the aggregates to break down and the soil particles to run together, so that, on drying, a smooth crust appears on the soil surface. Many soils contain a fairly large proportion of aggregates that

Plate 1 – The so-called 'dew-mounds' in the Negev Desert of Israel.
(a) – Aerial view of hills covered with stone-mounds.
(b) – Close-up view of individual stone-mounds.
Photo: N. Tadmor

(a)

(b)

Plate 2 – 'Desert Pavement'. *Photo: N. Tadmor*

Plate 3 – *Aristida scoparia* on shifting dunes in the Negev Desert. In the distant background, dense vegetation of *Artemisia monosperma* on arrested sand-fields. *Photo: N. Tadmor*

Plate 4 – Typical 'diffuse' vegetation on a slope and 'restricted' herbaceous vegetation in the wadi (Negev Desert). *Photo: N. Tadmor*

Plate 5 – A stand of Chihuahua Desert shrubs. The dominant species are Creosote-bush, Whitethorn and Tarbush. *Photo: E. M. Schmutz and R. F. Wagle*

Plate 6 – An untreated stand of oak-chaparral. The dominant species is Turbinella Oak. Openings similar to those shown in the right and left foreground occupy a sizeable portion of the area and contain little or no ground-cover.

Photo: E. M. Schmutz and R. F. Wagle

Plate 7 – Natufian blade-sickle.
By courtesy of the Israel Museum

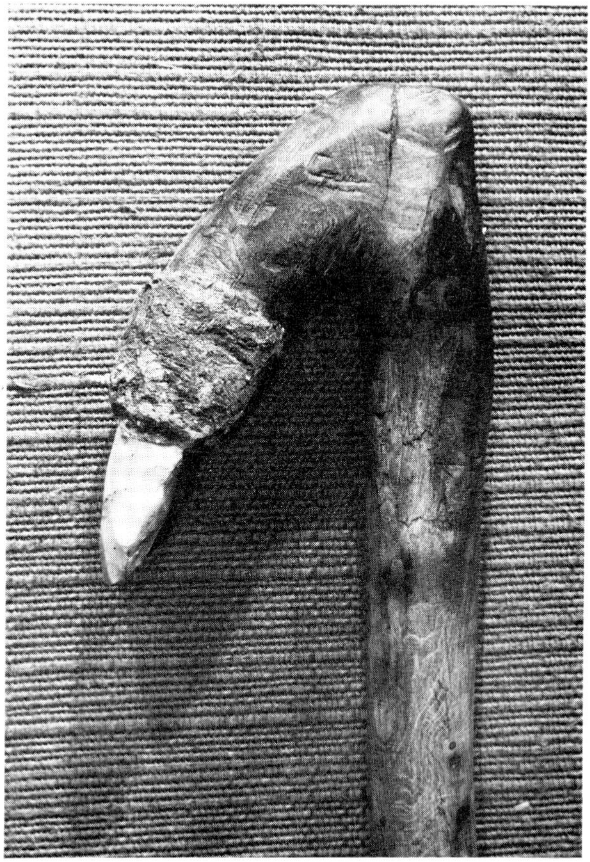

Plate 8 – Reproductions of original sketches by Guaman Poma de Ayala, depicting the agriculture of the Incas: (left) The sowing of maize, (right) Irrigating maize.
From *Races of Maize in Peru: Their Origins, Evolution and Classification.* Publication 915, Committee on Preservation of Indigenous Strains of Maize, National Academy of Sciences – National Research Council, Washington, D.C., 1961.

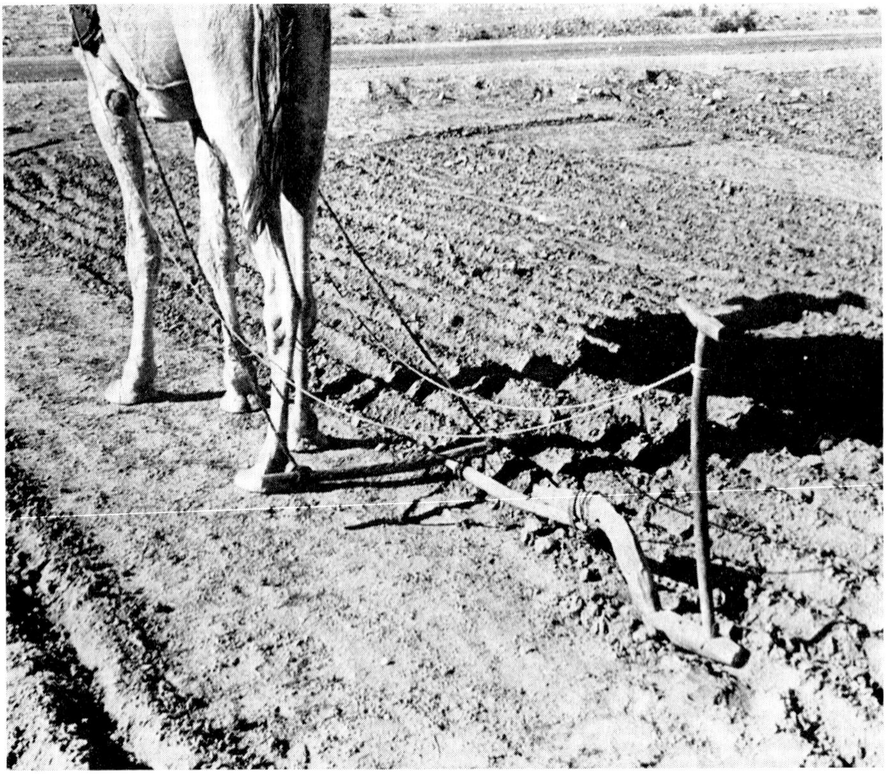

Plate 9 – Sowing an area in the Negev Desert wetted by a flash-flood. The seeds are scattered broadcast, and then covered by ploughing with a typical wooden plough of the Near East.

Photo: N. Tadmor

(a)

(b)

(c)

(d)

Plate 10 – Four types of grazing in the dry regions.
(a) On diffuse vegetation in the desert.
(b) On range in a semi-arid region.
(c) On self-seeding legumes in a ley-farming system adapted to dry regions.
(d) On stubble of a grain crop.
Photos: N. Tadmor

(a)

(b)

Plate 11 – Flood-water spreading in Nabatean Farming.
(a) Crop production in tributary valleys; the well-preserved low terraces for slowing down the flow of water are clearly seen in this aerial photo.
(b) Aerial view of Nabatean water dispersal system; note the so-called dew-mounds in the background (cf. P1.1).

(c)

(d)

(c) Reconstructed Nabatean farm, on the site of the original farm; note the areas in the background that produce the 'water crop'.
(d) Nabatean diversion dam in a wide wadi.
Photos: N. Tadmor

Plate 12 – Typical flood in a wadi of the Negev Desert, that had been previously dry for months. *Photo: N. Tadmor*

Plate 13 – Nabatean water-harvesting system: clearing the donor area of stones, in order to increase run-off. *Photo: N. Tadmor*

(a)

(b)

Plate 14 – Cisterns in the Negev Desert.
(a) Collecting system leading water from donor area to the cistern.
(b) Close-up view of cistern.
Photo: N. Tadmor

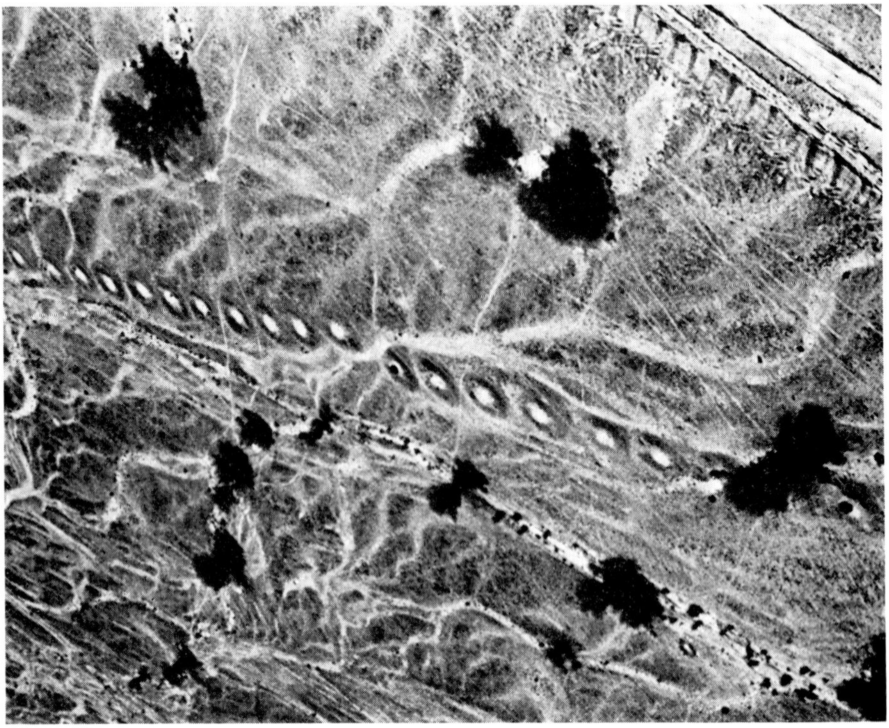

Plate 15 – Aerial view of line of quanat apertures.
Photo: N. Tadmor

Plate 16 – Border-strip irrigation.
By courtesy of Soil Conservation Service, Israel

(a)

(b)

Plate 17 – Furrow Irrigation.
(a) Primitive system; note the short irregular furrows, requiring considerable manual labour, and the irregular water distribution.
By courtesy of Soil Conservation Service, Israel
(b) Modern system; note capillary rise of moisture.
By courtesy of O. A. Lorenz, College of Agriculture, Davis, California

Plate 18 – 'Valley' self-propelled overhead sprinkler system pre-watering an area prior to sowing.

Photo: Holland Dreves Reilly, Inc.

Plate 19 – Maize plants suffering from a deficiency of moisture.

By courtesy of O. A. Lorenz, College of Agriculture, Davis, California

Plate 20 – Irrigation ditches lined with concrete, using portable slip-forms, conserve irrigation water and reduce erosion in gravity-irrigation systems.

By courtesy of the University of Wyoming

Plate 21 – Differences in salt tolerance between two closely related varieties of soybean (left and right).

By courtesy of G. H. Abel

Plate 22 – Two wheat varieties with sharply contrasted leaf-angles.
By courtesy of C. M. Donald, Waite Agricultural Research Institute, Australia

Plate 23 – Soil erosion due to faulty tillage practices.
By courtesy of Soil Conservation Service, Israel

Plate 24 – Applying anhydrous ammonia in the field.
By courtesy of 'John Deere Co.'; photo by Broyhill Co., Nebraska

Plate 25 – Fertilizer placement.
By courtesy of 'Better Crops with Plant Food' of the American Potash Institute

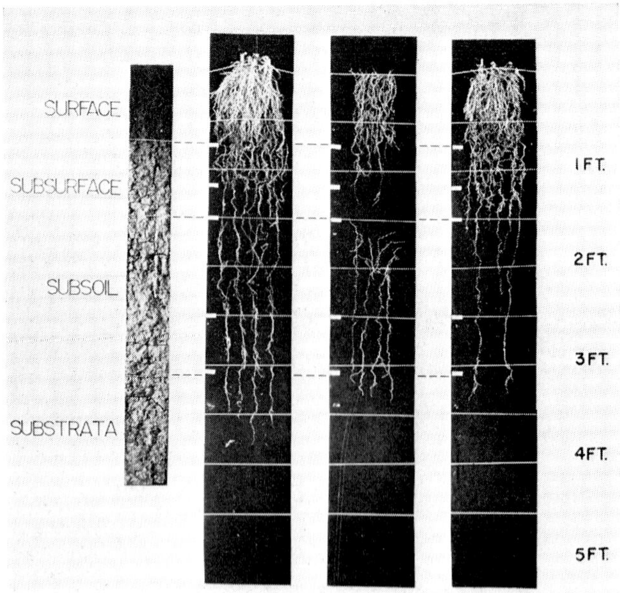

(a)

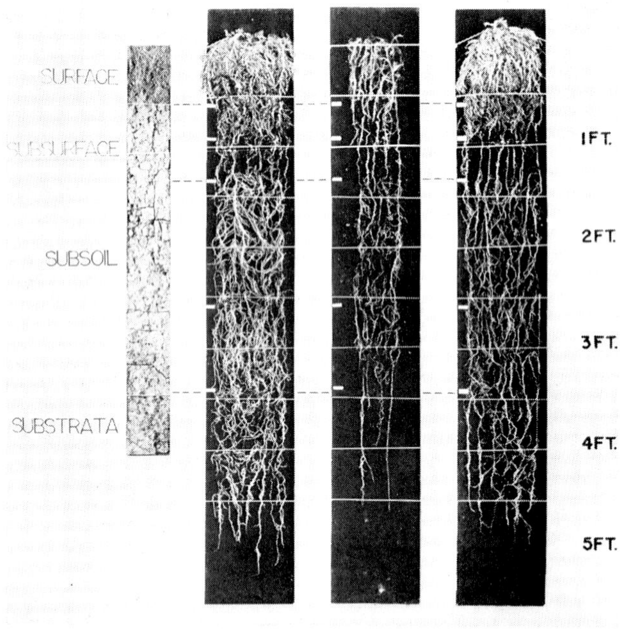

(b)

Plate 26 – Effect on maize root development of fertilizer application to a silt-loam.
(a) Unfertilized plots.
(b) Fertilized plots.
By courtesy of J. B. Fehrenbacker

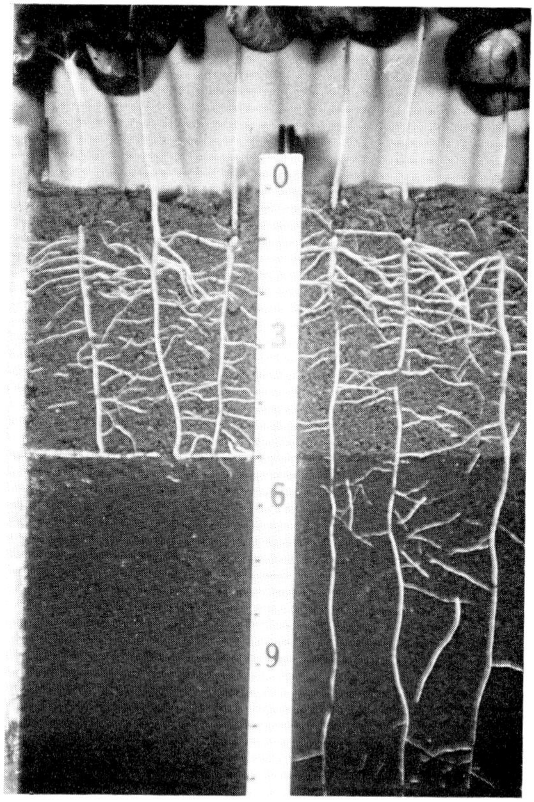

Plate 27 – Effect of soil compaction on cotton roots. In the left-hand soil, a hard pan is present at a depth of 5″ (12.5 cm) impeding root penetration. In the right-hand soil, there is no hard pan and roots penetrate freely beyond the ploughed layer.

Photo: J. McKinney, by courtesy of 'Crops and Soils'

Plate 28 – Land-leveller or float. Note the even surface after passage of the float.

By courtesy of Ministry of Agriculture, Israel

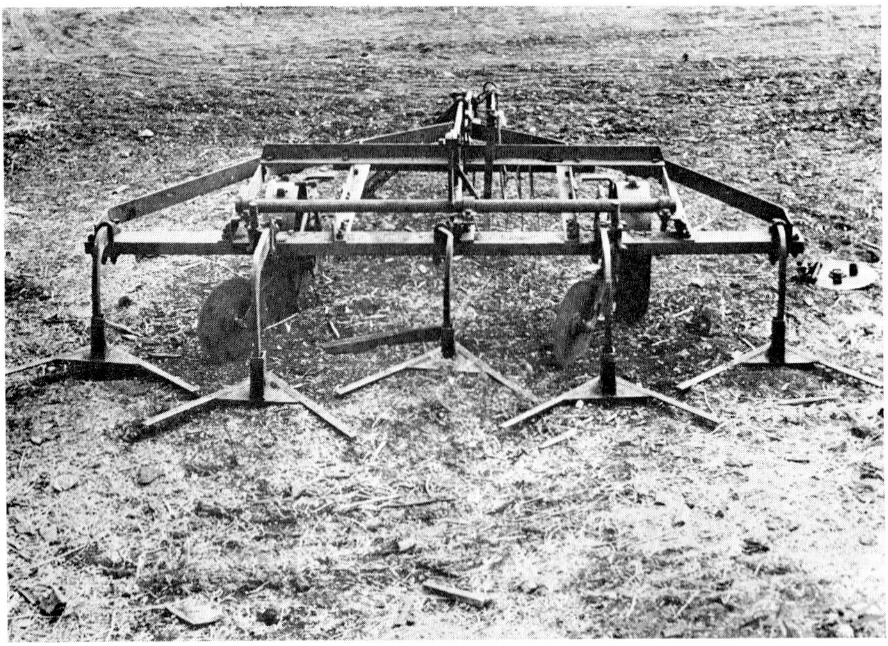

Plate 29 – Sweep-cultivator for stubble-mulch tillage.
By courtesy of Ministry of Agriculture, Israel

Plate 30 – The two soil zones (water management zone and seedling environment zone) as prepared for row-crops by minimum tillage.
By courtesy of W. E. Larson

Plate 31 – Plough planting.
By courtesy of Michigan State University

Plate 32 – Wheel-track planting.
By courtesy of U.S. Department of Agriculture

(a)

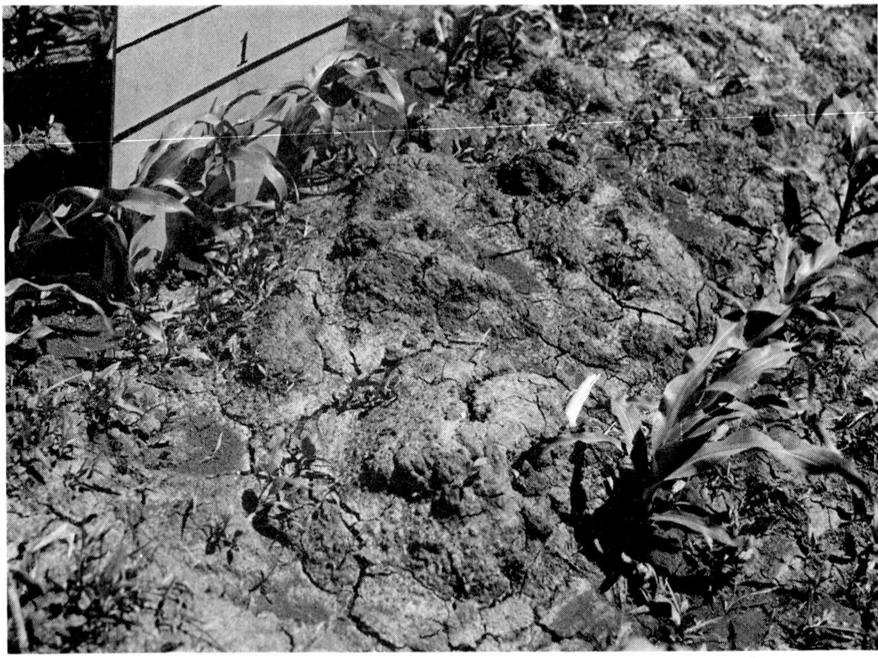

(b)

Plate 33 – Soil crusting on silt-loam as influenced by conventional (a) and wheel-track planting (b). Note how in the wheel-track planting the soil between the rows is rough and loose in contrast to the compacted soil of the conventionally tilled and planted inter-row space.

By courtesy of W. E. Larson

(a)

(b)

Plate 34 – Chemical fallow compared with conventional fallow.
(a) Chemical fallow (right) compared with conventional fallow (left).
(b) Maize growing after chemical fallow.
By courtesy of 'Crops and Soils'

Plate 35 – Locust swarm.
By courtesy of Ministry of Agriculture, Israel

Plate 36 – Biological control: a wasp, *Baphyplectes curculionis*, attacking larva of Lucerne Weevil.
By courtesy of 'Crops and Soils'

may retain their form for a long time—if they are not subjected to excessive stress when very wet or exceedingly dry. Such aggregates are said to be 'water-stable'. A soil with a stable structure can be tilled without damage within a far wider range of soil-moisture contents than a soil with a non-stable structure. The formation of crusts on the latter types of soil prevents or retards water infiltration, whilst the breakdown of the aggregates in the soil results in soil compaction and reduced pore space.

The stability of soil structure confers a further advantage; it increases considerably the ability of the soil to withstand erosion by wind or water. Primary soil particles are easily blown or washed away; when they are cemented together into larger, stable aggregates, their resistance to wind and water-induced movement is considerably increased.

With a good soil structure, moisture penetration is greatly facilitated; therefore less run-off occurs and the danger of erosion damage is considerably reduced. In Australia, a 50 per cent decline in aggregation in a loam soil was found to result in a reduced rate of moisture penetration: 1·25 mm in three hours instead of the previous 22·5 mm during the same time (Woodroffe, 1949). The stability of the soil structure and the stability of the soil itself are therefore very closely linked.

In general, the factors which affect soil structure adversely are more effective in doing so than are those that contribute to an improved structure or increase the soil's stability.

In soils that are poor in organic matter, soil aggregation may be due to physical factors only. For example, if dry clay-soil is gradually wetted and then ploughed at an optimum soil-moisture content, many aggregates will be formed. However, these aggregates will be broken down quite easily by further tillage operations, rainfall, or irrigation.

Factors Affecting Soil Structure

Kinds of Clay. The influence on its structure of the clay particles in the soil is determined mainly by the kind of exchangeable ions which predominate on their surface. The most marked effects, in this respect, are exercised by sodium and calcium ions.

Soils in which the clay is saturated with sodium form large, hard, compact clods on drying, with very few cracks. When these clods are rewetted, the sodium deflocculates the clay, and the clods break down into a plastic, formless paste.

By contrast, calcium-saturated clay, on drying, forms small friable clods with numerous cracks. When rewetted, the calcium prevents deflocculation; the aggregates swell but do not break down easily. These are the water-stable aggregates mentioned above.

Soil Organic Matter. Diverse groups of micro-organisms are active in the composition of organic matter in the soil. Fungi bind soil particles together by the mycelium filaments that they produce; one of the products of the breakdown of organic matter by other micro-organisms are sticky substances that cement soil particles together into aggregates. These substances are not entirely resistant to further breakdown, so that a constant supply of sticky substances is necessary for such maintenance of soil structure.

Experimental results indicate that such micro-organisms produce better aggregating substances at relatively low temperatures, resulting in a more stable structure. As temperature increases from 10 to 55°C, the soil-aggregating influence of organic residues was found to decrease. On the other hand, with increasing temperatures, less time is required for maximum aggregation to occur, but it is less stable (J. P. Martin & Craggs, 1946). It is assumed that the microbial population which is active at lower temperatures produces more effective aggregating substances than the organisms operating at higher temperatures, whilst the higher temperatures simply favour a more rapid decomposition of organic matter. In waterlogged soils, the rate of aggregation is considerably reduced, indicating that anaerobic organisms produce few aggregating substances, and that these are of low quality (J. P. Martin & Craggs, 1946).

In high-sodium soils, the addition of organic residues causes much less aggregation than in normal soils; probably the sodium disperses the inorganic soil colloids and the aggregating substances formed by the micro-organisms (Aldrich & Martin, 1954).

Soil Fauna. Burrowing and digging animals, and a great variety of insects and worms, have a considerable effect on soil structure. The most widely known and the most ubiquitous in moist climates are the earthworms, whose considerable role in improving soil structure was stated already by Darwin. However, their numbers are far smaller, and their effects of little significance, in the hot arid climates.

In areas with very sparse vegetation, as in the semi-deserts and along their fringes, ants and termites, by burrowing in the soil, can be as effective in improving soil structure as the earthworms are in the humid regions (Hopp & Slater, 1948). They break up the larger and coarser conglomerates mechanically; they digest organic matter and contribute various enzymes to the residues which affect their further decomposition (Pauli, 1967).

Grazing Animals. Grazing animals may accelerate the fertility-cycle considerably by increasing the rate of decomposition of plant residues following the return to the soil of faeces and urine and the resulting increased microbial activity. If, however, overgrazing occurs, plant residues and hence soil organic matter decrease; this results in increasing erosion from the denuded soil which in turn, further impairs the soil structure.

Crops. When virgin soils that had been covered by perennial grasses are brought under cultivation, soil structure usually declines rapidly. The rate of decline depends to a large extent on the kind of crops grown.

Crops may have diverse effects, both direct and indirect, on soil structure. Thus a more favourable and stable soil structure forms under perennial crops, in particular grasses, than under annual crops that are grown in rotation, as is shown by the following examples (Table XVII) from the results of Russian workers.

TABLE XVII

Effect of Cropping System on the Percentage of Water-stable Aggregates (1 > /4 mm) *in Irrigated Grey Desert Soils* (Geltzer, 1934)

Soil under annual crops in rotation	4
Soil under lucerne, after three years	20
Soil under lucerne, after six years	30
Perennial irrigated pasture	79

These results indicate that perennial grasses and, to a lesser extent, perennial legumes, are capable of improving the structure of a soil that has broken down under annual cropping. Similar results have been obtained by other workers: thus in Australia it was found that the percentage of aggregates larger than 0·2 mm decreased, in a virgin loam soil, from 52 to 24 per cent, after many years of cropping in a wheat-fallow rotation (Stephens, 1945).

In the Great Plains region of North America it was found that cropping systems consisting of continuous small grain, continuous row-crops, or cereal cropping alternating with fallow, showed a loss of approximately 80 per cent of the initial aggregation of the virgin Buffalo Grass range after approximately 40 years of arable cropping. On the other hand, seeding arable soil to Buffalo Grass doubled the amount of aggregation within five years (Olmstead, 1946) In South Africa, the structure of a soil that had been cropped continuously for 20 years was restored by a grass cover after seven years (Hyam, 1950).

Plant residues also supply organic matter to the soil. The relationships between plants, organic matter, and soil micro-organisms, were clarified in experiments carried out in the arid climate of New Mexico (Martin & Craggs, 1946). No aggregation of soil particles occurred in a sterilized soil in which roots of Sudan Grass proliferated. After inoculation of the soil by micro-organisms, small aggregates were formed which were subsequently bound together into larger aggregates by the rootlets of the Sudan Grass.

Microscopic examination of soil aggregates confirms that the smaller aggregates are formed by the cementing together of soil particles by colloidal materials, whilst larger aggregates are formed by the binding action of mycelia of Fungi and rootlets of higher plants. However, the main effect of crops on soil structure may be indirect and dependent on the amount of tillage involved. Perennial crops with little or no tillage have the best effect, closely planted crops with little inter-row cultivation do relatively little harm to soil structure, whilst the greatest damage to soil structure is caused by wide-spaced crops with frequent inter-row cultivation, such as maize and cotton.

Tillage Operations. The properties of clay change at different degrees of wetness. A dry clay is hard, compact, and breaks into clods with ploughing. If the clay is wetted gradually, the water molecules at first are adsorbed on the surface of the clay particles, acting as a lubricant and allowing them to slip easily against one another. The clay becomes soft and the clods break down easily. This is the ideal stage for tillage. If more water is added, the plasticity of the clay increases, it becomes sticky, and is then easily transformed by pressure into a plastic mass.

The moisture content at which a clay soil, on drying, ceases to stick to a metal rod pressed into the soil, is called the 'sticky point', and is optimal for tillage operations. For then the soil does not stick to the tillage implements and a minimum of damage is caused to soil structure by compaction. The pressure exerted by tillage equipment may even increase aggregation. However, any beneficial effects may be easily lost by subsequent tillage operations, such as disking, levelling, rolling, etc. If carried out in dry soil, these operations reduce a large proportion of the aggregates into fine dust; if the soil is wet, compaction and reduced pore space result.

Rainfall and irrigation. Raindrops hit the soil surface with considerable mechanical force, and can rapidly destroy the structure of the upper soil layer, creating a hard crust. This crust effectively seals the soil surface, and reduces the depth of water penetration and aeration. The overall effect of rainfall is therefore similar to the detrimental effects of tillage, the main difference being in the depth of the soil layer that is compacted.

Ellison (1952) has calculated that, for a rainfall of 75 mm per hour, the raindrops falling on one acre have a compacting effect on the soil that is equivalent to 29 average farm-tractors working on this area during the time of the rainfall!

The impact of raindrops against the soil destroys the aggregates in the surface soil, while the run-off water becomes turbid with suspended particles which clog the pores of the upper layers; thereupon permeability further decreases, with a concomitant increase in erosion damage (Vershinin *et al.*, 1959).

In South Australia, on a red-brown soil, the principal damaging effect of excessive rainfall in the month following sowing was found to be an increase, in apparent density, which amounted to 0·052 gm/cm^3 for each additional inch of rainfall. The corresponding decline in grain yields was approximately 680 kg/ha for each increase of 0·1 gm/cm^3 in apparent density (Millington, 1961).

Irrigation by surface methods (flooding, furrows, etc.) acts in a different way. After each irrigation, there is a considerable excess of water in the upper soil layer, during the 24–48 hours required for an equilibrium to be established until the soil reaches its full water-holding capacity. This excessive humidity can cause the breakdown of soil aggregates that are only partially stable, thereby compacting the soil.

Artificial Soil Conditioners. The influence on soil structure of most of the factors mentioned above is usually transitory. The discovery that soil micro-organisms can synthesize polysaccharides and other chemical compounds which enhance soil granulation, stimulated the search for synthetic materials which would act in a similar but more permanent manner (Harris *et al.*, 1966).

In recent years, synthetic polymers have been produced by the chemical industry, which are capable of stabilizing the structure of the soil for long periods, owing to the marked resistance of these products to microbial decomposition. One of the first and most widely tested of these 'soil conditioners' was the sodium salt of polyacrylonitrile, which is sold under various trade names, of which Krillium is probably the best known.

The three principal soil-conditioners to have been tested are: HPAN, supplied largely as a sodium polyacrylate; VAMA, a mixture of calcium hydroxide, a copolymer of vinyl acetate, and the partial ester of maleic acid; and IBMA, a copolymer of isobutylene and the half-ammonium-salt and half-amide of maleic acid.

The soil conditioners are water-soluble polyelectrolytes whose particles in solution carry a positive electrical charge, and therefore are attracted to the negative surface-charged clay particles, binding them together into aggregates. Anion exchange, hydrogen binding, and cationic bridge formations, are mentioned as possible mechanisms for clay-polyanion linkage (Harris *et al.*, 1966). The polyelectrolytes are relatively resistant to breakdown and their effect is therefore long-lasting.

The great hopes placed in the soil conditioners have not yet been substantiated in farming practice. They cannot, in themselves, improve soil structure, but merely tend to stabilize the situation as it exists at the time they are applied. If soil structure is poor to start with, the conditioner will perpetuate this situation. Structure improvement must therefore precede the use of the conditioners.

The effectiveness of soil conditioners depends on soil type, the crop to be grown, and the rates and methods of application. Conflicting results have been obtained, ranging from crop increases to decreased yields. Relatively heavy rates of application of the conditioners have frequently had adverse effects: e.g. poor germination, retarded flowering, and lowered yields. The quantities required for effective stabilization range from 1 000 to 1 500 kg per hectare, according to soil type, incorporated to a depth of 6–8 cm.

The soil conditioners are not likely to replace organic matter in soil management procedures. They are, however, an important potential supplement to the conventional methods. Regardless of their effectiveness, their cost at present precludes their large-scale use in farming practice. The only economically justified use developed up to the present, is for the prevention of crust formation above the seed-rows of high-value crops, whose seedlings are normally weak when sown in soils that tend to form a hard crust.

THE GASEOUS PHASE OR SOIL 'AIR'

The composition of soil 'air' is not markedly different from that of the free atmosphere, the only important difference being that the former is normally richer in carbon dioxide. However, the carbon dioxide content of the soil 'air' is far from constant. Additions of organic matter to the soil usually result in a rapid increase of carbon dioxide and a parallel decrease of soil oxygen; root respiration has similar effects. Oxygen deficiencies and excessive CO_2 in the root atmosphere have a depressive effect on the growth of most crop plants and their associated micro-organisms (Grable, 1966).

Carbon dioxide becomes toxic at concentrations above 1 per cent in the soil 'air'. Frequently, the toxicity of the carbon dioxide is more lethal to the root system of plants than is the relative lack of oxygen. When the oxygen content of the soil 'air' falls to below 10 per cent, this usually results in damage to the root system. When the level falls to 3 per cent, the roots of most cultivated plant species die (Grable, 1966). Soil 'air' usually contains sufficient oxygen for the vital processes of the plant and soil micro-organisms, but air as a whole may be in short supply in soils with a poor structure or those which are excessively wet.

Under waterlogged conditions, the escape of CO_2 and renewal of the air-supply ceases, and as a result undesirable biological activity occurs which may engender toxic concentrations of methane, hydrogen sulphide and other substances. Hence, root development slows down and may even cease. Meanwhile root respiration, which is essential for certain metabolic processes, is adversely affected, nutrient and water uptake being thereby curtailed.

Certain crops, such as rice, are capable of growing normally in inundated soils. As with other plants, their roots require oxygen for their normal functioning, but the oxygen in this case is absorbed by the aerial parts of the plant and transported to the roots through continuous intercellular spaces in the stems and roots (Barber *et al.*, 1963).

Part of the carbon dioxide in the soil combines with water to form carbonic acid (H_2CO_3). This is a weak acid which is nevertheless capable of dissolving certain mineral components of the soil and thereby making them available to the crop. However, a large proportion of the carbon dioxide produced in the soil escapes into the air above the soil and is assimilated by the green aerial parts of the plant.

In fertile soils that are rich in organic matter, and with a favourable moisture régime, soil micro-organisms free very large amounts of carbon dioxide. There is therefore a direct relationship between soil fertility and the amounts of carbon dioxide liberated from the soil.

The need for oxygen and the parallel sensitivity to excess carbon dioxide in the soil 'air' increase with increasing temperatures; these factors are therefore more critical in the hot arid areas than in the humid regions. The greater sensitivity to lack of oxygen at higher temperatures may be due to the increased intensity of respiration of roots (Vershinin *et al.*, 1959), or to the diminished solubility of oxygen in soil solutions at higher temperatures.

Chemical Soil Properties

The nature of the parent material of the soil, and the processes of weathering, determine the potential reservoir of plant nutrients.

The limited leaching to which arid soils have been subjected has resulted widely in the accumulation of mineral substances, including the important plant nutrients. These soils are therefore generally rich in calcium, potassium, magnesium, sodium, phosphorus and sulphur, and deficiencies of micro-elements are rare in them. However, the virtual absence of leaching has also resulted in many cases in the accumulation of salts that are detrimental to plant growth, such as chlorides and carbonates of sodium and magnesium. Where the soluble salt-content of the soil exceeds 0·5 per cent, soil productivity is adversely affected, while above 1–2 per cent the soil is virtually barren.

THE SOIL COLLOIDS

When a soil sample is agitated in water, the smallest particles remain in suspension. These are the soil colloids, consisting of an intimate mixture of mineral and organic particles, whose diameters generally lie between 0·5 micron and 1·0 millimicron. The inorganic colloids consist mainly of clay minerals, derived from the parent material during the weathering process; the organic colloids are complex organic materials, being the intermediate products in the process of decomposition of organic matter by soil micro-organisms.

The three principal groups of clay minerals which occur in various proportions in most soils are kaolinite, illite, and montmorillonite. The kaolinite group tends to predominate in mature, weathered soils and is therefore found mainly in humid, hot regions; the illite and montmorillonite clays predominate in younger soils and therefore tend to be more frequent in regions of moderate rainfall.

Kaolinite consists of silica and alumina sheets, in a 1:1 ratio, which form a non-expanding crystal lattice. Soils in which kaolinite predominates show little swelling or shrinking.

In illite, in which the silica and alumina sheets are in a 2:1 ratio, potassium bonds occur between the adjacent silica plates and prevent water molecules from expanding the structures.

In montmorillonite, the ratio of silica to alumina sheets is also 2:1, but the chemical bond between the plates is absent—so that, with hydration, this type of clay swells considerably, but shrinks again on drying. As a result, the soils in which montmorillonite predominates tend to develop large cracks on drying—particularly if they were compacted when still wet.

The important characteristic of the soil colloids is their enormous surface area exposed per unit volume, as a result of the small size of the individual particles. Montmorillonite particles, for example, have a surface area of 600–800 m^2/gm, and those of organic matter are similar: 500–800 m^2/gm. By contrast, the surface areas of illite (65–100 m^2/gm) and kaolinite (7–30 m^2/gm) are far smaller (Webber & Elrick, 1966).

The colloidal constituents of the soil are the seat of most of the chemical and physical reactions taking place in the soil.

SOIL REACTION—pH

pH is a numerical designation of relatively weak acidity or alkalinity in soil, as determined by the relative proportion of hydrogen and hydroxyl ions present.

In regions of low rainfall, the colloidal complex of the soil is generally dominated by the common basic cations—calcium, magnesium, and sodium. The colloidal complex will tend to remain high in adsorbed calcium ions as well as in other basic cations, and the soil solution will remain alkaline. In such soils, variable quantities of soluble bicarbonates, carbonates, biphosphates, and silicates, are often found which, upon hydrolysis and dissociation, yield hydroxyl ions in excess of hydrogen ions. As a result, calcareous soils may have reactions ranging from neutral to very alkaline (pH 7–11).

In regions of high rainfall, where leaching has drastically diminished the quantity of soluble and solid-phase salts, the hydrogen ions become dominant over the hydroxyl ions. The soil thus becomes acidic in spite of the presence of a certain amount of active calcium.

The soil reaction (pH) has a profound influence on many factors which determine the suitability of a soil for crop growth. In arid soils, excessive acidity is of rare occurrence, for reasons explained above. Excessive alkalinity, on the other hand, occurs quite frequently, and may have the following undesirable effects:

toxic effects on the root system;
some nutrients may become less soluble and less available; as a result, phosphate and minor-element deficiencies may occur;
certain desirable activities of micro-organisms may be inhibited; and
the amounts of exchangeable sodium may be sufficient for sodium carbonate to be present in the soil solution (sodic soils).

The use of pH to define soil conditions has serious limitations; thus accurate measurement of the pH of the soil is not possible, the correct interpretation of the measured pH is difficult, the pH varies considerably over a field at a given time, and it is also changing constantly with time (E. W. Russell, 1961).

ION ADSORPTION AND EXCHANGE

Adsorption

The clay fractions of soils are either colloidal or colloidal-like. While nearly all colloids have equivalent electrical charges for the same weight of colloid, each kind of soil colloid has its own adsorption capacity. This capacity is normally expressed in terms of milliequivalents per hundred grams of dry material (colloid or soil). The kaolinite colloids, owing to their lattice structure and chemical nature, generally have a relatively low adsorptive capacity: 3–15 meq/100 gm, whilst the adsorptive capacity of the montmorillonite and illite clays is many times greater (80–150 meq/100 gm). Organic colloids have a still higher adsorptive capacity, around 200–400 meq/100 gm (Webber & Elrick, 1966). Thus, the adsorptive capacities of soils vary with the amount and kinds of clay they contain, and with their organic matter content. The nature of cations that predominate in the colloidal complex of the soil determines the pH (soil reaction): if hydrogen is present in considerable amounts in the complex, the soil will be acidic; if calcium or magnesium predominates, the soils are in general neutral and provide a favourable milieu for most crops; if sodium and potassium predominate, the soils will be alkaline—a very frequent occurrence in arid and semi-arid soils.

Ionic Exchange

All soil solid materials exhibit the capacity to absorb gases, liquids, and other solids, to a certain degree. The amount of adsorption is directly related to the amount of surface present. The soil colloids are electrically charged; whether their charge is positive or negative, will depend on their chemical composition and on the nature of the soil solution. Under normal conditions, the soil colloids have a negative charge, which is neutralized by ions of the opposite charge—commonly by such cations as hydrogen, calcium, potassium, and sodium. These cations may be readily removed from the soil colloids by simple exchange with another cation in solution, and are therefore called 'exchangeable cations'. For example, if a potash fertilizer such as K_2SO_4 is added to a soil in which considerable calcium is associated with the colloids, the K ions will displace at least part of the Ca ions. The proportions of Ca ions displaced will depend on the amount of K_2SO_4 applied, or, in other words, on the concentration of the K ions in the soil solution.

Most cation-exchange reactions in the soil are reversible, though some ions, such as potassium and ammonium, when attached to specific exchange positions, are held more strongly than others. Calcium, on the other hand, is not held very strongly by the soil colloids and is easily replaced by plant nutrients, such as potassium or ammonium, when these are added to the soil in fertilizers. They, in turn, are held sufficiently strongly to prevent their loss by leaching, but still remain easily available to plants. The amount of ions that can be held on the surface of the soil particles is limited, so that the total amounts of bases in the soil that are available for adsorption may be greatly in excess of the amounts actually adsorbed. The capacity of the soil to hold and exchange cations is accordingly a measure of the ability of the soil to store plant nutrients and make them available to the growing plants, according to their requirements, and is therefore an important factor in soil fertility.

Soils of arid regions are generally *base-saturated*, i.e. most of the cation exchange capacity is occupied by calcium, magnesium, potassium, and sodium. Strongly alkaline soils are characterized by a relatively high ratio of monovalent bases, especially sodium, to bivalent bases.

Anion Exchange Capacity

Anions can also be held and exchanged by the soil—phosphates and sulphates more strongly than chlorides and nitrates (Tisdale & Nelson, 1966). Anion exchange capacity increases with decreasing pH, being the opposite of what occurs for cation exchange. For this reason, the agricultural soils of arid regions, with their relatively high pH, have a low capacity for anion absorption, and anions, with the exception of phosphates and, to a lesser degree, sulphates, are easily leached from the soil.

THE SOIL 'SOLUTION'

Certain plant nutrients are frequently present in the soil in quantities that greatly exceed the requirements of a growing crop. For example, a soil containing 0·1 per cent of P_2O_5 will have a total of approximately 14 tonnes of P_2O_5 per hectare in the root-zone, i.e. to a depth of one metre. The requirements of a single crop such as cotton do not exceed 50 kg of P_2O_5 per hectare, and yet they will not necessarily be met from the phosphorus reserves in the soil. This is indicated by the frequent considerable response of a growing crop of cotton to phosphorus fertilizer, demonstrating the low availability of the soil phosphorus to the plant.

The concentration of phosphate actually in the soil solution is usually low, and is practically independent of the moisture content of the soil. The availability of phosphorus to the crop will therefore depend on the speed with which the phosphorus removed from the soil solution is replaced from the soil reserves.

Most of the cations held by the exchangeable complex of the soil are only very slowly soluble. At a given time, only minute amounts of these nutrients are in solution in the soil-water, even in very fertile soils. The concentration of available plant nutrients in the soil solution is therefore another important component of soil fertility.

The application of thermodynamic principles has contributed to an understanding of the ionic relationships between plant and soil.

An exchangeable equilibrium is established between ions held by the exchange complex of the soil and the soil solution; ions migrating between the two phases form what is called 'the labile ionic pool' (R. S. Russell *et al.*, 1961). As ions are taken up by the plant from the soil solution, this equilibrium is upset; new equilibra are then established by ions moving from the surfaces into solution.

The soil solution is therefore the intermediary whereby nutrient ions move from the soil into the plant, and the rate of uptake of a nutrient is a function of the concentration of the ion in the soil solution.

The availability of cations to plants is controlled by the following factors:

(1) The concentration of ions in the soil solution, which depends on the nature of the colloidal fraction and on the degree of their base-saturation.
(2) The influence of one adsorbed ion on the release of another from the colloid surface. Polyvalent ions are held more strongly than univalent ions; hence, when the soil colloids hold large amounts of potassium and NH_4^{++} in proportion to the amounts of calcium and magnesium, the former will be replaced more easily than the latter and therefore plant deficiencies of calcium or magnesium may develop (Tisdale & Nelson, 1966).

By contrast, practically all the nitrate and chloride present in the soil is in the soil solution, so that the concentration of these anions varies inversely in relation to soil moisture content (E. W. Russell, 1961).

Whereas in humid regions the soil solution is very dilute and its osmotic pressure does not normally exceed one atmosphere even at wilting point

(E. W. Russell, 1961), the situation is entirely different in the irrigated soils of the arid and semi-arid lands, in which calcium and sodium salts, in particular, accumulate. Under these circumstances, osmotic pressure of the soil solution may have an appreciable effect on crop growth—especially with increasing soil-moisture tension.

ION ABSORPTION BY PLANTS

The three mechanisms whereby ions may enter the roots are: by diffusion, by exchange adsorption, and by the action of carriers or ion-binding compounds (Epstein, 1955). According to the carrier concept, ions combine with carrier molecules whose exact nature is not yet known, but which are able to traverse membranes that are not permeable to the free ions. After penetrating the cell, the ions are released from the carrier (ibid.). Movement of ions into the cell therefore requires an expenditure of energy by the root-cells, and the accumulation of ions can therefore occur *against* a concentration gradient. It is also a highly selective process: certain ions compete for the same carrier, whilst others use different carriers. For example, calcium competes with strontium for the same carrier, but not with potassium, as they have different carriers.

The soil solution may contain elements that are toxic to plants, or the concentration of the total salts in solution may be excessively high and therefore harmful to plants.

NUTRIENT ELEMENTS IN THE SOIL

Soil Nitrogen

Parent rock does not contain any nitrogen; the primary source of soil nitrogen is the inexhaustible supply obtained directly or indirectly from the atmosphere.* The amount of nitrogen in the ploughed layer of cultivated soils usually ranges from 0·02 to 0·4 per cent by weight (Black, 1968).

Most of the nitrogen in the soil is in organic form; it is generally assumed that organic matter contains about 5 per cent of nitrogen, of which amount only a fraction, becomes available yearly. The rate at which N becomes available depends on the rate of mineralization of the organic matter. This subject is discussed on pp. 330–2 and pp. 336–7.

The total amount of nitrogen in the soil is therefore not an indication of the amounts of this nutrient which are immediately available to a crop, but constitutes a reserve from which nitrogen may become available to plants—and not necessarily at a rate commensurate with the requirements of an actively growing crop.

* Fixation of nitrogen from the air is discussed on pp. 333–6.

The percentage of nitrogen in the soil tends to remain constant at a level which depends on the nature of the parent material, on the leaching characteristics of the soil (mainly determined by its texture), and on the management system adopted (Cooke, 1967).

Soil Phosphorus

Soil phosphorus is present in the soil in mineral and organic forms, usually in amounts far lower than those of nitrogen and potassium, and varying from 0·1 per cent to 0·4 per cent while rarely exceeding 0·5 per cent The quantity of phosphorus in the soil solution is always small, usually ranging from 0·1 p.p.m. to 0·5 p.p.m. in most soils. A level of 1 p.p.m. of P in the soil solution is generally considered adequate for plant growth (Seatz & Stanberry, 1963).

Most of the soil phosphorus is in inorganic form, and its original source is the apatite group of minerals. The organic compounds containing phosphorus are derived from the decay of animal and plant bodies.

Most of the soil phosphorus is tied up chemically in a form that is not available to the crop in the course of a growing-season; indeed the amount of available phosphorus is not necessarily closely related to the total soil phosphorus, *inter alia* because the chemical nature of the phosphate compounds is not the same in all soils. In California, it was found that phosphorus deficiencies were related to soil profile groups, as shown in Table XVIII.

TABLE XVIII

Relation of Soil Profile Group to Response of Grain to Phosphorus Fertilizers
(W. E. Martin & Mikkelson, 1960)

Soil profile group	*Percentage of soils deficient in P*	*Characteristics of soil*
Recent alluvial	19	Uniform texture, deep and permeable
Young alluvial	19	Uniform texture, deep and permeable
Older alluvial	30	Less well drained and permeable than above
Clay-pan	72	Strong accumulation at depth, poor drainage
Hard-pan	83	Cemented hard-pan layer, shallow, waterlogged
Upland	79	Developed *in situ* from weathering of parent rock

Soil Potassium

Of the major macronutrients, potassium is usually the most abundant in soils. In particular in arid regions, the presence in the soil of small fragments of little-weathered potassium-bearing primary minerals provides a considerable reserve of potassium (Babcock *et al.*, 1967).

Potassium occurs in the soil in a number of forms, though most of it is apt to be unavailable to plants even after years of cropping:

(a) *Water-soluble*—the potassium which is present in the soil solution. This is a very small fraction of the total, and even in fertile soils this form cannot supply the major requirements of a crop.

(b) *Exchangeable*—the potassium which is held by the exchange fraction of the soil. This is the important reservoir of readily available potassium, though together with water-soluble potassium it represents only a small part—commonly 1–2 per cent—of the total potassium of the soil (Arnold, 1960).

(c) *Non-exchangeable* but slowly available potassium. This comprises the native soil potassium from partly weathered minerals and forms the reserve from which the water-soluble potassium of the soil solution is gradually replaced; it therefore constitutes an essential element of the permanent fertility of the soil. The higher the temperature is, the greater will be the release of non-exchangeable K (Burns & Barber, 1961). Residual potassium from fertilizers is also in this form. The slowly available forms amount to 1–10 per cent of the total soil potassium (Tisdale & Nelson, 1966).

(d) *Inert*—present in the unweathered potassium-bearing parent minerals, and released at an extremely slow rate.

These groupings are somewhat arbitrary and the boundaries between them are not sharply delineated. The different forms are in equilibrium: as potassium is removed from the soil by plants and by leaching, a transfer occurs from the slowly available forms; on the other hand, especially after applications of large amounts of potassic fertilizers, some of the potassium may revert to slowly available forms.

The alternate drying and wetting of the soil, characteristic of irrigation agriculture, usually causes an increase of exchangeable potassium when the levels of soil potassium are low to medium. When the levels are high, the opposite may occur, though the reasons for this last phenomenon are not yet known (Tisdale & Nelson, 1966). Drying of both surface and subsoil can approximately double the exchangeable potassium content (Scott & Smith, 1957).

Calcium

The calcium content of soils varies more than does that of any other element. Calcium is a constituent of a number of primary rocks and minerals etc., which fall into five groups:

(a) calcium present in the mineral particles, which are nearly all of primary rock; (b) calcium carbonate, in general the most important source of calcium in soils; (c) calcium sulphate; (d) simple salts of calcium in the soil solution; and (e) exchangeable calcium.

Carbonate of calcium ($CaCO_3$) occurs alone or often with magnesium in many arid regions (calcareous soils). Concentrations of calcium sulphate are found only occasionally in some arid soils (gypsum soils), though beds of gypsum occur in numerous places. The most important form of calcium in soils as a whole is that associated with the exchangeable colloidal complex. Exchangeable calcium is considered readily available.

Calcium produces several specific effects, which result in the improvement of soil structure and in increased crop production. The leaching losses of calcium from soils depend on the quantities of calcium present, the forms in which they are held, the amount of percolating water, and the production of carbonic acid. In humid climates with high rainfall, leaching is active and the removal of calcium ions from the colloidal complex results in their substitution by hydrogen ions. In arid climates sodium ions frequently replace calcium ions on the colloids, yielding in soils of poor physical condition.

Magnesium

There is usually a very close relationship between calcium and magnesium in the soil. Magnesium carbonate is usually found in mixtures with calcium carbonate—particularly in dolomite ($CaCO_3.MgCO_3$). Magnesium deficiencies are in general rare in arid-land soils; fine-textured soils formed from parent rock with high magnesium content may contain several per cent of this element (Tisdale & Nelson, 1966).

Sodium

Large amounts of sodium may accumulate in the soils of arid and semi-arid regions. The significance of sodium accumulation has been discussed in Chapter IX.

Minor Elements

Deficiencies of micro-elements in arid zones are rare; however, in many soils which were originally considered as unproductive, it was found that a deficiency of one or more minor elements was the sole cause of their apparent infertility. Copper, zinc, and molybdenum deficiencies were identified in Australia, and supplying these elements in minute quantities transformed a previously infertile, valueless soil into a normally productive one.

Biotic Aspects of Soil Fertility

The soil habitat teems with a varied population of micro-organisms—Bacteria, Algae, Fungi, and Protozoa—whose abundance and considerable activities justify the term 'living soil'. Their numbers are not constant but may fluctuate hourly as the environment or food supply changes. The addition of organic matter to the soil, combined with favourable conditions of temperature and moisture, makes possible an enormous increase in numbers within a very short time; these disappear just as rapidly when the food supply is exhausted or the environment becomes otherwise unfavourable.

Certain groups of micro-organisms play extremely important roles in processes—some favourable, some unfavourable—that have a considerable bearing on soil fertility. The most important of these are described below.

Very little is, however, known about the microbial activity in arid soils, and the effects of long periods of extreme drought on the biological evolution of these soils (Aubert, 1962).

SOIL ORGANIC MATTER

The term 'organic matter' has a very broad meaning because it includes all materials of organic origin present in the soil, regardless of their origin or stage of decomposition. Thus, the term includes both fresh and highly decomposed crop residues and animal excretions, as well as the decomposing bodies of soil flora, fauna, and microbial components of the 'microcosm'.

Organic matter is not the same in all soils. The type of vegetation, the nature of the soil population, soil aeration and moisture conditions, climatic conditions, and management practices—all affect the kind and amount of source matter present in the soil. Soil organic matter is therefore truly a product of its environment.

DECOMPOSITION OF ORGANIC MATTER

In a normal, productive soil organic matter cannot accumulate. If the soil is not sterile, extremely alkaline, completely waterlogged, or excessively dry, every addition of organic matter will stimulate the activity of micro-organisms. These decompose the organic matter first into intermediary products—the humic compounds—which are fairly resistant to further decomposition and therefore form a more or less stable part of the colloidal organic matter in mineral soils. Ultimately, however, all organic material is converted into simple inorganic compounds, such as carbon dioxide, nitrates, and water. Different groups of micro-organisms are involved at each stage of this process, causing the decomposition of the various constituents of the organic residues at different rates.

The activities of the micro-organisms involved in the various stages of this so-called 'mineralization' of organic matter in the soil require certain essential environmental conditions: a favourable moisture régime, temperatures that are neither excessively high nor low, sufficient aeration, a balanced nutrient supply, etc. These requirements are identical to those needed for the normal development of most crop plants. Therefore, when conditions are favourable to crop plants, they are also favourable to the micro-organisms involved in the breakdown of organic matter. Under circumstances in which one or more essential requirements are in short supply, crop plants and micro-organisms may compete with each other for the limiting requirement—usually to the immediate detriment of the crop.

The main source of energy for soil micro-organisms consists of the carbohydrates of the organic matter; for the building of their body tissues they also require nitrogeneous compounds. If these are not available in sufficient amounts in relation to the carbonaceous materials of the added organic matter, the micro-organisms will draw on the available sources of nitrogen in the soil. As far as the crop plants are concerned, this nitrogen is immobilized until the transformation of the organic matter into humus is completed and the micro-organisms themselves die and decompose. Organic matter with a C:N ratio of 20–30 will itself supply all the nitrogen required for its decomposition; with a ratio wider than 30, soil nitrogen will be immobilized, whilst with a ratio narrower than 20, nitrogen will be released at a fairly early stage of decomposition (Figure 60 and cf. Tisdale & Nelson, 1966).

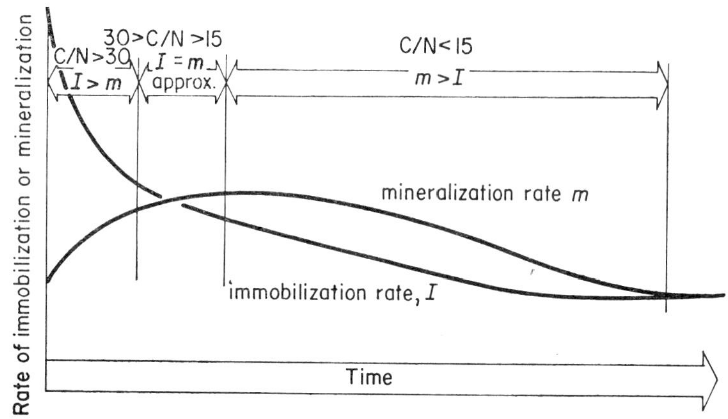

Figure 60. Relationship between carbon:nitrogen ratios and nitrogen availability during the decomposition of plant residues. C/N — carbon: nitrogen ratio
I = immobilization rate
m = mineralization rate
(From Broadbent (1957): by courtesy of the U.S. Department of Agriculture.)

This explains why the addition of undecomposed manure with a large proportion of straw, or the ploughing under of large quantities of cereal straw or stubble, with wide C:N ratios may cause a temporary deficiency of nitrogen, unless appreciable amounts of nitrogen fertilizers are added. Well-decomposed manures, on the other hand, in which the nitrogen:carbon ratio has been narrowed during decomposition outside the soil, do not have this effect. Similarly, the ploughing under of an immature stand of leguminous plants will rapidly release nitrogen, which is immediately available to the crop following.

The decomposition of low-nitrogen organic matter requires a larger amount of nitrogen in soils with high temperatures than in colder soils. A tonne of straw that is ploughed into the soil in winter, may require and 'fix' 8–9 kg of nitrogen, whilst the same quantity in summer may require 10–14 kg (Pinck *et al.*, 1946). If other nutrients, besides nitrogen, are in short supply in the organic material to be decomposed, these may also be 'fixed' by the micro-organisms. The ratio of carbon to nitrogen in the top-soil tends to be constant at approximately 10:1; in the subsoil the ratio is usually narrower, approaching 5:1. This may be due to fixed ammonium in the subsoil (Cooke, 1967).

The level of organic matter in the soil, and the rate of its decomposition, are markedly dependent on temperature and soil-moisture régime, which influence the development of the plants and the activities of the micro-organisms in the soil.

Jenny (1930) showed that every 10°C increase in the average temperature doubles the rate of decomposition of organic matter in soils when moisture conditions are favourable. The water supply affects the rate at which organic material is produced and also the rate of its decomposition, both increasing—up to a certain limit—with an improved soil-moisture régime.

The activity of the soil micro-organisms is influenced by the relation between soil temperature and soil moisture, the so-called 'hydrothermic quotient'.* For example, in a sandy clay soil of a semi-arid region under natural veld, the optimum conditions for microbial activity were found to be a hydrothermic quotient in the range of 1·9 to 2·4, at a soil depth of 10–15 cm. This represented soil temperatures ranging from 20 to 30°C, and a soil moisture ranging from 11 to 14 per cent (Pauli, 1967). Most soil organisms are incapable of activity at hydrothermic quotients below 1·5 or above 3·0 (ibid.), though there are some exceptions.

Semi-arid conditions are usually characterized by the alternation of a dry season and a rainy season, during which the combination of relatively high temperatures and a favourable moisture régime enables a considerable acceleration of the physical and biological factors that influence soil fertility. As a result, organic matter is rapidly mineralized down to water and carbon dioxide (Pauli, 1967).

NITROGEN FIXATION

Nitrogen fixation is the assimilation of free nitrogen from the air by soil organisms, and the formation of nitrogen compounds that eventually become available to plants.

* *Hydrothermic quotient.* Ratio of soil temperature (°C) to soil moisture (percentage weight-loss after drying) (Pauli, 1967).

Molecular nitrogen cannot be directly used by plants; first it has to be combined with other elements—a reaction which requires a very large expenditure of energy. The required energy can be supplied in nature by lightning; its main source, however, is the activity of certain groups of micro-organisms.

Several kinds of soil algae and bacteria are capable of fixing atmospheric nitrogen, either independently or in symbiosis with certain crop plants.

Non-symbiotic Nitrogen Fixation in Soil

Amongst the groups of micro-organisms that are capable of fixing nitrogen from the air independently, are *Azotobacter*, *Clostridium*, and several species of algae. Of these, *Azotobacter* is the most important. The fixed nitrogen becomes available to plants after the death of the micro-organisms.

In the U.S.A., it has been estimated that the amount of nitrogen fixed by these organisms in temperate climates may reach approximately 7 kg of nitrogen/ha/year, and more in warmer regions. Algae, especially the blue-greens (Cyanophyceae), are considered to be capable of adding appreciable amounts of nitrogen to arid-region soils (Fletcher & Martin, 1948). In particular, after irrigation, an algal layer appears on the soil surface, that remains active in nitrogen fixation until the soil dries. After several such cycles, the amounts of nitrogen added may reach several kilograms per hectare. The green algae that develop in paddy-rice fields are estimated to fix over 30 kg/ha/year (Cooke, 1967).

It is frequently assumed that the soils in arid climates have considerable non-symbiotic nitrogen-fixing ability and can therefore permanently maintain a constant level of available nitrogen under continuous wheat production, provided crop residues are available to the free-living nitrogen-fixing organisms. But this belief has not been confirmed by research. Thus in Australia it was found that, in the wheat-fallow rotation, non-symbiotic nitrogen fixation under the most favourable circumstances does not provide more than one-third of the nitrogen requirements of the wheat crop (Callaghan & Millington, 1956).

Investigations in California indicate that under fallow conditions only a few kilograms of nitrogen per hectare are added annually as a result of non-symbiotic nitrogen fixation (Delwiche & Wijler, 1956). The probable reason why no significant fixation of nitrogen occurs in semi-arid soils is that *Azotobacter* and similar organisms are efficient mainly under anaerobic conditions, such as prevail in soils which are saturated for long periods (Jensen, 1940).

Symbiotic Nitrogen Fixation

The fact that certain crops such as vetches, lucerne, and clovers, are capable of improving soil fertility, was already known in ancient times. In 1892, M. W. Beijerinck identified the specific organism which was capable of fixing nitrogen in symbiosis with a leguminous plant as a bacterium belonging to the genus *Rhizobium*. The rod-shaped bacteria are mobile in the soil; their mobility is, however, dependent on a sufficient supply of available phosphorus. The roots of leguminous plants excrete a substance which attracts the bacteria and promotes their rapid multiplication; the bacteria in turn produce a growth substance that enables them to penetrate into the roots through the root-hairs. They multiply rapidly and cause the formation of a swelling on the root; this is the typical nodule.

The bacteria obtain some of their nutritional requirements from the host plant: this provides them with the source of energy needed for fixing nitrogen. The nitrogenous substances formed by the bacteria are partly absorbed by the host plant, and partly excreted into the soil, where they may benefit non-leguminous plants growing in association with the legume. However, most of the nitrogen derived from fixation by the bacteria is liberated only *after* the breakdown of the roots and their nodules, and not by excretion from the active root system.

The relation between Bacteria and host plant, as described above, is typically symbiotic. However, there are circumstances under which the system is no longer one of mutual benefit.

Any deficiency in the supply of carbohydrates by the host plant to the nodule bacteria, as a result for example of low light intensity, will cause these to cease fixing nitrogen. The transport of sugars to the nodules breaks down if boron is deficient. The bacteria also require an adequate supply of calcium and phosphorus.

Fixation of nitrogen by clover was found in southern Australia to be proportional to the amount of superphosphate applied: approximately 80 kg of nitrogen was fixed for every 10 kg of phosphorus used (Donald, 1960). Whilst its exact role is not known, a minute supply of molybdenum is essential.

The bacteria will not be active in molecular nitrogen fixation if ample nitrate or ammonium is available in the soil. Nodule formation will also cease under these conditions.

Whenever conditions occur which limit or arrest molecular nitrogen fixation by the nodule bacteria, the relationship with the host plant ceases to be symbiotic, and becomes typically parasitic; the bacteria continue to obtain their nutrient supply from the host plant, without the latter benefiting in any way from the presence of the nodule bacteria.

The nodule bacteria are all of the genus *Rhizobium* but belong to a number of groups of strains, each of which infect certain host plants or groups of plants. In each group of bacterial strains are found individual ones that differ in the efficiency with which they fix nitrogen. The inefficient strains form very small nodules, few in number, which degenerate rapidly. Efficient strains can be isolated, multiplied, and used for seed- or soil-inoculation, thereby considerably increasing the quantity of molecular nitrogen obtained by a leguminous crop. The efficiency of nitrogen fixation also depends on the host plant and its ability to 'co-operate' with the bacteria; this ability is a hereditary trait.

There exists, therefore, the entire spectrum of relationships between host plant and bacteria, ranging from absolute parasitism to full symbiosis.

As in all bacteria, mutations may occur in *Rhizobium,* changing efficient strains into parasitic strains. Selected strains for inoculation should, therefore, in addition to being efficient in nitrogen fixation, also show great genetic stability and agressiveness in competition with native strains.

The *Rhizobium* bacteria, in their free-living stage, are subject to attack by various types of Fungi, by bacteria of other groups, and by bacteriophage. They are generally able to hold their own, but unfavourable environmental conditions can easily upset this balance, with disastrous results for the *Rhizobium* and their host plants.

Of particular significance for arid lands is the fact that adequate soil moisture is the first essential for the symbiotic relationship between the host plant and *Rhizobium*: when the moisture deficit in the plant becomes severe, the crop sheds its nodules (E. W. Russell, 1961). High soil temperatures also adversely affect nodule fixation. In Australia, the nodules of Subterranean Clover were found to fix nitrogen actively at 20°C, with an optimum at 25°C and almost complete cessation at 30°C (Meyer & Anderson, 1959). It is surprising that, notwithstanding the large number of factors which can affect the relationship between host plant and bacteria, symbiosis is the rule and parasitism the exception.

The host plant supplies to the bacteria from 15 to 20 per cent of the carbohydrates resulting from photosynthesis; in turn it obtains 90–100 per cent of the nitrogen fixed by the bacteria (Callaghan & Millington, 1956). The total amounts of nitrogen fixed by Rhizobia may be quite considerable, and range from 50 kg/ha/year for Subterranean Clover in New South Wales (Donald & Williams, 1954) to 250 kg/ha/year for lucerne in Arizona (Smith, 1944). An indirect indication of the level of nitrogen provided to the plant is given by the following example:

In a trial with peanuts grown for the first time in a loess soil in the Negev, in which specific symbiotic bacteria were entirely lacking, inoculated seed produced the same yield without nitrogenous fertilizers as non-inoculated seed grown with nitrogenous fertilizer (Shiffman, 1961). It can therefore be assumed that the Rhizobia fixed at least the equivalent of the amount of nitrogen applied in the fertilizer, namely, 180 kg N per hectare (Figure 61).

The amounts of nitrogen fixed in the host plant are no indication of the amount of residual nitrogen that may remain available to the following crop.

This latter amount depends primarily on the conditions under which the

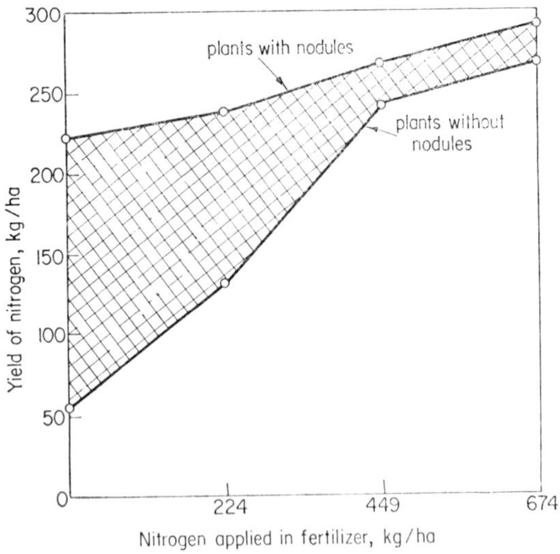

Figure 61. Yield of nitrogen in above-ground portions of nodulated and non-nodulated soybeans. The plants were grown on soil treated with ground maize to immobilize the soil nitrogen. The shaded area represents the difference in nitrogen uptake due to nodulation. (From Weber (1966): by courtesy of the American Society of Agronomy.)

host crop was grown, on the stage at which it was harvested, and on the conditions prevailing during the period between harvesting the legume and the following crop. Legumes harvested for seed will leave very little or no residual nitrogen behind in the soil. (This matter will be discussed in detail in Chapter XV.)

Rainfall may leach out much of the fixed nitrogen before the following crop can benefit from it.

NITRIFICATION

About 99 per cent of the nitrogen in the soil is bound in organic materials; to become available to plants, these organic compounds must first be mineralized. This mineralization occurs in three stages: aminization, ammonification, and nitrification. Each stage is the function of specific groups of micro-organisms, each group producing the substrate that is essential for the activity of the following group.

The rate of mineralization of the nitrogenous compounds may vary greatly, depending on the environmental factors affecting the activities of the micro-organisms involved in the process, and on the rate at which the nitrogen is taken up by the plants.

When crop production and, at the same time, the decomposition of the organic crop-residues, are limited by soil moisture, soil nitrogen is depleted relatively slowly. In the Great Plains of North America, the average loss in nitrogen over a period of 36 years was 39 per cent in all. (Hass *et al.*, 1957).

The rate of mineralization of organic nitrogen in the soil increases from the air-dry condition to field-capacity. The micro-organisms involved in nitrification continue to be active until near wilting point, whilst ammonification continues at even higher soil-moisture tensions (Robinson, 1957). Consequently, water deficiencies will limit crop growth more than they will limit nitrification, and as a result plants grown under these conditions will generally have a higher nitrogen content than those grown under moister conditions.

An opposite effect is obtained after the top-soil has dried out and the plant draws most of its water-supply from the subsoil, which is usually more deficient in nitrogen than is the top-soil. Under these circumstances, nitrogen content of the plant tissues will be relatively more reduced by water deficiencies than will the yield of dry matter—unless the plant has taken up an excess of nitrogen before the top layer has dried out (Black, 1966).

In the course of the growing-season of a crop, up to about 1–2 per cent of the nitrogen may be released from the organic compounds present in the soil (cf. p. 326) if moisture is not limiting. This will supply up to 80–100 kg N/ha in fertile soils. The decomposition of the residues of the previous crop may supply another 10 kg N/ha. However, the micro-organisms that decompose the plant residues may require about double this latter amount of nitrogen for their own metabolic processes; this equivalent, therefore, becomes temporarily unavailable to the cultivated crop.

The nitrates that have been formed move freely in the soil: downwards following rain or irrigation; upwards, by capillary action, during the drying out of the soil. Large amounts of nitrogen may be leached beyond the reach of the roots or out of the soil, if rainfall is excessive or irrigation is not well controlled. Of all plant nutrients, nitrogen is the most easily lost to the plants.

DENITRIFICATION

Certain groups of micro-organisms, which are active under anaerobic conditions and at high temperatures, are capable of reducing ammonia and nitrates into volatile compounds that escape into the atmosphere. Such denitrification will cause appreciable losses of soil nitrogen in poorly drained soils; it will also occur when organic matter that is high in nitrogen is added to wet alkaline soils under conditions of high temperatures, for example in paddy fields (E. W. Russell, 1961).

ANTIBIOSIS

The struggle for survival between the micro-inhabitants of the soil is no less severe than that proceeding above-ground among larger organisms. Normally, a balance between saprophytes and parasites is maintained in the soil. When ample organic matter is applied to the soil, saprophytes are generally favoured and they may compete successfully with parasites. For example, many soil bacteria have been shown to inhibit the growth of the pathogens causing take-all disease of wheat (Fellows, 1929), *Phymatotrichum* root-rot of cotton (King *et al.*, 1934), and the wilt diseases of a number of crops—such as cotton, melon, and flax—due to *Fusarium oxysporum* (Mitchell & Alexander, 1961). However, in the absence of a sufficient supply of dead organic matter, even saprophytes may attack living plant tissues.

Experience has shown that, if a plant which is susceptible to root disease is grown in a sterile soil and then inoculated with a pathogen that causes root-rot, it will be more severely infected than if the soil had not been previously sterilized. Root diseases are more prevalent in soils that are poor in saprophytic flora than in those with an abundant population of saprophytes. The saprophytes reduce the numbers of pathogens in the soil by (a) preying on the pathogens, (b) competing for nutrients, oxygen, etc., and (c) excreting substances that are toxic to the pathogens (antibiosis). For the antagonists to be effective in biological control, they have to build up sufficiently high populations, and environmental conditions must be favourable to them. The addition of appropriate crop residues is one of the means of achieving this aim. Yet results are not always consistent: e.g. the decomposition of the residues of a leguminous crop was found to depress the survival of *Helminthosporium sativum*, but to enhance that of *Ophiobolus graminis* (Garrett, 1956). Residues may even have entirely different effects on the same pathogen, depending on the condition and stage of their decomposition; the pathogen population may increase initially, but become more and more depressed as decomposition progresses—presumably as a result of nitrogen starvation due to the increased competition of the saprophytic population (Patrick & Tousson, 1965).

The addition of organic manures causes a rapid increase in the populations of certain predaceous Fungi which are capable of trapping and feeding upon eelworms, whose numbers are thereby reduced (Duddington, 1955).

Other Factors Affecting Soil Fertility

ORGANIC MANURES

The favourable influence of organic matter, in the form of crop residues, manure, or composts, on soil fertility, is due to the following effects:

1. During and after decomposition, essential plant nutrients, such as nitrogen, phosphorus, potassium, and micro-elements, are gradually released. Organic matter is therefore an important store of plant nutrients.
2. The humic compounds formed during the decomposition of organic matter may prevent the formation of certain mineral-phosphate complexes

in which the phosphorus is unavailable to plants. They may even reverse the process, by forming a soluble and available phosphate complex. The humic compounds may also increase the availability of many micro-elements (Pauli, 1967).

3. The colloids formed during decomposition are active in the cation exchange complex (C.E.C.) of the soil. In sandy soils, the exchange capacity is almost entirely due to organic matter.
4. The decomposition of organic matter improves soil structure—both directly, by providing sticky substances that are needed for aggregation of soil particles, and indirectly, by increasing the water-holding capacity of the soil, improving aeration, facilitating tillage operations, and reducing erosion hazards.
5. Organic matter serves as a source of energy and food for the many groups of micro-organisms involved in biological processes that are of significance to soil fertility.
6. The decomposition of organic matter in the soil releases considerable amounts of carbon dioxide, of which a large proportion escapes into the air. As the carbon dioxide is heavier than air, it concentrates in the atmosphere above the soil, and can be taken up by the leaves of plants during the day and assimilated in photosynthesis. Manuring the soil therefore increases the nutrient-engendering content of the atmosphere!

It accordingly follows that all the beneficial effects obtained from the addition of organic manures to the soil, are dependent on their decomposition under favourable conditions.

There is no point in attempting to increase the organic-matter content of the soil, as an objective in itself; but there is need for a regular supply of organic matter at a level that is commensurate with the rate of its biological decomposition.

The chernozem soils are often considered to be highly fertile, and yet crop yields on them are frequently disappointing. This is mainly because the decomposition of the high proportion of organic matter present in these soils is limited by the heat and drought of the late summer and the extreme cold of winter (Crowther, 1947).

However, not all decomposition of organic matter has favourable effects: whilst the decomposition of organic matter under aerobic conditions produces nitrates, sulphates, phosphorus, and compounds of calcium, magnesium, potassium, iron, etc., the dominant products of decomposition under anaerobic conditions are methane, hydrogen sulphide, ammonia, aldehydes, and ferrous iron (Vershinin *et al.*, 1959).

The amount of organic matter in the soils of arid regions is extremely low, being frequently limited to mere traces and rarely exceeding 1 per cent (Lemée, 1955), for the following reasons:

the sparse plant cover produces very little in the way of residues;
the dried residues are dispersed by the wind;
the rapid mineralization of the organic matter;
the small contribution of animal manure; and
the use of most of the organic residues for fuel.

Therefore, until the advent of chemical fertilizers, soil productivity of these lands, when irrigated, was mainly dependent on the maintenance of an equilibrium of organic matter at an adequate level, which, on decomposition, made available the nitrogen that was necessary for crop production. The use of organic manures, and the return of crop residues to the soil, made possible stable production when the moisture-supply was sufficient.

CHEMICAL FERTILIZERS

Over a century ago, J. Liebig propounded his theory that the only role of organic manures was to provide plant nutrients. In recent years, a school of thought has arisen, inspired by Howard (1940), Balfour (1945), and many others, that ascribes most of the ills of modern agriculture to the use of chemicals in general, and to that of chemical fertilizers in particular.

The detrimental effect on soil fertility that has been ascribed to chemical fertilizers is supposedly due to the poisoning of the soil micro- and macro-organisms, thereby causing an increase in the incidence of diseases and pests of plants, and indirectly affecting the health of the farm animals and humans consuming the products that have been grown with chemicals.

The addition of fertilizers to the soil has both beneficial and detrimental effects: on one hand it promotes plant growth by providing nutrients, but on the other it hinders plant growth by increasing the osmotic pressure of the soil solution and hence increasing the difficulties which the plant encounters in supplying its needs in water and nutrients. In arid regions, for dry land crops growing under conditions of high soil-moisture tension, this increase in osmotic pressure may be decisive under certain circumstances and more than outweigh the advantage of the additional nutrient supply.

When the long-term effects of various types of fertilizers on the physical and chemical properties of an irrigated soil are investigated, far-reaching differences may be observed. Aldrich & Martin (1954), when comparing different nitrogen carriers, found marked physical and chemical changes in the soil after sixteen years of differential treatment. In particular, the rate of water percolation was apt to be greatly affected. The use of sodium nitrate and ammonium sulphate brought about structural breakdown resulting in reduced macropore space. In the sodium nitrate plots, the poor physical conditions appeared to be due to an unfavourable calcium–sodium ratio. In the ammonium sulphate plots, the cause of the structural breakdown was apparently the dispersing action of the ammonium ion. The low pH, caused by the continuous use of ammonium sulphate and due to the accumulation of hydrogen in the exchangeable complex, inhibits the ability of soil organisms to nitrify the ammonium. No such unfavourable effects were found to result from the use of urea or calcium nitrate.

Where continuous use of large rates of application of certain types of fertilizer is apt to cause difficulties, the answer is not to reduce or abandon fertilizer application, but to change the type of fertilizer used or to apply appropriate soil amendments.

Salinity and alkalinity are major problems of crop production under irrigaation, and the opinion is frequently voiced that heavy fertilization increases these problems. When it is realized, however, that the quantity of salts added to the soil in irrigation water, even of satisfactory quality, easily amounts to 5 tonnes/ha/year, it is clear that the effect of fertilizers on salinity, under normal irrigation practice, is usually not of overriding importance, unless excessive quantities of unsuitable fertilizers are applied and provided periods of severe moisture stress are avoided.

Unbalanced fertilization may also cause difficulties: excess fertilization with one element may hinder the uptake of other essential nutrients.

It therefore follows that it is not the use of fertilizers *per se* that may have an undesirable effect on soil fertility, but their incorrect use. By choosing the appropriate carrier, avoiding excessive application or an unbalanced supply, and using proper application methods, it should be possible to avoid or mitigate the negative effects mentioned above.

In so far as organic matter is essential to soil fertility, fertilizers can play a considerable part in increasing the amounts of organic residues to be incorporated in the soil. Green-manures, grown with fertilizers, are frequently the only economical source of organic matter that is available.

Finally, the addition of fertilizers is frequently indispensable, when ploughing under bulky organic manures or crop residues, in order to avoid the setback to the cultivated crop which would otherwise result from the fixation of plant nutrients by the soil micro-organisms.

PESTICIDES, FUNGICIDES, AND HERBICIDES

An important characteristic of several commonly used chemicals for pest, disease, and weed control is their persistence in the soil. Soil fertility depends to a considerable extent on the maintenance of the equilibrium between the various components of the complex populations of micro-organisms in the soil. The disturbance of this precarious equilibrium by the residual effect of chemicals applied to the soil may affect soil fertility adversely (Audus, 1964).

There are vast differences in the persistence of chemicals in the soil: some are completely broken down within a few days after application, and others may persist in toxic concentrations for years. Persistence of chemical residues in the soil depends on many factors: the nature of the pesticide, the soil type, soil moisture and temperature, the rate of application, the formulation used, and the activity of micro-organisms (Lichtenstein, 1966). The stable chlorinated hydrocarbons may persist in the soil for years; the organic phosphorus compounds are usually decomposed rapidly, but occasionally they persist from one season to the next. Compounds incorporating copper, lead, and arsenic, persist for long periods (Weisner, 1963). D.D.T. benzine hexachloride, and Chlordene, amongst others, are toxic to the micro-organisms that are active in nitrification, and they can accumulate in the soil. However, usually no drastic effects are experienced with these pesticides if care is taken not to exceed the normal rates required for pest control, thereby avoiding a build-up of residues in the soil, such as may occur through excessive or imprudent use.

2,4-D, the ubiquitous weed-killer, decomposes rapidly in the soil, and has been shown to have a favourable effect on soil micro-organisms. Amitrole, Endothal, and Simazine, do not appear to have an appreciable effect on soil micro-organisms at normal application rates (Audus, 1964). Certain chlorinated organic compounds can be metabolized only by specific genera of soil organisms; e.g. Simazine is decomposed only by Fungi and Actinomycetes. The rate of biodegradation increases in soils that are rich in organic matter (ibid.).

From the foregoing it is clear that a slow, persistent build-up of toxic materials, resulting from the widespread use of agricultural chemicals, can have far-reaching effects on soil fertility. These will depend upon the physical, chemical and biological environment within the soil (Webber & Elrick, 1966). The processes will be different under arid and semi-arid conditions—with and without irrigation—as compared with those of humid regions. Knowledge of these effects is, however, still very limited.

DRYING THE SOIL

One of the important characteristics distinguishing the arid and semi-arid soils from those in humid temperate regions, is that the former periodically dry out to a considerable depth. This is accompanied by a rise in soil temperature to 50–60°C (Aubert, 1962). This periodical drying out and heating of the soil has profound chemical, physical and biological effects on soil fertility, which have some similarity to those ascribed to freezing in the colder and wetter regions. This is not surprising, as freezing is also a form of drying.

Exposing a ploughed soil to the effect of sun-rays for prolonged periods is an age-old practice in arid lands, and is still in present-day use both in countries using traditional agricultural methods and in those using modern ones.

The effect of drying and wetting on the soil structure depends on the type of soil, on the temperature of drying, on the degree of dryness achieved, and on the intensity of the wetting process.

The effect of drying the soil on its productivity was studied by Amiable (1925) in Tunisia, who found that yields of wheat were increased by 31–44 per cent following effective drying of the ploughed layer.

Under natural conditions, drying affects mainly the upper soil layer to a depth of 20–30 cm. Drying the soil has a number of effects which may be considered under the headings of chemical, physical and biological effects.

Chemical

The drying of the soil increases the availability of certain nutrients. Humus decomposition, in a moist, warm, and well-aerated soil, proceeds at a slow, albeit steady, rate (E. W. Russell, 1961). If the soil is first dried and then moistened, the decomposition of the organic matter is considerably accelerated, releasing increased amounts of nitrogen. The more thoroughly the soil is dried, the higher the temperature at which it is dried, and the longer it is kept dry, the greater will be the subsequent nitrogen flush on decomposition (Birch, 1959). In arid or semi-arid areas, in which intervals between irrigations or rains may be fairly long and the temperatures high, the upper soil layer may dry out completely during these intervals. Under these conditions, when the soil is again wetted, considerable mineral nitrogen may be released. It is to the increased amounts of available nitrogen that Birch (1960) ascribes the principal beneficial effects of drying.

Air-drying of a number of soils has been shown to result in an increase in potassium availability and uptake (Luebs *et al.*, 1956). However, Attoe (1946) found that, whilst the drying of soils that had not been fertilized with potassium generally resulted in a quite considerable increase in the content of exchangeable potassium, the opposite occurred in soils that had been fertilized with potassium prior to drying; in the latter case, drying increased the rate of fixation of potassium. This fixation did not prevent the added potassium, that had been fixed by drying, from being available for plant growth (Dowdy & Hutcheson, 1963). Drying of the soil had much less effect on P than on exchangeable K. Increases of P amounted to 10–11 per cent for fertilized plots and 18 per cent for unfertilized plots (Grava *et al.*, 1961).

Physical

Drying the soil, with subsequent gradual rewetting, improves soil structure and is capable of restoring a soil structure that has been degraded by excessive compaction. Woodburn (1944) found that the summer exposure of a dense and impermeable clay to several cycles of drying and wetting by sun and rain, respectively, caused extensive aggregate development. The effects of alternate drying and wetting on soil structure were greater than those of repeated cycles of freezing and thawing. The exact mechanisms whereby extreme drying improves aggregate stability are not yet understood (Harris *et al.*, 1966).

When soils dry gradually, they contract. This contraction is not uniform, and so cracks develop, sometimes to a considerable depth. The capacity of the soil to crack on drying depends mainly on its texture: the higher its content of soil colloids and the more uniform its texture is, the greater will be its tendency to crack on drying. The higher the content of colloids, the more the soil swells when wetted, and, consequently, the deeper the cracks become on drying. Adsorbed monovalent cations, especially sodium, have a similar effect (Vershinin *et al.*, 1959). The formation of shrinkage cracks on drying was found to be more extensive in soils with a high proportion of montmorillonite clay than in illite clay (Buehrer & Deming, 1961).

Cracks fulfil a number of important functions (Hardy & Derraugh, 1947): (a) they facilitate penetration in depth of rain or irrigation water; (b) the alternate opening and closing of the cracks facilitates the incorporation of organic residues into the lower soil layers; (c) they facilitate the breaking up of the land; (d) they enable the deep penetration of roots into clay soils; (e) they are beneficial to drainage, which may be of paramount importance when it becomes necessary to flush excess salts out of the root-zone; and (f) they help in drying the soil to a greater depth than would otherwise occur.

Soils in which deep cracks form are 'self-mulching': as a result of the action of wind, grazing animals and gravity, the cracks fill up progressively with soil particles from the surface. This leads in the course of time to a mixing of surface soil and subsoil, the effect being similar to that produced by burrowing animals and earthworms in soils of the humid regions.

Biological

A distinct shift in the balance of the soil population, between saprophytes and pathogens, occurs frequently in irrigated soils where the combination of high temperatures and favourable soil moisture frequently causes a proliferation of pathogens in the soil.

Drying the soil causes a marked reduction in the total number of microorganisms. This action is, however, selective in the sense that it favours the saprophytes at the expense of many of the soil pathogens. Many of the beneficial saprophytes form resistant spores, which enable them to survive the period of drying and increase rapidly on rewetting, whilst many of the pathogens do not have this advantage, and are therefore destroyed

Amongst the most resistant to drying are the nitrifying Bacteria; this is one reason why soils that have dried out during the summer months, show an explosive activity in nitrification with the first autumn rains, as discussed above (p. 343).

EROSION AND SOIL FERTILITY

Soil erosion is frequently the culmination of a process that begins with the loss of soil fertility and impaired soil structure (Plate 23). Soil erosion in turn further reduces soil fertility by selectively removing the smallest and lightest particles, thereby reducing the proportion of soil colloids and increasing that of the large, inert particles of sand. Thus a vicious circle is created, whereby the fertile top-soil is gradually removed, leaving the far less fertile subsoil exposed. The subsoil being relatively compact, water infiltration into it is slow and limited. Tillage becomes more difficult and less efficient. Biological activities in the soil are slowed down, as a result of impaired soil structure and deficiencies of available nutrients. Even long after the addition of fertilizers and manures, the subsoil remains less fertile than the top-soil that was removed by erosion.

The subsoil in turn is even more susceptible to erosion than the top-soil, because of its poor structure and low organic-matter content. Run-off increases, gullying results, and the field may soon be in a condition which makes cultivation uneconomical.

The damage is not confined to the eroded fields. Both wind and water transport the eroded soil selectively: the first to be deposited are the coarse sandy particles, which may cover to a great depth fields that were previously fertile, whilst the light fertile particles remain in suspension for great distances, and usually end up in the desert (loess soils) or in the oceans.

IRRIGATION

Irrigating arid soils is *not* the equivalent of creating a humid environment; irrigation has specific, far-reaching effects on soil fertility which have no counterpart in the conditions encountered in humid climates.

Irrigation makes possible considerable increases in crop production, but it affects soil fertility in a number of ways which may seriously impair its productivity on a sustained basis.

Irrigation water, unlike rain-water, contains quite considerable amounts of salts in solution. Certain of these salts, such as potassium sulphate or nitrates, can be of direct benefit to plants; others, such as calcium sulphate, can contribute to the improvement of soil sturcture. On the other hand, salts such as sodium chloride, or compounds containing boron, may have detrimental effects on the soil or the plant.

Irrigation water, even of high quality, contains appreciable amounts of undesirable salts. For example, water with a chlorine content of 200 p.p.m., which is considered a low and safe level, can add annually, under normal irrigation practice, up to 3 tonnes per hectare of NaCl (this does not even take into account other salts that are present in the water and inevitably get added to the soil).

It will be easily realized that constant additions of this magnitude can cause considerable changes in the soil's characteristics and fertility. The mechanism of these changes, the nature of the damage caused, the need for special precautions to be taken, and the special methods that can be applied to avoid permanent damage to irrigated soils resulting from salt accumulation, have been discussed in Chapter IX.

The negative effects of irrigation on soil fertility are not confined to the addition of harmful salts, especially when irrigation methods are primitive or faulty. The development of land for irrigation farming frequently makes large-scale levelling operations necessary, during which the subsoil may be exposed on fairly large areas. The subsoil is usually of very low productivity, because of poor physical condition and marked deficiencies in plant nutrients. For example, it has long been known that the semi-arid loess subsoils in Nebraska produced very poor crops of cereals, and that only legumes—provided they were supplied with phosphorus and potassium—could grow well under these conditions (Always *et al.*, 1917).

Whilst it is possible to build up the fertility of subsoils gradually, by growing a succession of legumes, this is usually a lengthy process: twenty years after a subsoil was exposed in Ohio, its colour was still noticeably lighter than that of the top-soil that had been treated in the same way. Also, under conditions of drought, the crops growing on the subsoil were affected much more quickly than those growing on top-soil (Bachtell *et al.*, 1956). The cumulative loss of income after 18 years of cropping due to the poor yields of the crops grown on the exposed subsoil, as compared to those grown on the soil in its original condition, was equivalent to the value of the land (ibid.).

Soil structure is adversely affected by the mechanical action of irrigation water, and by compaction due to the use of heavy tillage and harvesting machinery on moist soils. The effects on the dispersion of colloids, of leaching salts from sodic soils, and the disastrous effects of irrigation on soil structure and permeability, have already been discussed (pp. 249–51). Most surface methods of irrigation are conducive to soil erosion. In extreme cases, the land may become totally unfit for further irrigation farming (Penman *et al.*, 1967).

Changes in soil texture may result from the addition of silt carried by the irrigation water and from the downward movement of clay in the soil profile. In one area in southern Victoria, Australia, it was found that about 14 per cent of the clay content of the surface loam had been washed downwards by irrigation during a period of nine years (Penman, 1940). If the water is saline, sandy loams may become progressively heavier and less permeable, as occurred in the Salt River Valley of Arizona after twenty-eight years of irrigation (Pratt, 1959). The soil-water table may rise considerably, and restrict the volume of soil available to the root system. The problem becomes acute when the level of the water-table comes to within 2 metres of the soil surface. Waterlogging may become serious, especially where river irrigation is practised, as a result of the percolation of water from a dense network of rivers and canals. In the Punjab, in a single year (1942–43), more than 280 000 ha of land were rendered useless as a result of salt accumulation and waterlogging (Figure 62; Bharadwaj, 1961).

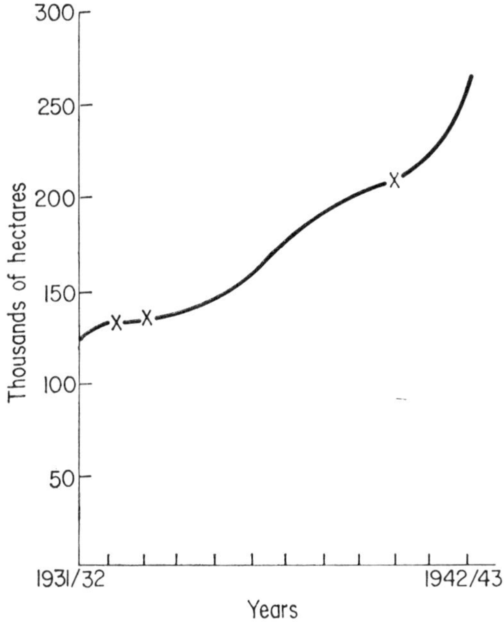

Figure 62. Areas damaged by salt accumulation and waterlogging in pre-partition Punjab. (After Bharadwaj (1961): by courtesy of the United Nations Educational, Scientific and Cultural Organization.)

There is small wonder, then, that many of the civilizations that were based on irrigation agriculture ended in disaster, and that large-scale failures, a few examples of which are given in Chapter IX, are not exceptional even in the twentieth century.

Irrigation farming may, however, also have favourable effects on soil fertility. The classical example is the effect of the silt-laden waters of the Nile on the fertility of the soils of the Nile Valley. Irrigation with silty water may considerably improve the texture of sandy soils, as has happened in many cases.

The use of large quantities of fertilizers that is implicit in successful irrigation farming, and the huge amounts of crop residues that remain after harvest, may enhance the fertility status of the irrigated soils. The combination of favourable temperatures and moisture conditions make possible a considerable increase in microbial activity. In the U.S.S.R., it was found that after two irrigations, the population of micro-organisms had increased more than six-fold (Penman *et al.*, 1967).

When a good crop rotation is used, which for example includes lucerne, irrigated soils may show increases in nitrogen and carbon; investigations in the U.S.A. have shown that, after 16 years, irrigated soils and subsoils contained, respectively, 44 and 58 per cent more nitrogen than did adjacent virgin soils and subsoils (Penman *et al.*, 1967). The less humus the soil contains at the time irrigation is initiated, the greater will the relative gains usually be that result from well-managed irrigation farming.

In summary it may be said that, with the present know-how and understanding of the problems involved, irrigated agriculture *can be permanent*. However, to ensure this requires both collective and individual efforts, continuously maintained. Irrigation and drainage systems have to be organized on a regional basis, and the individual farmer must employ irrigation techniques that are adapted to the soil type; he must, moveover, avoid over-irrigation, guard against salt accumulation, and practise good husbandry.

References

ALDRICH, D. G. & MARTIN, J. P. (1954). A chemical-microbiological study of the effects of exchangeable cations on soil aggregation. *Proc. Soil Sci. Soc. Am.*, **18**, 176–81.

ALWAYS, F. G., MCDOLE, G. T. & ROST, C. O. (1917). The loess soils of the Nebraska portion of the transition region. *Soil Sci.*, **3**, 9.

AMIABLE, J. V. (1925). La dessiccation des sols et l'accroissement de fertilité qui en résulte. *Ann. Serv. Bot. Tun.*, **3**(3), 219–41.

ARNOLD, P. W. (1960). Nature and mode of weathering of soil-potassium reserves. *J. Sci. Fd Agric.*, **11**, 285–92.

ATTOE, O. J. (1946). Potassium fixation and release in soils occurring under moist and drying conditions. *Proc. Soil Sci. Soc. Am.*, **11**, 145–9.

AUBERT, G. (1962). Arid zone soils. A study of their formation, characteristics, utilization, and conservation. *Arid Zone Res.*, **18**, 115–38.

AUDUS, L. J. (Ed.) (1964). *The Physiology and Biochemistry of Herbicides.* Academic Press, New York, xix + 555 pp., illustr.

BABCOCK, K. L., EGOROV, V. V. & KOVDA, V. A. (1967). Chemistry of saline and alkali soils of arid zones. pp. 175–217 in *Irrigation and Drainage of Arid Lands.* F.O.A./UNESCO, Paris, 663 pp., illustr.

BACHTELL, M. A., WILLARD, C. J. & TAYLOR, G. T. (1956). Building fertility in exposed subsoil. *Res. Bull. Ohio Agric. Exp. Stn.*, **782**, 35 pp.

BALFOUR, E. B. (1945). *The Living Soil.* Faber & Faber, London, 258 pp., illustr.

BARBER, S. A., WALKER, J. M. & VASEY, E. H. (1963). Mechanisms for the movement of plant nutrients from the soil and fertilizer to plant root. *Agric. Fd Chem.*, **11**, 204–7.

BHARADWAJ, O. P. (1961). The arid zone of India and Pakistan. *Arid Zone Res.*, **17**, 143–73.

BIRCH, H. F. (1959). The effect of soil drying on humus decomposition and nitrogen availability. *Pl. Soil*, **10**, 9–31.

—— (1960). Nitrification in soils after different periods of dryness. *Pl. Soil*, **12**, 81–96.

BLACK, C. A. (1966). Crop yields in relation to water supply and soil fertility. pp. 177–206 in *Plant Environment and Efficient Water Use* (Ed. W. H. Pierre, D. Kirkham, J. Pesek & R. Shaw) Amer. Soc. Agronomy and Soil Sci. Soc. Amer., Madison, Wisconsin, xii + 295 pp., illustr.

—— (1968). *Soil–Plant Relationships*. J. Wiley, New York, vii + 792 pp. illustr.

BROADBENT, F. E. (1957). Organic matter. pp. 151–7 in *Soil: Ybk. of Agriculture*. U.S. Dept. of Agric., xiii + 784 pp., illustr.

BUEHRER, T. F. & DEMING, J. M. (1961). Factors affecting aggregation and permeability of hardspot soils. *Soil Sci.*, **92**, 248–62.

BURNS, A. F. & BARBER, S. A. (1961). The effect of temperature and moisture on exchangeable potassium. *Proc. Soil Sci. Soc. Am*, **25**, 349–52.

CALLAGHAN, A. R. & MILLINGTON, A. J. (1956). *The Wheat Industry in Australia*. Augus & Robertson, Sydney, 486 pp., illustr.

COOKE, G. W. (1967). *The Control of Soil Fertility*. Crosby Lockwood, London, xvii + 526 pp., illustr.

CROWTHER, E. M. (1947). The Study of Soil Organic Matter and Organic Manures by Means of Field Experiments. pp. 123–39 in *C. R. Conf. de Pédologie Méditerranéenne*, Montpellier & Algiers, vi + 510 pp., illustr.

DELWICHE, C. C. & WIJLER, J. (1956). Non-symbiotic nitrogen fixation in soil. *Pl. Soil*, **7**, 113–29.

DONALD, C. M. (1960). The impact of cheap nitrogen. *Aust. J. Agric. Sci.*, **26**, 319–38.

—— & WILLIAMS, C. H. (1954). Fertility and productivity of a podzolic soil as influenced by subterranean clover (*Trifolium subterraneum* L.) and superphosphate. *Aust. J. Agric. Res.*, **5**, 664–87.

DOWDY, R. H. & HUTCHESON, T. B., jr. (1963). Effect of exchangeable potassium level and drying upon availability of potassium to plants. *Proc. Soil Sci. Soc. Am.*, **27**, 521–3.

DUDDINGTON, C. L. (1955). *The Friendly Fungi*. Faber & Faber, London, 188 pp., illustr.

ELLISON, W. D. (1952). Raindrop energy and soil erosion. *Emp. J. Exp. Agric.*, **20**, 81–97.

EPSTEIN, E. (1955). Passive permeation and active transport of ions in plant roots. *Pl. Physiol.*, **30**, 529–35.

FELLOWS, H. (1929). Studies of certain soil phases of the wheat take-all problem. *Phytopath.*, **19**, 103.

FLETCHER, J. E. & MARTIN, W. P. (1948). Some effects of algae and molds in the rain crusts of desert soils. *Ecology*, **29**, 95–100.

GARDNER, W. R. (1966). Soil-water movement and root absorption. pp. 127–49 in *Plant Environment and Efficient Water Use* (Ed. W. H. Pierre, D. Kirkham, J. Pesek & R. Shaw) Amer. Soc. Agronomy and Soil Sci. Soc. Amer., Madison, Wisconsin, xii + 295 pp., illustr.

GARRETT, S. D. (1956). *Biology of Root-Infecting Fungi*. Cambridge University Press, New York, xi + 292 pp., illustr.

GELTZER, F. Y. (1934). Effect of organic manures on the structure of grey desert soils. *Trans. Ist. Comm. Int. Soc. Soil Sci. Sov. Sect. A.*, **2**, 73–8.

GRABLE, A. R. (1966). Soil aeration and plant growth. *Adv. Agron.*, **18**, 58–106.

GRAVA, J., SPALDING, G. E. & CALDWELL, A. C. (1961). Effect of drying upon the amounts of easily extractable potassium and phosphorus in Nicollet clay loam. *Agron. J*, 54, 219–21.

HARDY, F. & DERRAUGH, L. F. (1947). The water and air relations of some Trinidad sugar-cane soils. *Trop. Agric. Trin.*, **24**, 76–87.

HARRIS, R. F., CHESTERS, G. & ALLEN, O. N. (1966). Dynamics of soil aggregation. *Adv. Agron.*, **18**, 107–69.

HASS, H. J., EVANS, C. E. & MILES, E. F. (1957). Nitrogen and carbon changes in Great Plains soils as influenced by cropping and soil treatments. *Tech. Bull. U.S. Dept. Agric.*, **1164**, 111 pp., illustr.

HOPP, H. & SLATER, C. S. (1948). Influence of earthworms on soil productivity. *Soil Sci.*, **68**, 421–8.

HOWARD, A. (1940). *An Agricultural Testament.* Oxford Univ. Press, London, xv + 253 pp., illustr.

HYAM, G. (1950). Crumb structure of the soil. *Fmg S. Afr.*, **25**, 119–20.

JACKS, G. V. (1963). The biological nature of soil productivity. *Soils Fertil.*, **26**, 147–50.

JENNY, H. (1930). A study on the influence of climate upon the nitrogen and organic matter content of the soil. *Bull. Agric. Res. Exp. Stn, Mo.*, **152**,

JENSEN, H. L. (1940). Contributions to the nitrogen economy of Australian wheat soils. *Linnean Soc. N. S. Wales*, **65**, 1–22.

KING, C. J., HOPE, C. & HEATON, E. D. (1934). Some microbial activities affected in manurial control of cotton root-rot. *J. Agric. Res.*, **49**, 1093–1107.

KOVDA, V. A. (1961). Principles of the theory and practice of reclamation and utilization of saline soils in the arid zones. *Arid Zone Res.*, **16**, 201–14.

KVASNIKOV, V. U. (1928). *Russian J. Agric. Sci.* **5**, 459, cited by Russell, E. W. (1961).

LEMÉE, G. (1955). Exposé général sur les rapports entre sol et végétation. *Arid Zone Res.*, **5**, 73–7.

LICHTENSTEIN, E. P. (1966). Persistence and degradation of pesticides in the environment. pp. 221–9 in *Scientific Aspects of Pest Control.* National Academy of Sciences, National Research Council, Washington, D.C., xi + 470 pp., illustr.

LUEBS, R. E., STANFORD, G. & SCOTT, A. D. (1956). Relation of available potassium to soil moisture. *Proc. Soil Sci. Soc. Am.*, **20**, 45–50.

MARTIN, J. P. & CRAGGS, B. A. (1946). Influence of temperature and moisture on the soil-aggregating effect of organic residues. *J. Am. Soc. Agron.*, **38**, 332–9.

——, ERVIN, J. O. & SHEPHARD, R. A. (1959). Decomposition and aggregating effect of fungus cell material in soil. *Proc. Soil Sci. Soc. Am.*, **23**, 217–20.

MARTIN, W. E. & MIKKELSON, D. S. (1960). Grain fertilization in California. *Bull. Calif. Agric. Exp. Stn*, **775**, 39 pp., illustr.

MEYER, D. R. & ANDERSON, A. J. (1959). Temperature and symbiotic nitrogen fixation. *Nature, London*, **183**, 61.

MILLINGTON, R. J. (1961). Relations between yield of wheat, soil factors, and rainfall. *Aust. J. Agric. Res.*, **12**, 397–408.

MITCHELL, R. & ALEXANDER, M. (1961). The mycolytic phenomenon and biological control of *Fusarium* in soil. *Nature, London*, **190**, 109–10.

OLMSTEAD, L. B. (1946). The effect of long-time cropping systems and tillage practices upon soil aggregation at Hays, Kansas. *Proc. Soil Sci. Soc. Am.*, **11**, 89–92.

PATRICK, Z. A. & TOUSSON, T. A. (1965). Plant residues and organic amendments in relation to biological control. pp. 440–57 in *Ecology of Soil-Borne Plant Pathogens* (Ed. K. F. BAKER, & W. C. SNYDER) Univ. of Calif. Press, Berkeley, 571 pp., illustr.

PAULI, F. W. (1967). *Soil Fertility*. Adam Hilger, London, xi + 204 pp., illustr.

PENMAN, F. (1940). Soil changes under irrigated pasture. *J. Dept. Agric. Victoria*, **37**, 83–100.

——, MINASHINA, N. G., KONONOVA, M. M. & MALINKIN, N. P. (1967). Some effects of irrigation and drainage on soils. pp. 509–62 in *Irrigation and Drainage of Arid Lands*. F.A.O./UNESCO, Paris, 663 pp., illustr.

PINCK, L. A., ALLISON, F. E. & GADDY, V. L. (1946). The nitrogen requirements in the utilization of carbonaceous residues in soil. *J. Am. Soc. Agron.*, **38**, 410–20.

PRATT, P. F. (1959). Chemical changes in an irrigated soil during 28 years of differential fertilization. *Hilgardia*, **28**, 381–420.

ROBINSON, J. B. D. (1957). The critical relationship between soil moisture content in the region of wilting point and the mineralization of natural soil nitrogen. *J. Agric. Sci.*' **49**(1), 100–5.

RUSSELL, E. W. (1961). *Soil Conditions and Plant Growth* (9th edn.) Longmans Green, London, xvi + 688 pp., illustr.

—— (1966). The soil environment of graminaceous crops. pp. 138–52 in *The Growth of Cereals and Grasses* (Ed. F. L. Milthorpe & J. D. Ivins) Butterworth, London, xii + 359 pp., illustr.

RUSSELL, R. S., BARBER, D. A. & NEWBOULD, P. (1961). Nutritional relationships between crops and soils. *Outlook on Agric.*, **3**, 103–110.

SCOTT, T. W, & SMITH, F. W. (1957). Effect of drying upon availability of potassium in Parsons silt loam surface soil and subsoil. *Agron. J.*, **49**, 377–81.

SEATZ, L. F. & STANBERRY, C. (1963). Advances in phosphate fertilization. pp. 155–80 in *Fertilizer Technology and Usage* (Ed. M. H. McVickar, G. L. Bridger & L. B. Nelson) Soil Sci. Soc. Am., Madison, Wisconsin, xv + 464 pp., illustr.

SHIFFMAN, J. (1961). Field experiments on inoculation of peanuts in northern Negev soils. *Israel J. Agric. Res.*, **11**, 151–8.

SMITH, H. V. (1944). A lysimeter study of the nitrogen balance in irrigated soils. *Tech. Bull. Ariz. Agric. Exp. Stn.*, **102**, 257 pp.

STEPHENS, C. G. (1945). A soil, land-use and erosion survey of part of county Victoria, S. Australia. *Bull. C. S. I. R. (Austr.)*, **188**.

TISDALE, S. L. & Nelson, W. L. (1966). *Soil Fertility and Fertilizers*. Macmillan, New York, vi + 694 pp., illustr.

VERSHININ, P. V., MELNIKOVA, M. K., MICHURIN, B. N., MOSHKOV, B. S., POYASOV, N. P. & CHUDNOVSKII, A. F. (1959). *Fundamentals of Agrophysics*. State Publ. House for Physics and Mathematics, Moscow. (Translated from Russian by the Israel Program of Scientific Translations, Jerusalem (1966)), xvii + 726 pp., illustr.

WEBBER, L. R. & ELRICK, D. E. (1966). Research needs for controlling soil pollution. *Agric. Sci. Rev.*, **4**(4), 10–20.

WEBER, C. R. (1966). Nodulating and nonnodulating soybean isolines: II. Response to applied nitrogen and modified soil conditions. *Agron. J.*, **58**, 46–9.

WEISNER, J. B. (1963). Use of pesticides. *Residue Rev.*, **6**, 1–22.

WOODBURN, R. (1944). Aggregation studies of Houston clay in Mississippi. *Proc. Soil Sci. Soc. Am.*, **9**, 30–6.

WOODROFFE, K. (1949). Investigations of the soil fertility problem in the wheat lands of South Australia. *Proc. Spec. Conf. Agric.*, Adelaide and Canberra.

YODER, R. E. (1937). The significance of soil structure in relation to the tilth problem. *Proc. Soil Sci. Soc. Am.*, **2**, 21–3.

XII

FERTILIZER USE IN DRY REGIONS

The elements that are essential in relatively large amounts to the growth of plants are carbon, hydrogen, oxygen, nitrogen, phosphorus, potassium, sulphur, calcium, and magnesium. The farmer is concerned mainly with the supply of nitrogen, phosphorus, and potassium, in the form of fertilizers. The other elements that are also essential to plant growth and required by plants in fairly large amounts are sulphur, calcium, and magnesium. These are, however, usually present in the soil in sufficient amounts for crop production, or are added incidentally by using commercial fertilizers supplying N, P, or K, or in the water used for irrigation.

The essential micro-nutrients iron, manganese, boron, zinc, copper, and molybdenum, are usually present in many arid soils in sufficient amounts to supply the needs of crop plants. The high proportion of calcium that is typical of many arid soils may, however, hinder absorption of iron and magnesium, and so cause deficiencies even when these two elements are present in the soil in sufficient amounts. Other micro-nutrients, such as sodium, cobalt, silicon, and vanadium, are not essential for all plants, but may be for some; sodium may play a useful role by partially replacing potassium.

The Dynamics of Response to Fertilizers in Dry Regions

In the non-irrigated soils of the arid regions, the weathering of the soil and the leaching of soluble salts are minimal. Nitrate content of the soil depends mainly on the amount of plant cover present. As this is usually very sparse, the decomposition of organic debris cannot provide large amounts of available nitrogen.

Most arid and semi-arid soils originally have large reserves of potassium and, to a much lesser degree, of phosphorus. Wherever cropping is practised, the soil phosphorus reserves have been considerably depleted and the amounts of available phosphorus are usually inadequate for satisfactory yields. Potassium reserves are usually adequate for rain-fed crops but may become deficient under irrigation.

The nutrient status of these soils, and hence the response of crops to fertilizers, does not remain static. Changes occur—either in the direction of continued depletion or of a build-up of soil fertility—according to the cropping system and the management practices adopted. Both these factors are dependent, in turn, on the moisture régime under which the crops are grown.

WITH RAINFALL TOO LIMITED FOR ARABLE CROPPING

These areas are generally used to support a very extensive pastoral economy. Even in developed countries such as Australia, fertilizers are not commonly used under these conditions.

RAIN-FED CROPS

In the semi-arid areas, in which rainfall can suffice for arable crop production, land which has previously lain undisturbed under scrub or other vegetation usually has sufficient natural fertility to produce cereal or forage crops for a more or less prolonged period without the addition of fertilizers. Eventually, however, yields begin to decline. In Australia, average wheat yields declined from 860 kg/ha in 1861–70 to 490 kg/ha in 1890–1900, as a result of the depletion of the initial supplies of nitrogen and phosphorus in the soil (Stephens & Donald, 1958).

In California, in regions where grain has been grown continuously for 50 years, production has frequently dropped to uneconomic levels, and the soils are referred to as 'worn-out grain lands'. Some are allowed to revert to range, but generally attempts are made to improve their productivity (Martin & Mikkelson, 1960).

Where traditional farming has been practised for centuries, most of the soils have reached a very low level of fertility, as a result of centuries of exhaustive cropping, with practically no returns to the soil excepting the meagre droppings of animals working on the land or grazing on the stubble of the field-crops. This is the case in most Mediterranean countries.

Whether the soils have been recently exhausted or have been cropped for centuries, the first essential step is to build up their fertility; this is done by the use of fertilizers, the introduction of legumes in the crop rotation, and periods of fallow. Because fertilizers are relatively inexpensive, are easy to apply, and give results that are usually visible to the eye, their use has been called a 'lead' practice which farmers adopt fairly early when moving from traditional farming methods to more productive systems (Williams & Couston, 1962).

The introduction of leguminous crops for green forage, hay, silage, or green-manure, into the crop rotation for building up fertility, may be absolutely dependent on the use of phosphatic fertilizers; without the latter, attempts at improved crop rotation are usually doomed to failure.

On these depleted soils, the response of leguminous crops to phosphatic fertilizers is most striking; without phosphorus, production is so low that there is almost no crop to harvest. However, with each cycle of the crop rotation, the response to phosphorus becomes less marked and more erratic though still usually giving significant increases in yield. In Australia, the results of research work on the residual value of phosphorus applied to leguminous pastures have led to the conclusion that generous applications of superphosphate should be made in the early years of pasture establishment, in order to build up the level of available phosphorus as rapidly as possible; subsequently, applications can be made every second or third year, instead of annually (Stephens & Donald, 1958). Nitrogen is usually ineffective for leguminous crops from the beginning, and remains so. However, when leguminous hay-crops are sown in admixture with a supporting cereal, yields are increased by nitrogenous fertilizers, though this increase is always at the expense of the quality of the hay.

On the same depleted soils, the grain cereals grown during the rainy season, respond to both phosphatic and nitrogenous fertilizers. When soils with an initially low phosphorus content receive regular applications of phosphatic fertilizers, the residual phosphate will increase the available phosphorus considerably. In South Australia, this increase was found to be linear in relation to the total amount of superphosphate applied (Woodroffe & Williams, 1953). It was found that in fields to which phosphorus, totalling 585 kg of P_2O_5 per hectare, had been applied, the residual value was sufficiently high to supply all requirements of wheat for five years, when it was grown in a wheat–barley–fallow rotation, without the need for further applications (Piper & Vries, 1964).

Similar results were obtained with barley on phosphorus-deficient soils (Fig. 63). It was found that for five successive years, yields of barley increased considerably as a result of superphosphate application; during the following five years, however, not only was there no increase in yields as a result of applications of superphosphate at the heavy rate that had been so effective in the preceding period, but yields were even depressed. This was due to the lowered nitrogen-level of the soil as a result of five years of cereal cropping. This probably resulted in a nitrogen–phosphorus imbalance, leading to reduced protein synthesis in the young plants (Woodroffe & Williams, 1953).

It is suggested that after a build-up of phosphorus reserves has been achieved during the early development period, relatively small applications of superphosphate to each cereal crop at sowing-time are sufficient. These supply an easily available source of phosphorus when the root system of the plants is still small, thereby stimulating the crop at an early stage and securing good

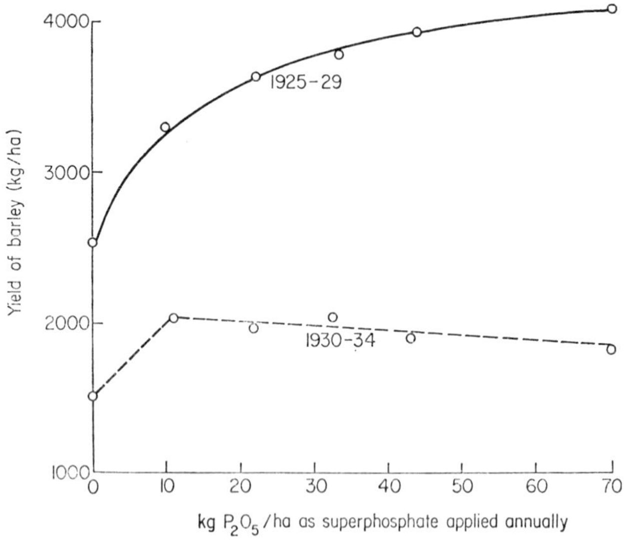

Figure 63. Response of barley to increasing rates of superphosphate application during two successive periods. During the initial period (1925–29) the phosphorus status of the soil was low, but available nitrogen status was relatively high. During the second period the soil phosphorus status had improved as a result of the high rates of superphosphate applied during the previous period, but available nitrogen had been lowered by cropping—hence overall yields during the second period were low and there was no response to superphosphate on areas that had previously received annually 20 kg/ha or more of P_2O_5. (After Woodroffe & Williams (1953): by courtesy of *Australian J. Agricultural Research.*)

establishment and better root penetration (Piper & Vries, 1964). The graph in Figure 64 illustrates dramatically the residual value of phosphorus applied in previous years, showing that it is possible to build up reserves that provide adequate P independently of current application (Donald, 1964).

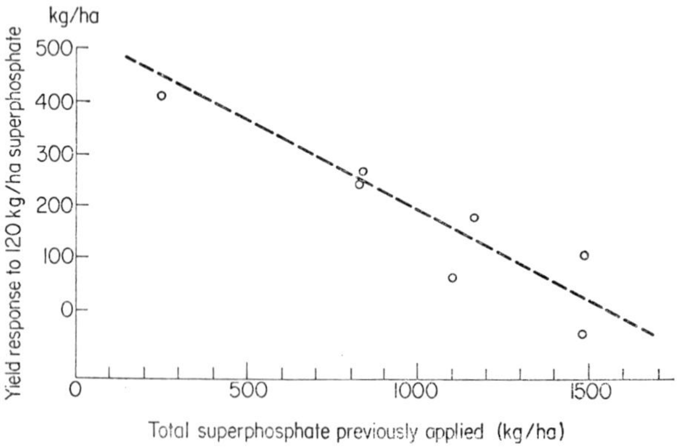

Figure 64. Effect of prior applications of superphosphate on the response by wheat to a current application of phosphorus in Western Australia. (After Donald, 1964.)

Experience in Israel shows that, once the initial severe phosphorus deficiency has been overcome, a single application of phosphorus to the leguminous crop in a three-course rotation (legume-wheat-sorghum) is sufficient to supply all the needs of the cereal crops in the rotation. Nitrogenous fertilizers continue to be effective on rain-grown cereals, even under conditions of improved fertility. The optimum time of application of nitrogen varies, as it is dependent on rainfall distribution.

CROPS GROWN ON STORED SOIL MOISTURE

In regions where a summer drought alternates with a winter-rainfall season, it is possible to grow summer crops whose entire moisture-supply is provided by water stored in the soil during the rainy season. Typical crops which give fairly good yields under these conditions are sorghums, cotton, sesame, and cucurbits.

Under these conditions, crops do not usually respond to fertilizers that are applied to them directly. Although early growth may be somewhat stimulated by such application, under the conditions of limited moisture supply that obtain, this is no advantage. The lack of sustained response is probably due to the fact that the fertilizers are applied in the top-soil. By the time the plants have reached a stage of development when nutrient requirements are maximal, this soil layer has completely dried out, and the nutrients contained in it are virtually ineffective. This view is supported by the fact that crops grown on stored soil-moisture give very poor yields on low-fertility soils, but react favourably to a fertility build-up resulting from fertilizing preceding crops.

The situation is different when a crop that is dependent on stored water also receives a limited amount of rainfall or a supplemental irrigation during its growth period. The results of 66 trials on wheat, grown in North Dakota on non-fallow land, show that the response to fertilizers increased with increasing amounts of stored water and of rainfall occurring during the growing-season. On the basis of these results, fertilizer recommendations are being made by regions, according to the probability of rainfall during the growing-season, and for each region they are related to the amount of stored moisture at the time of sowing (Weiser, 1960).

In Nebraska it is recommended to apply 25 kg N/ha for every 75 mm of available water in the 2-m soil profile (S. L. Tisdale & Nelson, 1966).

IRRIGATED CROPS

Irrigation becomes an effective tool, for increased production maintained on a high level, only when applied in conjunction with the most intensive cultural practices, of which an appropriate fertilizer régime is one of the most important.

Increased needs for fertilizers under irrigation are due to:

1. *A more intensive cropping schedule:* In an intensive crop rotation under irrigation, eight or more crops may be harvested in the course of five years. It is evident that, under these conditions, nutrient removal may assume considerable proportions
2. *Higher yields produced:* In Israel, yields of 20000 feed units* annually from irrigated pasture, 120 tonnes/ha of lucerne, 450–500 tonnes of stock beets, 100–120 tonnes of forage maize, 4500–5000 kg of cottonseed, or 80–100 tonnes of sugar-beets—are common. The level of production that is possible under irrigation may increase the crop's requirements for nutrients up to twenty-fold. This is possible only if the soil is rich in nutrients—derived either from its own reserves or from fertilizers that have been judiciously applied by the farmer.
3. *Leaching of nutrients due to irrigation:* Salinity and alkalinity are major problems of irrigation farming. The removal of excess salts by leaching is an almost essential feature of good farming practice: neglect of this process brings on salinity problems. The leaching process, however, is indiscriminate and also removes soluble nutrients from the soil, thereby enhancing the deficits resulting from increased plant production. Of the principal nutrients, the most affected is nitrogen. Losses of potassium by leaching are far less marked, except in sandy soils. Turk *et al.* (1943) showed that potassium held in exchangeable form leaches relatively slowly.

With the transition from rain-fed to irrigated agriculture, it is essential that the farmer break away from the traditional fertilizer practices evolved under dry land farming, and adopt an entirely new approach that must be appropriate to the new situation.

A farmer who is accustomed to the fertilizer rates applied to rain-fed crops will, at first, usually apply to irrigated crops, rates that are at best fifty to one hundred per cent higher than those which he used previously. The responses obtained from these amounts are generally so unspectacular as to create the impression that further increases of yield from fertilizer application are not to be expected. It is only when rates of applications are increased three to fourfold or even more, that the potentialities of heavy fertilizer application for increasing the efficiency of water-use are fully realized. In many crops—such as maize for grain or forage production, fodder beets, irrigated pastures, and many vegetable crops—really astonishing yield increases are obtained by heavy fertilizer dressings. The most consistently spectacular responses are obtained from nitrogenous fertilizers. Rates of up to 200 kg of N/ha for maize, 300 kg for fodder beets, and 500 kg and more for irrigated pastures, can give economic returns (I. Arnon, 1963).

* *Feed Unit:* In the Scandinavian feed-unit system of valuing feeds, the value of 1 kg of barley grain is taken as the standard for measuring the relative values of different feeds.

The high rates of nitrogen applied call almost automatically for an increase in phosphatic fertilizers. Most of the soils of the dry regions are poor in phosphorus. However, response to phosphorus application is usually erratic, and largely dependent on fertilizers that had been applied to previous crops in the rotation. Only a relatively small fraction, usually less than 25 per cent, of a phosphatic fertilizer is used by the crop to which it is applied; the excess is, however, not lost but contributes to the building up of phosphorous reserves in the soil (cf. pp. 355–7). It is therefore not surprising that after several consecutive applications at the rate of 150–200 kg of P_2O_5 per ha, the response of crops to this element tends to become erratic. It accordingly appears logical to adjust the rates of phosphatic fertilizers, not to the requirements of the individual crops, but to those of the crop rotation as a whole.

Most of the soils of the arid regions are known to be relatively rich in potassium. With the low yields that are usually obtained in these regions under dry land cropping, liberation of fixed potassium from the soil has usually been able to keep pace with the removal of this element by the crops.

When irrigation is first practised on these soils, response to potassium fertilizers is usually meagre or non-existent. However, it has been found that this situation is not necessarily static. In an experiment on irrigated pastures in Israel, in which annual rates of up to 3000 kg per hectare of sulphate of ammonia gave economical returns, clover, sown into the high-nitrogen plots, showed distinctive signs of potash starvation, which were relieved by potash fertilization. Similar responses have been observed in various other crops—e.g. cotton and sugar-beets—and it is evident that the soil reserves of K are not always able to keep pace with the high rates of plant production that are possible under irrigation.

Plant Nutrients

Although we dealt briefly in the last chapter with the main nutrient elements in the soil, we must now consider them in more detail—with special reference to their significance to crops and their application in fertilizers in dry regions.

NITROGEN

Role

Nitrogen plays a vital role in all the living tissues of the plant, of which it constitutes up to 1–4 per cent of the dry-weight. No other element has such an effect in promoting vigorous plant growth. A deficiency of other elements may have as great a detrimental effect as a deficiency of nitrogen, but the response of most crops—excepting legumes—to nitrogen supplied as a fertilizer is usually more striking visually than is that to other nutrients. Nitrogen is also the element which is most frequently in short supply in the soil.

The forms in which plants usually absorb nitrogen from the soil are nitrate

(NO_3^-) and ammonium (NH_4^+) ions. In warm, well-aerated and slightly alkaline soils, absorption of the nitrate form will usually predominate (S. L. Tisdale & Nelson, 1966). The ions absorbed are first reduced within the plant to more elemental forms — $-N^{=}$, $-NH^-$, or $-NH_2$ — which are subsequently built into more complex compounds and ultimately into protein. Nitrogen is also part of the all-important chlorophyll molecule.

Abundant protein tends to increase the size of the leaves and accordingly to bring about an increase in carbohydrate synthesis (ibid.). A high nitrogen-supply favours the conversion of the carbohydrates into proteins, which in turn promotes the formation of protoplasm. Protoplasm, being highly hydrated, is conducive to the succulent growth that is typical of plants growing on a high N-régime.

Nitrogen Requirement of Crops and Limitations in Use

Nitrogen requirements vary considerably at different stages of development of the plant: minimal in the early stages, the requirements increase as the rate of growth accelerates, to reach a peak in most crops during the period between the onset of flowering and early fruit or grain formation.

There are quite a number of cases in which the application of the full amounts of nitrogen fertilizers, that are required for maximum yields, may have undesirable side-effects. In small-grain cereals, excessive nitrogen increases the proportion of straw to grain, and in combination with succulent growth, is conducive to 'lodging' of the tall but weak straw. In cotton, the excessive vegetative growth hinders mechanical harvesting and makes the crop more susceptible to insect attack. In sugar-beets, high levels of nitrogen reduce the sugar content and increase the amount of noxious nitrogen in the roots, which impedes sugar crystallization. In malt barley, malting quality, as measured by the percentage of malt extract, will usually be reduced if nitrogen is applied for maximum grain-yields. The undesirable effects of high levels of N fertilization can be alleviated by proper timing of application and, particularly, by a well-balanced supply of potassium and phosphorus.

Commercial Nitrogenous Fertilizers

Commercial nitrogenous fertilizers can be classified in four groups: ammoniacal compounds, nitrate salts, amides, and protein fertilizers.

The bulk of the chemical nitrogen compounds are ammonia derivatives, produced synthetically—mainly by reacting nitrogen and hydrogen gases. The basic compound NH_3 can be used either directly or after transformation into many other commercial fertilizers.

Ammoniacal Fertilizers. These have a specific advantage for agricultural use in the arid regions: the NH_4^+ ion is retained by the colloidal fraction of the soil, and is therefore less liable to leaching in irrigation agriculture than are the nitrates.

A few of the most important ammoniacal fertilizers are: ammonium sulphate (20·5 per cent N, 23·4 per cent S), ammonium nitrate (32–33 per cent N), ammonium chloride (26–28 per cent N), anhydrous ammonia (82 per cent N), aqua ammonia, and solutions of ammonia and other nitrogen salts (24–49 per cent N). When ammonium salts are applied to the surface of alkaline or calcareous soils, very unstable ammonium carbonate or ammonium bicarbonate salts are formed. These decompose, and so free ammonia may be lost. The higher the temperature is, the greater are the losses. Losses by volatilization in alkaline soils, following the application of aqua ammonia, may exceed 50 per cent (Humbert & Ayers, 1957).

(a) *Ammonium Sulphate.* This is the nitrogenous fertilizer that is most widely used in the arid regions. Its marked tendency to acidify the soil is a disadvantage in temperate and tropical climates, whose soils tend toward a low pH in any case, and its continued use under these conditions makes periodic liming essential. However, this is usually not the case in the soils of arid regions, and the sulphur provided by the fertilizer may frequently be beneficial as a nutrient in its own right.

(b) *Ammonium Chloride.* This is not widely used except for fertilizing paddy rice, where it replaces ammonium sulphate. The latter is undesirable because under anaerobic soil conditions, the sulphate may be reduced to the toxic compound hydrogen sulphide (H_2S).

(c) *Liquid Fertilizers.* In recent years, the use of anhydrous ammonia and aqua ammonia has expanded considerably in a number of countries. As the ammonia is obtained in the first stage of the manufacture of nitrogenous fertilizers, it can be delivered at a much lower price within a reasonable distance from where it is manufactured.* It then has further advantages in that no physical effort by the farmer is involved; instead of having to empty heavy sacks into his fertilizer distributor, he moves liquid fertilizer from the transport tanks to the field-applicator by pump or compressed air. Moreover the fertilizer distribution in the field, not being affected by wind, is also more uniform than is the application of solid fertilizers.

Nitrogenous liquid fertilizers can be divided into three types, according to the amount of pressure required in the tanks, and the equipment used for storage, transport, and application:

* In the U.S.A., anhydrous ammonia and aqua ammonia can be supplied at a little over half the price of the nitrogen in ammonium nitrate (Cooke, 1967).

(i) *High Pressure: Anhydrous Ammonia:* Anhydrous ammonia contains the highest percentage of nitrogen (82 per cent) of any of the commercial fertilizers available. Relatively high pressure is required to maintain the ammonia as a liquid, and the higher the temperature is, the greater will be the pressure required. In recent years the use of refrigerating equipment to liquify the ammonia gas and keep it liquid has made possible the replacement of high-pressure tanks, such as that shown in Plate 24.

When no longer under pressure, ammonia gas volatilizes rapidly; nitrogen losses in the field can, however, be avoided by proper application methods. In heavy and medium-textured soils, applying the ammonia to the soil by injection, through tubes running down the rear of a blade-type applicator, to a depth of 15–18 cm, will effectively prevent losses of gaseous ammonia. In lighter soils, the depth of application must be increased slightly.

The ammonia is effectively retained by the colloidal fraction of the soil. Losses may, however, occur if the physical condition of the soil during application is not satisfactory—especially if it is not in good tilth and contains many clods.

In irrigation agriculture, ammonia can be applied by metering into the irrigation water—except where overhead irrigation systems are used.

As regards *the reaction of anhydrous ammonia in the soil,* in wet soils we have:

$$NH_3 + H_2O \longrightarrow NH_4OH$$

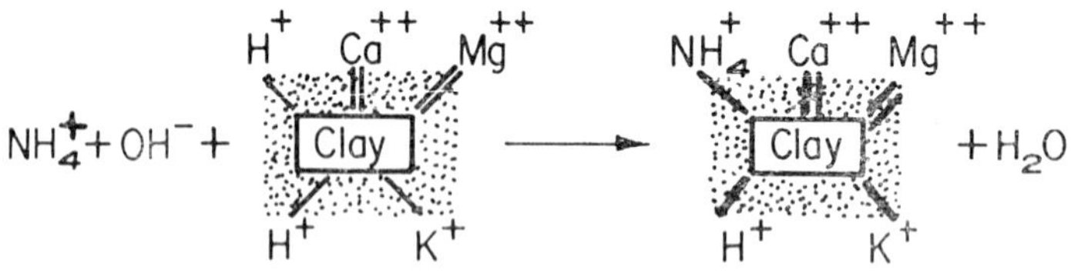

This reaction explains why ammonia is more strongly held in acidic soils than in basic ones.

In dry soils:

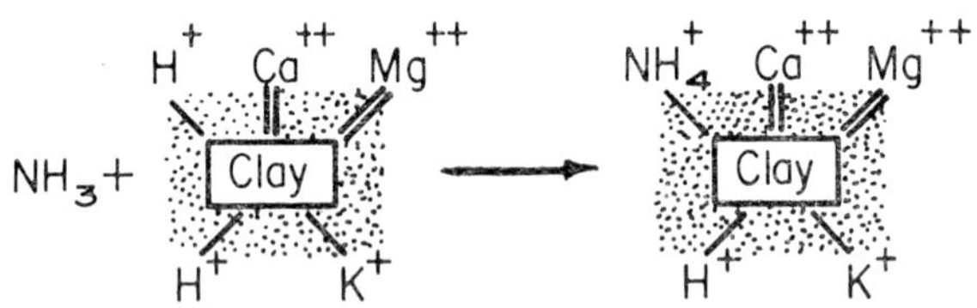

In dry soils, in addition to its adsorption by neutralization of exchangeable hydrogen, as shown schematically above, the anhydrous ammonia may also be adsorbed by Van der Waal's forces—intermolecular forces of attraction such as those which hold dust to a wooden surface (Kurtz & Smith, 1966).

As free ammonia is toxic to micro-organisms, their numbers in the soil are initially depressed following its application; however, nitrification is subsequently stimulated, probably as a result of the more rapid recovery of nitrifying Bacteria as compared with soil Fungi (Eno & Blue, 1954).

Free ammonia is also toxic to higher plants. Damage to their germinating seeds and to seedlings can be avoided by placing the band of ammonia 8–10 cm below, and to the side of, the seed-row.

For close-planted crops, such as wheat, it is desirable to apply the ammonia fertilizer at right angles to the direction of sowing. As a further precaution, applying the fertilizer about two weeks ahead of sowing is indicated.

(ii) *Low Pressure: Aqua Ammonia:* Aqua ammonia is a solution of ammonia in water, with a nitrogen content of at least 20 per cent. A certain amount of pressure is required to offset the natural tendency of the aqueous ammonia to change into anhydrous ammonia. The application of aqua ammonia can be made closer to the surface than that of anhydrous ammonia, but must be at a depth of not less than 10–15 cm.

(iii) *Non-Pressure Solutions:* Solutions containing dissolved ammonium nitrate or urea, but no free ammonia, can be applied with very simple equipment in a variety of ways—on the soil surface, injected into the soil, or metered into the irrigation water.

Nitrate Salts

Fertilizers belonging to this group are very soluble, and their nitrogen is readily absorbed by the plants without undergoing any transformation. They are, however, very easily leached from the soil.

The principal nitrate sources are sodium nitrate (16 per cent N), potassium nitrate (13·8 per cent K, 36·5 per cent N), and calcium nitrate (15·5 per cent N). They are not acid-forming, and have no special advantages for crop cultivation in arid regions. The most widely available form—sodium nitrate—is even undesirable in irrigation agriculture, because of the sodium component. Ammonium nitrate has already been mentioned in the former group of ammonized fertilizers.

Amide Nitrogen Fertilizers

The principal fertilizers in this group are calcium cyanamide (20–22 per cent N) and urea (46 per cent N).

The main justification for the use of urea is its high nitrogen content—the highest of any solid nitrogen fertilizer available—which makes it attractive for areas where the cost of transport is an important factor in calculating fertilizer costs, and for application by aeroplane. Urea is rapidly transformed in the soil, by the enzyme urease, into ammonia and carbon dioxide, and ultimately behaves like the ammoniacal sources of nitrogen.

Protein Fertilizers

Protein fertilizers, such as dried blood or dried fish-meal, are of historical interest only.

In recent years, much attention has been devoted to the development of a new class of fertilizers with a high content (about 38 per cent) of nitrogen which has low initial solubility and is released in the soil only gradually and over a long period. These fertilizers, which are reaction products of urea and formaldehyde, resist leaching and release nitrogen throughout the period of growth of the plant. They are therefore well-suited to irrigated crops; their present high cost, however, precludes large-scale use for the time being.

Relative Value of Various Fertilizer Carriers of Nitrogen

Most of the research on the relative effectiveness of the various forms of nitrogen supplied in commercial fertilizers has been carried out in humid regions. Studies comparing different nitrogen carriers have, for the most part shown no consistent differences in their effects on the yields or composition of individual crops.

Whatever nitrogen carrier is applied to the soil, within a relatively short time the nitrogen will be present in two forms only: $(NH_4)^+$, which is rapidly attracted to and adsorbed by the negatively charged colloids of the soil, and NO_3^-, which is not held by the soil colloids. The ammonium ion may also be 'fixed' by being held in the lattice structure of certain soil minerals.

I. Arnon *et al.* (1963), comparing the effects of various sources of combined nitrogen in dry, liquid, and gaseous forms, on forage and grain crops, as well as on stock and sugar-beets, all grown in Israel, found that the liquid form showed a clear advantage over ammonium sulphate when applied to green fodder; effects were identical in the production of grass, but the liquid form was less effective than ammonium sulphate in its influence on fodder and sugar-beets. However, the differences either way did not exceed 12–15 per cent. No significant differences were observed on the effects of different sources of nitrogen on dry matter, protein, or sugar content.

The choice of the commercial fertilizer to be used will depend firstly on the effective cost, *in the field*, per unit of nitrogen, then on the physical characteristics of the fertilizer, which affect the ease of its application (hygroscopicity, granulation, liquid or gaseous forms), and finally on its specific effects on soil properties. *A priori*, for the arid-region soils, preference will generally be given to acid-forming fertilizers, whilst those adding sodium are best avoided. In the warm and moist soils of so much irrigation farming, there is practically no difference in the relative rapidity with which the nitrate and the ammoniacal forms of N become available to plants. On the other hand, the ammoniacal form is less subject to leaching—though the importance of this aspect should not be exaggerated, as it is rapidly nitrified under the favourable temperature and moisture conditions that are characteristic of irrigation agriculture conditions. High temperatures as well as drying of the soil favour losses of NH_3 from ammonium fertilizers and urea when these are applied to alkaline and calcareous soils with a low cation-exchange capacity.

With overhead irrigation, which makes it possible to confine added water to the root-zone only, losses of nitrogen by leaching are practically eliminated, whatever the form of nitrogen used (Nielson & Banko, 1960).

PHOSPHORUS

Role

The quantities of phosphorus that are present in the tissues of most plants are much smaller than those of the other two major nutrients, nitrogen and potassium, being only about one-tenth of the nitrogen and one-fifth of the potassium.

Most of the phosphorus is absorbed by the plant in the form of orthophosphate ions—mainly the primary orthophosphate anion $H_2PO_4^-$ and, to a much lesser degree, the secondary orthophosphate ion $HPO_4^=$ (S. L. Tisdale & Nelson, 1966). The proportion of the latter used by the plant is, however, favoured by the higher pH values which characterize arid-region soils.

During growth, phosphorus concentrates mainly in the plant tissues in which cell division is proceeding actively. It is an essential constituent of nucleic acid, of phytin, and of phospholipids. An adequate supply of easily available phosphorus is of great importance during the early stages of growth when the limited root system is not yet capable of drawing sufficiently on the phosphorus reserves of the soil and cannot effectively compete with soil micro-organisms for the available phosphorus.

A deficiency at this stage adversely affects the laying down of the primordia for the reproductive parts. High doses of phosphorus, when available to young maize plants, were also found to enhance the rate of nitrogen metabolism (Pleshkov, 1956). Adequate phosphorus, supplied at a later stage, therefore cannot remedy the adverse effects of a deficiency at the critical earlier stage.

Fixation of Phosphorus

The process whereby soluble phosphates, as applied in fertilizers, are changed into less soluble forms in the soil, is called 'fixation' or 'reversion'. The nature of the fixation may affect the efficiency of different forms of fertilizer phosphorus differently, according to soil type.

Phosphate fixation in alkaline and calcareous soils is mainly due to the formation of insoluble phosphate compounds of calcium; some fixation may also be due to Al^{+++} or $Al(OH_3)$ in the clay fraction of alkaline soils and, to a minor degree, to the activity of micro-organisms (Hemwall, 1957).

Two factors influence the supply of phosphorus to the crop:

(a) The concentration of phosphorus in the soil solution; this is mainly important for the young plant.
(b) The concentration of phosphorus in the soil, and its capacity to ionize and thereby maintain the level of phosphorus in the soil solution. This is the main source of supply for the developed plant.

In soils in which 'fixation' of phosphorus occurs, only a small proportion of the fertilizer phosphate (usually not more than 5–15 per cent) may be taken up by the crop to which it is applied. However, as we have already seen (pp. 355–7), the remaining phosphate is not lost. A distinction must be made between non-solubility and non-availability of the phosphorus, as the rapid fixation does not prevent the fertilizer phosphorus from contributing gradually to the replacement of phosphorus in the soil solution.

The fertilizer phosphorus that is not taken up by the crop to which it is applied should be considered as a capital investment; remaining as it does in the 'labile ionic pool', its presence will enable the next addition of phosphate to increase the potential to a greater extent than would have been the case otherwise (Russell *et al.*, 1961).

Relative Contributions of Fertilizer Phosphorus and Soil Phosphorus

In view of the rapid fixation of any phosphorus applied in fertilizers, much work has been done with radioactive isotopes in order to distinguish between the effects of fertilizer phosphate and phosphorus that is already present in the soil (Figure 65).

The relative proportion of the phosphorus requirements of the plant that are derived from the fertilizers applied directly to the crop, is influenced by the kind of crop and its stage of development, as well as by the soil type, soil phosphorus level, and environmental conditions (Woodroffe & Williams, 1953).

By using fertilizer phosphorus labelled with the radioisotope P^{32}, it was demonstrated that the main contribution of fertilizer phosphorus is to the young plant. The ready availability of the fertilizer phosphorus encourages rapid root-proliferation, following which the plant is better able to draw on the phosphate reserves of the soil and thus become progressively less dependent on the fertilizer phosphorus. The rate of P diffusion to the root is far slower than that of K and especially N; therefore, total uptake of P depends largely on the total root surface-area of the plants and hence on the rate of

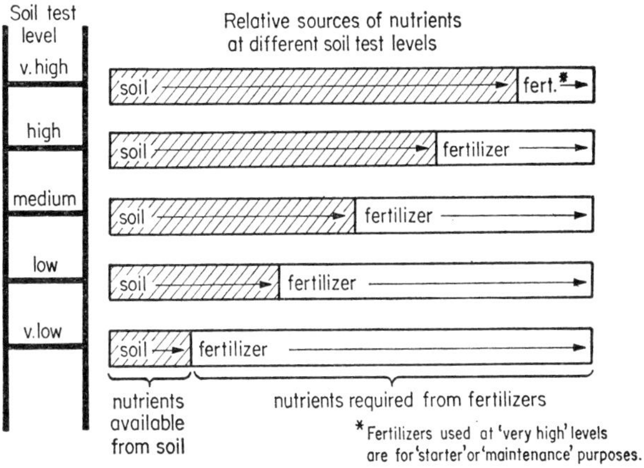

Figure 65. Relative sources of nutrients at different soil test levels. (By courtesy *Crops & Soils*.)

root elongation: as the root system develops, the proportion of phosphorus derived from the soil reserves increases progressively. The absolute requirements for phosphorus are also greatest in the young plant: in many crops, by the time they have achieved 20 per cent of their total growth, they have already taken up 50 per cent of their total phosphorus requirement (Hall *et al.*, 1953).

It has also been shown that, the higher the rates of phosphate fertilizer that are applied, the less will be the phosphate contributed to the plant from the soil reserves. When phosphorus fertilizer was applied to lucerne at the rate of 50 kg/ha P_2O_5, the crop derived only 15–18 per cent of its total phosphorus requirement from the fertilizer; when the rate of application was increased threefold, the contribution of fertilizer phosphorus also increased to 45·9 per cent (Jacob *et al.* 1949). In other words, the relative amounts of phosphorus absorbed by the plant from the soil and from fertilizer is approximately proportional to the supply available from each of these sources.

The availability of P is also dependent on soil temperature, as plants are more capable of taking up P from a warm soil than from a cold one hence the need for adjusting rates of P fertilization according to the growing-seasons of the crop (Seatz & Stanberry, 1963).

With approaching maturity, the phosphorus in the plant migrates to its fruits and seeds; therefore, harvest of the produce constitutes a serious drain on the phosphorus reserves of the soil. The amount of phosphorus that needs to be applied to a phosphorus-deficient soil is usually far in excess of the immediate requirements of the plant, because of the rapid fixation of the phosphorus in the soil. Larger doses are therefore needed, in order to saturate the fixing power of the soil and leave an excess available to the crop. For example, land-levelling required prior to irrigation may expose calcareous or extremely infertile subsoils that are poor in phosphorus. It has been shown that extraordinarily large amounts of phosphorus—up to 140 kg of P per ha—may increase the productivity of these soils markedly, even to the same level as that of the unexposed soil. This effect may already be evident in the first season after exposure, provided N and Zn are also added (James *et al.*, 1967). On the other hand, in well-designed crop rotations, in which phosphate fertilizers have been applied regularly, reserves of phosphorus are gradually built up from the residual fertilizer, from organic manures, and from the crop residues.

It has been shown that plots which were fertilized annually at the rate of 400 kg/ha of single superphosphate, had, after three years, doubled the amount of available phosphate as compared with unfertilized plots. However, each additional increment of fertilizer increases the residual effect—so that, in a well-fertilized crop rotation, the response of individual crops to further phosphate fertilization may become progressively less (cf. pp. 355–7).

The availability of the phosphorus in the soil is also influenced by the water régime, as was shown in an experiment in Utah in which the level of soil moisture had as much effect on the availability of phosphorus to an irrigated sugar-beet crop, as had the rate of fertilizer applied and the method of its placement (Haddock, 1949).

Phosphorus Fertilizer Application

Owing to the rapid fixation of phosphorus, there is very little natural downward movement of this element in the soil. With irrigation, the downward movement of phosphorus is increased, the depth of penetration being proportional to the amount of water applied (Jordan *et al.*, 1952). However, even with irrigation, the downward movement is limited. It has been found that after 28 years, 80 per cent of the P fertilizer applied to an irrigated soil remained in the top 30 cm of the soil and very little had penetrated beyond 60 cm. It is therefore frequently recommended to incorporate phosphoric fertilizer into the soil by ploughing. This has several disadvantages, however: the emerging seedling does not find sufficient available phosphorus and is hindered in its development, and by the time its root system has developed sufficiently to exploit the soil efficiently, the phosphorus is no longer readily available. Moreover, phosphate applied in this way is exposed to a great surface of contact between soil and fertilizer, and fixation is rapid and complete.

It has repeatedly been shown, under conditions of ample soil moisture, that placing water-soluble phosphates, such as superphosphate, in bands in the vicinity of the seed-row, considerably reduces fixation, so that the seedlings are able to derive their phosphorus requirements from the fertilizer and develop rapidly. In arid-zone soils that are low in available phosphorus, banding may stimulate early root-growth and thereby enable the plant to resist drought conditions later in the season (Kafkafi, 1963). Broadcast application, on the other hand, ensures greater contact between the fertilizer and the feeder roots of the plant, and therefore ensures more efficient uptake as the plant develops and its root system expands. Water-insoluble fertilizers benefit most from thorough mixing with the soil.

Granular P fertilizers are often recommended for general farm practice because of reduced fixation due to the smaller surface-area involved; however, on alkaline soils, where the immobilization of P by Al and Fe is of minor importance, finely powdered fertilizers with highly soluble P are more effective than granulated fertilizers (Hagin, 1958).

On soils that are very deficient in phosphorus, band fertilization will not produce maximum yields. The general level of phosphorus in the soil should first be built up by broadcast applications and incorporation in the top layer of the soil. Subsequent row-applications will then generally be effective.

Perennial crops, such as lucerne and permanent pastures, are a special problem. The phosphate can only be broadcast on the soil surface and, at the most, lightly disked in. Experience has shown that downward movement of phosphorus, under these conditions, does not exceed a few centimetres, even after a year or longer. However, it has been proven conclusively that top-dressed phosphates are highly beneficial to crops that require annual applications of phosphorus. In parts of Australia, it was found that broadcasting superphosphate on the surface of pastures was just as beneficial as drilling it in (Ignatieff & Page, 1958).

Phosphate Fertilizers

In the 19th century the main source of fertilizer phosphorus was finely ground animal bones. Liebig (1843) was able to show that treating the bones with sulphuric acid increased the availability of the phosphorus. It was J. B. Lawes in England, however, who developed the treatment of rock phosphate by sulphuric acid into an industrial process for the production of superphosphate.

The main phosphate fertilizers are:

Phosphoric Acid. H_3PO_4, containing 55 per cent of P_2O_5, is used mainly in the manufacture of triple superphosphate or ammonium phosphates. It can however, be applied directly to the soil by injection, or be added to irrigation water, and is particularly effective in the alkaline and calcareous soils that are typical of arid lands.

The Superphosphates. These are the most important commercial phosphate fertilizers, and consist of calcium orthophosphate. The three commercial fertilizers are ordinary or single superphosphate (16–22 per cent of P_2O_5), enriched superphosphates (25–30 per cent of P_2O_5), and triple superphosphates (44–52 per cent of P_2O_5). The superphosphates have no appreciable effect on soil pH. The single superphosphate consists of a mixture of monocalcium phosphate and gypsum in approximately equal amounts. The gypsum may be beneficial in its own right, especially on alkaline irrigated soils. The triple superphosphate is advantageous in areas in which transport costs are significant: in granular form, it is convenient to handle and fixation is retarded. However, under dry conditions, the powdered forms usually give the best results (S. L. Tisdale & Nelson, 1966).

Ammonium Phosphates. These are very popular in many regions because their high content of essential plant nutrients makes their transport, storage, and handling, more economical than is the case with most of their rivals. However, on alkaline soils they may release free ammonia, which is toxic to young seedlings. Proper precautions should therefore be taken in regard to placement in relation to the seed-rows. Ammonium phosphates have an acidifying effect on the soil.

Heat-treated Phosphates. A number of different types of phosphate fertilizers have been developed by heating rock phosphate; these include defluorinated rock phosphate, Rhenania phosphate, calcium metaphosphate, basic slag, and others. Such fertilizers are generally almost insoluble, and so are not suitable for the alkaline or calcareous soils of arid lands.

POTASSIUM

Role

The amounts of potassium that are absorbed by the plant as ionic K^+ are larger than those of other nutrient cations. Potassium content is about 0·5 to 2·5 per cent of the dry-weight of the plant; 2 per cent is generally considered an adequate level for near-maximum growth (Nelson & Stanford, 1958). Potassium has a high degree of mobility in the plant, moving to the meristematic tissues whenever it is required there.

Though potassium is not a constituent of important plant components such as protoplasm, fats, and cellulose, an essential role has, at one time or another been ascribed to this element in almost every important physiological process of the plant (Mulder, 1950); it thereby directly determines the rate of growth and yields. Photosynthesis may be decreased, and respiration increased, under conditions of potassium deficiency.

The main processes affected by potassium supply are:

1. Synthesis of simple sugars and starch.
2. Translocation of carbohydrates.
3. Reduction of nitrates and synthesis of proteins.
4. Normal cell division.
5. Maintainance of plant turgor.

As potassium is involved in so many vital processes, an adequate supply to the plant of available potassium is essential. Resistance to certain plant diseases, such as powdery mildew of wheat and root-rot of lucerne, is thereby increased. Potassium also plays an important role in the water economy of the plant. It increases the moisture content of plant tissues and promotes the turgidity of plant cells. The application of adequate amounts of K fertilizer was found to be more effective in improving the moisture status of sugar-cane, than was a reduction of the interval between irrigations (Barber & Humbert, 1963). Soil moisture is used more efficiently when K is plentiful, especially in periods of drought, than when it is deficient. Potassium deficiencies prevent the normal ripening of cereals and result in shrivelled grain, while applications are particularly effective on crops that are grown for starch or for sugar.

The most important contribution of potassium fertilizers in irrigation agriculture is to offset to some degree the negative effects of the heavy applications of nitrogen that are needed for high yields.

Excess amounts of potassium may also have undesirable effects. The absorption of other ions, in particular of calcium and magnesium, may be impeded, and as a result deficiency symptoms may appear.

Fixation of Potassium

When a potassium fertilizer is added to the soil, a part of the potassium is adsorbed by the soil colloids, in more or less unavailable compounds. The part so 'fixed' is usually directly proportional to the amount of colloidal matter in the soil, being greatest in heavy soils, and least in light soils. It is assumed that the potassium ions enter into inter-layer lattice positions, where they become fixed, after displacing equivalent amounts of calcium, magnesium or sodium. The exchange occurs most readily in soils in which calcium, magnesium, or sodium, predominate—a situation that is characteristic of many arid lands.

Fixation is not generally as acute a problem for potassium as for phosphorus; it helps to reduce leaching losses and luxury consumption. Fixed potassium can be considered as a reserve supply whose availability to plants is intermediate between that of exchangeable and naturally non-exchangeable forms.

As a rule, potassium uptake is facilitated by increases in soil temperature (Miller & Army, 1954).

Potassium Fertilizers

All commercial potassium fertilizers are water-soluble. The chief sources of potassium are potassium chloride (KCl) and potassium sulphate (K_2SO_4).

Potassium Chloride. This salt, called commercially 'muriate of potash', contains the equivalent of 60–63 per cent of K_2O (according to its purity) and is the most widely used form of potassium fertilizer.

Potassium Sulphate. This salt contains 50–53 per cent of K_2O. Except for certain crops, such as tobacco, which are susceptible to the Cl-ion, there is no essential difference between the behaviour of these two fertilizers in the soil or in their effect on crops. Under certain conditions, the accompanying sulphur may be an advantage to be taken into consideration. In general, however, the choice will depend on the relative cost of the K^+ ion in the two fertilizers.

Potassium Nitrate. This salt, which contains 44 per cent K_2O and 13 per cent of N, is an excellent but expensive fertilizer which is used on a relatively restricted scale.

SECONDARY AND MICRONUTRIENTS

Increasing yields due to irrigation, more productive varieties, and improved technology, have increased the rate of depletion of secondary and micronutrients. Another circumstance contributing to the more frequent occurrence of micronutrient deficiencies nowadays than formerly is that high-test, relatively pure fertilizers are increasingly replacing the bulkier nutrient carriers that were previously marketed, and that usually included some secondary and micronutrients as incidental impurities. A further factor in this connection is the decline in the use of farmyard manure for field-crops. Finally, the replacement by mineral nitrogen, of much of the nitrogen that is symbiotically fixed by leguminous plants, has reduced the role of legumes as mobilizers of soil micronutrients that are otherwise unavailable to other plants. Deficiencies are frequent in the alkaline and calcareous soils which are prevalent in many of the dry regions, and in which the availability of boron, iron and manganese, is relatively low (Jewitt, 1966).

Calcium

Calcium is absorbed by the plant as the ion Ca^{++}, and is in equilibrium with magnesium, potassium, sodium, and boron. Any upset of this balance has undesirable effects on vital physiological plant processes. These disturbances are not necessarily due to a deficiency of calcium but often it seems to a lack of nutrient balance. No clearly defined physiological functions can be ascribed to calcium; but it is a constituent of calcium pectase, which serves as a cementing material between cell-walls.

Most arid-zone soils are sufficiently rich in calcium, and many fertilizers in any case contain significant amounts of calcium for the needs of plants. However, gypsum ($CaSO_4$) may be required in relatively large amounts in alkali soils, in order to improve the aggregation of the soil colloids through the replacement of sodium by calcium in the exchange complex.

Magnesium

Magnesium is absorbed by plants as the ion Mg^{++}. Magnesium is involved in the formation of chlorophyll of which it is the only mineral constituent. It is also involved in the metabolism of phosphorus, and is required for the activation of many enzymes that are involved in carbohydrate metabolism. It has a significant effect on the oil content of a number of crops. Fortunately, there is generally no deficiency of magnesium in arid-zone lands.

Sodium

Sodium is absorbed by plants as the ion Na^{+}. It is not an essential element, except possibly for a limited number of crops that have been domesticated from wild plants that grew near the sea—such as sugar-beets. It can replace potassium to some extent, and increase crop-yields when there is a relative deficiency of the latter. There is usually no shortage of sodium in arid-land soils, and enormous amounts are added to them by irrigation.

Sulphur

Sulphur is absorbed by the plant mainly as the sulphate ion ($SO_4{}^{--}$), which is subsequently reduced within the plant. Although sulphur is not a constituent of chlorophyll, it plays an important role in chlorophyll production; moreover it is required for the synthesis of a number of amino-acids which contain sulphur, and is also a constituent of certain vitamins An adequate supply of sulphur increases the oil content of linseed, soybeans, and peanuts.

There is usually no shortage of sulphur in arid-zone soils. Many of the commercial fertilizers incidentally supply significant quantities of sulphur, and irrigation waters also usually have a high content of sulphates, which in itself may be ample for crop requirements.

Micronutrients

Most of the micronutrients are components of some key substance in plant metabolism, or are essential to the functioning of an enzyme system. Therefore, though the quantities of the essential minor elements required for normal plant growth, are extremely small, as compared with the major nutrients, a deficiency of a micronutrient may have extremely disturbing effects on vital plant processes.

When a deficiency occurs, a very small amount of a chemical containing the micronutrient concerned may suffice to restore normal plant growth; however, an excess application of certain minor nutrients may be even more dangerous to the plant than the deficiency it was supposed to cure. For several of the micronutrients, such as iron, manganese, copper, and zinc, their concentration in the plant in relation to other heavy metals is of greater importance than the absolute amounts present (S. L. Tisdale & Nelson, 1966).

Boron. This mineral element can be absorbed in a number of ionic forms. Its role is not clear, though it is thought to be necessary for some aspect of cell-wall formation. Dicotyledenous plants generally have a higher requirement for boron than do the monocotyledons (Gerloff, 1963). Boron toxicity assumes great practical importance when farmers irrigate protractedly with water having a high boron content.

The amount of boron required to alleviate a deficiency in one crop, such as lucerne, may be toxic to a more sensitive crop such as field beans (S. L. Tisdale & Nelson, 1966). Deficiencies of boron are generally rare in arid soils, and available B is well distributed through most arid-soil profiles (Berger & Pratt, 1963). When such deficiencies do occur, they can be remedied by the application of borax, applied to the soil at rates of 5–10 kg/ha. However, boron becomes rapidly fixed when applied to an alkaline soil; therefore borax has to be applied annually in small dressings, to boron-sensitive crops such as lucerne, sugar-beets and cotton, when they are grown on boron-deficient soils (Cook & Davis, 1957).

Iron. This metallic element is involved in the activation of a number of enzyme systems, while a continuous supply is necessary for the production of chlorophyll. Iron is absorbed from the soil in ionic form or in complex organic salts. Chlorosis is the typical symptom of its deficiency in plants. This does not necessarily indicate a deficiency of this nutrient in the soil, but is proof of the inability of the plant to absorb soil iron, which happens most frequently on calcareous and alkaline soils. Poorly aerated soils and high water-tables are also conducive to making the soil iron unavailable to plants. In irrigated soils, Fe deficiencies may be alleviated by improved drainage and by increasing the intervals between irrigations (Berger & Pratt, 1963). Crops differ in their susceptibility to iron chlorosis, one of the field-crops that is most prone to it being sorghum.

Copper. This metallic element is absorbed by plants as the cupric ion Cu^{++}.

It functions as an enzyme in reduction and oxidation systems of the plant, is involved in the activation of several enzymes, and possibly also in the light reactions of the plant. Its salts are apt to be poisonous at higher concentrations, and some are used as fungicides.

The copper content of most soils is very low, but a deficiency leading to crop problems is extremely rare (Mulder, 1950). Copper deficiencies have however, been widely recorded in southern Australia on a considerable variety of soils (Stephens & Donald, 1958).

Manganese. This metallic element is absorbed as the ion Mn^{++}; it is relatively immobile and tends to concentrate in the leaves of the plant. Manganese functions in the activation of a number of important enzymes, and plant growth is frequently impaired by a deficiency or an excess of manganese. As occurs with iron, a deficiency of manganese causes chlorosis; however, the leaves are somewhat darker in this latter case. In certain plants, such as cotton, an excess of manganese causes physiological disorders.

Manganese deficiencies occur quite frequently on calcareous and alkaline soils, as well as on sandy soils.

Zinc. This metallic element is absorbed by plants as the ion Zn^{++}. It plays a role in the production or destruction of a number of auxins, and a deficiency of it causes imbalance in the auxins of the plant. It is an activator of a number of enzymes. Crops differ in their susceptibility to zinc deficiencies; for example, maize will show marked zinc deficiency under conditions in which alfalfa is able to grow normally (D. I. Arnon, 1958).

Deficiencies of zinc in field-crops are relatively rare, though they may appear occasionally on alkaline soils, on calcareous soils, and on those which are excessively rich in phosphorus. Such deficiencies frequently occur in California in areas in fields that have been levelled for irrigation (A. L. Brown & Krantz, 1960). The deficiency can be overcome by spraying a crop, such as lucerne, with a 0·5 per cent solution of zinc sulphate, or, alternatively, by applying zinc sulphate to the soil at rates of 45 to 90 kg/ha (Powers & Pang, 1947).

Cobalt. This metallic element is essential for plants that rely on nitrogen fixation (S. L. Tisdale & Nelson, 1966), as it is involved in the formation of the haemoglobin needed for that process. To a certain extent, it may be replaced by vanadium (ibid.).

In Australia, cobalt deficiencies occur on fairly extensive areas.

Chlorine. This halide element is absorbed as Cl^-. For many years it was not considered essential for plants, but more recently it has been shown that this view was mistaken, and that chlorine is essential for all higher plants (Johnson *et al.*, 1957). Its function is not yet clear. However, growth increases of 80 per cent have been reported in cotton supplied with 3 meq Cl per litre in nutrient solution (Eaton, 1966).

The Cl content of arid soils is usually high, because of limited leaching and reletively high salt-content, while under irrigation, Cl salts usually predominate. The problem is therefore not one of Cl deficiency, but of direct or indirect damage to crops through its excess, which can be very detrimental to a number of plants.

Molybdenum. This metallic element is probably adsorbed as the ion MoO_4^{--}. It is required only in minute quantities, but deficiencies have been reported for a number of crops, and especially legumes—first in Australia, and subsequently in many other areas of the world (Berger & Pratt, 1963). It is required by Rhizobia for nitrogen fixation and by non-legumes for nitrate reduction (S. L. Tisdale & Nelson, 1966), and hence is important in protein synthesis.

Mo can be applied by soaking the seeds of crop plants, before sowing, in a solution of sodium molybdate.

Determining Fertilizer Requirements

Diagnostic methods, which would make it possible to advise the farmer on the kinds and amounts of fertilizers to apply to each crop grown under the specific conditions of the individual farm, are of considerable economic importance. The problems involved have been under study for many years, and a number of different approaches have been developed.

SOIL TESTS

There is no difficulty in determining the soil's content of essential elements by chemical methods. However, it is not the amounts of nutrients present in the soil that are of significance to crop production, but their availability in the proper proportions throughout the growth of the plant. This is a function not only of the soil but also of the crop, its environment, and management. It is therefore not surprising that no chemical tests have yet been devised that are able to provide an accurate assessment of the amounts of essential elements which are available to the plant in the soil, and hence the amounts that need to be added. In spite of these difficulties, by correlating the results of soil tests with the crop responses obtained in fertilizer experiments in the field, it has been possible to develop analytical methods which make it possible to predict *approximate* fertilizer needs. These tests are, however, empirical, and appropriate only to the specific soil-types for which they were developed. Their interpretation requires calibration with fertilizer field experiments, and their reliability depends on the practical experience of the interpreter. Most of the original tests have been developed for the acidic soils of the humid temperate regions and are not suitable for arid-region soils with their tendency to alkalinity. Much of the work has, therefore, to be repeated in arid regions.

There also arise numerous practical problems of sampling, analysis, and interpretation. Applying the results of soil tests is further complicated by the fact that the availability of soil nutrients is not the same for all crops; different crops are also able to obtain their nutrients from different soil-depths. Accordingly, the interpretation of the chemical tests will be different for different crops (and even for varieties of the same crop), and will vary according to their places in the crop rotation. Nitrogen availability in particular is influenced by complex edaphic, climatic, and biotic factors that frequently make a chemical soil test meaningless.

A rise in temperature of the soil may speed up nitrification and thereby release relatively large amounts of nitrogen; leaching and denitrification, on the other hand, may cause rapid losses of nitrates. Tests for potential nitrogen availability have been developed, but these methods have been adopted to only a very limited extent.

It is mainly for phosphorus and potassium that soil tests have been developed, which, after correlation with crop response data and a great deal of practical experience, can be used for fertilizer recommendations; the probability of profitable response increases with lowering of the nutrient level, as determined by analysis (Figure 66). Using the results of the soil test, and taking

Figure 66. Effect of level of fertility on the probability of obtaining a profitable response from a given nutrient. (From Fitts (1955): by courtesy of American Potash Institute.)

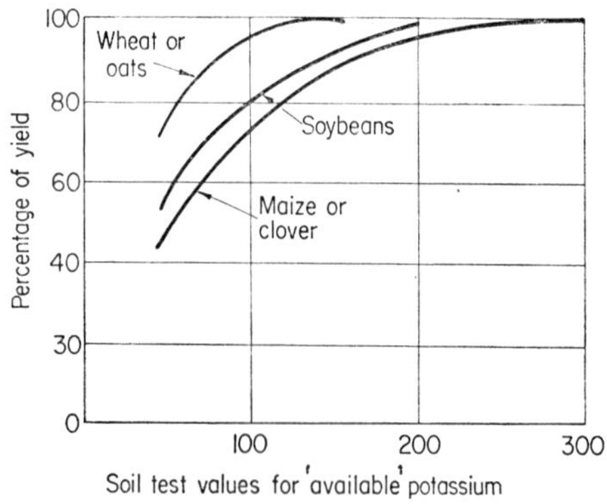

Figure 67. Relationship between percentage of yields without added potassium for different crops and soil test values for available potassium (where yield with adequate added potassium is 100 per cent). (From Viets & Hanway (1957): by courtesy of the U.S. Department of Agriculture.)

into account the nature of the crop, the physical conditions of the soil, the management of the preceding crops and their residues, together with cultivation methods, moisture régime, etc., an experienced diagnostician may be able to predict the probability of profitable crop response to fertilizers, and the approximate amounts of them that should be applied (Figure 67).

On the basis of the soil tests, the soils are classified into from three to five levels in relation to phosphorus and potassium availability. The accuracy of the tests does not generally justify more than three levels: *low*, at which an application of about double the amount of nutrients removed by each cropping is recommended in order to allow for some build-up in the soil in addition to the immediate crop requirement; *medium*, at which the amount recommended is slightly higher than that removed for phosphorus, and slightly lower than that removed for potassium; *high*, at which less than the rate of removal is recommended (Nelson & Stanford, 1958).

In some regions in the U.S.A., recommendations on fertilizer applications are being made by computerized services, which take into account soil type, yield potential, cropping sequence, subsoil moisture, the amount and frequency of precipitation, and managerial ability of the farmer, as well as the actual results of the soil tests. Any results of new research are immediately included in the programming. By these means, one hundred soil samples can be processed per minute (Walsh, 1964).

BIOLOGICAL TESTS

A number of biological tests have been developed for diagnosing fertilizer requirements by using higher plants, soil fungi, or microbes, as indicators. One method consists of growing a suitable indicator plant in the greenhouse, in pots containing a relatively small amount of the soil to be tested. The plants exhaust the soil rapidly, nutrient deficiencies being thereby revealed. Analysing the plants after harvesting, or comparing the growth of the indicator plants in containers that have received various combinations of nutrients, with the unfertilized control, make possible conclusions regarding the fertilizer requirements of the soil that is being tested. Another approach is to prepare a number of cultures from the soil to be tested, inoculate them with the appropriate micro-organism—such as *Azotobacter*, *Aspergillus niger*, etc.—and then study the response of the micro-organisms to the soil, with and without added phosphorus and potassium.

The biological methods give a better indication of the general availability of plant nutrients than do chemical soil tests, but have the same drawbacks in that they cannot take into account individual crop response, the physical condition of the soil, environmental conditions, etc.

PLANT TISSUE TESTS

As the plant itself is an integrator of all the factors affecting nutrient availability, it has long been thought that an analysis of plant tissue should reflect accurately the nutrient status of the soil.

At first the ash of the entire plant was analysed, but it was found that there were many cases in which no correlation could be established between the amounts of the essential elements in the ash and the response of the crop to fertilizers. Subsequently, it was found that analysing the tissue of a metabolically active part of the plant, such as a fresh leaf, gave a generally accurate estimate of the nutrient status of the plant at the time of testing—provided it was not growing under conditions of acute deficiency. When there is a severe deficiency of one or more nutrients, growth is stunted and the plant tissues may show a higher *concentration* of the nutrient(s) than those of a plant growing under more favourable conditions.

The tissue tests give rapid and generally fairly accurate results. The interpretation of leaf analysis data is based on the assumption that, when the concentration of a certain element in the tissues of the plant is below a certain *critical level*, the uptake of that element from the soil is insufficient.

Because a deficiency of a single nutrient may affect the uptake of non-deficient nutrients, the ratio between elements is frequently used instead of the critical level of a single nutrient.

The main disadvantage of tissue tests is that the results are obtained at a time when it is too late to correct a deficiency of phosphorus or potassium for the current crop, though a top-dressing with nitrogen might still be feasible. Plant tissue tests are therefore used for perennial crops and for high-value crops such as cotton and sugar-beets, to reveal nitrogen deficiencies. With cotton, excellent correlations were obtained in 90-day-old plants between nitrogen, phosphorus, potassium, calcium and magnesium levels in the leaf-stalks and the amounts of these elements available in the soil, at a time when no deficiency symptoms could be seen (Joham, 1951).

DEFICIENCY SYMPTOMS

Certain nutrient deficiencies cause characteristic symptoms to appear in the plant which make a visual diagnosis possible. The principal symptoms are severe stunting, leaf or stem discoloration, and delayed or advanced maturity.

The characteristic hunger-sign for nitrogen—yellowish-green leaves—is well known. It is a familiar form of 'chlorosis'. The older leaves are the first to show these chlorotic symptoms, which spread progressively to the younger leaves of the plant. Nitrogen-deficiency symptoms may be related to a deficient water régime: the upper soil-layer, which contains most of the available nitrogen, dries out rapidly between irrigations. In this case the hunger signs do not indicate a deficiency of nitrogen in the soil, but the inability of the plant to absorb nitrogen from the dry soil-layer.

Phosphorus deficiency is indicated by a dark-green colour of the seedlings, and a purplish tint of the stems and leaves. The purple coloration is due to a decrease in protein synthesis as a result of P deficiency, with an attendant increase in sugar concentration in the plant tissues. Higher sugar concentrations promote the production of anthocyanin, the source of the purple coloration. However, the symptoms usually appear at a time when it is already too late to correct the situation.

A deficiency of available potassium is characterized by stunted growth, brown, necrotic leaf-margins, and a tendency to 'lodge' because of a poorly developed root system. In lucerne, the leaves show characteristic white spots.

Chlorosis is, in many crops, an indication of a shortage of available magnesium or iron rather than of nitrogen.

The above are but a few examples of visual deficiency symptoms. Special coloured plates have been prepared, showing deficiency symptoms in a number of crops, such as those in Hambidge (1949).

Using deficiency symptoms as a diagnostic tool is simple and frequently effective; however, it has a number of drawbacks. Thus the deficiency symptoms are an indication that the plant is already under severe stress, and yields may already be affected even if nutrients are applied immediately. On the other hand, nutrient deficiencies may be insufficiently severe to cause visual symptoms, whilst still affecting yields adversely. Such a level of deficiency is called 'hidden hunger' (Garrard, 1958). Moreover, the symptoms may be misleading: different nutrient deficiencies may show similar symptoms, while similar symptoms may also be caused by waterlogging, low temperatures, disease, or damage by insects, etc. Altogether, much experience is needed for a correct assessment of nutrient supply based on visual symptoms.

FIELD EXPERIMENTS

Field experiments, that are carried out according to certain rules which ensure that a valid statistical analysis of the results can be made, are still the most satisfactory method for ascertaining crop response to fertilizers in a given environment, and are also essential for calibrating the soil and plant tests mentioned above.

The main drawback of field experiments is that, since the nutrient level in the soil is not constant, the results obtained reflect a one-time situation, that is not likely to be repeated in time or space. In order to enable generalized conclusions to be drawn from field experiments, these have to be repeated over a period of years—on all the important crops, in different regions, and under various moisture régimes.

Factors Affecting the Efficiency of Fertilizer Application

SOIL-MOISTURE RÉGIMES

In arid agriculture, the basic problem in plant nutrition is that of adjusting fertilizer applications to the moisture régime under which the plants are expected to grow. This is true for both rain-fed and irrigated crops. Basically, the problem is one of nutrient-soil moisture interactions. At one end of the spectrum, under conditions of sparse rainfall, is the need to limit fertilizer application to rates which will not promote more growth than the available soil moisture can sustain until harvest—or in other words, to prevent upsetting the very delicate and critical balance between vegetative and reproductive growth under conditions of limited moisture. At the other end of the spectrum when the farmer controls the supply of water to his crops, his aim is to ensure a level of nutrient supply that will enable the plant to make full and efficient use of the favourable moisture conditions which it enjoys.

Soil moisture affects the efficiency of fertilizer use mainly in two ways: by improving the uptake of nutrients and by increasing dry-matter production. When considering the effect of soil-moisture level on the uptake of nutrients by plants, indirect effects of soil moisture must also be allowed for—such as the effects on the physiological activities of the plant, on soil aeration, and on the osmotic pressure of the soil 'solution' (D. A. Brown *et al.*, 1960).

Effect of Soil Moisture on Nutrient Uptake

Three mechanisms are involved in determining the effect of soil moisture on nutrient uptake:

1. A favourable water-supply increases the mass and distribution of the root system; actively growing roots will take up more nutrients by exploring a greater volume and depth of soil than do others. When soil moisture is

inadequate, phosphorus uptake is impeded—probably because of reduced growth of roots towards the sources of phosphorus in the soil (Cooke, 1967). Conversely, nutrient deficiencies can restrict root development, so that uptake of water and nutrients can be impaired—especially from the lower soil-levels.

2. The flow of soil-water transports the more soluble, and therefore mobile, nutrients to the roots (nitrates, sulphates, etc.).
3. The less mobile nutrients, such as phosphorus and potassium, reach plant roots partly through diffusion. As plant roots take up the nutrients in their immediate vicinity, these may be replaced by ions that diffuse slowly, from areas of higher concentrations to the depleted zone around the roots. Though the distances involved are very small, at most 5–6 mm, this is an important mechanism of nutrient supply to the roots. The diffusion occurs through the films of water surrounding the soil particles. As the soil becomes progressively drier, the water films around the soil particles become thinner, thereby reducing the rate of diffusion of nutrients to the roots.

When soil moisture increases from wilting point to field-capacity, almost linear increases in the absorption of N, P, K and Ca, by cotton and soybean plants, have been observed (D. A. Brown *et al.*, 1960). Similar results have been obtained by Watanabe *et al.* (1960), working with maize seedlings (Figure 68). Uptake of P decreased to 80 per cent at 1 bar,* and to 50 per cent at 3 bars moisture tension relative to uptake at $\frac{1}{3}$ bar. We may note that the uptake of P follows a curve closely resembling the moisture desorption curves for the same soil. The uptake of nitrogen, which is more mobile than potassium and phosphorus, is less affected by soil-moisture stress than that of the two latter macronutrients.

Several investigators, working under a wide variety of conditions, report that potassium availability increases as soil-moisture level increases. The results obtained by Barber (1959), who found a correlation of 0·97 between relative yield without potassium and the amount of precipitation during the active growing period of maize (June–August), are shown in Figure 69.

It is assumed that in years of high rainfall, the top-soil layer, which is relatively rich in available potassium, remains more continuously moist than lower layers, and therefore that the potassium supply in it is sustained at a satisfactory level. In work on fertilizer uptake by grain sorghum under dry-land conditions in the southern Great Plains region of North America, it was found that phosphorus uptake practically ceased when the soil-moisture content approached wilting point. The amount of phosphorus that was absorbed

* *Bar*. A unit of pressure in the metric system, equal to one million dynes per cm^2 and slightly less than one atmosphere.

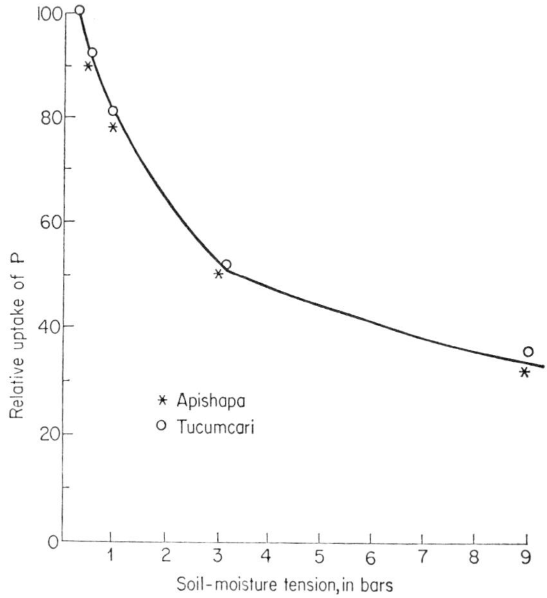

Figure 68. Relative uptake of phosphorus by maize seedlings in relation to soil-moisture tension for two soils, named 'Apishapa' and 'Tucumcari'. (After Watanabe *et al.* (1960): by courtesy of Elsevier Publishing Company.)

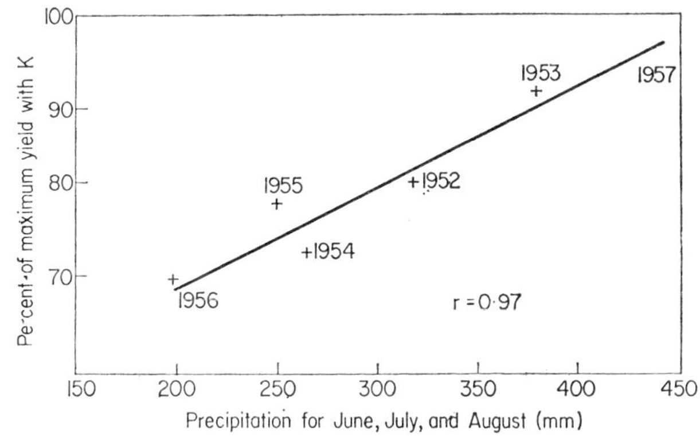

Figure 69. Relationship between the precipitation for June, July and August, and the relative yield of maize without potassium. (After Barber (1959): by courtesy of the American Society of Agronomy.)

as long as soil moisture was not limiting, did not suffice for maximum dry-matter production. Nitrogen absorption was less readily influenced by the depletion of soil moisture, and probably continued even after the soil had reached the wilting point (Eck & Fanning, 1961).

The general conclusion can therefore be drawn that when the soil-moisture tension is low, the abilities of the plant to absorb the nutrients, and of the soil to supply them, are optimal, and therefore nutrient availability is at its highest level (Viets, 1967).

Increased Dry-matter Production

When the water-supply is deficient the stomata close, uptake of carbon dioxide is reduced, and photosynthesis decreases. Less dry matter is produced and fertilizer nutrients are less efficiently utilized than when water is plentiful. This interaction of moisture-supply and nutrient-supply is reciprocal.

The interrelation between soil moisture and fertilizers (or their effects on yields) has been most intensively studied. Only a few typical examples for some crops can be cited. In experiments carried out in Mexico, irrigating an improved disease-resistant variety of wheat, grown on a deep alluvial soil that was deficient in nitrogen, increased yields by only 5 per cent, although soil moisture was maintained at the optimal level. When nitrogen was added to the driest soil, yields were increased to 3·7 times the former level, but when the same amount of nitrogen was added in combination with optimum soil-moisture conditions, yields were increased almost sevenfold (Richardson, 1960).

The following results, obtained by Parks & Knetsch (1959) for maize, are typical of many experiments of this kind. When irrigation water was applied without modification of current practices, including nitrogen fertilization at 67 kg/ha, there was little or no increase in grain yields. As the fertilizer applications were increased, two important relationships were found: (a) limited supplies of irrigation water were used more efficiently, and (b) the increased supply of irrigation water was used with even greater efficiency. The highest yield obtained in this experiment was 8 190 kg/ha. In Figure 70 the effect of the typical interactions between nitrogen fertilization and soil-moisture régime on the yield of maize grain is strikingly evident. These results were obtained on a loess soil with a low level of inherent fertility (Shimshi, 1966).

On cotton, a well-defined interaction between soil moisture and fertilizers has been established (Scarsbrook *et al.*, 1959). At low moisture-levels there was no response to fertilizing with nitrogen at rates above 67 kg of N per ha. At the intermediate moisture-level, there was a response to 134 kg of N per ha, and at the highest moisture-treatments, cotton responded to all the levels of N tested—including 268 kg/ha. Conversely, the nitrogen applied increased the efficiency of water utilization. At the highest moisture-level tested, the cotton lint produced per mm of water applied was about doubled by applying 268 kg of N per ha, as compared with 67 kg N per ha (Table XIX).

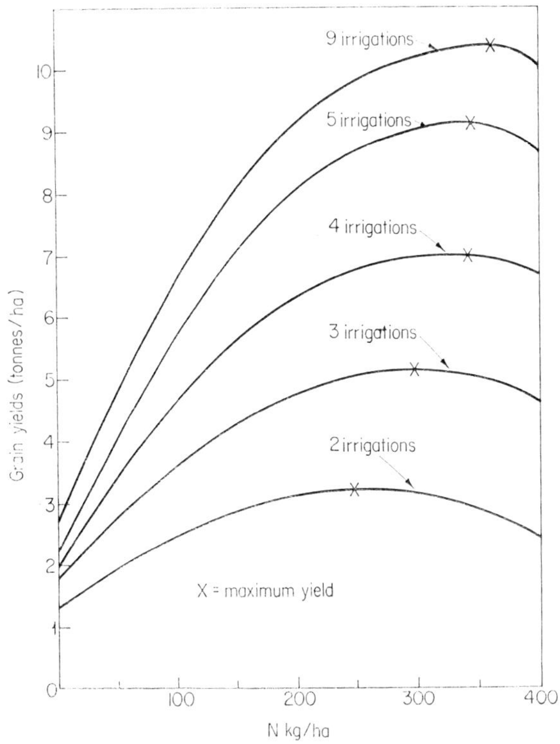

Figure 70. Effect of nitrogen fertilization at different soil-moisture régimes on grain yields of maize. (After Shimshi, 1966.)

TABLE XIX

Effect of Nitrogen Fertilization on Efficiency of Water Utilization in Cotton

Rate of N application (kg/ha)	*kg of lint produced per mm of water*
67	0·52
134	0·72
201	0·81
268	1·02

Haddock (1949), working on sugar-beets, found that when soil fertility was held constant, irrigation could increase yields by 17·5 tonnes/ha. Conversely, when soil-moisture stress was held constant, fertilizers were able to increase yields by 15 tonnes/ha. Improving the soil-moisture relationship and soil-fertility conditions simultaneously, increased yield by 30 tonnes of beets per ha.

Stanberry *et al.* (1955) studied the effects of three moisture régimes and 10 rates and/or frequencies of phosphate applications on the yields of lucerne in

Arizona. They found that a progressive increase in lucerne production, with added increments of P_2O_5, occurred for each moisture treatment, and that, for a given amount of phosphate applied, yields increased as soil-moisture tension was decreased. The yield of hay over a three-year period from low-fertility, low-moisture plots was 16·1 tonnes/ha, compared with 36·1 tonnes/ha in high-fertility, high-moisture plots.

In Nebraska, under conditions in which moisture was *not* limiting to yields (over 600 mm of rainfall), water-use efficiency of grain-crops (wheat, oats, maize, sorghum) was found to increase by an average of 29 per cent following judicious use of fertilizers. The fertilized crop consumed only an extra 25 mm of water (Olson *et al.*, 1964).

In other trials in Nebraska, it was found that maximum yields on low-nitrogen soils resulted from the application of 22 kg of N per ha for every 75 mm of available pre-planting moisture (Figure 71; Ramig & Rhoades, 1963).

Many more experiments of this kind have been reported in the literature, with extraordinarily consistent results, showing that nutrient and water requirements are intimately linked, and that fertilization increases the efficiency with which crops use available water; furthermore, favourable water régimes make possible fertilizer applications at levels which would be entirely ineffective under conditions of moisture stress (Figure 72).

Fertilizer Application under Conditions of Low Moisture-supply

The Problem. A special problem arises on the fringe of the semi-arid regions in which rainfall is normally low and droughts are frequent. In these regions the effects of fertilizers under varied conditions of precipitation have not been studied as intensively as in the humid regions; many of the fertilizer experiments that have been carried out in semi-arid regions have not been sufficiently reliable for predicting crop response to fertilizers. The results of the fertilizer trials are erratic, and often fail to correlate with chemical soil tests (Power *et al.*, 1961*b*). Therefore, no clear-cut answer has yet been formulated to the question of whether and when fertilizer application is justified under conditions of low moisture-supply.

The following example gives an indication of the difficulties with which farmers and researchers are faced in deciding on fertilizer policy under these conditions. In one season, yields of wheat were increased significantly by fertilizer application in Oklahoma in seven locations out of eight; in the following season, which was much drier, a yield increase with fertilizer application was obtained in only one location out of these eight (Eck & Stewart, 1954). Long-term weather records may be helpful for calculating the probability of a good or a dry season, so that the farmer may know what are the 'odds' of fertilizer applications being profitable.

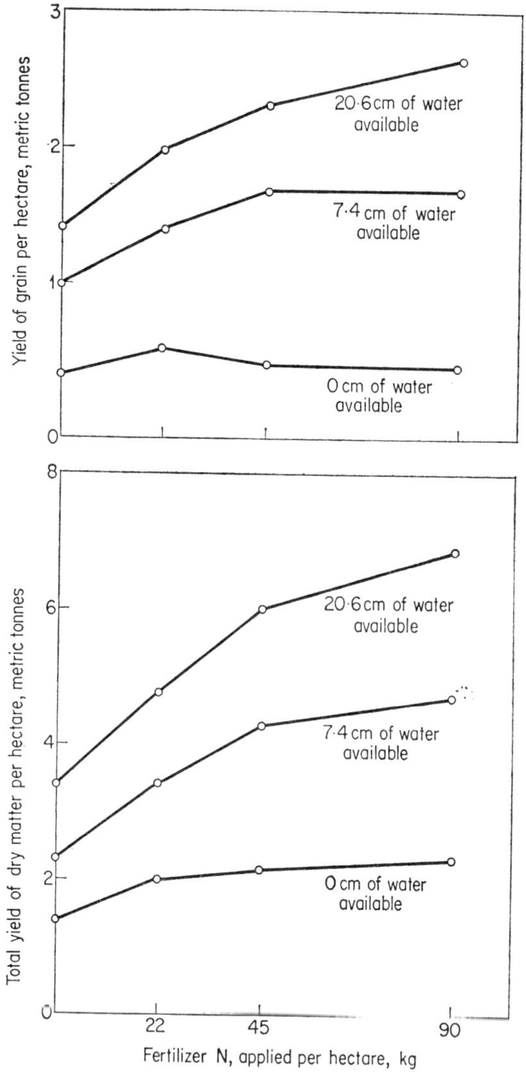

Figure 71. Yield of grain and total dry matter of winter wheat in western Nebraska as influenced by rates of nitrogen application and the amount of available soil-stored water at sowing time. The different quantities of water were supplied by pre-irrigation, and correspond to zero, 61 cm, and 183 cm, of soil wetted to field-capacity, respectively. (After Ramig & Rhoades (1963): by courtesy of the American Society of Agronomy.)

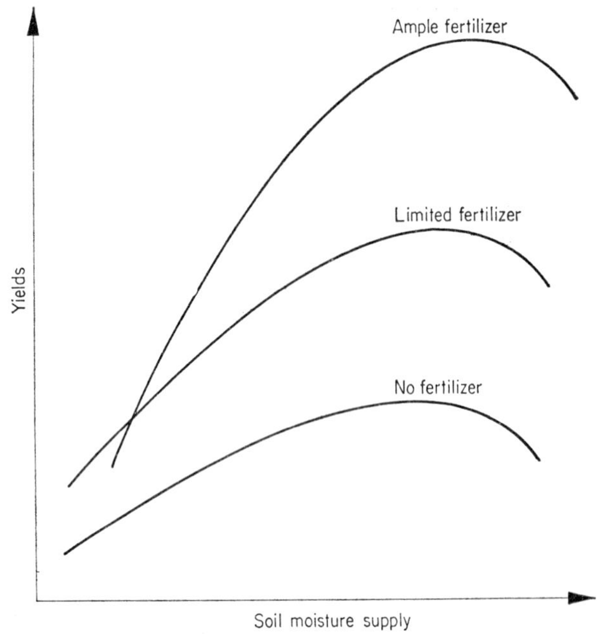

Figure 72. Schematic graph showing generalized crop response to levels of fertilization in relation to soil-moisture régimes.

Less guesswork is involved when the crop is not dependent exclusively on precipitation that falls during the growing-season. In research carried out in Kansas on sorghum, for example, it was found that the greater the depth of soil moisture occurring at seeding time, the greater was the effect of nitrogen on yields (Orazem & Herring, 1958). The depth of soil moisture at seeding time can be determined without difficulty, and this information will therefore reduce the amount of uncertainty involved in deciding on whether to apply fertilizers, and, if this decision is in the affirmative, at what rates.

Under the distinctive climatic pattern of the dry regions, with their relatively limited and erratic rainfall and long drought-periods, certain patterns of fertilizer use have evolved. Farmers have frequently been disappointed, under conditions of limited moisture-supply, by crop response to fertilizers; in certain cases their use was actually detrimental to yields.

Under such conditions, an extremely cautious fertilizer policy has evolved, and even progressive farmers usually employ very limited rates of application. However, when judiciously applied, fertilizers make possible increased utilization of the limited moisture available to plants (pp. 390–7).

Minimum Precipitation Required. When soil moisture is definitely inadequate, crops may fail to respond to fertilizers, or the response may be too small to be economic. The minimum amount of rainfall needed to ensure a fairly satisfactory yield-level of grain-crops, which in turn makes possible an economc return on the fertilizers applied, depends on many factors, such as the efficiency of the rainfall, moisture reserves in the soil, preceding crop history, temperature, evaporation, etc. An empirical and somewhat arbitrary figure is 300 mm of rainfall in a winter-rainfall region, combined with soil-moisture reserves from fallow or else a moisture-conserving crop. Table XX gives examples.

TABLE XX

Relation of Rainfall and Crop Sequence to Response of Winter Grain to Nitrogen Fertilization (Martin & Mikkelsen, 1960)

	Per cent of tests responding to nitrogen with seasonal rainfall of:		
	< 250 mm	250–300 mm	> 300 mm
Annual barley	20	46	96
Fallow barley	30	35	56
Fallow wheat	31	50	56
Average	27	44	67

These results show clearly the effect of both rainfall and fallow on the crops' response to fertilizers. With less than 250 mm of rainfall, moisture is the only limiting factor, and less than one-third of the tests showed a response to nitrogen—and even then fertilization was not profitable. When rainfall was in excess of 300 mm, grain yields were almost always increased under continuous cropping; however, when the crops were grown after fallow, which allowed a build-up of available nitrogen, not much more than half the tests responded to fertilization, notwithstanding ample soil moisture.

Timing and Balanced Fertilizer Application. Fertilizers, especially nitrogen, may stimulate early growth and thereby exhaust the soil-moisture supply before the period of maximum water requirement of the crop. This is illustrated by the results of research in California, summarized in Table XXI.

TABLE XXI

Cereal Grain Response to Applied Nitrogen under Different Moisture Conditions in California in 1960 (Luebs & Laag, 1964)

Location	Crop	Available soil moisture at planting (mm)	Rainfall after planting (mm)	Soil nitrogen in 0–15 cm depth at planting (kg/ha)	Yield of grain without N (kg/ha)	Additional yield due to N (kg/ha)
1	barley	155	158	24·2	1584	–253
2	barley	25	50	4·4	319	0
3	barley	85	88	6·6	385	+121
4	wheat	175	123	8·8	1166	+594

At location 1, the moisture-supply was fairly constant but initial soil nitrates were high; nitrogen application resulted in reduced yields. At location 2, the moisture was deficient throughout the growing period and yields were low, applied nitrogen having no effect. At location 3, initial nitrogen was relatively low and moisture content moderate, rainfall during the season being also low; yields were low, but addition of nitrogen increased them by 31 per cent. The greatest response to added nitrogen was, however, at location 4, where soil nitrates were low at the time of planting, and moisture conditions were fairly constant throughout the growing period.

Occasionally, early stimulation of crop growth may enable a more rapid and deeper penetration of the roots into the subsoil, so that the fertilized crop is able to draw more effectively on subsoil reserves of water that have been accumulated during a previous period of fallow; or possibly the stimulation may favour a more active root system, with greater water-extraction ability (Smith, 1954). Phosphorus has been found to improve yields of wheat under conditions of relatively low rainfall, provided the drought is not too severe, by improving early growth and vigour, favouring increased root-development, and advancing the maturity of the crop (Norum, 1963).

Under the semi-arid conditions of Montana, the response of wheat to P fertilization has been found to be inconsistent and of variable magnitude, though yield increases have frequently been recorded. On soils that are very deficient in P, yield increases were obtained even when the moisture-supply at seeding time was low. Yield increases were in direct proportion to available soil-moisture at seeding time (Power *et al.*, 1961*a*).

Similar research results were obtained with phosphorus in South Australia. In a year of extreme drought, on a phosphate-deficient red-brown earth, yields of wheat were extremely low, and yet response to phosphoric fertilizers, in relative terms, was very striking, as is shown in Table XXII.

TABLE XXII

Effect of Phosphorus Application on Yield of Wheat grown under Drought Conditions

(Piper & Vries, 1964)

	Yield (kg/ha)	*In percentage of control*
Control	8·1	100
12·4 kg P_2O_5/ha	23·6	289
24·8 kg P_2O_5/ha	46·6	574

Nitrogen can also be effective under conditions of limited moisture. In Nebraska, wheat fertilized with nitrogen was found to remove 25–50 mm more water from the 2-m soil profile than did the non-fertilized wheat. The additional soil moisture was removed at tensions greater than 15 atmospheres, presumably as a result of the increased root growth stimulated by the nitrogen (Ramig & Rhoades, 1963).

Generally speaking, a balanced nutrient-supply is beneficial even under conditions of limited rainfall, as it actually enables the crop to make more efficient use of the limited soil moisture available. By applying only a phosphoric fertilizer at the time of sowing, and withholding the nitrogenous fertilizer until the crop, by its appearance, shows the need for a stimulant, excessive initial growth can be avoided.

This was found to be the case in Nebraska, where water-use efficiency in rain-fed grain-crops was increased by an average of 29 per cent, when nitrogen application was delayed so as to coincide with the period of maximum requirement. Excessive vegetative development of the young crop was thereby avoided, and maximum efficiency in water utilization resulted (Olson *et al.*, 1964).

When nitrogenous fertilizers are applied in this way, they actually advance crop maturity by as much as 10–14 days, and unfavourable crop reactions occur only when excessive and unbalanced nitrogenous fertilizer is applied. However, when the rainfall is insufficient to ensure a satisfactory crop, the additional yield resulting from the fertilizer may be too small to be economically justified.

The use of fertilizers on forage crops is less critical than on grain-crops under dry conditions. Their growing period is shorter, and any stimulation of growth improves yields—irrespective of when the stimulation occurs

Many of the forage crops are legumes, to which only phosphoric fertilizers need be applied. As has already been pointed out, except in cases in which drought is so severe as to prevent normal absorption of the nutrients, phosphorus is always beneficial in improving drought-resistance.

In soils which are deficient in potassium, and which are the exception in semi-arid regions, it was found that large responses to fertilizer potassium are obtained when rainfall is low. The responses to the fertilizer decreased as rainfall increased, presumably because of increased availability of native soil K (Paauw, 1958).

Trends. In the dry regions of the U.S.A., the general trend is for a more and more widespread use of fertilizers, even where it was formerly taken for granted that fertilizer use was detrimental (Viets, 1967). This is due, in part, to nutrient deficiencies becoming more acute, as the original native fertility of the soil is gradually depleted through continuous cropping. To the north, in Canada, soils in southern Saskatchewan have been cultivated for about fifty years and are still fertile; the depletion of the plant nutrients has been slow, largely because moisture deficiency limits crop yields (Warder *et al.*, 1963). In recent years, however, large responses to phosphate and nitrogenous fertilizers are being obtained, even with a limited moisture-supply—especially on soils which are cropped continuously to wheat (ibid.). On soils that are low in available phosphorus, consistent responses from P fertilizer are obtained, even under variable and limited soil-moisture conditions (ibid.).

An additional factor in promoting these favourable responses is that moisture losses have been reduced by improved tillage practices and chemical weed-control, so that a more favourable moisture régime increases crop responses to fertilizers.

Soil Moisture and Micronutrient Uptake. Little is known of the effect of the moisture régime on the uptake of micronutrients, although certain findings indicate that it may be important. For example, it is known that drought may cause temporary deficiencies of boron. This may be due to the reduced decomposition of organic matter which normally releases boron, or to reduced boron uptake as a result of the drying out of the upper soil layers in which the main activity of the roots is concentrated.

LIGHT INTENSITY

A very close relationship between light intensity and the response to various levels of nitrogen fertilization has been demonstrated in rice (Figure 73; cf. Stansel *et al.*, 1965). As light intensity was reduced, yields become lower at all the rates of nitrogen application that were tested. Light determined the amount of N that could be effectively used: with 100 per cent light, all rates of N, including the highest (134 kg/ha), were effective in increasing yields.

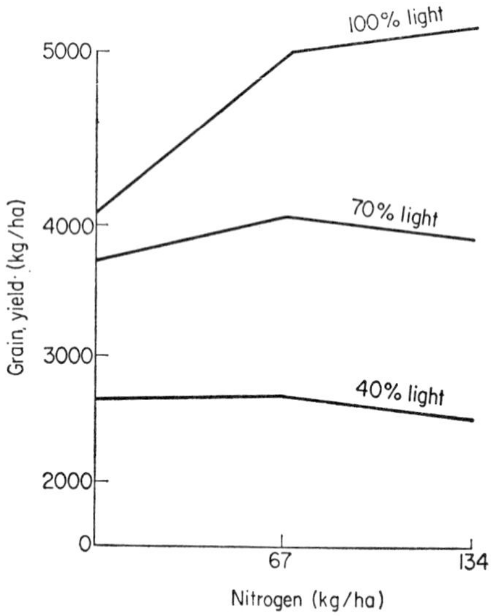

Figure 73. Rice yields at different light intensities and nitrogen fertility levels. (From Jackson (1967): by courtesy of *World Farming*.) Based on data from Stansel *et al.* (1965).

When light was reduced to 70 per cent, yields were increased by N up to 67 kg/ha, though a higher rate of application reduced yields. With 40 per cent light, the application of N at a low rate had no effect on yield and the heavy rate caused slightly reduced yields.

The later in the life-cycle light intensity is reduced, the larger will be the reduction in grain yield recorded. The effects of light and nitrogen were found to be similar in a number of widely divergent locations (which included California and Texas), three different varieties, and two seasons, showing that light may influence nitrogen utilization and yield over a wide range of environments (Stansel *et al.*, 1965).

SOIL AERATION

An expenditure of energy is required for the plant to take in ions from the external medium against a concentration gradient. This energy is provided by root respiration which, in turn, requires oxygen. Therefore, factors that impair the aeration of the soil, such as waterlogging, soil compaction, etc., will have an adverse effect on the uptake of nutrients—particularly from clay soils.

HEREDITARY CHARACTERISTICS

Not only are there considerable differences between plant species in their ability to take up mineral nutrients from the soil, but there may be marked differences between cultivars of the same crops in this respect, possibly based on differences in the root system. It has been shown that the higher the cation exchange capacity of a plant's roots is, the more effective will the plant be in using soil phosphorus (Drake & Steckel, 1955).

Different cultivars, grown under identical conditions of nutrient-supply, may have greater differences in mineral content than the differences occurring within a cultivar receiving different amounts of fertilizers. Strains and cultivars within the same species may also differ in their susceptibility to toxicity from high concentrations of an element (Gerloff, 1963).

In a comparison of two Australian wheat varieties, Bencubbin and Charter, it was found that Bencubbin has a greater tolerance of low soil-phosphorus status than Charter, and has the capacity to give a greater yield of grain though of lower phosphorus content. The difference between the two varieties may be ascribed to the fact that Bencubbin was bred in an area with low soil-phosphorus levels (Lipsett, 1964).

A differential varietal response to phosphate fertilizers was found in barley growing in Saskatchewan (Mitchell, 1957). The non-responsive variety was apparently better able than others to absorb phosphate when this was available in limited quantities, but failed to maintain a high absorption rate when the supply was ample.

Rice varieties of the species *Oryza japonica* give a striking response to increasing fertilization in their grain yield, whilst the *O. indica* varieties are better adapted to low levels of fertility, and give the highest yields of all varieties when no fertilizers are applied. At high levels of nitrogen, the *O. indica* varieties showed a great increase in sterility (Angladette, 1966).

A differential response to different nitrogen levels has also been found between maize hybrids, and has been ascribed to the higher level of nitrate reduction activity of the more efficient hybrid (Hageman *et al.*, 1961).

Two strains of cotton, Acala 4–42 and Acala C 15–17, were found to differ considerably in their susceptibility to potassium deficiency, the former presenting a normal appearance at potassium levels at which the latter showed the characteristic leaf symptoms associated with potassium deficiency (Gutstein, 1963).

Certain strains of soybeans are more susceptible to iron deficiency than others; this trait has been ascribed to a single recessive gene (J. C. Brown *et al.*, 1967). In India, local varieties of sorghum were found to be relatively insensitive to applications of N and to variation in the density of their populations, whilst hybrids were very sensitive to both factors (Doggett, 1965).

Many of the varieties that have been developed through natural or oriented selection, under the conditions of low fertility that are characteristic of primitive agriculture, are simply not capable of making efficient use of fertilizers; the economic use of fertilizers may therefore be dependent on replacing the primitive varieties by others which are capable of responding to high fertility levels. In general, varieties with a wide range of adaptation do not tend to show significant variety × fertilizer interactions (S. L. Tisdale & Nelson, 1966).

PLANT POPULATION

The optimum number of plants per unit area depends on a number of factors, among which availability of plant nutrients is of considerable importance. When plants are widely spaced, they may be able to obtain, from the relatively large volume of soil available to the individual plant, an adequate supply of plant nutrients. However, when moisture is not limited, high yields can be achieved only at plant densities at which competition between plants for nutrients is relatively severe; an adequate supply of nutrients is therefore essential under these conditions.

TIMING OF FERTILIZER APPLICATION

The proper timing of fertilizer application aims at obtaining the maximum response from the crop, by making the nutrients available at the time and rate at which they are most needed. However, the greater the number of fertilizer applications there are during the growing period, the greater will be the costs of application.

Proper timing of the fertilizer application may sometimes have an effect as great as the actual amounts applied. The need to supply phosphoric fertilizers early in the life of the plant has already been stressed. The practice of applying fertilizers in split-applications—part at the time of sowing and the rest at different growth-periods—has become part of accepted agricultural practice. In particular, under irrigation, and especially for N fertilizers, this practice appears to be based on sound theoretical considerations. However, many experiments on maize, cotton, sugar-beets, groundnuts, and other crops, grown under overhead irrigation, have shown no advantage whatsoever from split-applications as compared with a single application prior to sowing (I. Arnon, 1963). This is probably due to the more complete control of water-supply that is made possible by overhead irrigation, as compared with furrow irrigation or flooding. Downward movement of the fertilizers, as far as it occurs, corresponds to the development of the root system—and should be more beneficial than negative, as long as the fertilizers are not leached beyond the reach of the roots.

Another possible reason for the lack of effect of split-applications could be the high rates of nitrogen applied to most crops that are grown under intensive irrigation practice. In research carried out in Nebraska with irrigated maize, it was found that the time of nitrogen application was extremely important when low rates were used, but had little effect when rates in excess of 90 kg of nitrogen per hectare were applied (Olson *et al.*, 1960).

The whole amount of potassium required by the crop is usually given in a single application prior to, or at, the time of sowing. Losses by leaching are minimal, and even the potassium that becomes immobilized in the clay lattices of the soil is gradually released to the crop.

Phosphorus is fixed very shortly after application, and is then gradually released to the crop. Split-applications are therefore hardly more effective than single, relatively heavy dressings before sowing.

TECHNIQUES OF FERTILIZER APPLICATION

The principal methods of applying fertilizers are *directly to the soil*, by broadcasting or by localized placement, *indirectly*, through irrigation water, and *directly to the plant*, by spraying on the foliage. The choice of method depends on a number of factors, such as the kind of crop and the stage of development, the characteristics of the fertilizer and the amounts required, the fixing power of the soil, the size of the area to be fertilized, and the season of the year (e.g. whether the fields are accessible to ground machinery or not).

BROADCASTING

Broadcasting of fertilizers may be carried out before ploughing the land, before sowing, or on the growing crop. Broadcasting followed by ploughing makes possible an even distribution of the fertilizer throughout the ploughed layer, in which plant roots are most active. This method may be of particular importance for incorporating part of the phosphorus and potassium in soils that are deficient in these elements, as the first step in a build-up of sufficient reserves. It is also advisable when there is need for heavy rates of application which might harm the crop if they were applied in localized placement. It has been found that in alkaline soils, the thorough incorporation of the phosphorus fertilizer is of particular importance (Stanberry, 1959).

Broadcasting before sowing is used mainly for easily leached nitrogenous fertilizers, or in cases in which a light levelling of the soil follows the ploughing. The fertilizer is then disked or harrowed in. However, for summer-growing crops it may be advisable to incorporate nitrogen fertilizers at a greater depth, in a more continuously moist zone, so that the nitrogen remains available during the period between rains or irrigations when the top level of soil dries out.

Localized Placement

The volume of soil that is in intimate contact with the roots and root-hairs usually varies between 0·1 per cent and 5 per cent of the whole (Wiersum, 1962). This indicates the importance of water and nutrient movement by diffusion toward the roots, and of fertilizer placement for nutrients—especially those with low mobility, such as phosphorus. The merits of localized placement for phosphorus fertilizers have already been discussed (pp. 368–9). Localized placement is usually more effective than broadcasting fertilizers, especially when moderate amounts of phosphorus and/or potassium are to be applied to widely spaced row-crops. Experience has also shown that applying a small portion of the total nitrogen requirement in the proximity of the seed-row, greatly enhances seedling growth and increases overall efficiency of the nitrogen fertilizer (Grunes, 1959).

Special equipment has been devised that can place the fertilizer band at the sides and slightly below the seed-row (Plate 25).

The effective volume of the fertilizer band is not more than one-hundredth to one-thousandth of the total volume of soil occupied by the roots—depending on the stage of development of the crop (Ohlrogge *et al.*, 1957). However, the most extensive root systems develop in soil in which plant nutrients are most abundant (Wilkinson & Ohlrogge, 1962).

Notwithstanding the fact that only a small fraction of the root's surface is in contact with the banded fertilizers, the rapid proliferation of the roots in the fertilizer band enables there to be a high recovery-rate of the plant nutrients in the band. This proliferation of roots is, however, dependent on the presence of ammonium nitrogen *and* phosphorus in the band. The combination of nitrogen and phosphorus in the fertilizer bands not only led to greatly increased numbers of roots, but these last were also finer and more hair-like, so that a considerable increase in root surface area resulted (Duncan & Ohlrogge, 1958). When either nitrogen or phosphorus is applied alone in the band, recovery is much lower—even when the second element is abundant in the rest of the root system.

As a result of competition between the cations and anions of the fertilizers in the band, and their reactions with soil ions, the source of nitrogen may affect the availability of the phosphorus (Starostka & Hill, 1955). This may explain why ammonium sulphate was found to improve phosphorus uptake whereas calcium nitrate had the opposite effect (Figure 74; Ohlrogge *et al.*, 1957).

Another point of interest is that, although the concentration of salts in the band may exceed the accepted limits beyond which damage is normally caused to the plants, the fact that only a very small part of the root system is involved probably explains why generally no harmful effects are experienced (Duncan & Ohlrogge, 1958).

However, placing fertilizers in too-close proximity to seeds may cause a

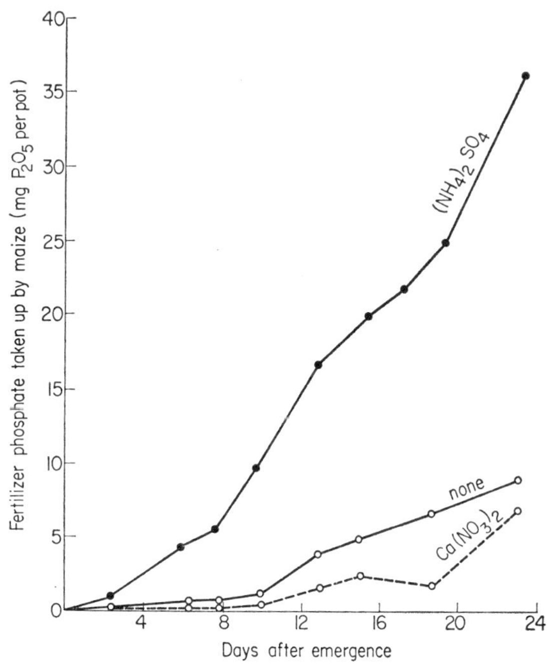

Figure 74. Effect of source of nitrogen on the uptake of phosphorus from a fertilizer band. The nitrogen-containing compound is mixed into the fertilizer band. (From Ohlrogge *et al.* (1957): by courtesy of the American Potash Institute.)

delay in germination or even reduce emergence markedly. This danger is enhanced under conditions of soil-moisture stress.

Numerous field experiments have shown that the most satisfactory results are obtained when the fertilizer band is 5–7 cm from the seed-row, and about 5 cm beneath it; this placement ensures efficient utilization of the fertilizer, and minimizes the risk of damage to the germinating seed and the seedling.

Broadcasting on the standing crop (top-dressing) is mainly used for applying nitrogen to the crop during the period of peak requirements, or after heavy rains have leached the soil nitrogen beyond the root system.

In recent years, top-dressing of fertilizers with a high nitrogen content (e.g. urea) by aircraft, has been used extensively—especially when the soil is too wet for conventional fertilizer equipment, or when large areas have to be treated in a relatively short time. It has often been found convenient to combine aerial fertilizer application with insect or weed control.

Applying Fertilizers in Irrigation Water

Applying fertilizers in irrigation water is very attractive to the farmer as it makes for economy in labour, savings in equipment, and faster crop-responses. The method is becoming increasingly popular in many countries. The application of phosphoric acid and nitrogen in irrigation water is being used quite extensively. Special metering devices are available for applying the desired rate of fertilizer. However, where water distribution is not uniform, fertilizer application in solution will further aggravate the negative results of faulty irrigation techniques.

When overhead irrigation is used, the fertilizer solution should be added to the water at the beginning of the irrigation period; it is then, subsequently, washed off the plants by clear water, and damage to the plant tissues is thereby avoided.

Foliar Application of Fertilizers

Experiments have shown that phosphorus, nitrogen, and potassium, in solution, when sprayed on the foliage, are easily absorbed and spread rapidly to all parts of the plant. Spraying fertilizer solutions directly on the foliage of a crop has the advantage of avoiding the problems of fixation, loss of availability, and losses by leaching, which occur when fertilizers are applied to the soil.

The main drawback of this method is that, in order to apply a major nutrient in a quantity which can have a significant effect on the crop, a large amount of dilute solution is needed—more than the foliage is capable of absorbing—while increasing the concentration of the solution causes scorching of the foliage and damages the crop. The other alternative is to spray a dilute solution frequently, but this will make fertilizing labour-consuming and costly.

These limitations do not apply to micronutrients. Deficiencies of iron, manganese and zinc, are frequent in alkaline soils, and are not necessarily due to an absence of these elements in the soil, but rather to the chemical reactions which make them unavailable to the plant. Applying these minor nutrients to the soil therefore does not usually relieve the deficiency, whereas spraying directly on the foliage is usually highly effective.

Fertilizing the Subsoil

In irrigation farming the upper 15–20 cm of the soil are periodically dried out during the intervals between irrigations. It is evident that nutrients, normally applied in the upper horizon of the soil, will pass through similar cycles of periodical unavailability. Thus, water applications not only improve the moisture status of the soil layer in which the most active and abundant roots are usually found, but also make available nutrients that had become positionally unavailable to plants.

Deeper placement of fertilizers in irrigation farming has the advantage of avoiding these periodical cycles of availability and unavailability of nutrients due to soil dryness.

Subsoil moisture plays an important role in irrigation farming through serving as a steady source of water-supply, thereby offsetting the moisture fluctuations of the top-soil. It is good irrigation practice to wet to its full depth, even prior to sowing, the whole soil-horizon which will eventually be penetrated by the root system. This practice will, however, be ineffective if conditions are not favourable for the penetration of roots into the subsoil. Experience has shown that simply improving the physical conditions of the subsoil—even the breaking up of impervious layers between soil and subsoil —is not sufficient, though it may be noted that adequate soil fertility may enable plants to overcome compact layers.

Fertility levels are, on the whole, lower in subsoils than in surface soils—especially as regards readily available nutrients (Winters & Simonson, 1951).

Gliemeroth (1953) has shown that the fertility of the subsoil may greatly affect the utilization of subsoil water. He found that subsoil fertilization doubled the proportion of roots occurring in the soil layer at a depth of 36–54 cm as compared with more shallow fertilization.

These results have been confirmed by other workers. Kohnke and Bertrand (1956) found that maize roots did not penetrate deeply into a compacted silty clay-loam, mainly because of lack of oxygen. Subsoiling* alone improved root development under these conditions, but fertilization of the subsoil gave even more striking results.

The adage that the livelihood of farmers is derived from the 18–20 cm of

* Subsoiling . . . loosening the soil to an appreciable depth without inversion (see pp. 426–7).

top-soil is certainly not justified in irrigation farming; the full possibilities of subsoil fertilization have not yet been developed in practice, or even thoroughly investigated.

However, it can be safely concluded that whilst keeping the entire soil-profile moist is good farming practice, this will not be fully effective unless a proper moisture-fertility balance is maintained.

INTERACTIONS BETWEEN NUTRIENT ELEMENTS

The statement is commonplace that a balanced supply of nutrients is essential for optimum production, yet it is unfortunately true that this is not always reflected in agricultural practice. Especially in irrigation agriculture, the high fertilizer rates that are normally used to ensure high yields, increase the likelihood of creating an imbalance. It is also a moot question as to whether a balance among only the three main macronutrients—N, P and K—is adequate.

'Plant growth is a function of two variables: of nutrition intensity and balance' (Shear *et al.*, 1946). Not only is maximum yield dependent on optimum intensity occurring in combination with a proper balance of all the nutrients, but the effect, on yield and quality characteristics, of changing the level of supply of one or more nutrients, will clearly depend on the amount of imbalance created in relation to other essential elements. In many cases, the minimum amount of a nutrient for normal growth is not an absolute value, but depends on the relative amounts of certain other elements that are available.

The physiological effects of the macronutrients cannot be considered in isolation from one another. If there is a shortage of potassium, carbohydrate production is limited; as a result of N-deficiency, only part of the organic acids in the plant can be built into amino-acids and proteins; phosphorus has an essential role in the intermediary stages of plant metabolism. Therefore, the concentration of a nutrient in the soil, in relation to certain others, may have a greater influence on plant growth than the absolute amounts available at a given time.

Irrigation farming is characterized by the high rates of fertilizers that are applied, especially in the case of nitrogen. However, the high nitrogen rates needed for high yields frequently have adverse effects on certain quality characteristics. In many cases, adequate supplies of potassium may counteract, at least partly, the adverse effects on quality of high N-levels.

Plant nutrients may either aid or compete with one another for entry into the plant, and improve or impair their respective availability; e.g. high levels of N nutrition may affect K intake adversely. Ehrendorfer (1958) found that when N application was increased threefold, the K content of the plant was reduced by half. Various indirect effects, due to reducing the K content below a certain level, as a result of heavy applications of nitrogen, may therefore be exercised on various physiological processes.

The uptake of fertilizer phosphorus by wheat and oats was found to be stimulated up to twofold by nitrogen fertilization; this was attributed to the stimulating effect of nitrogen on root activity (Olson & Dreier, 1956).

K-fertilization at very high levels may cause an imbalance in the supply of other essential nutrients, thereby indirectly causing some physiological deficiencies in plants. As the cation content of a plant is generally almost constant, an excessive increase in K-supply will reduce uptake of Ca and Mg (Barker & Bradfield, 1963), and an adequate potassium supply will limit sodium uptake. When irrigation water high in magnesium was used, it was found that poor growth of plants resulted from a decreased uptake of potassium (Ignatieff & Page, 1958).

The uptake of zinc from the soil is improved by the use of ammonium sulphate; sodium nitrate, on the other hand, causes a decrease in zinc uptake. This differential response has been ascribed to the effect of the nitrogenous fertilizers on soil pH.

Boron improves the absorption of calcium; phosphorus improves the absorption of molybdenum. Excessive concentrations of phosphorus may cause a deficiency of zinc; high levels of calcium may indirectly cause chlorosis as a result of iron deficiency, whilst heavy potassium rates may improve iron availability.

The balance between nutrients may be quite satisfactory at a given level of yields, yet prove inadequate when a higher yield level is made possible; e.g. by irrigation. This is a frequent occurrence with potassium in arid lands.

Additional Effects of Fertilizers on Plants

PLANT COMPOSITION

Proteins

The vital role of proteins in human nutrition, and the disastrous effects on human health of an inadequate balance between proteins and carbohydrates in the diet, are well known. Accordingly the possibility of increasing, by the judicious use of fertilizers, the absolute and relative amounts of proteins produced, is of major importance—especially for those populations whose low standard of living makes them dependent largely on plant proteins.

The overall picture that emerges from a considerable volume of research is that only after plants have used available nitrogen required for growth processes conducive to maximum yield, do fairly large doses of nitrogenous fertilizers increase protein content. An example is shown in Figure 75 (cf. Fernandez & Laird, 1959). It will be noted that there is first a drop in protein content when nitrogen is applied at a relatively low level; this is due to the

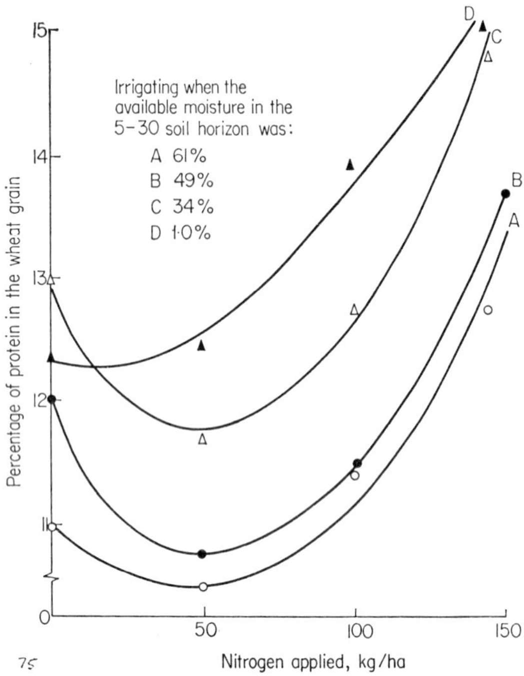

Figure 75. Relationship between the percentage of protein in the whole grain of wheat and the amount of applied nitrogen under different soil-moisture conditions. (After Fernandez & Laird (1959): by courtesy of the American Society of Agronomy.)

steep increase in yield which causes a 'dilution effect'. It is only after the basic requirements of nitrogen have been met, that protein content increases with increased nitrogen. This increase is usually accompanied by a reduced biological value* of the protein which is not, however, sufficient to offset the advantage of increased protein production.

Intensive protein synthesis requires, in addition to fairly high levels of nitrogen, a balanced supply of potassium and phosphorus. The general effect of potassium, except when it is supplied at excessively high levels, is to increase the effectiveness of nitrogen and, in conjunction with phosphorus, to counteract the unfavourable effect of high N-fertilization on the quality of the protein.

Nitrate Accumulation in Plant Tissues

Under conditions of limited soil-moisture supply and low atmospheric humidity, the plant may take up more nitrate than it requires, as a result of

* *Biological value.* The degree to which the protein is able to replace 'body protein' that has been degraded in the course of functioning of the organism.

the high concentration of the soil solution and its high rate of transpiration. The plant is generally not adversely affected by the excess nitrogen, but it may be harmful and even lethal to farm animals that eat forage with high nitrate content.

Unbalanced nutrition resulting from high rates of nitrogen application under conditions of P deficiency, is also conducive to excessive nitrate accumulation in the plant tissues.

Carbohydrates and Lipids

An adequate supply of potassium is essential for condensing sugar into starch, but the relatively high applications of potassium as well as of nitrogen, reduce the starch concentration in the edible parts of the plant, as a dilution effect.

The effect of fertilizers on sugar content has not been found to be consistent, probably because of the transitional nature of sugars in the metabolism of the plant.

Potassium was found to have a beneficial effect on the oil content of fruits and seeds, possibly because of its role in balancing the carbohydrate–nitrogen relationships in plants.

Mineral Content

The mineral constituents of foods can vary considerably, both in quantity and in their proportions to one another.

Many factors, of which fertilizers are the most important, control the uptake of minerals by plants. The observed differences in mineral content of various parts of the plant are not always understood (Lawton & Cook, 1954). When nutritive elements are applied to a low-fertility soil, the percentages of the respective elements in the plant tends to increase until a complete, well-balanced treatment is achieved. Subsequently, the more luxuriant growth generally causes a reduction, through dilution, in the levels of these elements.

The potassium content of plants is more directly influenced by the potassium content of the soil solution than is the case with other elements, because it is not converted into organic substances in the manner of most of the others.

K-content of the seed of most crops is usually far less affected by K-fertilization than is that of the vegetative parts. Stems of maize and wheat may contain three times as much potassium as the grain (Stubblefield & Turk, 1940), and variations of K-content under the influence of manuring are far more evident in the stems than in the grain.

Phosphorus Fertilization

Whilst there is a considerable body of research on the effect of phosphorus fertilizers on the mineral composition of forage crops, in which an increase of P-content as a result of P-fertilization has been conclusively demonstrated (Nehring, 1965), much less work on this subject has been carried out on food-crops. However here, too, there is a general tendency for phosphorus applications to result in an increase in P-content of the produce. This was found to be true in cereals (Gupta & Das, 1954; Merker, 1956; Wenzel, 1957), in which the major proportion of the phosphorus is in the form of phytin and only a small amount is present in the inorganic form.

Climate and soil fertility may mask the effect of phosphoric fertilizers, and heavy nitrogen applications, in conjunction with phosphorus, may reduce the P-content of the crop as a result of dilution due to increased growth.

Maturity

A delay in attaining maturity of the crop may have considerable economic significance to the farmer. In rain-fed cereal crops it may prevent the grain from ripening fully before the available soil moisture is exhausted, so resulting in shrivelled grain. In an irrigated crop such as cotton, harvesting difficulties and impaired quality of the lint may result from a delaying of maturity until after the early rains.

It is frequently stated that nitrogen fertilizers cause a delay in the attainment of maturity of most crops. Actually, if there has been a deficiency in available nitrogen, an adequate supply of this nutrient can advance maturity of wheat by as much as two weeks. Similar results were obtained with maize (Pesek *et al.*, 1955). Properly balanced nitrogen applications, given at the right time and in the right amounts, will generally advance maturity, rather than otherwise. It is the heavy applications of nitrogen, out of balance with the supply of other major nutrients, that delay maturity.

It is well known that phosphorus helps to hasten maturity in most crops, and in particular to offset the opposite tendency of heavy nitrogen applications to delay maturity.

FERTILIZERS IN RELATION TO DISEASE AND PEST INCIDENCE

The nutrient status of a crop may have a considerable effect on the ability of the plants to withstand disease, and the resistance of undernourished plants to disease may be considerably weakened. The high fertilizer rates associated with irrigation are bound to affect the reaction of cultivated plants to disease and insects, and may even help to control one particular disease while encouraging another. Excessive nitrogen frequently increases susceptibility to diseases. On the other hand, it has been found (Lachover & Avizohar-Hershenson, 1960) that nitrogen fertilizers, in particular in the form of aqueous ammonia and applied at high rates, have a marked influence on

Sclerotium rolfsii, an extremely important disease that causes considerable damage to many crops in the irrigated soils of the Mediterranean region.

Walker (1946) has shown that the effect of several wilt diseases—which are also of great importance in irrigated soils—can be greatly reduced by appropriate potash fertilization in soils that are deficient in this element. H. B. Tisdale and Dick (1942) stated that heavy applications of potash could be very beneficial for the control of cotton wilt; however, varieties differed in their response to such treatments. Potash deficiency also increases susceptibility to powdery mildew of wheat. Applications of potassium fertilizers were found to reduce the number of cysts of the eelworm *Heterodera schachtii* on the roots of sugar-beets (Curtis, 1964).

Insects are not affected in the same way as higher animals are by mineral deficiencies in the plants on which they feed. Haseman (1946) found that, in a number of cases, a shortage of certain minerals actually proved beneficial rather than detrimental to insects. It was found that chinch-bugs mature faster, live longer, and are more prolific, when feeding on maize grown under conditions of N-deficiency than on normal maize. Thrips have also shown differential preference for plants with a low level of nitrogen nutrition. Wheat-stem sawfly damage was increased by potassium deficiency.

Haseman (1946) claims that, as insects generally require smaller amounts of minerals relative to their weight than do plants, it is quite possible that plants with a low mineral content may be more favourable hosts than those with a higher mineral content. Depletion of soil nutrients, associated with irrigation, either directly through leaching or indirectly through increased yields, may therefore increase susceptibility to insect damage, unless offset by appropriate fertilization. On the other hand, heavy fertilization, by increasing vegetative growth, may favour the increase of certain insects, as is the case with the bollworm.

EFFECT OF FERTILIZERS ON ROOT DEVELOPMENT

Any influence on root development by fertilizers has relevance to the water relationships of the plant (Figure 76). As water movement in unsaturated soils is both limited and slow, an extensive proliferation of roots is essential to enable the plant to utilize soil moisture effectively. The extent of the root system also determines the depth of the reservoir from which plants can absorb water, and hence their ability to make use of water that is stored in the subsoil.

A high soil-fertility is usually conducive to the proliferation of the root system; penetration in depth is frequently limited as a result of an unsatisfactory nutrient status underneath the ploughed layer (Plate 26).

Fox & Lipps (1955) showed that, for lucerne, a favourable nutrient régime

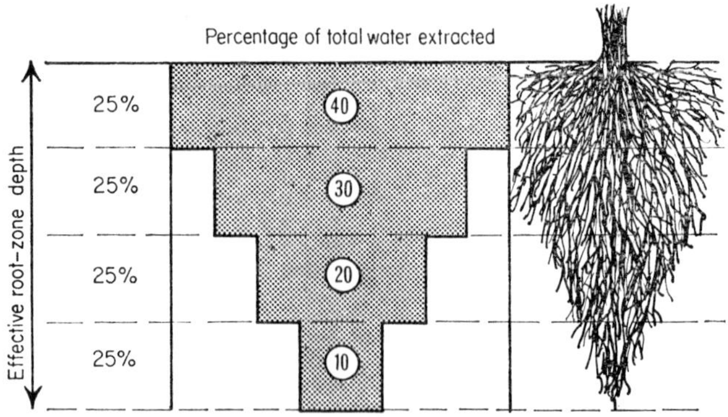

Figure 76. Effect of fertilizers on the root system. The top quarter of the root-zone is the first to be exhausted of available moisture. Adequate fertilization helps the plant to develop a deeper root system, and enables it to use moisture from the lower root-zone. (By courtesy of the U.S. Department of Agriculture.)

is even more important than a favourable physical environment, for the deep rooting of the plant and for satisfactory nodulation.

Fehrenbacher & Rust (1956) reported that, in a well-fertilized rotation, maize roots penetrated to a depth considerably greater than in unfertilized controls.

Fertilizers were found to increase the root development of grasses, which resulted in more efficient extraction of soil moisture at high tensions and at greater depths (Smika *et al.*, 1965).

It is well known that heavy nitrogen fertilization tends to increase the top-root ratio, as under these conditions only a small quantity of carbohydrates is translocated to the roots. However, nitrogen does not always have this effect, and Kmoch *et al.* (1957), working with wheat grown on four moisture levels and at three nitrogen rates, found that the added nitrogen increased root weights at all moisture levels and at nearly all soil depths, thereby permitting more effective utilization of the soil moisture. In all probability the problem is one of nutrient balance (Duncan & Ohlrogge, 1958).

The effect of potassium fertilizers on low-potassium soils, of decreasing the lodging of maize due to weakness of the root system is well known. Hoffer & Krantz (1949) have shown that, as a result of potassium deficiency, iron accumulates at the nodes of the stalk, thereby interfering with the transloca-tion of nutrients to the roots; the roots become weakened and susceptible to root-rots—hence the subsequent lodging.

The application of phosphorus fertilizers frequently favours the develop-ment of the root system in various crops and may thereby increase the ratio of roots to shoots.

EFFECT OF FERTILIZERS ON OSMOTIC PRESSURE OF SOIL SOLUTION

Growth of many crops is inversely proportional to the osmotic pressure of the soil solution between the limits of 2 and approximately 10 atmospheres (Wadleigh & Ayers, 1945), because water becomes progressively less available to plants as the osmotic pressure of the soil solution increases.

In humid regions, fertilizers will rarely have a markedly adverse effect on crop growth as a result of an increase in the osmotic pressure of the soil solution. However, the situation is different in dry regions; as the soil dries, the osmotic pressure of the soil solution increases, and will in time reach a level that adversely affects plant growth. This effect is felt sooner, the higher the concentration of salts in the solution. Soluble salts from fertilizers may accumulate, especially in semi-arid climates in which low rainfall and high evapotranspiration are the rule. Even under irrigation, moisture levels fluctuate from high to low between irrigations, and an increase in osmotic pressure of the soil, resulting from the heavy applications of fertilizers characteristic of irrigation agriculture, may have a detrimental effect on crop production, if periods of severe moisture stress are allowed to occur.

The foregoing is no reason for renouncing the benefits of fertilizer application, but further underlines the need for adjusting fertilizer supply to the moisture régime of the soil.

Fertilizers differ considerably in their effect on the salt concentration of the soil solution. High-analysis fertilizers generally have a lower salt-index* per unit of plant nutrients than do lower-analysis fertilizers. For example, the salt-index per unit of plant nutrients for three representative nitrogen carriers (Rader *et al.*, 1943) is as shown in Table XXIII.

TABLE XXIII

Salt-index of Three Nitrogen Carriers

	Salt-index	*in per cent*
Sodium nitrate	6·060	100
Ammonium sulphate	3·253	53·6
Anhydrous ammonia	0·572	9·4

In brief, low-analysis potassium and inorganic nitrogen carriers will generally cause the greatest increase in osmotic pressure of the soil solution, whilst phosphate carriers, urea, and ammonia, have relatively little effect in this respect (Rader *et al.*, 1943).

* *Salt-index.* Ratio of increase in osmotic pressure produced by a fertilizer, in relation to that produced by the same weight of sodium nitrate based on a relative value of 100.

Economic Aspects of Fertilizer Use

The supplying of fertilizers to crops is one of the easiest and most economical ways to overcome an important limiting factor in crop production. However, the fertilizer rates that give the highest yields are not necessarily the most profitable ones (Figure 77). The response to increasing amounts of fertilizers

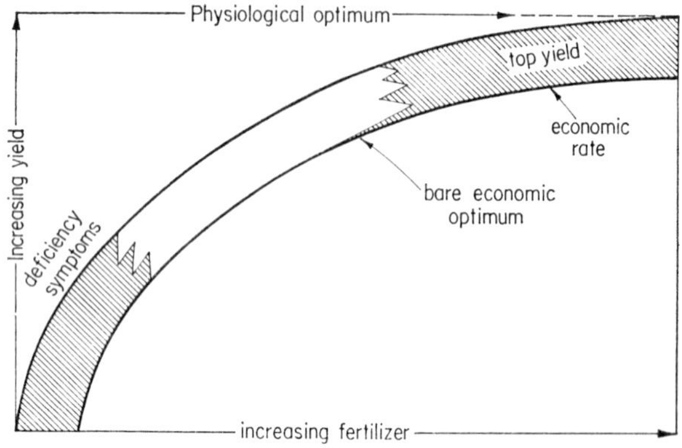

Figure 77. Physiological optimum and economic optimum rates of fertilizer application. (After Garrard (1958): by courtesy of the American Potash Institute.)

follows the law of diminishing returns, and a point is invariably reached at which an increment of fertilizer will increase costs more than it does the return. Where a nutrient deficiency is severe, the first units of nutrient added may cause a very small yield response; with additional increments the rate of response increases until an inflection point is reached. This is followed by a maximum, after which the yields decrease with additional fertilizer increments. In this case the overall yield-response curve is sigmoid.

Many methods have been devised to analyse the results of fertilizer experiments from the point of view of maximum economic return. Where the fertilizer experiments have been planned in such a way that a response curve can be drawn, the point of maximum return can be determined graphically, as in Figure 78.

All this appears at first sight to be fairly simple and straightforward; in practice, however, supplying fertilizers in the most economical way is a complicated problem, and fertilizer practice is still highly empirical. The following are some of the difficulties encountered in an economic analysis of fertilizer profitabilty:

1. The calculation of the most profitable rate of fertilization is based on data obtained from field experiments; these results need not necessarily apply in the following season, when the same crop is grown on a different field

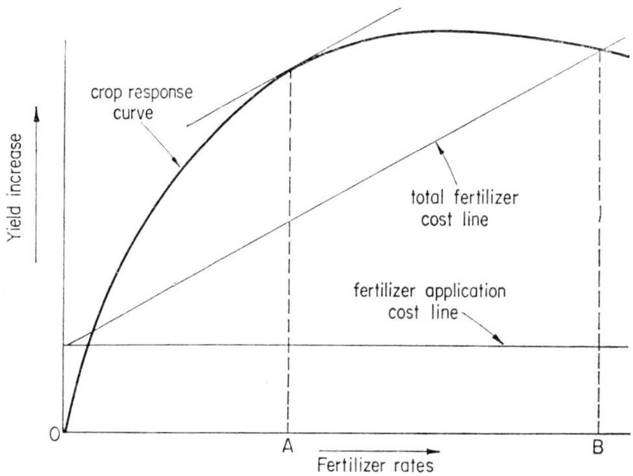

Figure 78. Generalized crop response curve with associated cost lines shown in terms of quantity of crop increase required to pay for the fertilizer. Maximum total profit occurs at point A, where returns minus costs are highest. Point B is a break-even point, where the value of the yield increase just equals the cost of fertilizer and its application. (From Stritzel (1963): by courtesy of Iowa State University.)

or a different crop is grown on the same field. The response of the crop to fertilizer is dependent on many environmental factors which cannot be forecast, and on agronomic practices which may differ from farm to farm—such as varieties, plant population, maintenance of fertility level, etc. (Figure 79). A difference in the amount of rainfall or in its distribution, or the incidence of diseases or pests or other similar factors, can make all the difference in the response of the crop to fertilizers.

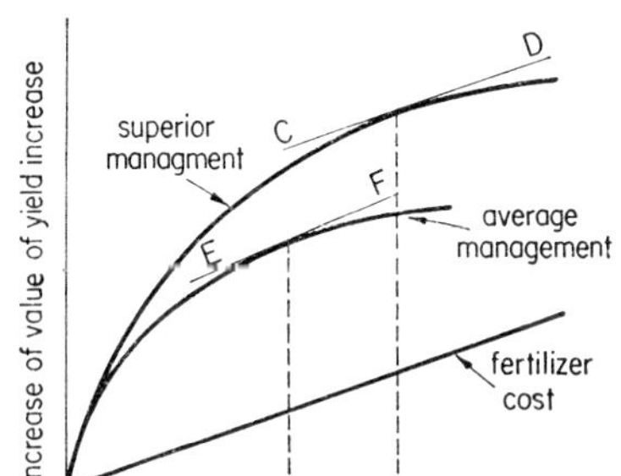

Figure 79. Effect of management level on maximum profit rates of fertilizer application. The vertical lines from the points A and B to the fertilizer rate line indicate the points that give maximum profit for each management level. Clearly, the more efficient the management, the higher the amounts of fertilizer that can be profitably applied. (From Stritzel (1963): by courtesy of Iowa State University.)

2. The determination of profitability of fertilizers as described above takes into account only the effect of the fertilizer on the yield of the crop to which it is applied. There are many supplementary effects that are not taken into account. Quality effects are disregarded, as well as residual effects on soil and subsequent crops. The interactions between different nutrients may give a higher crop response than the sum of responses of each nutrient separately.
3. Increased yields due to the use of fertilizers will increase the costs of harvesting, handling, and transport. Bumper crops due to fertilizers may cause a drop in prices.

The optimum economic level as determined above corresponds to the point at which the cost of the last increment equals the additional value of the crop in question. This depends, however, on whether the farmer has sufficient capital to exploit all the profitable investment potentialities available—a situation that is indeed rare in agricultural enterprises, particularly in developing countries where capital scarcity is the rule.

It is clear that an accurate determination of the most profitable rate of fertilizer application cannot be made in advance of sowing a crop, and even post-factum calculations are only approximations.

For the farmer in a developed country in the temperate regions, using advanced agricultural technology, the inability to predict accurately the most profitable rate of fertilizer application is not a serious handicap in the use of fertilizers. For fertilizers are cheap, and do not generally represent more than 5–6 per cent of the cost of production (Bublot, 1967). High yields are obtained as a result of costly inputs in labour, mechanization, and pest, weed and disease control, but are also dependent on adequate and high levels of fertilizer application. Even if the farmer applies amounts in excess of immediate requirement, this is not money lost, but will be mostly retrieved by increased yields of subsequent crops.

The situation is entirely different for a farmer in a developing country, especially in the arid regions. Fertilizer application is usually the first step taken to increase yields, whilst agricultural technology is still backward, varieties are primitive, and yields generally low. Fertilizers are relatively expensive, because they usually have to be imported. Even a low rate of fertilization may account for a relatively high proportion of the total costs of production. The risk of no response, or even of a negative response, if rainfall is insufficient or badly distributed, is great. Even when rainfall is satisfactory, an increase in yield of as much as 30 per cent in relation to the unfertilized crop may still be insufficient to cover the cost of fertilization. Small wonder, therefore, that under these circumstances, even farmers who wish to improve their production methods, adopt a very cautious approach to fertilizer application and apply only moderate amounts. However, fertilizer use is an essential ingredient of increased productivity; the financial risks involved can be reduced only if all the other essential factors of production are adopted—such as an improved crop rotation for increasing soil fertility and improving the moisture régime, soil and water conservation methods, disease, pest and weed control, and also improved varieties.

Under irrigation, the farmer has potential control of one of the main production factors—water—and his situation thus becomes more like that of the farmer in the temperate regions. Labour-intensive and capital-intensive crops are grown, and fertilizers account for only a relatively small fraction of the total costs of production. The rates that have to be applied are high, but the risk of their not being profitable is much smaller than that of the potential losses due to low yields if nutrients are limiting. Under these circumstances, fertilizer application may be considered as an insurance measure. In view of the high costs involved in irrigation farming, high yields are an economic necessity, and the farmer actually cannot afford *not* to apply high rates of fertilizers.

References

ANGLADETTE, A. (1966). *Le Riz*, G. P. Maisonneuve & Larose, Paris, 930 pp., illustr.

ARNON, D. I. (1958). Trace elements. pp. 1–31 in *Proc. Conf. Wooster, Ohio*, 1957, 410 pp., illustr.

ARNON, I. (1963). The transition from primitive to intensive agriculture in a Mediterranean environment. *World Crops*, **15**(4), 126–34.

——, BEREVSKY, A., BERKOVITZ, J., DOR, Z., HOROWITZ, M., MENKES, I. & LACHOVER, D. (1963). [Comparative value of nitrogen carriers in several crops—in Hebrew] *Ktavim*, **13**, 151–65.

BARBER, S. A. (1959). Relation of fertilizer placement to nutrient uptake and crop yield. II. Effects of row potassium soil-level and precipitation. *Agron. J.*, **51**, 97–9.

—— & HUMBERT, R. P. (1963). Advances in knowledge of potassium relationships in the soil and plant. pp. 231–59 in *Fertilizer Technology and Usage* (Ed. M. H. McVickar, G. L. Bridger & L. B. Nelson) Soil Sci. Soc. Am., Madison, Wisconsin, xv + 464 pp., illustr.

BARKER, A. V. & BRADFIELD, R. (1963). Effect of potassium and nitrogen on the free amino-acid content of corn plants. *Agron. J.*, **55**, 465–70.

BERGER, K. C. & PRATT, P. F. (1963). Advances in secondary and micronutrient fertilization. pp. 287–329 in *Fertilizer Technology and Usage* (Ed. M. H. McVickar, G. L. Bridger & L. B. Nelson) Soil Sci. Soc. Am., Madison, Wisconsin, xv + 464 pp., illustr.

BROWN, A. L. & KRANTZ, B. A. (1960). Zinc deficiency in California soils. *Calif. Agric.*, **14**(6), 8–9.

BROWN, D. A., Place, G. A. & PETTIET, J. V. (1960). The effect of soil moisture upon cation exchange in soils and nutrient uptake by plants. *Trans. 7th Int. Congr. Soil Sci.*, **3**, 443–9.

Brown, J. C., Weber, C. R. & Caldwell, B. E. (1967). Efficient and inefficient use of iron by two soybean genotypes and their isolines. *Agron. J.*, **59**(5), 459–62.

Bublot, G. (1967). The profitability of fertilizers. *Phosphorus in Agric.*, **21**(47), 13–18.

Cook, R. L. & Davis, J. F. (1957). The residual effect of fertilizer. *Adv. Agron.*, **9**, 205–16.

Cooke, G. W. (1967). *The Control of Soil Fertility*. Crosby Lockwood, London, xvii + 526 pp., illustr.

Curtis, G. L. (1964). The effect of potassium chloride on the infestation of sugar-beet by Beet Eelworm. *Ann. App. Biol.*, **54**, 269–80.

De Turk, E. E. *See* Turk, E. E. De.

Doggett, H. (1965). Hybrid grain sorghum from eastern and southern Africa. *Field Crops Abstr.*, **18**(2), 71–6.

Donald, C. M. (1964). Phosphorus in Australian agriculture. *J. Aust. Inst. Agric. Sci.*, **30**, 75–105.

Drake, M. & Steckel, J. E. (1955). Solubilization of soil and rock phosphate as related to root cation exchange capacity. *Proc. Soil Sci. Soc. Am.*, **19**, 449–50.

Duncan, W. G. & Ohlrogge, A. J. (1958). Principles of nutrient uptake from fertilizer bands. II. Root development. *Agron. J.*, **50**, 605–8.

Eaton, F. M. (1966). Chlorine. pp. 98–135 in *Diagnostic Criteria for Plants and Soils* (Ed. H. D. Chapman) University of California, v + 793 pp., illustr.

Eck, H. V. & Fanning, C. (1961). Placement of fertilizer in relation to soil moisture supply. *Agron. J.*, **53**, 335–8.

—— & Stewart, B. A. (1954). Wheat fertilization studies in western Oklahoma. *Bull. Okl. Agr. Exp. Stn.*, **432**, 16 pp., illustr.

Ehrendorfer, K. (1958). Untersuchungen über Angebot und Aufnahme der Haupnährstoffe. *Z. Acker-u. Pfl. Bau.*, **101**, 105, 301.

Eno, C. F. & Blue, W. G. (1954). The effect of anhydrous ammonia on nitrification and the microbial population in sandy soils. *Proc. Soil Sci, Soc. Am.*, **18**, 178–81.

Fehrenbacher, J. B. & Rust, R. H. (1956). Corn root penetration in soils derived from various textures of Wisconsin-age glacial till. *Soil Sci.*, **82**, 369–78.

Fernandez, R. G. & Laird. R. J. (1959). Yield and protein content of wheat in Central Mexico as affected by available soil moisture and nitrogen fertilization. *Agron. J.*, **51**, 33–6.

Fitts, J. W. (1955). Using soil tests to predict a probable response from fertilizer application. *Bett. Crops*, **39**(3), 17–20.

Fox, R. L. & Lipps, R. S. (1955). Influence of soil profile characteristics upon the distribution of roots and nodules of alfalfa and sweetclover. *Agron. J.*, **47**, 361–7.

Garrard, H. L. (1958). Do your crops have hidden hunger? *Bett. Crops*, **42**(5), 4–12.

Gerloff, G. C. (1963). Comparative mineral nutrition of plants. *A. Rev. Pl. Physiol.*, **14**, 107–24.

Gliemeroth, G. (1953). Bearbeitung und Düngung des Unterbodens in ihrer Wirkung auf Wurzelentwicklung, Stoffaufnahme und Pflanzenleistung. *Z. Acker-u. Pfl. Bau*, **96**, 1–44.

Grunes, D. L. (1959). Effect of nitrogen on the availability of soil and fertilizer phosphorus to plants. *Adv. Agron.*, **11**, 369–96.

Gupta, Y. P., & Das, N. B. (1954). The quality of wheat as affected by manures and fertilizers. *J. Indian Soc. Soil Sci.*, **2**, 121–5.

GUTSTEIN, I. (1963). [The response of cotton varieties to potassium fertilization—in Hebrew] *Hassadeh*, **44**, 144–8.

HADDOCK, J. L. (1949). The influence of plant population, soil moisture and nitrogen fertilization on the sugar content and yield of sugar-beets. *Agron. J.*, **41**, 79–84.

HAGEMAN, R. H., FLESHER, D. & GITTER, A. (1961). Diurnal variation and other light effects influencing the activity of nitrate reductase and nitrogen metabolism in corn. *Crop Sci.*, **1**, 201–4.

HAGIN, J. (1958). Availability of dicalcium phosphate to plants when applied in various forms. *Pl. Soil*, **10**, 101–13.

HALL, N. S., CHANDLER, W. V., BAVEL, C. H. M., REID, P. H. & ANDERSON, J. H. (1953). A tracer technique to measure growth and activity of plant root systems. *Tech. Bull. N. Carol. Exp. Stn*, **101**, 40 pp., illustr.

HAMBIDGE, G. (Ed.) (1949). *Hunger Signs in Crops*. American Society of Agronomy and Nat. Fert. Assoc., Washington D.C., xvi + 390 pp., illustr.

HASEMAN, L. (1946). Influence of soil minerals on insects. *J. Econ. Ent.*, **39**, 8–11.

HEMWALL, J. B. (1957). The fixation of phosphorus by soils. *Adv. Agron.*, **9**, 95–112.

HOFFER, G. N. & KRANTZ, B. A. (1949). Deficiency symptoms of corn and small grain. pp. 59–84 in *Hunger Signs in Crops*. American Society of Agronomy and Nat. Fert. Assoc., Washington D.C., xvi + 390 pp., illustr.

HUMBERT, R. P. & AYERS, A. S. (1957). The use of aqua ammonia in the Hawaiian sugar industry. *Proc. Soil Sci. Soc. Am.*, **21**, 312–16.

IGNATIEFF, V. & PAGE, H. J. (1958). *Efficient Use of Fertilizers*. F.A.O., Rome, xxi + 335 pp., illustr.

JACKSON, E. A. (1967). Tropical rice: the quest for higher yields. *Wld. Fmg.*,**8**(8), 14–16, 40–1.

JACOB, W. C., MIDDLEM, C. H., VAN, NELSON, W. L., WELCH, C. D. & HALL, N. S. (1949). Utilization of phosphorus by potatoes. *Soil Sci.*, **68**, 113–20.

JAMES, D. W., LEGGETT, G. E. & DOW, A. T. (1967). Phosphorus fertility relationships of central Washington irrigated soils. *Bull. Washington Agric. Expt. Stn.*, **688**, 18 pp., illustr.

JEWITT, T. N. (1966). Soils of arid lands. pp. 103–24 in *Arid Lands: A Geographical Appraisal* (Ed. E. S. Hills) UNESCO, Paris, xviii + 461 pp., illustr.

JOHAM, H. E. (1951). The nutritional status of the cotton plant as indicated by tissue tests. *Pl. Physiol.*, **26**, 76–89.

JOHNSON, C. M., STOUT, P. R., BROYER, T. C. & CARLTON, A. B. (1957). Comparative chlorine requirements of different plant species. *Pl. Soil*, **8**, 337–53.

JORDAN, J., SIMKINS, O., COREY, G., KNIGHT, R. & BARKER, G. O. (1952). Uptake and movement of fertilizer phosphorus. *Soil Sci.*, **73**, 305–13.

KAFKAFI, U. (1963). [*The Effect of Phosphorus Fertilizer Placement on the Efficiency of Phosphorus Uptake by the Plant*—in Hebrew.] Ph.D. Thesis, Hebrew University of Jerusalem, ii + 155 pp., illustr. (mimeographed).

KMOCH, H. G., RAMIG, R. E., FOX, R. L. & KOEHLER, F. E. (1957). Root development of winter wheat as influenced by soil moisture and nitrogen fertilization. *Agron. J.*, **49**, 20–5.

KOHNKE, H. & BERTRAND, A. R. (1956). Fertilizing the subsoil for better water utilization. *Proc. Soil Sci. Soc. Am.*, **20**, 581–6.

KURTZ, L. T. & SMITH, G. E. (1966). Nitrogen fertility requirements. pp. 195–235. in: *Advances in Corn Production* (Ed. W. H. Pierre, S. A. Aldrick & W. P. Martin) Iowa State University Press, Ames, vi + 476 pp., illustr.

LACHOVER, D. & AVIZOHAR-HERSHENSON, Z. (1960). The use of aqua ammonia for the control of *Sclerotium rolfsii* and its residual value, in Israel soils. *Israel J. Agric. Res.*, **11**, 141–50.

LAWTON, K. & COOK, R. L. (1954). Potassium in plant nutrition. *Adv. Agron.*, **6**, 235–303.

LIEBIG, J. VON (1843). *Chemistry in its Application to Agriculture and Physiology.* (Translated from German.) J. M. Campbell, Philadelphia, iv + 135 pp.

LIPSETT, J. (1964). The phosphorus content and yield of grain of different wheat varieties in relation to phosphorus deficiency. *Aust. J. Agric. Res.*, **15**, 1–8.

LUEBS, R. E. & LAAG, A. E. (1964). Tillage and nitrogen for dryland grain in a winter rainfall climate. *Bull. Calif. Agric. Exp. Stn*, **805**, 34 pp., illustr.

MARTIN, W. E. & MIKKELSEN, S. D. (1960). Grain fertilization in California. *Bull. Calif. Agric. Exp. Stn*, **775**, 39 pp., illustr.

MERKER, J. (1956). Untersuchungen an den Ernten und den Böden des Versuches 'Ewiger Roggenbau'. *Halle (Jaale) Kühn-Arch.*, **70**, 153–215.

MILLER, E. V. & ARMY, T. J. (1954). Effect of environment on the potassium and sodium contents of plants. *USDA & USDC Southern Coop. Serv. Bull.*, **36**, 118–54.

MITCHELL, J. (1957). A review of tracer studies in Saskatchewan on the utilization of phosphate by grain crops. *J. Soil Sci.*, **8**, 73–85.

MULDER, F. (1950). Mineral nutrition of plants. *An. Rev. Pl. Physiol.*, **1**, 1–24.

NEHRING, K. (1965). Düngung, Qualität und Futterwert. Pp. 1268–354 in *Handbuch der Pflanzenernährung und Düngung* (Ed. H. Linser) Springer Verlag, Wien, Band III/2, xviii + 1916 pp., illustr.

NELSON, W. L. & STANFORD, G. (1958). Changing concepts of plant nutrient behavior and fertilizer use. *Adv. Agron.*, **10**, 67–141.

NIELSON, R. F. & BANKO, L. A. (1960). A new look at nitrate movement in soils. *Fm. Home Sci.*, **21**(1), 2–4.

NORUM, E. B. (1963). Fertilized grain stretches moisture. *Bett. Crops*, **47**(1), 40–4.

OHLROGGE, A. J. MILLER, M. H. & DUNCAN, W. G. (1957). Getting effective nutrient use from fertilizer bands. *Bett. Crops*, **41**(6), 26–30.

OLSON, R. A. & DREIER, A. F. (1956). Nitrogen, a key factor in fertilizer phosphorus efficiency. *Proc. Soil Sci. Soc. Am.*, **20**, 509–14.

——, LAMKE, W. E. & RHOADES, H. F. (1960). Time of N application is important. *Nebraska Exp. Stn Quart.*, **7**(1), 10–11.

——, THOMPSON, C. A., GRABOUSKI, P. H., STUKENHOLTZ, D. D., FRANK, K. D. & DREIER, A. F. (1964). Water requirement of grain crops as modified by fertilizer use. *Agron. J.*, **56**, 427–32.

ORAZEM, F. & HERRING, R. B. (1958). Economic aspects of the effects of fertilizers, soil moisture and rainfall on the yields of grain sorghum in the sandy lands of south-west Kansas. *J. Fm. Econ.*, **40**, 697–708.

PAAUW, F. VAN DER (1958). Relation between the potash requirement of crops and meteorological conditions. *Pl. Soil*, **9**, 245–68.

PARKS, W. L. & KNETSCH, J. L. (1959). Corn yields as influenced by nitrogen level and drought intensity. *Agron. J.*, **51**, 363–4

PESEK, J. T., NICHOLSON, R. P. & SPIES, C. (1955). What about fertilizers in dry years? *Iowa Farm Sci.*, **9**(10), 3–6.

PIPER, C. S. & VRIES, M. P. C. DE (1964). The residual value of superphosphate on a red-brown earth in South Australia. *Aust. J. Agric. Res.*, **15**, 234–72.

PLESHKOV, B. P. (1956). The effect of phosphorus fertilizers on the nitrogen metabolism of plants. *Doklady Moskov. Selskokhoz. Akad. K. A. Timiryazeva*, **25**, 158–64. (Also in *Chem Abstr.* (1958) **52**, 3044.)

POWER, J. F., BROWN, P. L., ARMY, T. J. & KLAGES, M. G. (1961*a*). Phosphorus responses by dryland spring wheat as influenced by moisture supplies. *Agron. J.*, **53**, 106–8.

——, GRUNES, D. L. & REICHMAN, G. A. (1961*b*). The influence of phosphorus fertilization and moisture on growth and nutrient absorption by spring wheat. I. Plant growth, nitrogen uptake and moisture use. *Proc. Soil Sci. Soc., Am.* **25**, 207–10.

POWERS, W. L. & PANG, T. S. (1947). Status of zinc in relation to Oregon soil fertility. *Soil Sci.*, **64**, 29–36.

RADER, L. F., WHITE, L. M. & WHITTAKER, C. W. (1943). The salt-index—a measure of the effects of fertilizers on the concentration of soil solution. *Soil Sci.*, **55**, 201–18.

RAMIG, R. E. & RHOADES, H. F. (1963). Interrelationships of soil moisture level at planting time and nitrogen fertilization on winter wheat production. *Agron. J.*, **55**, 123–7.

RICHARDSON, H. L. (1960). Increased world food supplies through greater crop production. *Outlook on Agric.*, **3**, 9–22.

RUSSELL, R. S., BARBER, D. A. & NEWBOULD, P. (1961). Nutritional relationships between crops and the soil. *Outlook on Agric.*, **3**, 103–10.

SCARSBROOK, C. E., BENNETT, O. L. & PEARSON, R. W. (1959). The interaction of nitrogen and moisture on cotton yields and other characteristics. *Agron. J.*, **51**, 718–21.

SEATZ, L. F. & STANBERRY, C. (1963). Advances in phosphate fertilization. pp. 155–80 in *Fertilizer Technology and Usage* (Ed. M. H. McVickar, B. L. Bridger & L. B. Nelson) Soil Sci. Soc. Am., Madison, Wisconsin, xv + 464 pp., illustr.

SHEAR, C. B., CRANE, H. L. & MYERS, A. T. (1946). Nutrient element balance: a fundamental concept in plant nutrition. *Am. Soc. Hort. Sci.*, **47**, 239–48.

SHIMSHI, D. (1966). [The effect of nitrogen fertilization on certain aspects of soil-water-plant relationships—in Hebrew.] Ph.D. thesis, Hebrew University of Jerusalem, Israel, viii + 145 pp., illustr. (mimeographed).

SMIKA, D. E., HAAS, H. J. & POWER, J. F. (1965). Effects of moisture and nitrogen fertilizer on growth and water use by grass. *Agron. J.*, **57**, 483–6.

SMITH, G. E. (1954). Soil fertility—the basis for high crop production. *Bett. Crops*, **38**, 22–30.

STANBERRY, C. O. (1959). Commercial fertilizers in neutral and alkaline soils of the West. *Reclamation Era*, November 1959.

——, CONVERSE, C. D., HAISE, H. R. & KELLEY, O. J. (1955). Effect of moisture and P variables on alfalfa hay production on the Yuma Mesa. *Proc. Soil Sci. Soc. Am.*, **19**, 303–10.

STANSEL, J. W., BOLLICH, C. N., THYSELL, J. R. & HALL, V. L. (1965). The influence of light intensity and nitrogen fertility on rice yields and components of yields. *Rice J.*, **68**(4), 34–5, 49.

STAROSTKA, R. W. & HILL, W. L. (1955). Influence of soluble salts on the solubility of, and plant responses to dicalcium phosphate. *Proc. Soil Sci. Soc. Am.*, **19**, 193–8.

STEPHENS, C. G. & DONALD, C. M. (1958). Australian soils and their responses to fertilizers. *Adv. Agron.*, **10**, 167–256.

STRITZEL, J. A. (1963). What is a soil test? *Iowa Farm Sci.*, **17**(12), 6–8.

STUBBLEFIELD, F. M. & TURK, E. E. DE (1940). The composition of corn, oats and wheat as influenced by soil, soil treatment, seasonal conditions and growth. *Proc. Soil Sci. Soc. Am.*, **5**, 120–4.

TISDALE, H. B. & DICK, J. B. (1942). Cotton wilt in Alabama as affected by potash supplements and as related to varietal behavior and other important agronomic problems. *J. Am. Soc. Agron.*, **34**, 405–26.

TISDALE, S. L. & NELSON, W. L. (1966). *Soil Fertility and Fertilizers.* Macmillan, New York, vi + 694 pp., illustr.

TURK, E. E. DE, HOLBERT, J. R. & HOWK B. W. (1935). Chemical transformation of phosphorus in the growing corn plant, with results on two first generation crosses. *J. Agric. Res.*, **46**, 121–41.

—— WOOD, L. K. & BRAY, R. H. (1943). Potash fixation in corn-belt soils. *Soil Sci.*, **55**, 1–22.

VAN DER PAAUW, F. *See* PAAUW, F. VAN DER.

VIETS, F. G. Jr. (1967). Nutrient availability in relation to soil water. pp. 458–67 in *Irrigation of Agricultural Lands* (Ed. R. M. Hagan, R. H. Haise, & T. W. Edminster) Agronomy Series; Amer. Soc. Agron., Madison, Wisconsin, **11**, xxxi + 1180 pp., illustr.

—— & HANWAY, J. J. (1957). How to determine nutrient needs. pp. 172–84 in *Soil, The Yearbook of Agriculture.* U.S. Department of Agriculture, Washington D.C., xiii + 784 pp., illustr.

VON LIEBIG, J. *See* LIEBIG, J. VON.

WADLEIGH, C. H. & AYERS, A. D. (1945). Growth and biochemical composition of bean plants as conditioned by soil–moisture tension and salt concentration. *Pl. Physiol.*, **20**, 106–32.

WALKER, J. C. (1946). Soil management and plant nutrition in relation to disease development. *Soil Sci.*, **61**, 47–54.

WALSH, L. M. (1964). Computers tailormake recommendations. *Crops Soils*, **17**(1), 15–17.

WARDER, F. G., LEHANC, J. J. & STAPLE, W. J. (1963). The effect of fertilizer on growth, nutrient uptake and moisture use of wheat on two soils in south-western Saskatchewan. *Can. J. Soil Sci.*, **43**, 107–16.

WATANABE, F. S., OLSEN, S. R. & DANIELSON, R. E. (1960). Phosphorus availability as related to soil moisture. pp. 450–6 in *Trans. 7th Intern. Cong. Soil Sci.*, Vol. 3, 637 pp., illustr.

WEISER, V. (1960). *North Dakota Fertilizer Guide.* North Dakota Ext. Circ. A–350, 6 pp.

WENZEL, W. (1957). Einfluss der Versorgung der Planzen mit Phosphosäure and ihren Gehalt an P–haltigen Verbindungen. *Z. Pflanzenernährung, Dung. Bodenkde*, **79**, 247–60.

WIERSUM, L. K. (1962). Bodenstructuur, bewortling en plantenvoeding. *Landbouwk. Tijdschr.*, **74**, 961–72.

WILKINSON, S. R. & OHLROGGE, A. J. (1962). Principles of nutrient uptake from fertilizer bands. I. Mechanism responsible for intensive root development in fertilized zones. *Agron. J.*, **54**, 288–91.

WILLIAMS, M. S. & COUSTON, J. W. (1962). *Crop Production Levels and Fertilizer Use.* F.A.O., Rome, 48 pp., illustr.

WINTERS, E. & SIMONSON, R. W. (1951). The subsoil. *Adv. Agron.*, **3**, 2–86.

WOODROFFE, K. & WILLIAMS, C. H. (1953). The residual effect of superphosphate on soils cultivated for wheat in South Australia. *Aust. J. Agric. Res.*, **4**, 127–50.

XIII

TILLAGE

The earliest form of tillage was probably practised at the close of the Neolithic period. Small plots of ground were cleared, often with the aid of fire, and then dug with a 'digging stick'—the probable ancestor of the plough (Bishop, 1938). The digging stick in the course of time developed into the foot-plough, which was provided with a rest on which the cultivator pressed with his foot. At a later stage, one or more individuals would pull the tool with cords, whilst another would hold and push it. When animals replaced man-power, the traction plough was evolved. The first proof of its use is found in a Sumerian seal of about 3500 B.C. The plough depicted on the seal is strikingly similar to the wooden plough used to this day by the felaheen* in the Near East. In the Egyptian pyramids, 4000-year-old tablets depicting wooden ploughs drawn by cows have been found.

At a somewhat later stage, an upright tube, open at both ends, was attached to the plough. An attendant would drop the seed through a funnel placed on the tube—the earliest example of the 'once-over' principle of sowing and tilling simultaneously. This method is still in use in the Near East.

Greek and Roman farmers were convinced that loosening and mixing the soil improved its fertility. The Middle Ages was a period of stagnation in agriculture. In the 18th century, Tull (1731) proposed the theory that plants feed on soil particles, and the more the soil is cultivated, the more particles can be absorbed by the plant roots. The 19th century added its spate of theories concerning the role of soil tillage. A few superstitions still prevailed; in 1891 Walden still stated categorically that the use of tillage implements was more effective in feeding plants than was the addition of fertilizers to the soil.

The composite picture that emerged by the beginning of the 20th century

* Felaheen . . . Arab farmers.

indicates a fairly good understanding of the objectives and effects of tillage: aeration of the soil, promoting increased availability of nutrients, better root penetration, weed control, breaking up compacted layers, etc.

The general attitude during the 19th century was therefore that the importance of tillage operations could hardly be overstated. This attitude, equating a maximum number of tillage operations with good husbandry, was further strengthened by the fact that the cost of upkeep of draft animals was practically the same, whether they were working in the field or were stalled; farm labour was also cheap and plentiful.

The importance formerly attached to cultivation as an objective *per se*, is vividly reflected in the following quotation from the *Standard Cyclopedia of Horticulture*: 'It would have been a sorry thing for agriculture if there had been no weeds. They make us stir the soil, and stirring the soil is the foundation of good farming' (Bailey, 1950). In short, weeds justify their existence simply by obliging the farmer to cultivate the soil!

It was only following gradual replacement of draft animals by tractors, that these traditional concepts of tillage were submitted to close scrutiny. For cost of tillage operations had become a substantial part of the overall production costs. Changes in the number and depth of soil cultivations could make possible considerable savings in production costs. Many experiments on the effects of tillage on crop production have been carried out since the beginning of the present century, and the results obtained were most surprising to the early investigators. The overall picture that emerged was that the need and justification for numerous tillage operations had been grossly exaggerated in the past. As a result of the objective reappraisal of the effects of soil cultivation, a new understanding, new attitudes, and new methods, have evolved.

Traditional Tillage Methods in Dry Regions

The wooden plough, developed in the Near East several thousand years ago, cannot penetrate the dry, hard soil that develops there during the summer months. The farmer must therefore wait until the rains have moistened and softened the soil, before he can begin to cultivate. The wooden plough loosens the moist soil, without inversion. Weeds and stubble remain on the soil surface.

Seeds of *winter cereals* and other rain-fed crops are sown broadcast, before ploughing. No other tillage operations are carried out. For summer crops the soil is re-ploughed two or more times during the winter. The light plough and the light animals that pull it make it possible to exploit every interval between rains without causing excessive compaction of the soil. Weeds are controlled, though not very effectively; many weeds take root again fairly rapidly.

The seeds of *summer crops* are dropped into the furrow behind the plough, to be covered by the next passage of the plough, or else they are sown through a funnel attached to the frame of the plough.

This type of ploughing is the only tillage operation carried out; the work is arduous and slow. The period during which tillage is possible is strictly limited, the primitive plough being incapable of penetrating the soil when this has dried and become hard. For these reasons, the concept of the desirability of frequent tillage operations could not develop under these conditions, as it did in the temperate humid regions. On the contrary, traditional ploughing in the drier regions, as developed thousands of years ago, is more similar to the modern trend of 'minimum tillage' than to the concepts prevailing until quite recently in humid climates with developed agriculture.

Modern Tillage Methods in Arid Soils

With the advent of modern equipment, an entirely new concept of tillage methods developed in the dry regions—different from that of modern tillage in temperate humid climates, and from traditional tillage in arid zones. Heavy equipment, drawn by tractors, made possible the cultivation of dry or almost dry soils. The effects achieved are entirely different from those of cultivating moist soil.

Most of the classical concepts concerning tillage were first developed in north-west Europe, and later in the U.S.A., and they were naturally adapted to the temperate regions. Under the cool and humid conditions of these regions the soil is almost permanently moist; also rainfall is very rarely torrential and therefore is not conducive to erosion. Basic cultivation has been based on the mould-board plough, which cuts and inverts a furrow, completely covering all weeds, trash, stubble, and manure, and leaving a clean surface. Subsequently, disking, harrowing, and rolling, are needed to pulverize the furrow, compact the soil, and prepare a seed-bed.

When these methods were transferred to arid lands, whether for the development of areas that had previously not been cultivated or to replace traditional cultivation methods, results were frequently disastrous.

Optimum moisture conditions for cultivation are not encountered frequently in dry regions: during the rainy season, torrential downpours occur frequently and the soil may be too wet for heavy equipment; during the dry season, the soil generally dries out almost completely. European-style ploughing shatters the soil, breaking the land into clods whose size and hardness depend mainly on the compaction caused by previous farming operations on land that was still moist; the heavy and frequent disking needed to break these clods pulverizes the soil into dust, breaking down soil aggregates. The soil is therefore easily eroded by water in winter and by wind in summer; soil compaction may cause hard-pans and reduce water infiltration.

Objectives of Tillage

IMPROVEMENT OF SOIL STRUCTURE

In Chapter XI, the possibilities of improving soil structure by tillage operations, carried out under conditions of optimal soil moisture, were described (p. 318). However, ideal conditions of soil moisture during tillage operations are the exception rather than the rule, and the effect of tillage on soil structure is more frequently detrimental than beneficial, resulting in soil compaction when the soil is too wet and in soil pulverization when it is too dry. Additional cultivation operations are therefore frequently needed in order to correct defects caused by previous cultivations.

MOISTURE CONSERVATION

The effects of tillage methods on crop growth and yields are to a large degree attributable to differences in soil-moisture régimes which in turn are due to improved infiltration and conservation of water.

Water is lost by surface run-off, by infiltration beyond the root-zone, through evaporation from the soil surface, and through transpiration by plants.

Tillage can be effective in reducing surface run-off if it is carried out according to approved soil-conservation practices. These improve infiltration and may double the amount of water stored in the soil.

Rough surfaces induced by tillage can increase infiltration of water into the soil by reducing run-off and minimizing crust formation. These effects can be further enhanced by a mulch of crop residues which protects the soil from raindrop impact.

The influence of soil compaction on loss of water by evaporation depends on the moisture content of the soil. At high moisture-levels, conduction of water to the soil surface is greater in a compacted soil than in a loose one. When the soil surface is dry, however, water moves as vapour through the pore spaces, and evaporation will then be greater in a loose soil. Therefore, tillage may increase water losses due to evaporation by loosening the top layer of soil, but, by effectively destroying weeds, it may considerably reduce losses caused by transpiration (Figure 80).

Crop response to tillage will depend on the amounts and seasonal distribution of the rainfall. In seasons during which precipitation provides adequate water to the crop throughout the growing period, differences in soil-moisture régime due to tillage have little or no effect on yields. Conversely, tillage methods that increase available soil moisture are beneficial to the crop in years of less than optimal rainfall. For example, an investigation in tillage methods in Iowa (Amemiya, 1968), listed maize consistently out-yielded conventionally tilled maize in the six of eleven years that the crop showed signs of severe water stress. The yield increases due to listing (which was

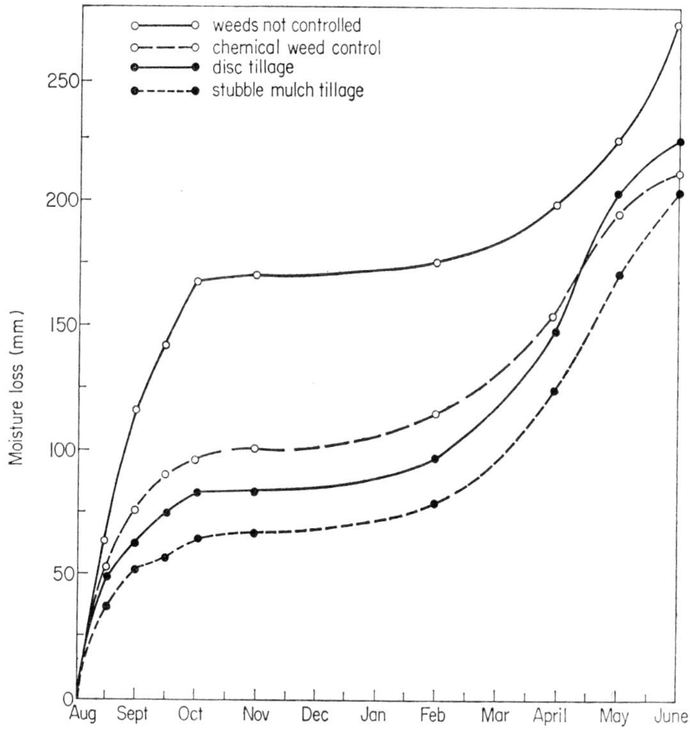

Figure 80. Cumulative moisture losses under different methods of weed control and tillage, in the Great Plains region of the U.S.A. (From Wiese & Army (1958): by courtesy of the American Society of Agronomy.)

conducive to water conservation) ranged from 308 kg/ha to as much as 2588 kg/ha; the average increase being 1524 kg/ha. In the other five years of the investigations, yields were above average because of favourable precipitation patterns, and there was little difference in yields among different tillage methods.

The effects of the tillage treatments were most marked in years in which there was little soil-stored moisture at sowing time and when a large proportion of the early season precipitation occurred during the first six to seven weeks after planting.

SOIL DRYING

In the arid regions, a prolonged rainless period makes it possible to dry out the soil completely. The alternate drying and rewetting of the soil has a marked beneficial effect on soil fertility as described in Chapter XI (pp. 342–3).

The heat of the sun is not sufficient to dry the soil to the full depth of the rooting system of most cultivated crops, as the top layer, when once it has dried out, protects the lower soil levels. Complete drying out to a considerable depth can be achieved by growing a crop that reaches maturity during the dry season, followed by deep-ploughing that turns up huge, hard clods that are then thoroughly baked in the sun. These clods break down subsequently under the effect of light rains or irrigations, thereby restoring the structure of even badly compacted soils.

SOIL AERATION

Air movement is usually satisfactory in soils with a good structure. Badly compacted soils tend to crack on drying during the dry season; the deep fissures improve aeration to a considerable depth.

SOIL PERMEABILITY

The heavy equipment used in modern irrigation farming, such as wheel tractors, disks, subsoilers, and landplanes, etc., causes soil compaction, sometimes to a depth of 50 cm, and thereby adversely affects water penetration (Figure 81; Doneen & Henderson, 1953). This is accompanied by reduced root-development and a striking reduction in plant growth.

In some cases compacted layers have been found to remain dry, even when the soil above the compacted layer is saturated with water. In these cases, waterlogged conditions above the compacted layer are an additional detriment to the crop (Raney *et al.*, 1955).

In order to improve water infiltration, compacted layers must be eliminated. This is achieved by tillage equipment which shatters the soil below the compacted zone. A depth of 30 cm is usually sufficient, but if the previous crop was harvested when the soil was still fairly wet, a depth of 50 cm or more may be necessary.

Improved infiltration may be essential as a preliminary to leaching of excess salts. The effects of tillage on infiltration are very transient. Rainfall, irrigation, and cultural operations, compact the soil rapidly and the rates of water infiltration are soon reduced again to their former level.

ROOT PENETRATION

Tillage usually affects only one-tenth of the soil volume in which the root systems of most crops develop, and therefore has only a limited effect on root penetration. Even subsoiling has not been found to improve root penetration markedly. A plentiful supply of nutrients and soil moisture at the lower soil levels is considerably more effective in promoting deep root penetration than tillage operations, unless a hard pan is present. In this case, its breaking up by ploughing or subsoiling is usually essential in order to improve root penetration (Plate 27).

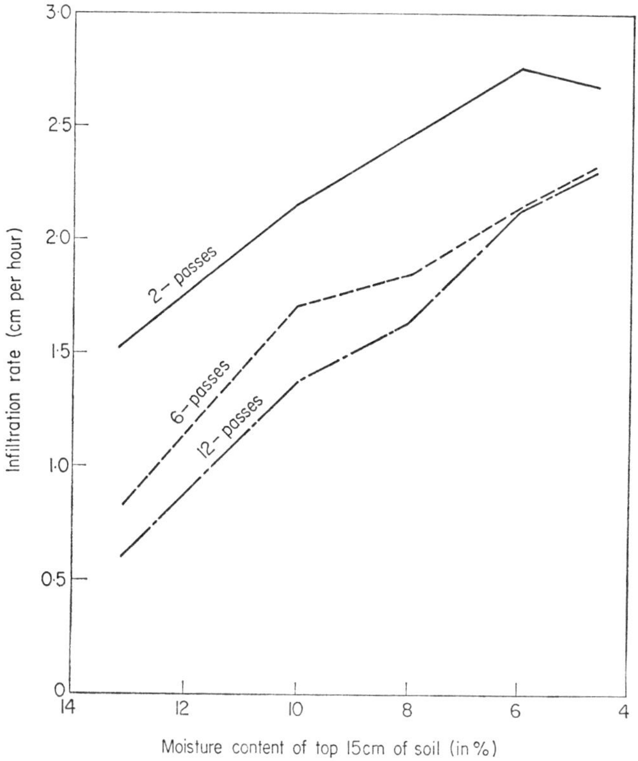

Figure 81. Effect of compaction by tractor tracks on infiltration, in relation to soil-moisture content at the time of tractor passes. Note that even two passes soon after irrigation markedly reduced the infiltration rate. (From Doneen & Henderson (1953): by courtesy of the American Society of Agriculture Engineers.)

SEED-BED PREPARATION

One of the earliest functions of soil tillage was to prepare a seed-bed and to cover the seeds. An initial ploughing is not particularly effective in this respect; it results in a too-loose soil, with high evaporation losses. After it, seeds cannot be placed at uniform depth, contact between seeds and soil particles is not sufficiently intimate to ensure movement of moisture to the seed, and the seed-bed dries out before the seeds have germinated. Seed emergence is therefore irregular, and stand establishment is incomplete. A number of additional tillage operations are usually needed to offset some of the negative effects of the ploughing—to break down the clods that have been turned up, to compact the soil that has been excessively loosened, to break up the plough-sole, and to level the soil surface.

A well-prepared seed-bed, free from weeds, is extremely important in giving a good start to the crop.

SOIL INVERSION

Under certain circumstances, soil inversion may be a desirable objective. It may bring to the surface fertile soil with good structure, covering the top layer that has lost its structure and is poor in nutrients. It may mix together surface soil and deeper layers which differ in texture, thereby improving the physical characteristics of the top-soil. It may help to bury the seeds of weeds—particularly numerous in the top few centimetres of the soil—at a depth which effectively prevents their germination. It may help soil sanitation by a thorough burying of crop residues which may carry insects at various stages of development or the spores of plant pathogens. For example, ploughing under the stubble of barley is an effective means of controlling the spread of Barley Stripe (*Helminthosporium* spp.), while burying the stubble of wheat prevents re-infestation by the Wheat Leaf-miner (*Scythris temperatella*). Complete incorporation of cotton stalks is a sanitary measure imposed by law in most cotton-growing countries, as an effective measure for controlling the Pink Bollworm (*Platyedra gossypiella*).

WEED CONTROL

Weed control is probably the most beneficial and universal role of tillage. Experiments have shown repeatedly that, on soils which are free from weeds, crop yields are not markedly increased by tillage. With proper timing, it is possible to control weeds efficiently and economically when they are small, by using light equipment for cultivation.

Deep-ploughing, carried out during the summer months in dry soil, is probably the most effective means of controlling perennial weeds such as 'couch grass' (*Cynodon dactylon*) and Johnson Grass (*Sorghum halepense*), two almost ubiquitous weeds of the warm semi-arid areas.

EFFECTS OF TILLAGE ON SOIL TEMPERATURES

Tilling a soil, by increasing its air content, decreases its thermal conductivity and hence its ability to absorb heat to any depth. As a result, the air layer above and near the ground absorbs more of the sun's heat, and this increases the lag of soil temperature behind air temperature (Thornthwaite, 1958).

Depth of Cultivation: Deep-ploughing

Increasing the depth of ploughing and other tillage operations has a very marked effect on costs: for every additional centimetre of increased depth, the amount of soil to be moved or turned is increased by about 150 tonnes per hectare! The enormous additional effort required cannot be accomplished without a considerable expenditure of fuel.

Much research, aimed at determining the most effective depth of tillage, has therefore been carried out—mostly in temperate climates. It was generally found that increasing the depth of ploughing beyond the minimum needed for a specific purpose—weed control, seed-bed preparation, shattering a plough-sole, etc.—did not improve yields.

Similar results were obtained in experiments in dry regions, when working with moist soil. Moisture conservation, an extremely important objective in dry land farming, is adversely affected by the depth of tillage—especially if the tilled soil layer is inverted.

'Moisture conservation varies inversely as the depth of dry soil at the surface, when rainfall occurs' (Staple, 1964). As soils dry rapidly to the depth of cultivation, shallow tillage (up to 7 to 10 cm maximum) is less wasteful of moisture that is deeper tillage.

The dry surface soil has to be wetted by rain to field-capacity before any moisture can penetrate to at least 10–12 cm, at which depth it will be relatively safe from evaporation, and will contribute effectively to crop production. By avoiding tillage implements that invert the soil, sub-surface moisture is not exposed and moisture losses are reduced.

However, the cultivation of dry, or nearly dry, soil, is typical of arid lands, and has no counterpart in temperate-region cultivation. It has specific objectives and specific effects, and depth of ploughing has relevance to these effects. Deep tillage, when carried out under the proper conditions, has marked beneficial effects that are difficult to achieve by other methods; these effects are important in dry land agriculture and still more so in irrigated agriculture. These specific effects are: sanitation, improvement of soil structure, and control of perennial weeds.

In order to achieve these beneficial results the soil has to be completely dried out after the deep-ploughing. The success of the deep-ploughing depends, therefore, on the soil management during the year preceding the deep-ploughing, and on the type of crop grown. The most desirable crop is a winter cereal that withdraws all the available moisture by the time it is mature. A reasonable amount of soil compaction during seed-bed preparation and sowing of the winter crop is desirable, as it favours the formation of clods when the land is deep-ploughed in the following summer. Deep-ploughing should not start before the land has dried out sufficiently, but should not be too late in the season as otherwise the time available for drying and weathering the clods will be too short. At least three months between the completion of ploughing and onset of the rains (or irrigation) should be allowed for this purpose.

The depth of ploughing will vary from 50 to 70 cm—the lesser depth being for dry land farming, the greater depth for irrigated soils. This ploughing turns up huge, hard clods—a sight that temperate-zone farmers will find it difficult to associate with their traditional concepts of soil tilth! No attempt should be made to reduce these clods by cultivation. This is not only expensive and unpractical, but it will actually defeat the purpose of the ploughing. The clods must be baked by the sun and completely dried out. As a result, most parasites and pathogens are destroyed, and the rhizomes and other underground parts, of perennial weeds dry out and die. The weathering of the clods starts under the influence of superficial moistening by dew; their breakdown is, however, dependent on the first rains of the season or on irrigation—preferably by overhead sprinklers. The clods absorb the water like a sponge, whereupon the air that is trapped within the clods creates internal pressures and tensions which fissure the clods; after a short time these crumble down into a fine tilth, and a light tillage implement, such as a harrow or cultivator, is sufficient to prepare an excellent seed-bed.

Deep-ploughing is a very expensive operation, requiring special heavy apparatus; in modern dryland and irrigated farming it has, however, been found to be an important practice for maintaining a high level of yields. Practical experience has shown that a time interval of five to six years between one deep-ploughing and the next, strikes a proper balance between the cost and the benefits of the operation.

Subsoiling and 'Vertical Mulching'

There are many cases in which soil inversion—the typical function of the plough in temperate regions—is unnecessary, and sometimes undesirable. For example, when a soil has been carefully levelled for irrigation, conventional ploughing will undo much of the levelling. After ploughing, it may become necessary to disk and land-plane the field in order to re-establish the level surface required; these are costly operations and also pulverize the soil excessively. In irrigation farming, on land that is reasonably clean of weeds and crop residues, ploughing can frequently be replaced by 'subsoiling', e.g. loosening the soil to an appreciable depth without inversion. The subsoiler consists of 2 or 3 strong vertical shanks, with chisels, mounted on a cultivator frame, operating to a depth of 60–70 cm, that cut a narrow slot through the soil, and shatter the subsoil. Subsoiling is probably the most effective method of breaking up compacted or cemented layers and thereby improving root penetration, aeration, and water penetration. To be effective, it must be done in dry soil. If the subsoil is wet, instead of the desirable shattering effect, the soil will smear and become compacted along the track of the subsoiler shanks.

Unfortunately, the effects of subsoiling are not long-lasting, the slots left by the subsoiler closing and becoming sealed off within a few months. In order to prolong the beneficial effect of the subsoiler, a method called 'vertical mulching' has been developed. The object of this is to fill the slots with organic material and thereby keep them open and functioning for longer periods. By modifying a conventional subsoiler and attaching a flail-type chopper to the unit, suitable organic material is chopped and introduced into the slots that are opened by the subsoiler. It was found that only dry carbonaceous material that decomposed slowly, such as cereal stubble, maize stalks, corn cobs, cotton burrs, wood chips, etc., was suitable (Spain & McCune, 1956).

Lucerne yields on irrigated soils with a low infiltration rate were increased in California as a result of vertical mulching, in which rice hulls were used as the organic material (Curley *et al.*, 1959).

For vertical mulching to be effective, the compacted layer with low permeability must be sufficiently shallow to be shattered by the subsoiler, and there must be a permeable layer under the compacted layer (Curley *et al.*, 1959).

Ridge Cultivation

The frequent traffic of machinery and equipment, especially in irrigated fields, causes a breakdown of soil structure in the top-soil layer, and considerable compaction of the lower layers. As a result, it is difficult to prepare a good seed-bed, germination is adversely affected, and irregular stands are obtained.

One of the methods developed to overcome these drawbacks is 'ridge cultivation'. The principle of the method is to establish permanent 'tracks' in the field along which the tractors travel during the entire period of soil preparation and inter-row cultivation, and to create a deep-tillage zone between the tracks, which will then serve as the seed-bed and growing area for the crop.

It has been shown that the first time the tractor crosses a ploughed field, the soil is packed ten times as much as by subsequent passages (Cook *et al.*, 1958). Much compaction of the soil can therefore be avoided by driving along old tracks.

Ridge cultivation was originally developed for preparing seed-beds in fields that were to be irrigated by furrow irrigation; it was soon found, however, that the method has merits which are quite independent of the irrigation system adopted. The furrows between the ridges provided efficient drainage during the rainy periods; in early spring the ridges dried more rapidly, and the soil warmed up sooner, than in land cultivated 'on the flat', in the conventional manner.

All the unfavourable effects of compaction following tillage operations remained confined to the furrows between the ridges; as a result, only one-quarter, approximately, of the field was adversely affected, whilst roots found favourable conditions in the remaining three-quarters, in which soil structure was well preserved and ensured favourable conditions of aeration and moisture.

There are four stages at which ridge cultivation can be initiated:

1. Immediately after harvesting the previous crop.
2. When preparing the land, before winter.
3. During winter cultivation.
4. In spring, for seed-bed preparation.

If it is decided to start with ridge cultivation immediately after harvesting the preceding crop, the method used has been called 'precision tillage' (Figure 82).

In this case, the first operation consists of chopping up the residues of the previous crop. Then follows use of a tool consisting of two 75-cm shanks equipped with chisels, followed by three bed-listers, all mounted on a cultivator frame. In a single operation, the soil is sub-tilled to a depth of 50–65 cm by the chisels, whilst the listers place the ridges accurately above the zone loosened by the chisels, effectively covering the residues of the previous crop.

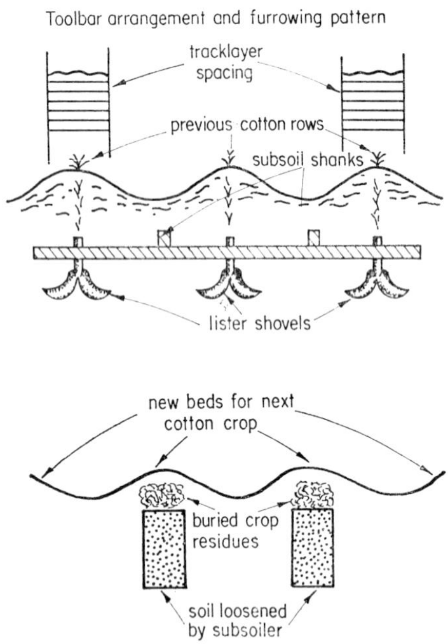

Figure 82. Schematic representation of precision tillage. (From Johnson (1963): by courtesy of *World Farming*.)

During the winter months, the ridges are reshaped and weeds are killed by a special implement called a disk-bed shaper, the wheels of the tractor always travelling in the furrows. If the field is too wet for cultivation, weeds are controlled by spraying.

Precision tillage has been developed at the Shafter Station in the San Joaquin Valley in California, for planting cotton continuously in the same field. Results show that root penetration is improved, yields are increased, and the cotton is earlier-maturing when 'precision' tilled, as compared with the conventional tillage methods, while there is a decrease of over 60 per cent in the total energy requirements of the various operations indicated in Table XXIV (cf. Stockton *et al.*, 1964).

TABLE XXIV

Operations involved in Preparing a Seed-bed for Cotton by Two Tillage Methods (Stockton *et al.*, 1964)

	Conventional tillage	*Precision tillage*
Stalk disposal	x	x
Double-disk harrowing	x	
Ploughing	x	
Disk harrowing	x	
Listing for irrigation	x	
Precision tillage	–	x
Nematocide application	x	x
Planting	x	x

Land-Levelling

Land-levelling is of great importance in irrigation farming. Its purpose is to obtain an even distribution of water in the field. Water should be applied at intervals and in quantities that are calculated to restore the water content of the soil in the root-zone to field-capacity. This is impossible to achieve for the field as a whole, unless its surface is sufficiently level and uniform. Assuming that water is being supplied at the optimal rate, both high spots and low spots will reduce yields—the first receiving too little water, the latter too much.

Whilst careful and accurate levelling is essential for land that is irrigated by flooding or furrow irrigation, a reasonably even surface is also important when overhead irrigation is practised, as otherwise water will run off the high spots and collect on the low spots. Levelling must therefore be considered as an essential element in land preparation. Levelling operations should, however, be restricted to the absolute minimum possible; for as they have to be carried out in reasonably dry soil, there is a tendency for soil aggregates to be broken down by them, and for soil structure to be thereby impaired (Plate 28).

DEPRESSION STORAGE

An uneven micro-relief may make it possible to store temporarily an appreciable amount of water during a rain or irrigation, so that it can later percolate into the soil. This transient storage capacity has been called 'depression storage' (Larson, 1962). Rough tillage methods, if done on the contour, could have a potential storage for a 50 to 75 mm fall of rain, as compared to the 25 mm or less that can be stored with a smooth soil surface as usually prepared by conventional tillage practice (ibid.).

Seed-bed Preparation

The germinating seed and the young seedling have a number of requirements which make certain demands, that may be conflicting, on the desired character of the seed-bed: these will now be considered.

Germination requires close contact between seed and soil particles, or otherwise water will not move rapidly enough into the seeds and germination will be delayed; free access of air to the germinating seed is also required, in view of the high oxygen requirements for the vital processes involved in germination and early growth.

The seedling must develop and reach the soil surface and light as soon as possible, to begin growth that will be independent of the nutrient reserves in the seed. The seed cannot be placed too deeply as otherwise its store of energy may not suffice for growth to the surface, and the seedling, if it emerges, will be weakened and hence unduly susceptible to attack by diseases and pests; nor can sowing be too shallow, or the seed-bed may dry out too quickly to allow the seed to absorb sufficient water for germination. The development of the seedling will also be adversely affected, if its roots have to penetrate an excessively compacted soil, or large clods; numerous air-pockets are also detrimental.

A seed-bed should provide the essential conditions for good germination, and a favourable environment for the emerging seedling. These are, however, often conflicting requirements; thus a firm seed-bed may be needed to ensure close contact between seed and soil, whereas a loose seed-bed may be desirable for good aeration. The smaller the seed is, the greater will be its sensitivity to seed-bed conditions and the greater the need for meticulous seed-bed preparation.

Ploughing a too dry, too wet or compacted soil will leave many large, hard clods on the surface. Attempts to reduce these clods to a fine tilth by mechanical means, will usually result in a breakdown of the soil into dust, and the crusting of the soil surface after the first rain or irrigation.

In semi-arid and arid regions, the rapid drying of the soil surface favours the development of hard crusts which impede emergence. At the same time, because the soil dries so rapidly, the rate of seedling emergence assumes considerable importance.

Excessive packing of the soil surface will also increase the soil's crust strength and further impede emergence of the seedlings, thereby causing a spotty, unsatisfactory stand, and frequently resulting in a need for reseeding. Thus it has been shown, for example, that an increase in crust strength from 108 to 273 millibars* results in a decrease in emergence of bean seedlings from 100 per cent to nil (Richards, 1953). A light rain or irrigation will soften the clods and enable their breakdown by a single light harrowing or cultivation.

Insufficient compaction of the seed-bed will leave many large airpockets in it, which interfere with germination and the root development of the seedling. On the other hand, excessive compaction impedes soil-moisture movement and therefore reduces the moisture-supply available to the seed. The poor aeration of an excessively compacted seed-bed is also detrimental to germination. However, adequate compaction of the seed-bed is effective in improving emergence, only if sufficient moisture for germination is available below the seed (Stout *et al.*, 1961).

Packing the soil at the surface may favour the formation of a crust and impede seedling emergence; it is therefore recommended to compact the soil at seed level, leaving the soil above the seed in a loose condition. This can be achieved, for example, by the use of a narrow press-wheel attached to the seed drill that presses the seeds into the soil, and then covers the seed with loose soil. The surface soil should not be broken down too finely, but left in a cloddy condition. Such a seed-bed does not crust easily and water penetration into it will be satisfactory.

Newer Concepts in Tillage Methods

STUBBLE-MULCH FARMING

The traditional methods of tillage, developed in temperate moist climates and based on the mould-board plough, when adopted indiscriminately in arid-land cultivations have frequently had disastrous effects, mainly by increasing erosion. The best-known example is that of the 'dust-bowl' of the central United States, which was created as a result of ploughing the original prairie for the growing of cereals, using clean-tillage methods, until wind erosion assumed devastating proportions. Under the conditions of limited rainfall, maintaining a clean fallow was considered essential to build up sufficient moisture in the soil for the cereal crop. The frequent cultivations that were needed to keep the fallow weed-free, pulverized the surface soil; the degraded and unprotected top-soil was then easily eroded by wind and rain.

The traditional concept of 'clean' cultivation, in which the farmer equated

* *Millibar.* A unit of pressure of 100 dynes per square centimetre.

good farming practice with the complete burying of crop residues and weeds, had to be abandoned. Instead, a new approach has developed, which aims at keeping the soil protected at all times, whether by a growing crop, or by crop residues left on the surface during fallow periods. The name of this new system is 'stubble-mulch farming'.

In experiments in Nebraska it was found that the rates of erosion of a mulched surface were about one-fifth of those resulting from 'clean tillage' with mould-board ploughing (Zingg & Whitefield, 1957). Wind-erosion was also substantially reduced: residues of various amounts, types, and orientation, reduced the force of the wind immediately above the soil's surface by from 5 to 99 per cent (Zingg, 1954). The crop (and weed) residues left on the surface of the soil not only offered protection against erosion, but improved water penetration and reduced losses from evaporation.

In a three-years grain rotation in Nebraska, a clean-ploughed loess soil lost 2 to 6 times as much run-off water and 4 to 8 times as much soil by erosion as did stubble-mulched land (Duley, 1960).

Within certain limits, the control of run-off and erosion is proportional to the amounts of crop residues left on the soil's surface; small amounts are more effective against wind-erosion than against water-erosion (Zingg & Whitefield, 1957). When the total straw production of the crop does not exceed one tonne per hectare, very little protection can be expected from the residues left on the surface (Luebs & Laag, 1964).

Stubble-mulch tillage was found to improve moisture conditions in the seed-zone significantly, when compared with disk tillage. In the central Great Plains (Kansas), soil moisture at seeding time was significantly greater under stubble-mulch than on clean-tilled plots in 3 out of 6 years, in a fallow-wheat crop rotation. This improvement occurred in years with high-intensity rainfall or when 4 to 6 days of consecutive precipitation occurred more frequently than usual (Harris, 1963). On the other hand, stubble-mulch tillage had no influence on soil-moisture storage at a depth of 1·30 m (Army *et al.*, 1961).

Mulches have also been found effective in reducing salinity. Thus low precipitation, high evaporation, high water-tables, and a compacted soil, generally cause salt increases in the upper soil layer under bare fallow. Seasonal precipitation has to be in excess of 250 mm to reduce salinity. Using a straw-mulch was found to result in appreciable reduction in salinity, even with below-normal precipitation (Benz *et al.*, 1967).

Better stands of wheat have been found to develop under stubble-mulching as a result of an improved soil-moisture régime (Zingg & Whitefield, 1957). However, in certain seasons the exclusive use of subsurface tillage during the fallow period may produce a surface that is too cloddy for seed-bed preparation and moisture conservation (Staple, 1964).

Mulches may reduce soil temperatures. For a period of 3 to 6 months after applying a straw-mulch of 5–8 tonnes/ha, soil temperatures were reduced by as much as 3 to 6°C at 2·5-cm depth and by 2 to 4°C at a 10-cm depth (McCalla & Duley, 1946).

Heavy mulches (20 tonnes of straw per hectare) may cause a still more considerable reduction of soil temperatures; a lowering of temperature of 17·7°C to a depth of 25 cm, due to such a heavy mulch, has been recorded (McCalla & Duley, 1946). However, under conditions of subsurface tillage, the *normal* amounts of crop residues have only a very limited effect on soil temperatures—especially after a few months of decomposition, when differences may amount to no more than 1 to 2°C during the warmest parts of the day (ibid.). Soil temperature differences due to mulching are still more reduced when the crop shades the soil effectively.

The high C/N ratio of the crop residues may depress nitrification especially in the early growth period and thereby adversely affect the growth of the crop and, ultimately, its yield.

Research data appear to indicate that the response to stubble-mulch tillage is intimately linked with nitrogen availability, as is indicated by the greater response to nitrogen of cereals growing on stubble-mulch tilled plots as compared with conventional tillage, and also by the reduction in protein content of the grain and the reduced straw production of mulched plots. Soil analysis also showed reduced nitrates under stubble-mulching (Luebs & Laag, 1964).

The lower rate of nitrification under stubble-mulch, as compared to 'clean-tilled' plots, may be due to lower temperatures, to less effective aeration, and most important, to nitrogen fixation by the crop residues, which decomposed more slowly on the soil surface then when incorporated in the soil (Figure 83).

The obvious conclusion to be drawn from these observations is that a favourable effect from stubble-mulching can be expected only if nitrogen applications are made in the quantities and at the times required to offset the negative effect of the mulch on available nitrate supply.

Stubble-mulch tillage, however, presents practical problems: the residues left on the surface interfere with seed-bed preparation, weed control and sowing operations. The traditional tillage and sowing equipment are not suitable under these conditions, and a whole new class of equipment had to be developed before the new system became practical. Subsurface cutting implements were developed which loosen the soil and cut the roots of weeds, leaving the residues on the surface as undisturbed as possible. The typical tillage implement for this purpose is the 'sweep-cultivator' (Plate 29), the sweeps consisting of large inverted V-shaped blades, attached to a cultivator frame and operating parallel to the soil surface at a depth of 10–15 cm. The sweeps slightly lift and shatter the soil.

For mechanical weed-control, the two main types of implements used are the 'blade cultivator', with a single wide blade on a two-wheel carriage, and

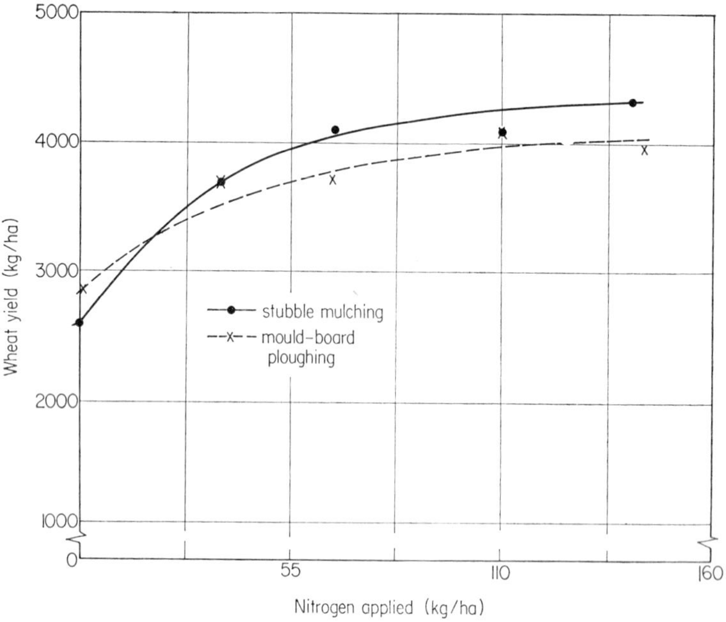

Figure 83. Average trend of wheat yields under stubble-mulching and mould-board ploughing in relation to amount of nitrogen applied. Note that stubble-mulching gives superior yields to mould-board ploughing only when adequate amounts of nitrogen are applied. (Zingg & Whitefield (1957): by courtesy of the U.S. Department of Agriculture.)

the 'rod-weeder', in which a slowly revolving horizontal rod, operating at a depth of 10 cm or more, compacts the subsurface and moves clods and weeds to the surface.

For seed-bed preparation, treaders—a kind of rotary hoe run in reverse—which break the soil crust, compact the subsurface, and kill small weeds, are used. Special seeding equipment has been devised in which disk-furrow openers clear narrow paths in the stubble, in which the seeds can be placed.

Stubble-mulch farming has been found to be most effective and beneficial in regions and seasons of low rainfall, and is critically needed where wind-erosion presents a considerable hazard. Under conditions of higher rainfall, negative results have been reported; apparently, the stubble-mulch is then conducive to excessive soil moisture, with its resultant ill-effects.

MINIMUM TILLAGE

Research in recent years has repeatedly shown that frequent tillage operations are rarely beneficial, and frequently detrimental, in addition to being costly. It is a paradox that the heavy equipment used for tillage, cultivation, and harvesting, damages the soil structure and increases soil compaction.

The problems are most acute under conditions of intensive cropping, involving frequent tillage operations and much movement of machinery—especially on soils with a relatively low organic-matter content (Free, 1960). Experience in orchards has even shown that abandoning tillage entirely can improve soil structure and water permeability as compared with the traditional cultivation methods. The root system of a crop growing in an undisturbed soil tends to improve soil structure: the roots penetrate the soil, causing cracks to form and clods to break down into crumbs, while the fine rootlets bind together small soil particles into larger crumbs. When the roots die and decompose, they leave innumerable channels which improve movement of air and water in the soil.

Crop residues remaining on the surface of the soil protect the soil structure of the upper layer. Non-tillage is, however, generally not a practical proposition for annual crops, though the development of herbicides, such as the bipyridils that kill weeds and become herbicidally inactive before the sowing of the crop, sometimes makes it possible to drill crops without any previous cultivation.

A world-wide tendency to reduce tillage operations to a minimum has been in evidence in recent years. This involves two complementary approaches: to omit any tillage operations whose cost exceeds its contributions to increased productivity or efficiency of production, and to combine agricultural operations.

Minimum tillage has been made practical and economical by the development of special equipment for combined tillage and sowing operations, and by the enormous progress achieved in chemical weed-control, both selective and general, which has rendered superfluous many previously indispensable tillage operations. Minimum tillage frequently gives as good as, or even better yields than, the conventional methods; the advantage in yield of maize obtained with minimum tillage was found to be even greater in years of low rainfall than in years of heavy rainfall (Doren & Ryder, 1962). These favourable results are ascribed to improved soil physical conditions, including higher rate of infiltration, less soil resistance to penetration, and less soil compaction (Figure 84), or to indirect effects, including fewer weeds, more root growth, better vegetative development, and less lodging (Rao *et al.*, 1960).

These advantages were more apparent on coarse and medium-textured soils than on the more heavy clay soils. In general, it takes from two to three years before the favourable effects of minimum tillage are fully apparent. The improvement of soil compaction, due to previous conventional tillage methods, is a gradual process.

In general, germination is less even with minimum tillage than with conventional tillage. Moreover it should be noted that there is more danger of failures in crop establishment if the soil is worked when too dry than when it is too wet.

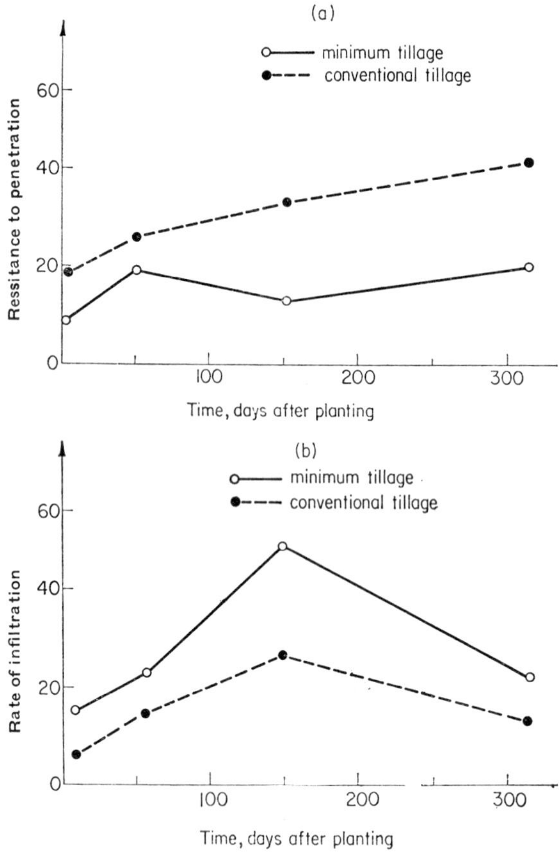

Figure 84. Comparison of conventional tillage and minimum tillage. Seasonal changes in the maize row:
(a) in soil resistance to root penetration
(b) in rate of water infiltration
(Rao *et al.* (1960): by courtesy of the American Society of Agronomy.)

'Minimum tillage' can be defined as a method aimed at reducing tillage to the minimum necessary for ensuring a good seed-bed, rapid germination, a satisfactory stand, and favourable growing conditions.

In recent years, studies of soil structure in the root-zone of maize have shown that the optimum conditions in the row are different from what is optimum in the region between the rows. From this we may conclude that fields in which row-crops are grown should be managed as two distinct zones, called by Larson (1962) seedling environment zone and water management zone and by Shear (1965) seed-bed zone and root-bed zone, respectively (Plate 30).

As tillage requirements are not the same for all crops or soils, a number of different approaches and systems have been developed, to achieve the aims of minimum tillage.

Plough Planting

In this method, only a single trip over the field is required. The tractor pulls a plough and a planter, simultaneously (Plate 31). The seed-row is centred on the furrow slice. The area between the rows remains rough, and weeds do not germinate easily in it.

A plough-packer, attached to the frame of the plough, makes possible 'once-over' soil preparation. Costs of seed-bed preparation were found to be considerably reduced, whilst yields were not reduced, as compared with those obtained by conventional methods (Cook & Peikert, 1950). The main disadvantage of the method is that planting is slowed down, so that sowing may be delayed beyond the optimum period.

Till-planting

A special 'till-planter' has been developed, that prepares a seed-bed and sows two rows in one operation. The seed-bed is prepared by an implement equipped with a narrow and deep-penetrating sweep, a wider and shallower sweep, and sections of a rotary hoe. The strip between the row is not disturbed (Aldrich, 1956).

Wheel-track Planting

In this method, the field is ploughed as usual, but instead of using the conventional methods of seed-bed preparation, the seeds are planted in a seed-bed prepared by the wheels of the tractor (Plate 32). Experience has shown that tractor wheels make a firm, satisfactory seed-bed even on cloddy soil, ensuring rapid and even germination.

The soil between the rows remains rough and loose, and is therefore better able to absorb moisture and to reduce run-off. Weed seeds lie dormant in the loose soil, until rain falls or the field is irrigated (Plate 33).

Wheel-track planting was found to save about 40 per cent of tillage costs (Peterson *et al.*, 1958). By adding two extra rear wheels to the tractor, it is possible to use a 4-row planter. Experience has shown that the optimal soil moisture for ploughing is also optimal for seed germination (Hansen *et al.*, 1958). Any delay between ploughing and sowing always reduces soil moisture, which may become too low for seed germination. In order to prevent the top-soil from drying out and becoming hard before sowing, it is essential to plough and plant the field in one and the same day.

Minimum tillage may entail certain changes in crop management: for example, conventional tillage, by promoting aeration, increases the rate of decomposition of organic matter in the soil, and hence the release of nitrogen to crop plants. Under conditions of minimum tillage, it may be necessary to offset the reduced supply of nitrogen from the soil reserves, by increased mineral fertilization.

Certain problems related to the use of minimum tillage have not yet been resolved. Herbicide costs are still frequently in excess of tillage costs. Too little is known of the long-term effects of herbicide residues in the soil. The possibility that herbicide-resistant strains of weeds may develop has also to be considered.

Inter-row Cultivation

Inter-row cultivation is supposed to be beneficial by destroying weeds, by improving soil aeration, and by reducing losses of moisture from the soil.

Numerous experiments carried out in the field to determine the effect of inter-row cultivation on yields and on moisture conservation, have shown conclusively that the stirring of the soil was beneficial mainly in controlling weeds.

Besides weed control, cultivation may occasionally have additional beneficial effects. As a result of improved soil porosity (Musgrave & Free, 1936), infiltration is increased and soil loss reduced after cultivating a crusted soil surface, such as usually forms after irrigation or heavy rainfall on certain soil types (Borst & Woodburn, 1942). Infiltration rates have been increased sixfold by loosening 0–8 cm of crusted surface soil (Duley, 1939). The best results are obtained when the rough configuration of the soil surface is preserved.

Improved nitrification in the upper soil layer, and somewhat lower temperatures in the loose soil, also result from cultivation, as compared with an uncultivated soil.

An important benefit from maintaining a loose layer on the surface of the soil is obtained on soils that tend to shrink and crack deeply on drying. Such cracks, which penetrate deeply, may cause the loss of considerable amounts of moisture throughout much of the root-zone. In Texas, it was found that the evaporation from the side-walls of shallow shrinkage cracks in a clay soil varied from 33 to 91 per cent of that from a comparable area of surface soil (Adams & Hanks, 1964). These cracks are filled and covered by the loose layer that is formed during cultivation of the soil surface.

The general rule, nevertheless, is to keep the number of cultivations to the minimum required for weed control, and to cultivate at the most appropriate time for this purpose. Cultivation should be as shallow as is compatible with this purpose, in order to reduce damage to the roots of the crop as far as possible.

Summary

It has been shown in this chapter that most tillage operations have both beneficial and detrimental effects.

Ploughing, the traditional and most basic tillage operation in temperate climates, inverts the soil, covers manure and crop residues, and destroys weeds, many insects, and disease agents. Frequent ploughing, however, creates a plough-sole which impedes infiltration and root penetration; the very 'cleanness' achieved is, in the dry regions, a factor for increased soil-loss with disastrous results.

Subsoiling improves infiltration, reduces erosion, and breaks up hard-pans; it may, however, favour the spread of diseases carried from one crop to the next on the crop residues or stubble, as well as encourage the development of perennial weeds.

Too frequent tillage of *dry* soils causes a breakdown of soil aggregates, resulting in poor soil structure, crusted soil surfaces, and reduced resistance to erosion.

Tillage of *wet soils* caused excessive compaction, resulting in poor infiltration and decreased aeration of the soil.

Deep-ploughing is very effective in the destruction of perennial weeds, and in ensuring a complete drying of the soil. The latter is an important factor in achieving improved soil structure and sanitation of soils in arid areas particularly in those that are under irrigation. Deep ploughing is, however, extremely expensive; it leaves an uneven soil surface requiring renewed levelling, and may reduce soil fertility by inverting the most fertile layer and replacing it with biologically inactive soil.

From the foregoing, two essential conclusions can be drawn:

1. Tillage operations should be reduced to the absolute minimum required for achieving the farmer's purpose.
2. By alternating different methods of tillage, and applying each method under the conditions in which its beneficial effect is maximized, it is possible to reduce the detrimental effects of tillage to a bare minimum. A 'rotation' of tillage operations may therefore be as important as a rotation of crops, and the two are best integrated.

References

Adams, J. E. & Hanks, R. J. (1964). Evaporation from soil shrinkage cracks. *Proc. Soil Sci. Soc. Am.*, **28**, 281–4.

Aldrich, S. R. (1956). New styles in seed beds. *What's New in Crops & Soils*, **9**(3), 12–15.

AMEMIYA, M. (1968). Tillage-soil water relations of corn as influenced by weather. *Agron. J.*, **60**, 534–7.

ARMY, T. J., WIESE, A. F. & HANKS, R. J. (1961). Effect of tillage and chemical weed control practices on soil moisture losses during the fallow period. *Proc. Soil Sci. Soc. Am.*, **25**, 410–13.

BAILEY, L. H. (1950). *The Standard Cyclopedia of Horticulture.* Macmillan, New York, vol. III. vi + 1 216 pp., illustr.

BENZ, L. C., SANDOVAL, F. M. & WILLIS, W. O. (1967). Soil salinity changes with fallow and a straw mulch of fallow. *Soil Sci.*, **104**, 63–8.

BISHOP, C. W. (1938). *Origin and Early Diffusion of the Traction Plough.* Smithsonian Instutition, Washington D.C., Publ. **3477**, 16 pp., illustr.

BORST, H. L. & WOODBURN, R. (1942). Effect of mulches and surface conditions on the water relations and erosion of *Muskingum* soil. *Tech. Bull. U.S. Dept. Agric.*, **825**, 16 pp., illustr.

COOK, R. L., MCCOLLY, H. F., ROBERTSON, L. S. & HANSEN, C. M. (1958). Minimum tillage. *Ext. Bull. Mich. St. Univ.*, **352**, 23 pp.

—— & PEIKERT, F. W. (1950). A comparison of tillage implements. *Agric. Engng*, **31**, 211–14.

CURLEY, R. B., KRANTZ, B. A., VOMOCIL, J. A. & CHANCELLER, W. J. (1959). *Vertical Mulching one Possible Solution to Soil Compaction Problems.* Calif. Agric. Ext. Serv.

DONEEN, L. D. & HENDERSON, D. W. (1953). Compaction of irrigated soils by tractors. *Agric. Engng*, **34**, 94–5, 102.

DOREN, D. M. VAN & RYDER, G. J. (1962). Factors affecting use of minimum tillage for corn. *Agron. J.*, **54**, 447–50.

DULEY, F. L. (1939). Surface factors affecting the rate of intake of water by soils. *Proc. Soil Sci. Soc. Am.*, **4**, 60–4.

—— (1960). Yields in different cropping systems and fertilizer tests under stubble mulching and plowing in eastern Nebraska. *Res. Bull. Nebr. Agric. Exp. Stn*, **190**, 53 pp.

FREE, G. R. (1960). Minimum tillage for soil and water conservation. *Agric. Engng*, **41**, 96–9, 103.

HANSEN, C. M., ROBERTSON, L. S. & GRIGSBY, B. H. (1958). The practicality of plowing for and planting corn in one operation. *Q. Bull. Mich. Agric. Exp. Stn*, **40**, 549–54.

HARRIS, W. W. (1963). Effect of residue management, rotations and nitrogen fertilizer on small grain production in north-western Kansas. *Agron. J.*, **55**, 281–4.

JOHNSON, C. (1963). Cotton production. *Wld Fmg*, **5**(4), 14–16.

LARSON, W. E. (1962). Tillage requirements for corn. *J. Soil Wat. Conserv.*, **17**, 3–7.

LUEBS, R. E. & LAAG, A. E. (1964). Tillage and nitrogen for dry-land grain in a winter rainfall climate. *Bull. Calif. Agric. Exp. Stn*, **805**, 34 pp., illustr.

MCCALLA, I. M. & DULEY, F. L. (1946). Effect of crop residues on soil temperature. *J. Am. Soc. Agron.*, **38**, 75–89.

MUSGRAVE, G. W. & FREE, G. R. (1936). Some factors which modify the rate and total amount of infiltration of field soils. *J. Am. Soc. Agron.*, **28**, 727–39.

PETERSON, A. E., GERGE, O. K., MURDICK, S. T. & PETERSON, D. R. (1958). Wheel-track corn planting. *Circ. Wisc. Ext. Serv.*, **559**.

RANEY, W. A., EDMINSTER, T. W. & ALLAWAY, W. H. (1955). Current status of research in soil compaction. *Proc. Soil Sci. Soc. Am.*, **19**, 423–8.

RAO, A. A. SWAMY, HAY, R. C. & BATEMAN, H. P. (1960). Effect of minimum tillage on physical properties of soils and crop response. *Trans. Am. Soc. Agric. Engnrs*, **3**(2), 8–10.

RICHARDS, L. A. (1953). Modulus of rupture as an index of crusting of soil. *Proc. Soil Sci. Soc. Am.*, **17**, 321–3.

SHEAR, G. M. (1965). The role of herbicides in no-tillage crop production. *Agric. Chem.*, **20**, 31–6.

SPAIN, J. M. & MCCUNE, D. L. (1956). Something new in subsoiling. *Agron. J.*, **48**, 192–3.

STAPLE, W. J. (1964). Dryland agriculture and water conservation. pp. 15–30 in *Research on Water*. Soil Sci. Soc. Am., Madison, Wisconsin, vi + 121 pp., illustr.

STOCKTON, J. R., CARTER, L. M. & PAXMAN, G. (1964). Precision tillage for cotton. *Calif. Agric.*, **18**(2), 8–10.

STOUT, M. A., BUCHELE, W. F. & SNYDER, F. W. (1961). Effect of soil compaction on seedling emergence under simulated field conditions. *Agric. Engng*, **42**(2), 68–87.

SWAMY RAO, A. A. *See* RAO, A. A. SWAMY.

THORNTHWAITE, C. W. (1958). Introduction to arid zone climatology. *Arid Zone Res.*, **11**, 15–22.

TULL, J. (1731). *Horse Hoeing Husbandry*. London.

VAN DOREN, D. M., JR. *See* DOREN, D. M. VAN, JR.

WIESE, A. F. & ARMY T. J. (1958). Effect of tillage and chemical weed control practices on soil moisture storage and losses. *Agron. J.*, **50**, 465–8.

ZINGG, A. W. (1954). The wind erosion problem in the Great Plains. *Trans. Am. Geophys. Un.*, **35**, 252–8.

—— & WHITEFIELD, C. J. (1957). Stubble-mulch farming in the western states. *Tech. Bull. U.S. Dept. Agric.*, **1166**, 56 pp., illustr.

XIV

PLANT POPULATION AND DISTRIBUTION PATTERNS

Reducing plant density under dry conditions in order to achieve a balance between the amount of plant cover and the limited soil moisture, has long been an accepted cultural technique. It is only in recent years that agronomists have become fully aware that the higher potential yields made possible by the favourable water régimes provided by irrigation, the high soil-fertility levels resulting from heavy applications of fertilizers, and the genetic potentials of new varieties and hybrids, could be achieved only after appropriate adjustments of the plant population. Research on adjusting population levels to all other factors of production has therefore become very extensive in recent years.

General Plant Density–Yield Relationships

total dry-matter production (biological yield)

When plants are widely spaced, dry-matter yields will at first tend to increase linearly with increases in plant density. This indicates that no appreciable competition is occurring between neighbouring plants. Further increases in density bring diminishing returns, indicating that competition between plants is making itself felt, and that the production of the individual plant is being reduced; this reduction is, however, still being more than compensated for by the increase in the number of plants.

A further increase in density results in a 'plateau' of practically constant yield of dry matter, indicating that an equilibrium has been reached at which the increase in production due to increased plant numbers compensates almost exactly for the reduction in the production of the individual plant (Figure 85).

To obtain the maximum yield possible, it is essential for the crop to utilize as efficiently as possible all available production factors: water, nutrients, light and CO_2.

The number of plants required per unit area to achieve this aim will depend on the nature of the crop and on its environment. This number cannot be too small, or not all the production potential will be fully utilized: nor can it be too large, or excessive plant competition will reduce the overall efficiency of the crop.

Increasing the number of plants per unit area, reduces the volume of air and soil that the individual plant can exploit, and therefore increases competition between plants for soil nutrients, moisture, CO_2 and light. Competition for physical space, however, is rarely limiting (Black, 1966), though such competition among roots may occasionally occur when they have to penetrate dense layers or hard clods through which there are relatively few channels (Donald, 1963).

Maximum exploitation of the factors needed for growth is achieved only when the plant population exercises maximum pressure on all the above factors (Donald, 1963). As a result, the individual plants are under relatively severe stress because of inter-plant competition. In Subterranean Clover, for example, the total dry matter produced per plant at a wide spacing (6 plants/m^2) was 34 gm, whilst at the density which produced maximum yields (1 500 plants/m^2) the individual plant produced only 0·6 gm, or less than 2 per cent of the former amount (Donald, 1954). Thus we come to the somewhat paradoxical conclusion that maximum yields are obtained from plant populations which do not allow the individual plants to achieve their maximum potential.

ECONOMIC YIELD

Many studies have shown that the relationship of yield of seed or fruit (economic yield) to density is different from that of total dry-matter production (biological yield). In the latter, total production yields rise with increasing density to a maximum, which then remains practically constant (Donald, 1963). At this density, any increase in total dry-matter yield per unit area due to an additional number of plants, is offset by an equal loss due to the decrease in weight per plant. This relationship also applies to other vegetative yields, such as total yield of potato tubers (Holliday, 1960).

The relationship of seed production to population density is different. As plant density increases, the yield of seed increases to a maximum, which remains constant within a certain range, and then declines more or less steeply as population pressures increase still further—even when moisture and nutrients are not limiting (Holliday, 1960). With maize, for example, the decline in yields is fairly steep when the optimum density is exceeded, whilst with cotton—a crop in which individual plants have a far greater ability to adjust to population density—the graph depicting yield of cottonseed exhibits a fairly wide plateau of optimum densities (Figure 85).

Donald (1963) points out that the two curves of seed yield and biological

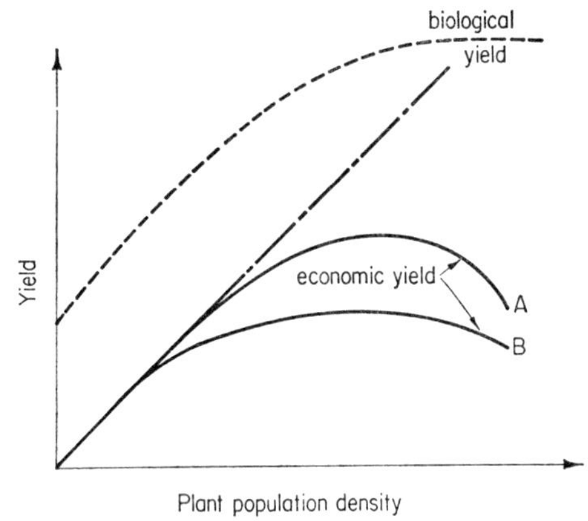

Figure 85. Generalized curves showing effect of plant population on biological yield and on economic yield. A: maize (grain). B: (cottonseed).

yield are interdependent, and that peak seed yields occur approximately at the same density at which the biological yield starts levelling off. He therefore suggests that 'the minimum density giving the ceiling biological yield may also be the density giving the maximum seed yield'.

In the young plant, most of the net assimilates are used for the production of additional leaves. With the appearance of the inflorescence, competition for assimilates between the leaves and the inflorescence takes place. Then, when seed formation starts, most of the assimilates move to the grain.

Thus there exists an internal competition, within the individual plant, between vegetative and reproductive parts. This internal competition becomes more severe as competition between plants increases, e.g. with increasing plant population pressures.

As plant density is increased, changes may occur in the allocation of assimilates to different parts of the plant, as a result of which a greater proportion of the plants (or of the reproductive parts of a individual plant), may become barren. Grain production then shows a decline in yields, whereas the total dry-matter production may remain constant (Harper, 1961).

In maize, it has been demonstrated that the reduction in grain yield resulting from excessive plant density is due primarily to the increase in the percentage of barren plants (Prine, 1966). It was found that if the stand was thinned shortly before silking, the average number of ears per plant was the same as in a widely spaced stand. When thinning was done during the ten days after the first plant had silked, the number of ears was markedly reduced in comparison with widely spaced plants. Making more light available after silking had no effect. In maize, therefore, there is a critical period for light in relation to grain formation; stressed plants aborted their ears from three days before silking until about eight days after the emergence of the silk.

As a result of the differential effect of competition on biological and economic yield, the amount of stover (maize straw) required to produce a kilogram of grain decreases with increasing plant density up to a certain level; a further increase in density beyond this level then results in a lower grain/stover ratio (Dungan *et al.*, 1958).

COMPONENTS OF YIELD

As planting density increases, most components of yield of the individual plant are reduced. However, there are cases in which certain components do not follow this downward trend. For example, in Subterranean Clover it was found that, while the number of inflorescences per plant and the number of seeds per plant decreased progressively with increasing density, the number of seeds *per inflorescence* and the weight per seed rose to a peak value as density increased, up to a certain critical level, and fell only beyond this level of density. This was, incidentally, the level that corresponded to maximum seed yield per unit area (Donald, 1954).

Donald (1963) suggests the following explanation: at wide spacings there is little competition between plants at the early stages of growth, and a large number of primordia are initiated in early plant growth. As growth proceeds, inter-plant competition increases progressively. The load of inflorescences reduces the number of seeds per inflorescence and reduces seed weight, compared with somewhat denser stands. In moderately dense stands, inter-plant competition may be sufficiently severe at the time of flower initiation to reduce the number of floral primordia to a level at which the load on the individual plant is not excessive and maximum efficiency in seed production, as expressed by seed number and size, is achieved.

In very dense stands, both inter-plant and intra-plant competition are sufficiently severe to reduce all components of yield.

An interesting feature is that among the several highly plastic components of plant yield, the weight of the individual seed is only rarely influenced by changes in plant density. Harper (1961) considers this to be an internal or physiological homeostasis with respect to the organ that is essential for reproduction and dispersal. However, as we have already seen, there may be very considerable differences in the manner in which different crops react to plant density in other connections (ibid.).

Crops with *indeterminate growth* systems, such as *Vicia faba*, 'in which the flowering apices do not arise directly from the major vegetative apices', respond to density mainly by changes in the *number* of parts formed. On the other hand crops with *determinate growth*, such as sunflowers, respond mainly by changes in the *size* of the parts.

Crops in which a determinate phase follows an indeterminate phase (after vernalization and photoperiodic response), such as winter wheat, may respond by changes in the number of parts during the indeterminate phase, and by changes in the size of parts during the determinate phase.

Interrelationships between Plant Density and Factors Affecting Production

MOISTURE-SUPPLY

Although it is well known that the optimum density for a crop has to be lower than normal under conditions of a limited water-supply, the obverse—namely, that plant populations must be increased in order to exploit fully the favourable conditions provided by irrigation—has frequently been overlooked.

NUTRIENTS

All available evidence indicates that, at higher levels of nutrient-supply, more plants per unit area are required to exploit fully the higher soil fertility potential and thereby to produce maximum yields, than is the case at lower levels of supply. Conversely, as plant density increases up to a certain limit, the crop will continue to respond to higher levels of added nutrients.

In a study on maize concerning the relationship between density of population and different levels of nitrogen supply, it was found that the higher the level of nitrogen was, the greater would be the plant population required for achieving maximum yields (Figure 86). With a low supply of nitrogen the maximum yield of grain was 4720 kg/ha, which was obtained with a population of 3 plants/m^2; with a medium nitrogen level, the maximum yield was 5800 kg/ha, obtained with 4 plants/m^2; with a high nitrogen level, the highest yield was 7430 kg/ha, which required a density of 5 plants/m^2 (Lang *et al.*, 1956). At low levels of nutrient-supply, dense populations may have specific adverse effects on crop plants. In spacing experiments with rye-grass (*Lolium rigidum*) in Australia, it was found that when the nitrogen-supply was limited, the nitrogen content of the grass fell from 1·8 per cent for widely spaced plants to 1 per cent for extremely crowded plants (Donald, 1954). In addition, the *total* nitrogen taken up by the sward was less with crowded plants than with more widely spaced plants. Donald (1963) suggests that the plants submitted to excessively severe competition had a poorer root system and were less efficient in the uptake of nutrients from the soil than were less stressed plants.

LIGHT

Water régime and nutrient-supply to the crop can be regulated to enable the maximum number of plants to achieve their full yielding potential; however,

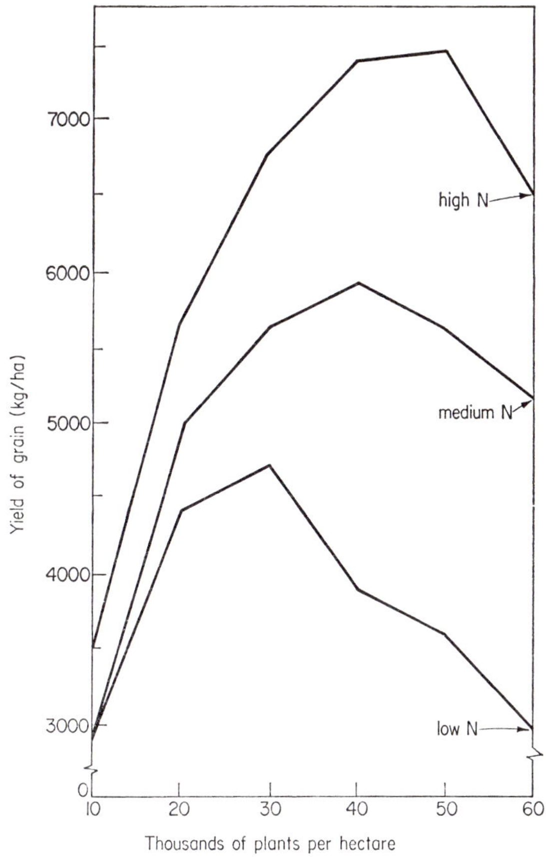

Figure 86. Effect of plant population at various nitrogen levels on grain yield of maize. (Lang *et al.* (1956): by courtesy of the American Society of Agronomy.)

increasing the number of plants per unit area will reduce the amount of light available to the individual plant, due to greater shading.

With irrigation and modern cultural practices, light is most frequently the factor limiting productivity, so that yields ultimately depend on the efficiency with which light is intercepted by the crop.

There is a basic difference in competition between plants for light and for other factors needed for growth. Moisture, nutrients, and CO_2 are withdrawn by the plants from a common pool. In plant populations of a single crop, assuming that differences in competitive ability between individuals of a similar genotype are not very great, the plants will share equally in the supply, until it is exhausted. By contrast, there is no pool of light that becomes exhausted as a result of withdrawal by the plant community (Donald, 1963).

A sparse stand, by withdrawing less water from the common pool at an early stage of growth, may 'save' water that will be available to the crop at a later, more sensitive stage. Any available light that is not intercepted by the canopy at a given moment, is lost to the crop, 'Light energy is instantaneously available and it must be instantaneously intercepted' (Donald, 1963).

Competition for light is universal and is absent only for a short period after emergence, or in the very wide spacings that are characteristic of plant communities which have become adjusted to arid conditions. In these last, the leaf cover never reaches a density which ensures the most effective utilization of radiation (Blackman & Black, 1959). In contrast with competition for water or nutrients, which increases in severity with decreasing availability, competition for light in a dense population will be equally acute under conditions of high light intensity and of overcast skies (Donald, 1963). Competition between plants is not a static process, and the intensity of competition will depend on the time when it occurs. In the early growth stages, there will be little mutual shading and even at relatively low light-intensities the plant will be able to photosynthesize with full efficiency.

As the plants develop, mutual shading increases, and light begins to become a limiting factor. When plants reach the flowering stage, they may even lose weight on cloudy days (Jackson, 1963).

Competition for light may be between plants, when one casts a shadow over any other, or within a plant, when one leaf shades another.

Essentially, there are two basic ways of ensuring maximum interception of light by the canopy: by improving the foliage pattern, and by increasing plant density.

(*a*) *Improving the Foliage Pattern*. The most efficient plant canopy is not that with the most leaves, but that in which the inter-plant and intra-plant competition for light is reduced to a minimum, so that the canopy as a whole intercepts a maximum of light. This aspect has been treated in detail in Chapter X (pp. 286–9).

(*b*) *Increasing Plant Density*. The light intensity required to give a maximum rate of photosynthesis by the individual leaf is about 1500–2000 foot-candles, whilst light intensities in the field may reach 10000 foot-candles (Donald, 1963). For example, in a sward of Bermuda Grass (*Cynodon dactylon*), isolated leaves were light-saturated at 2500 to 3000 foot-candles, whilst the light required to saturate the sward as a whole ranged up to 7000 foot-candles, depending on height and foliage pattern (Alexander & McCloud, 1962).

It is therefore possible to reduce the distance between individual plants, so as to increase the efficiency of the plant canopy *as a whole* in utilizing the available sunlight. However, there is an optimal ratio of leaf area to ground surface (LAI) beyond which increasing plant density will increase excessively the proportion of the foliage which is below the compensation point.

Hodgson & Blackman (1956) found that in *Vicia faba*, with 11–12 plants/m^2, light intensity at the base of the plants was above the compensation point, whilst with a density of 55–65 plants/m^2, during the flowering phase, the level of light intensity at the base of the plants fell as low as 0·03 per cent of daylight and an estimated 38 per cent of the leaves of the individual plants were at or below the compensation point.

An interesting experiment by Miller (1956) illustrates why, up to a certain plant density, the reduction in production of the individual plant is more than compensated for by the increase in the number of plants (Figure 87). A

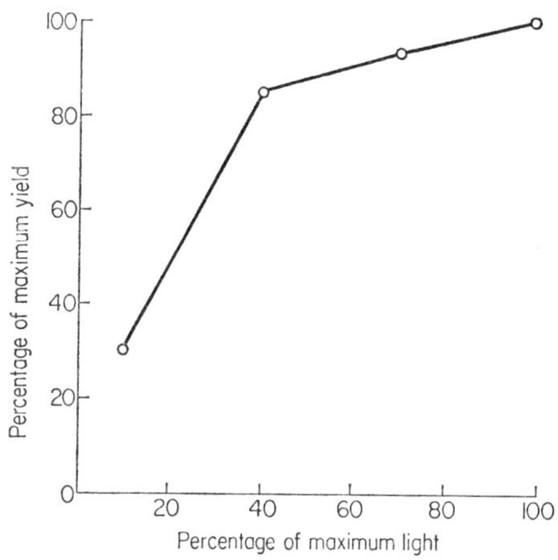

Figure 87. Grain yield of maize in relation to light. (Based on data from Miller, 1956.)

reduction of up to 40 per cent of the total sunlight causes a reduction of only 14 per cent of the yield of the individual maize plant; however, further reductions in sunlight cause a steep decline in yield.

Wit (1960) has shown that there is a linear relation between the reciprocal of yield and the area available per plant. The slope of this line decreases with increasing yield when the potential for yield is high. This could be expected, for a single plant will be able to cover more area the more its growth is promoted by favourable growing conditions.

The general conclusion that can be drawn is that the more favourable the conditions are under which a crop is grown, the greater will be the plant density required to exploit fully the potential and to achieve maximum yield; conversely, higher-than-average plant densities will require more favourable growing conditions.

Breeding for Adaptation to High Plant Densities

When production is not limited by moisture and nutrient-supply, and maximum interception of available light is achieved by an optimum plant population, the only means of increasing productivity further is to use genotypes with a greater adaptation to high plant densities.

Improvements in cultural techniques, and in particular irrigation and nitrogen fertilization, have led to the adoption of increased sowing rates. It was observed that not all the crop varieties which had been bred for the standard plant densities formerly employed, responded in the same manner to increased population pressures. Hybrids have been found to require more plants per ha than open-pollinated varieties, in order to reach the point of maximum yield per unit area. However, there are also differences in this respect between hybrids, and increasing attention is being given to the selection of hybrids that perform well in dense populations (Stinson & Moss, 1960).

Different maize hybrids were found to respond differently to plant density. The single cross Hy2 × Oh7 yielded well at high population-densities, whilst WF9 × C103 yielded best at a lower population-density (Lang *et al.*, 1956). Similar results were obtained in Oklahoma, where significant differences in grain yield were recorded among six hybrids in their response to plant density (Rutger & Crowder, 1967).

In comparing the normal Hy2 maize hybrid with a semi-dwarf mutant, it was found that the compact type yielded slightly more at high population-densities than did the normal type at any population-density tested. Furthermore, hybrids carrying the same dwarfing genes reacted in a similar way to population pressures (Nelson & Ohlrogge, 1957). At low plant densities, some maize hybrids produce several stems and multiple ears; others produce a single stem and one, or at the most two, ears per stem. It was observed that the hybrids with a high tendency to tiller and produce multiple ears, are those that produce the lowest percentage of barren stalks at high densities. Conversely, varieties that are single-stemmed at low densities, show the highest sterility in denser stands (Lang *et al.*, 1956).

The general conclusions to be drawn from breeding work with maize for high performance in thick plantings are that (a) hybrids differ in their capacity to utilize light for grain production; (b) hybrids that are most tolerant to shade are those that are best adapted to dense stands; and (c) the incidence of barren stalks is greater in the intolerant than in the tolerant varieties, when both are grown in shade (Stinson & Moss, 1960). Another hereditary characteristic of significance is the ability of the plant to resist lodging at high plant densities.

Similar differences in tolerance to dense plantings have been observed in various other crops. Rossiter (1959) found that the ranking of strains of Subterranean Clover was entirely different when they were grown at wide spacings as compared with greater crop density. In widely spaced plants the later varieties produced the highest yields of seeds, whilst the opposite was true for the dense populations. In this case, the main factor limiting yield was soil moisture, which was sufficient for widely spaced plants but became progressively more limiting with later flowering.

Bunch types of peanuts, beans, and peas, grow in an erect manner, and the pods are concentrated in a relatively small area. To produce maximum yields these varieties must accordingly be more closely planted than the runner types with their prostrate growth.

In order to evaluate the potential yielding ability of different varieties, the plant population should be high enough to exert a near-maximum demand on the fertility of the soil (Duncan, 1954).

Other Effects of Plant Densities

ROOT SYSTEMS

The extent and distribution pattern of root systems are markedly influenced by plant density. Very widely spaced plants of maize, which were presumably free from competition, developed a circular distribution of the roots, with a radius of about 75 cm. At high densities, the root systems of the individual plants interpenetrated to a high degree. At a 2·5-cm spacing within rows,* portions of as many as 32 overlapping root systems could be recorded at any one point beneath or near the row of plants, whilst at an 80-cm spacing there were only two overlapping root systems (Haynes & Sayre, 1956).

PLANT HEIGHT

With increasing density and competition for light, plant height may be markedly increased. For example, the average heights of wheat plants grown in rows spaced 30, 15 and 7·5 cm, apart were 63, 75 and 85 cm, respectively at the age of 16 weeks (Wassermann, 1963).

LODGING

The risk of lodging is markedly increased by high population levels (Dungan *et al.*, 1958).

PLANT COMPOSITION

Denser plant populations were found to decrease the protein and oil contents of maize hybrids (Lang *et al.*, 1956).

PLANT MORTALITY

At very high densities, competition may be so severe among the plants of a

* Each experimental plot consisted of a single row; the plots were spaced 250 cm apart, as a safeguard against between-row influence.

population that a proportion of them will die. In wheat, at a density of 1078 plants/m^2, only 40 per cent of the plants survived to the age of 6 months (Puckridge, 1962). This 'self-thinning' did not reduce plant-density to that giving the highest yield of grain per unit area. Thus, 'plant survival had precedence over total seed production per unit area' (Donald, 1963).

DISEASE INCIDENCE

When sorghum seeds were sown more closely in the rows, the more rapid spread of seedling diseases was assumed to be one of the reasons for the relatively smaller number of plants established, as compared with sowing more sparsely (Porter *et al.*, 1960).

There are occasional instances in which very close planting is effective in reducing the incidence of disease—as is the case for rosette disease of peanuts.

MATURITY

In maize hybrids, silking was found to be delayed by one day for each additional plant per m^2 (Lang *et al.*, 1956).

PLANTING PATTERNS

The following planting patterns might conceivably influence yields: square arrangements, rectangular arrangements, very elongated rectangles (wide row-interval with very small intervals within rows), irregular spacings in the row (pockets or hills within the row), and the direction of the rows.

SQUARE ARRANGEMENTS

It would be reasonable to expect that square arrangements of plants would be more efficient in the utilization of the light, water and nutrients, available to the individual plant, than would be a rectangular arrangement, as square arrangements would reduce to a minimum the competitive effect of neighbouring plants on one another. For wheat, Holliday (1963) has shown that decreasing row-widths below the standard 15–20 cm, i.e. reducing rectangularity, generally increases yield slightly, whilst increasing row-widths has a depressing effect on yields.

RECTANGULAR ARRANGEMENTS

The question of rectangularity has great practical significance. When prehistoric Man attached a hollow reed to his primitive plough, through which seed was fed during tillage, rectangular patterns of stand establishment became standard agricultural practice

This method of sowing had several advantages over broadcast sowing: a considerable saving in seed, a more uniform emergence and more complete stand, and easier weeding and harvesting. The distance of 30 to 40 cm between rows was very convenient, both for hand- and hoe-weeding, and is characteristic of traditional farming to this day. It was only when animal-drawn hoes were introduced for inter-row cultivation, that inter-row distances had to be increased to 75–100 cm. This led to a reduction in yields which, however, was not very marked, because of the relatively low yielding potentials. With the improvements in agricultural techniques and the introduction of varieties with high-yield potential, increases in plant population became essential. With increasing density of plant populations, in order to adjust to higher levels of potential yield, it was found in a number of crops that decreasing the larger dimension (inter-row distances) was more beneficial than decreasing the smaller dimension (intra-row distances) in rectangular spatial arrangements (Holliday, 1963). Fortunately, with the advent of selective herbicides, and in particular pre-plant and pre-emergence applications of them, mechanical inter-row cultivation was no longer indispensable and narrower rows are now a practical proposition. One of the most recent trends in crop production is a return to the inter-row distances that were characteristic of primitive farming!

However, extremely narrow rows would be necessary to achieve square spatial arrangements with the plant-density required for maximum yields of many crops, such as the small-grain cereals. This would entail considerable technical difficulties in most crops.

With wheat and similar crops, sowing machines with very narrow-spaced coulters cause difficulties during sowing and have a low efficiency due to the choking of the spaces between the coulters with crop residues or moist soil. Mechanical row-planters, for crops such as groundnuts and cotton, cannot work effectively in rows less than 60 cm apart. When furrow irrigation is practised, ridging for irrigation sets a limit of about 70–80 cm to the minimum practicable spacing between rows. Mechanical harvesting of certain crops, such as sugar-beets or cotton, also requires substantial inter-row spacings.

Much research has shown that quite considerable variations in planting patterns have relatively little influence on yield per unit area for a wide range of crops (Harper, 1961). With Broad Beans (*Vicia faba*), at equal numbers of plants per unit area, row spacings of 17, 34 and 51 cm had no influence on yields per ha (Hodgson & Blackman, 1956).

VERY ELONGATED RECTANGLES

Extreme rectangularity may be detrimental to yield, but usually some degree of rectangularity may be just as satisfactory as square-planting (Holliday, 1963). In general, the plant population has a greater effect on yield than does the spatial arrangement.

Limited Moisture-supply

Plant population must be adjusted to available soil-moisture levels, either within rows or between rows. Increasing the distance between the plants within the row in order to adjust to low moisture-levels generally defeats its own purpose: the young plants, with little or no intra-row competition, show excessive vegetative development; soil moisture is rapidly depleted, and the plants are unable to form satisfactory ears or to mature their grain normally. The alternative is to space plants more closely within the row, and to increase the distance between rows in order to compensate for low moisture-levels. In the wide-spaced rows, the soil-moisture supply is not exhausted as rapidly as in narrow rows (Brown & Shrader, 1959). Intra-row competition prevents excessive vegetative growth, and the laterally developing roots have to grow farther to reach moisture. They therefore continue to find available moisture between the rows later in the season, when it can be used for grain production—provided the distance between the rows has been well adjusted to the available soil moisture. The heads are small but relatively numerous under these conditions, and maturity is more uniform than with plantings in close-spaced rows.

It might be assumed that the wide row-spacings, by exposing large areas of bare soil to radiation, would increase moisture losses due to evaporation thereby defeating the purpose of the wider spacings. However, under dryland conditions, evaporation is influenced more by the moisture-supply at the soil surface than by radiation. Therefore, once the upper soil layer has dried, further moisture losses by evaporation become negligible. Under these conditions, wide rows are not more conducive to greater water-loss by evaporation than are narrow rows.

Moisture-supply not Limiting

Spatial arrangements may affect the efficiency of radiant energy interception or of moisture utilization.

Radiant Energy Interception. The main effect of spacing on yield is believed to be largely due to differences in radiant energy distribution. Closer and more uniform plant distribution should increase the proportion of radiant energy which is intercepted by the plants and reduce that reaching the soil surface. In experiments with maize, it was shown that the ratio of the net radiation at the ground to that above the crop was lower, the higher the rate of planting became. With the same population, the closer the inter-row spacing was, the lower became the ratio (Yao & Shaw, 1964*b*). It has been estimated that 60-cm inter-row spacings might increase the energy available for photosynthesis by 15 to 20 per cent compared with 100-cm rows (Denmead *et al.*, 1962).

Moisture Utilization. Tanner & Lemon (1962) have shown that, when soil moisture is available and a substantial crop-cover shades the ground, most of the net radiation is used for evapotranspiration. Therefore, differences in net radiation resulting from different plant populations and row spacings should also result in differences in water-use and water-use efficiency.

In trials carried out by Yao & Shaw (1964*a*) it was found that, in maize fields when inter-row spacings were maintained constant, doubling the plant-density increased water-use, but the increase in water-use was much smaller than the increase in plant-density. The narrowest row-spacing (52 cm) used significantly less water than did wider spacings, with the differences in water-use between treatments becoming more apparent as the season progressed. The highest efficiency of water-use was achieved at the narrowest spacings with the double plant-density.

With sorghum grown with two or more irrigations, narrow rows produced considerably more grain than did wider rows with identical plant populations. Soil moisture was extracted more rapidly, after irrigation, directly under plants than midway between rows, for rows more than 35 cm apart. It is therefore assumed that a more uniform extraction of soil moisture under narrow rows led to their greater grain-producing ability when water was not limiting (Grimes & Musick, 1960).

IRREGULAR SPACINGS IN THE ROW

In order to improve the efficiency of cultivation for the control of weeds, crops may be sown in such a way that rows of equal width run in two directions. This is achieved by *check-row planting*, which allows cross-cultivation. This method, which considerably slows sowing operations, has lost much of its utility with the advent of improved weed-control practices, both mechanical and chemical.

There are relatively few experiments comparing uniformly distributed plants in the row with hill or pocket groupings. Uniform spacing generally gives a greater yield than hill groupings under favourable soil-moisture conditions. However, when moisture is limiting, the advantage may be small or nil (Dungan *et al.*, 1958). Donald (1963) suggests that the advantage of the uniform spacing under favourable conditions is due to reduced competition for light; when soil moisture is lacking, light is no longer the limiting factor, and the advantage of uniform spacing is lost.

ROW DIRECTION

Very few investigations on the effect of row direction on yields have been made. The few results have been surprisingly similar with different crops grown under widely different conditions.

In Australia, yield increases in dry-matter production by wheat, of 6·5 and 11 per cent, respectively, were obtained in two experiments with N–S sowings as compared to E–W sowings (Santhirasegaram, 1962). In the U.S.S.R. spring wheat sown N–S produced from 100 to 300 kg/ha more than when sown E–W. It is assumed that sunlight and moisture were used more fully in the N–S than in the E–W rows, and that the plants in the N–S rows were the less subject to excess heat (Perekaljskii, 1951).

In India, it has been found that maize planted in a north–south direction yielded significantly more grain and forage than when it was planted in an east–west direction (Dungan *et al.*, 1955). Similarly, in Illinois, spring oats drilled in a north–south direction yielded slightly more grain than when it was drilled in an east–west direction; the advantage of the north–south row increased as inter-row spacings increased (Pendleton & Dungan, 1958).

Whilst a full explanation for these results is not available, it is possible that sunlight and the amount of reflected light are influenced by row direction, with N–S rows enjoying better lighting than the northern exposure of the E–W rows.

References

Alexander, C. W. & McCloud, D. E. (1962). CO_2 uptake (net photosynthesis) as influenced by light intensity of isolated Bermuda Grass leaves contrasted to that of swards under various clipping régimes. *Crop Sci.*, **2**, 132–5.

Black, J. N. (1966). Competition within grass and cereal communities. pp. 167–77 in *The Growth of Cereals and Grasses* (Ed. F. L. Milthorpe & J. D. Ivins) Butterworths, London, xii + 359 pp., illustr.

Blackman, G. E. & Black, J. N. (1959). Physiological and ecological studies in the analysis of plant environment. *Ann. Bot.*, **23**, 51–63.

Brown, B. L. & Shrader, W. D. (1959). Grain yields, evapotranspiration and water-use efficiency of grain sorghum under different cultural practices. *Agron. J.*, **51**, 339–43.

Denmead, O. T., Fritscher, L. J. & Shaw, R. M. (1962). Spatial distribution of net radiation in a corn field. *Agron. J.*, **54**, 505–10.

de Wit, C. D. *See* Wit, C. D. de.

Donald, C. M. (1954). Competition among pasture plants: the influence of density of flowering and seed production in annual pasture plants. *Aust. J. Agric. Res.*, **5**, 585–97.

—— (1963). Competition among crop and pasture plants. *Adv. Agron.*, **15**, 1–118.

Duncan, E. R. (1954). Influences of varying plant population, soil fertility and hybridity on corn yields. *Proc. Soil Sci. Soc. Am.*, **18**, 437–40.

Dungan, G. H., Lang, A. L. & Pendleton, J. W. (1958). Corn plant population in relation to soil productivity. *Adv. Agron.*, **10**, 435–73.

——, Sisodia, U. S. & Singh, G. D. (1955). The benefit of sowing maize for fodder in north and south lines. *Allahabad Farmer* (*India*), **29**, 8–13.

Grimes, D. W. & Musick, J. T. (1960). Effect of plant spacing, fertility and irrigation managements on grain sorghum production. *Agron. J.*, **52**, 647–50.

Harper, J. L. (1961). Approaches to the study of plant competition. pp. 1–39 in *Mechanisms in Biological Competition* (Ed. F. L. Milthorpe) 15th Symp. Soc. Exp. Biol., ix + 365 pp., illustr.

HAYNES, J. L. & Sayre, J. D. (1956). Response of corn to within-row competition. *Agron. J.*, **48**, 362–4.
HODGSON, G. L. & BLACKMAN, G. E. (1956). An analysis of the influence of plant density on the growth of *Vicia faba*. *J. Exp. Bot.*, **7**, 147–65.
HOLLIDAY, R. (1960). Plant population and crop yield. *Fld Crop Abstr.*, **13**, 159–67, 247–54.
—— (1963). The effect of row width on the yield of cereals. *Fld Crop Abstr.*, **16**, 71–81.
JACKSON, J. E. (1963). Relationship of relative leaf growth rate to net assimilation rate. *Nature, London*, **200**, 909.
LANG, A. L., PENDLETON, J. W. & DUNGAN, G. H. (1956). Influence of population and nitrogen levels on yield and protein and oil contents of nine corn hybrids. *Agron. J.*, **48**, 284–9.
MILLER, R. T. (1956). M.Sc. Thesis, Univ. of Illinois Graduate Coll., Urbana, Ill.; quoted by Dungan *et al.* (1958).
NELSON, O. E. & OHLROGGE, A. J. (1957). Differential response to population pressures by normal and dwarf lines of maize. *Science*, **125**, 1200.
PENDLETON, J. W. & DUNGAN, G. H. (1958). Effect of row direction on spring oat yield. *Agron. J.*, **50**, 341–3.
PEREKALJSKII, F. M. (1951). [Direction of the drills of a sown area—in Russian] *Selek Semenovod*, **18**, 70–3.
PORTER, K. B., JENSEN, M. E. & SLETTEN, W. A. (1960). The effect of row spacing, fertilizer and planting rate on the yield and water-use of irrigated grain sorghum. *Agron. J.*, **52**, 431–4.
PRINE, G. M. (1966). Light at silking affects ear number of south's corn. *Crops & Soils*, **19**(3), 16.
PUCKRIDGE, D. W. (1962). Thesis, Dept of Agronomy, Univ. of Adelaide, Adelaide, Australia; quoted by Donald, C. M. (1963).
ROSSITER, R. C. (1959). The influence of maturity grading on total yield and seed production in strains of *Trifolium subterraneum* L. grown as single plants and in swards. *Aust. J. Agric. Res.*, **10**, 305–21.
RUTGER, J. N. & CROWDER, L. V. (1967). Effect of high plant-density on silage and grain yields of six corn hybrids. *Crop Sci.*, **7**, 182–4.
SANTHIRASEGARAM, K. (1962). Thesis, Waite Inst., Univ. of Adelaide, Adelaide; quoted by Donald, C. M. (1963).
STINSON, H. T. & MOSS, D. N. (1960). Some effects of shade upon corn hybrids, tolerant and intolerant of dense planting. *Agron.*; *J.* **52**, 482–4.
TANNER, C. & LEMON, E. (1962). Radiant energy utilized in evapotranspiration. *Agron. J.*, **54**, 207–12.
WASSERMANN, V. D. (1963). Thesis, Dept of Agronomy, Univ. of Adelaide, Australia; quoted by Donald, C. M. (1963).
WIT, C. D. DE (1960). On competition. *Versl. Landbouwk. Onderz. Wagenin.*, **66** (8) 1–82.
YAO, A. Y. M. & SHAW, R. H. (1964*a*). Effect of plant population and planting pattern of corn on water-use and yield. *Agron. J.*, **56**, 147–52.
—— & —— (1964*b*). Effect of plant population and planting pattern of corn on net radiation. *Agron. J.*, **56**, 165–9.

XV

CROPPING SYSTEMS AND ROTATIONS

Objectives of Crop Rotation

A well-planned crop rotation should have the following advantages over monoculture or a haphazard succession of crops; it should:

1. Maintain, and even improve, soil fertility.
2. Prevent the build-up of pests, weeds, and soil-borne diseases.
3. Control soil-erosion.
4. Ensure a balanced programme of work throughout the year.
5. Prevent or limit periods of peak requirements of irrigation water.
6. Conserve moisture from one season to the next.

Principles of Crop Rotation

The traditional principles on which the planning of crop rotations is based are:

1. Alternate growing of crops with differential ability to absorb nutrients from the soil or having different depths of rooting systems.
2. Alternating crops susceptible to certain diseases, with those that are resistant.
3. A planned succession of crops that takes into account any detrimental or beneficial effects of one crop on the following crop. These effects may be due to toxic substances, effects on nutrients-supply, build-up of organic matter, soil structure, soil micro-organisms, or residual soil moisture.
4. Alternating soil-exhausting crops with crops that contribute to the improvement of soil fertility.
5. Alternating crops with different peak requirements of labour, water, etc.

Are Crop Rotations Obsolescent?

Crop rotations were once considered to be the essential basis of a stable agriculture. They were the only available means of preventing the deterioration of soil fertility and of achieving relatively high and stable yields. By contrast, monoculture of the so-called cash-crops was considered the epitome of backward, self-defeating farming methods which resulted in an exhausted and usually eroded soil.

The technological innovations in agriculture justify a re-appraisal of the need and justifications of crop rotation in the following connections:

1. *Soil fertility:* At a time when chemical fertilizers were non-existent, nitrogen fixation and the application of farmyard manures were the only ways of ensuring a high level of soil fertility and providing essential nutrients for crops. At present, chemical fertilizers are generally plentiful and inexpensive; by contrast, the costs of growing legumes and handling stable manure have increased steeply. Heavy applications of fertilizers also increase the amounts of organic residues that can be returned to the soil.
2. *Prevention of build-up of soil-borne diseases, pests and weeds:* The increasing array of fungicides, pesticides, and herbicides, available to the farmer, has provided him with tools for protecting his crops which are far more effective than the simple rotation of crops. There are a number of serious soil-borne diseases for which crop rotation never was effective.
3. *Control of soil erosion:* Minimum tillage, stubble-mulch farming, and sowing on the contour, make it possible to prevent soil erosion, even in monoculture.
4. *Ensurance of a balanced programme of work throughout the year:* This was one of the great advantages of crop rotation when most farm operations were carried out with animal power and manual labour. With the mechanization of agriculture and the use of increasingly specialized, sophisticated, and expensive, machinery, only the farmer who grows at the most two crops, can afford the expense of purchasing and maintaining the equipment that is required to grow a crop economically. Furthermore, the farmer growing only one or two crops becomes a specialist in them, with a high level of know-how and competence. In these respects the farmer growing a number of crops in rotation is at a disadvantage.

As a result of recent innovations in agricultural technology, including those mentioned above, the historical view that a planned rotation of crops is essential for the maintenance of soil fertility and efficient water-use is no longer generally accepted. The new attitude is summarized by Pendleton (1966) as follows: 'The trend towards monoculture has not resulted in great losses of water, soil and yields, as predicted not too many years ago. To a large degree, this may be attributable to technological advancement. This new era of "crop specialization" has resulted in more total food or fibre from a given farm, and therefore more efficient use of water.'

LIMITATIONS OF MONOCULTURE

Before rotations can be abandoned as an essential element in good farming, a number of limitations must still be overcome. For rain-fed crops, under semi-arid conditions, deficient soil moisture limits the amounts of chemical fertilizers that can be applied and, especially for nitrogen, may even make their use impracticable. Low yield-levels also make the use of heavy fertilizer doses prohibitive. In the dicussion on cropping systems adapted to these conditions (*see* pp. 462–6), it will be shown that rotations including legumes are still frequently the most economical solution to this problem.

Fertilizers are not cheap or easily available everywhere, and are especially costly in the less developed semi-arid regions that are remote from industrial centres. A crop that supplies forage *and* nitrogen may be able to justify itself economically. The organic nitrogen from symbiotic fixation may be more suitable than mineral nitrogen under conditions of limited moisture supply, being released more gradually, to become available at a time when conditions are favourable both for microbial activity and plant growth. Monoculture also tends to upset nutrient balance in soils by drawing heavily on certain nutrients and leaving excessive amounts of others unused.

Under continuous irrigation, there is a simultaneous build-up of diseases, pests and weeds, and a breakdown of the soil's structure. Whilst many effective fungicides, pesticides, and herbicides, are available none of these has yet enabled full control of the diseases, insects and weeds, that are apt to be encountered in irrigated fields. Their maximum effectiveness is achieved only when they are integrated in a system of crop management which includes crop rotation. In arid regions a root-rot of cotton, caused by *Phymatotrichum omnivorum*, is very common. The incidence of this disease can be considerably alleviated by including cereals in the crop rotation. In Texas, by sowing cereals in a four-year crop rotation, seedling mortality of cotton was reduced from 40 to 5 per cent (Curl, 1963). The complete eradication of this disease, by rotation, is not possible, as the Fungus can parasitize some 2000 species of weeds and cultivated crops (ibid.).

The most effective way of revitalizing the soil, achieving sanitation, and improving soil structure, is the inclusion in the rotation of rain-fed or lightly irrigated crops that completely dry out the soil. If, in dry-land agriculture, legumes play a major role in improving soil fertility, in irrigated agriculture it is the cereals which are the key to improved sanitation and soil structure.

The efficacy of crop rotation in reducing the severity of a number of soil-borne diseases, is ascribed not only to the 'starvation' of the pathogen in the absence of a suitable host, but also to the stimulation of antagonistic properties of the soil microflora, by the non-hosts and their residues (Patrick & Tousson, 1965).

Just as the repeated cropping of a susceptible host causes a build-up of the population of a plant pathogen, it is assumed that the repeated turning under of certain crop residues may favour the build-up of a suitable antagonist. In addition, the host may be less susceptible to attack under the more favourable conditions of soil fertility resulting from the decomposition of organic matter.

The need for specialized, expensive equipment is an important argument that cannot be overlooked, but must be considered within an overall economic appraisal of the cropping systems. This problem can also be overcome by employing contractors for certain operations such as mechanical harvesting. The most desirable crop rotation is the one that gives the maximum profit, on a sustained basis with the least fluctuations, over a period of years.

When a crop, such as a legume, that has a favourable effect on the yields of the succeeding crop, is included in the rotation, the total yield of the benefiting crop may be increased up to a point, although the actual area devoted to the crop is reduced. This is the so-called complementary effect—indicating that the inclusion in the rotation of crops that raise soil fertility or improve sanitation does not necessarily reduce the income from cash-producing crops.

Against the economic advantages that are claimed for monoculture, are the lesser fluctuations and smaller risks involved in growing a number of crops, as price and weather fluctuations do not affect all crops equally. Thus farmers will have to balance greater stability in income, and thus lower risks, against the possibly higher income obtainable from a single crop.

The farmer who grows the same crop continuously cannot be sure that his yield levels are not declining gradually. Such a decline will be masked by annual yield fluctuations due to environmental factors. Worse still, yields may drop suddenly and steeply, as the result of a build-up of disease organisms to epidemic proportions. Even in a crop such as cotton, which is relatively insensitive to continuous cropping, yields may decline after a period of time. In Australia, cotton grown continuously on previously virgin land, maintained its yield level unimpaired for eleven consecutive years; after that, yields declined gradually, irrespective of the rates of fertilizer application (Phillips, 1963).

The 'soil-toxin theory', which assumes that plant roots produce phytotoxic substances leading to so-called soil-sickness, has been debated for many years. A recent review of the available evidence indicates that there are indeed conditions under which phytotoxic substances are secreted by the roots and underground stems of certain plants, or are produced during the decomposition of their residues. Information on the nature of these substances is, however, still contradictory or lacking (Patrick & Tousson, 1965).

Certain crops, such as flax (*Linum* spp.), Guayule (*Parthenum argentatum*), and Brome Grass (*Bromus inermis*), are particularly subject to 'soil-sickness'; stands become sparser and yields decline when the crop is grown continuously in the same soil. The conditions are usually species-specific, and other crops grow well in the same soil. It is possible that other or additional factors, besides phytotoxic substances, are involved (Patrick & Tousson, 1965).

Where irrigation farming is practised, the possibility of using irrigation water efficiently for many months of the year, as is possible when several crops are grown, may be the overriding economic factor. With a single crop, it is the peak period of water requirement that will limit the area that it is possible to cultivate with the amount of water available.

Summing up, it can be stated that the farmer is no longer as dependent on crop rotation for maintaining high levels of production as he was in the past. However, a well-planned crop rotation still has important functions in soil sanitation and fertility, weed control, regulation of labour and water-use, and stability and security of income, so that in many cases it can confer economic advantages over monoculture.

Types of Crop Rotation Adapted to Arid and Semi-arid Regions

Whilst crop rotations may vary considerably in different regions, depending on tradition, the type of crops grown, local economic factors, etc., certain prototypes have developed which will be discussed here, as well as the fundamental principles on which they are based.

WINTER-RAINFALL REGIONS WITH MILD WINTERS

(a) *Transition Regions Between Aridity and Semi-aridity (250–400 mm rainfall)*

In the transition region between aridity and semi-aridity, where the average precipitation fluctuates between 250 and 400 mm, rain-fed crop production becomes possible, but is fraught with great dangers. In favourable seasons, when precipitation is above average, exceptionally good yields of grain can be obtained as a result of fertility build-up during the seasons of low rainfall and crop failure. Large areas are usually ploughed up and sown, to become dust-bowls when there is a succession of drought years.

The principal crop rotations followed in these areas are:

1. *Continuous wheat* (or barley), with crop failures during the dry years serving as unplanned fallow periods. If good yields are to be obtained during the favourable seasons, heavy nitrogen fertilizer applications are essential. Inputs are therefore high, with a low probability of success.
2. *Cereal-fallow:* In Australia, in the wheat-growing regions with an annual precipitation of 250 to 375 mm, experience shows that the fallow-wheat

sequence usually assures the farmer of a crop, whereas continuous wheat would generally be a complete failure (Callaghan & Millington, 1956).

In the low-rainfall regions, a succession of drought years is a frequent occurrence, and therefore even the additional moisture stored by the fallow may be insufficient to ensure a good crop. Cultivated fallow may promote an initial stimulation of yields, but in the long run it usually has adverse effects of which wind and water erosion are the ultimate outcome. This system also requires fairly high inputs (tillage, fertilizers, etc.) with, however, uncertain returns.

3. *Leguminous leys:* In some countries with advanced agriculture, rotations have been adopted in which cultivated fallow is replaced by leguminous leys. A ley rotation adapted to semi-arid conditions is based on alternating periods of two or more years of annual, self-seeding legumes—grazed by cattle or sheep—with one or more years of grain-crops.

This system has a built-in auto-regulatory device ensuring a yearly, albeit fluctuating, income. In the good rainfall seasons, the legume and the stubble of the cereal crop produce year-round grazing (provided the number of head of stock is properly adjusted to the carrying capacity), whilst good grain yields are harvested. In the poor rainfall seasons, the relatively low production of the legumes is supplemented by grazing the cereal fields which have no prospect of producing a satisfactory grain crop. What would have been a total loss in an exclusively grain-producing system, will give a fair yield of forage—even in conditions under which no grain is formed. The legumes may produce from 100 to 500 feed units per hectare even with a yearly rainfall of no more than 60–100 mm; with 200 mm, which is insufficient for grain production, the cereal may still produce 100 to 1300 FU/ha. In a good year, the legumes will produce around 2000 FU/ha and the grain stubble will provide another 300–500 FU/ha.

This integrated system of animal husbandry and cereal cropping also has the advantage of relatively low inputs: the annual legumes are self-seeding, so that no tillage, or only a minimum, is required. Only phosphatic fertilizers are needed, and if applied in excess of current requirements, the residues are not wasted. The legumes provide all the nitrogen that is needed by the cereal crop, but in a very good year additional nitrogen may be top-dressed. The cereal can be seeded into the legume stand by a chisel-drill without additional tillage; excessive legume volunteer plants in the cereal can be controlled by herbicides. Grain volunteer plants in the legumes provide balanced grazing. Perennial weeds such as Johnson Grass and Couch Grass, are kept in check both by the grazing and by the competition of the cereal crop.

The effectiveness of the whole system depends on how animal numbers are in balance with the fluctuating forage supply. The means to achieve this will be discussed in the companion volume.

It is estimated that in Australia, the use of clover leys in the cereal areas has contributed to a 50 per cent increase in yields of grain (Donald, 1960) with practically no application of fertilizer nitrogen. The fertility levels of originally poor soils have been dramatically raised to moderate and even high levels of fertility by sowing Subterranean Clover fertilized with superphosphate, and, where deficiencies of micronutrients occur, with sources of copper, zinc and molybdenum.

The effect of this cropping system on soil fertility is illustrated by the figures in Table XXV.

TABLE XXV

Effect of Inclusion of a Fertilized Clover Ley on Soil Fertility
(Donald & Williams, 1954)

	N (per cent)	P (per cent)	S (per cent)
Original soil	0·076	0·012	0·010
Subterranean Clover (with fertilizer)	0·184	0·028	0·024

(b) *Semi-arid Regions with 400 mm Rainfall and Above*

Traditional Crop Rotations. The main objective of arable farmers in these regions is usually the production of cereals, preferably wheat. In the more problematic areas, wheat is replaced by barley.

Farmers are usually aware that continuous cereal-growing invites disaster, and therefore most of the wheat—or barley in the drier regions—is grown in alternate years. As normally only a single crop can be grown each year, the two-course rotation predominates throughout the region. Some of the variations of these rotations are: wheat–weed fallow; wheat–cultivated fallow; wheat–pulses (beans, lentils, chick-peas); wheat–cultivated summer crops; and wheat–barley. When three- or four-course rotations are practised, they usually take the form of wheat–barley-fallow, or of wheat–pulses–barley fallow. A common factor of the rotations is that they do not include any crops for improving soil fertility, which is usually at a very low level. The farmer does not even adhere rigorously to any of these rotations. If the rains are early and abundant, and the season bodes well, he will usually be tempted to sow a wider area to wheat than was allocated in the rotation.

Cultivated summer crops have the advantage of aiding in the control of weeds, which may present serious problems in continuous wheat-growing. However, when the summer crop has a detrimental effect on the moisture status of the soil, it may offset any beneficial effect in regions in which soil moisture is the limiting factor. For this reason, in western Kansas, yields from wheat that has been sown immediately after the harvesting of cultivated summer crops (sorghum, maize, etc.) are frequently lower than those obtained from 'continuous' wheat. Between the harvesting and seeding of two successive wheat crops, three months elapse, during which time cultivated fallow will conserve soil moisture that will remain available to the wheat crop (Laude *et al.*, 1955).

Improved Crop Rotation. The traditional crop rotations described above are all exhaustive and make no contribution to soil fertility. The basic problem is, therefore, to devise a type of crop rotation that will raise the level of soil fertility, thus making it possible for the following crop to benefit fully from the favourable moisture régime prevailing during its growing period.

It is frequently assumed that pulses are desirable preceding crops for winter cereals, and it was originally thought that extending the area under pulses would have a beneficial effect on soil fertility. Many Ministries of Agriculture in the semi-arid regions have adopted the inclusion of pulses in the rotation as an official policy aimed at improving overall productivity. However, results are always disappointing, as harvesting a seed crop from the legume usually results in a drop of at least 30 per cent in the yield of the following wheat crop, as compared with wheat following fallow. The legume removes more nitrogen than the root nodules produce. The soil is also left in a very unfavourable physical condition, and weed infestation in the subsequent cereal crop is heavier than usual.

In research on the effect of leguminous crops in the rotation, Lyon (1936) has shown that, when cereals for grain are grown after pulses (beans, soybean, etc.), yields were only 47 per cent of those obtained after legumes that had been grown for hay (red clover, alsike clover) (average results of 8 years). In soils in which cereal grain-crops and leguminous hay crops alternate, there was after eight years, an increment of 500 to 650 kg/ha of nitrogen; whilst after alternating cereal and leguminous grain-crops, there was even a small deficit, of 50 to 110 kg/ha of N, after the same trial period.

Leguminous crops that are not allowed to mature seed, but are used for pasture, soiling, hay, silage, or for turning under as green-manure, have been shown to be very important for improving soil fertility. They also enhance the response of the winter cereals to other factors affecting production.

At first, green-manures were considered the ideal solution for rapidly raising the level of soil fertility. On the whole, the results obtained in semi-arid conditions have been disappointing, and green-manures have generally shown little, if any, benefit over results derived from leguminous forage crops. Where green-manures have been beneficial, the increase in yield rarely justified the loss of a cropping season or the risks and expenditure involved. However, legumes harvested for forage, before producing seeds, have proved very effective in increasing yields.

Under conditions in which the inclusion of legumes in the crop rotation increased the yields of the subsequent cereal crop, residual soil moisture after removal of the legume may be high, and not much less than that after fallow, as is shown in Table XXVI.

TABLE XXVI

Effect of Fallow and Legumes in the Rotation, on Amount of Residual Moisture at Time of Sowing the Following Wheat Crop
(Arnon, 1961)

Preceding crop	*Amount of residual moisture in soil (30 to 120 cm depth)*	
	m^3 per ha	*in percentage of control*
Fallow	3500	100
Green-manure (fenugreek)	3500	100
Silage (horse-beans)	3230	92
Hay (vetches)	2950	84
Grain (fenugreek)	2520	72

The earlier the legume was harvested, the higher was the amount of residual moisture (Figure 88). Green-manure and silage were as effective or nearly as effective in conserving moisture as was cultivated fallow. By contrast, when a deep-rooted legume crop such as lucerne or sweet clover is turned under for green-manure, the soil has usually dried out to a depth of several feet.

The Value of Fallow in the Crop Rotation. Fallow as a means of increasing soil moisture in the higher rainfall regions is not of overriding importance, and as a means of maintaining soil fertility it is not very effective. In southern California, from 30 to 40 per cent of the organic matter has been lost from the surface 25 cm after 50 to 70 years of grain cropping, mostly in the alternate fallow-grain system (Luebs & Laag, 1964). In south Australia, the inherent fertility of the red-brown soils was quite rapidly depleted by the bare fallow–wheat rotation which was at first widely practised. The situation improved only after the introduction of Subterranean Clover pastures into the rotation, which rapidly restored the productivity of worn-out wheat soils.

Simply resting the soil, even for a period of only one year ('weedy' fallow) is detrimental to yields and a very poor antecedent for the winter cereals. The depressant effect is due mainly to weed competition, for a cultivated fallow does not show the same negative influence.

Part of the value of fallow is that satisfactory moisture conditions may coincide with temperatures that are favourable for nitrification, so that sufficient amounts of nitrate nitrogen may be made available for the following crop. Fallow is also effective in accumulating nitrate in the soil. In a semi-arid winter-rainfall area, fallow increased available nitrogen from five- to sixfold over recent cropping (Littlejohn, 1956). However, the benefits of cultivated fallow in regions of adequate rainfall are usually overrated and, except in years of unusually low rainfall, the yields obtained after a leguminous hay-crop are generally higher than those obtained after a cultivated fallow.

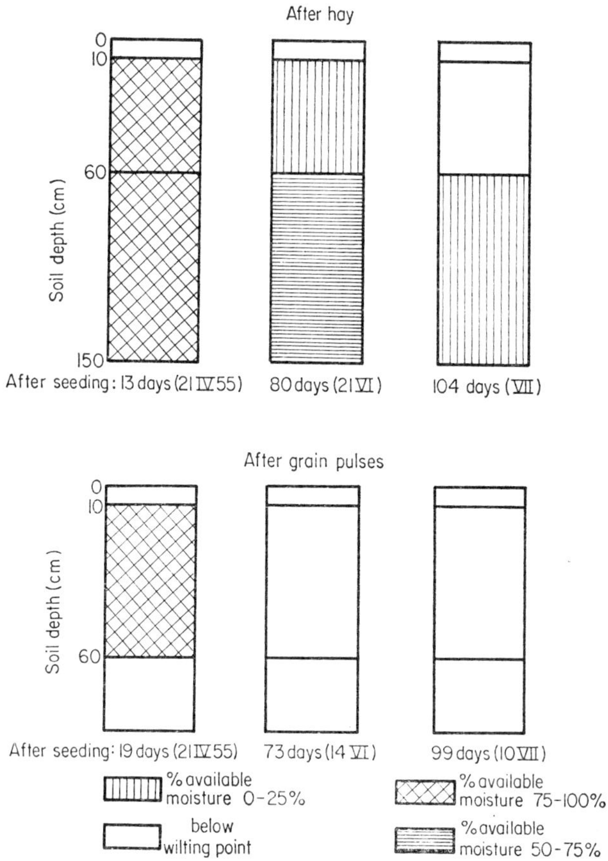

Figure 88. Effect of the stage at which a legume is harvested on the amount of residual soil-moisture available to the following sorghum crop (Arnon, 1961).

In summary it may be said that the most promising approach towards raising the level of soil fertility in semi-arid regions of sufficient rainfall (400 mm and above) is the inclusion in the rotation of a leguminous forage crop. When the legume is cut before seed is formed, the amount of plant nutrients removed from the soil is relatively small, whilst the soil is enriched in nitrogen and organic matter. Weeds are cut before flowering and are therefore well controlled. The soil can be ploughed when still moist and an excellent tilth is obtained.

As regards the types of rotation, the areas devoted to a soil-improving crop may vary from one-third to one-sixth of the total area, with a tendency to devote as large an area as possible to the winter cereal in the crop rotation. Hence we get a plethora of three-, four-, five-, and six-year crop rotations, with various sequences of crops, in which the winter cereals occupy an important position. The best overall results in Israel have been obtained with the following three-course rotation:

First-year—winter cereals (wheat, barley, or oats).
Second year—summer crops (sorghum, maize, sunflowers, sesame, or chick-peas).
Third year—hay crops (vetch, annual clover, etc.).

A different sequence is sometimes adopted, in which the winter cereal follows the summer crop. This is justified when fields are infested by grassy annuals, such as *Phalaris* and wild oats. In such a case it is the summer crop which is assigned the best place in the rotation, thereby benefiting from residual moisture after the legume and from the favourable nitrogen status resulting from the latter.

SUMMER-RAINFALL REGIONS IN THE MIDDLE LATITUDES

Continuous cereal growing

In the semi-arid regions of the world in which grasslands have been ploughed up for crop production in modern times (as in the Great Plains of the U.S.A.) and an adequate moisture supply is assured, the general tendency is to practise monoculture, with wheat grown continuously for long periods. In the areas with relatively favourable precipitation, alternate wheat-fallow is practised only occasionally—either when weed infestation has become a serious problem, or when exceptionally deficient soil-moisture at sowing time makes wheat sowing too hazardous.

Experience has shown that the accumulated fertility of soils which have always been under native grass before they were ploughed up, enables there to be continuous wheat cropping without declining yields for thirty years or more—provided weeds are kept under control (Harper, 1959). However, during this period, at least half the original organic matter and nitrogen will have disappeared, and so it is difficult to prophecy whether (and, if so, when) continued monoculture will start to have undesirable results. Even if a satisfactory level of soil fertility is maintained, sooner or later a build-up of grassy weeds occurs, which makes wheat-growing unprofitable. Under these conditions, a 3-year rotation of fallow–wheat–sorghum was found to give good weed control, as well as providing for efficient use of water and nitrogen and giving satisfactory protection against wind erosion. It also gave a higher gross income than a fallow–wheat–wheat rotation (Ramig & Smika, 1964).

Alternating Wheat and Fallow

In the rainfall-deficient areas, the main importance of cropping sequence is to provide the minimal quantity of moisture required by the wheat crop;

hence alternate wheat–fallow is almost inevitably the standard practice. By sacrificing a crop, moisture is conserved from one season to the next, so that the combined precipitation of two rainy seasons may suffice for producing one satisfactory crop. Fallow is considered essential for wheat production in the summer-rainfall areas of the middle-latitudes when annual precipitation is less than 350–375 mm (Mathews, 1951).

Fallow is effective in conserving soil moisture only if the amount, and, in particular, the distribution, of rainfall during the fallow period is such as to permit moisture penetration in depth. Practically all available moisture in the upper 20 cm of the soil will be lost by evaporation. Therefore, total seasonal rainfall must not only be sufficient, but individual rains must be sufficient to penetrate in depth. As losses by evaporation are confined mainly to the upper 10–20 cm of the soil surface, a far greater proportion of the total precipitation is lost from a relatively large number of light rains than from a smaller number of heavy rains. In an average season in southern Saskatchewan, less than 10 showers contribute to water storage in fallowed fields from May to October. In a wet season, twice as many showers contribute, but in drought years the storage is almost nil (Staple, 1964). The relative water losses from a layer of fallow soil due to evaporation and drainage, respectively, are shown in Figure 89.

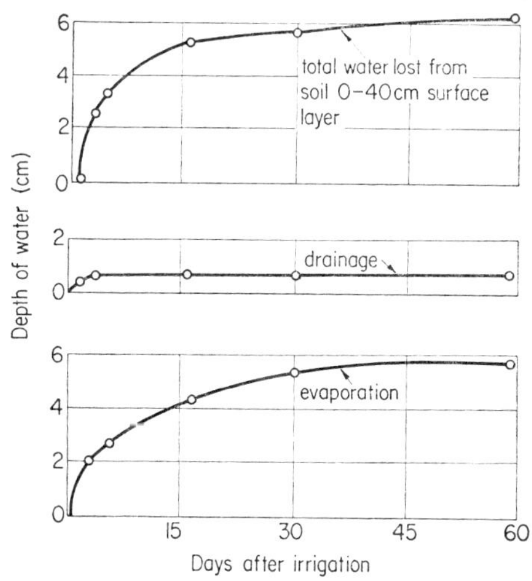

Figure 89. Accumulated loss of water from a layer of fallow soil. The accumulated evaporation from this soil during the two summer months is shown on the lowest curve. The top curve shows the total depth of water leaving the 0–40 cm soil layer during the same period. The middle curve is the difference between the two former curves and represents downward drainage, which—as can be seen—is minimal in comparison with loss by evaporation. (From Richards & Richards (1957): by courtesy of the U.S. Department of Agriculture.)

Conditions for infiltration must be satisfactory, and run-off must be minimized by proper cultivation measures. Moreover, the loss of the stored moisture must be minimized by effective weed-control.

However, even with the best management practices, fallow is not efficient in moisture storage, as the following examples demonstrate. In the Great Plains of North America, under an alternate crop–fallow system, the average amount of moisture conserved in the soil during the fallow year was only 16·3 per cent of the total precipitation under conditions in which losses due to run-off and deep percolation were negligible and weeds were well controlled (Mathews & Army, 1960). In North Dakota, in a region with an average annual precipitation of 400 mm, the average moisture storage efficiency over a 19-year period was 20 per cent (C. E. Evans & Lemon, 1957).

However inefficient, the additional 80–100 mm or so of water stored in the soil by fallowing may make the difference between success and complete failure of the succeeding crop in regions with marginal precipitation. Long-term weather records indicate that, in the southern high plains of the United States, a wheat–fallow rotation will produce yields in excess of 700 kg/ha in approximately eight summers out of ten, whilst with continuous cropping to wheat the odds of obtaining such yields are in only five seasons out of ten (Army *et al.*, 1959).

In a recent study in Nebraska, comparing a fallow–wheat rotation with continuous winter wheat under semi-arid conditions, it was shown that fallow–wheat average yields were over three times as great as average continuous wheat yields. The fallow–wheat rotation also provided stable production with no crop failures, while continuous wheat failed to produce a crop in about one year out of three. Average water-use efficiency for fallow–wheat was 80 per cent greater than for continuous wheat, when annual precipitation was between 246 and 430 mm, indicating that the fallow–wheat was the more efficient of the two methods tested, in producing winter wheat with rainfall of less than 430 mm. At least 580 mm rainfall were required in a summer-rainfall area for continuous wheat with nitrogen fertilizer to use water as efficiently as fallow–wheat without fertilizer. Nitrogen content of wheat grown on fallow was 0·22 per cent higher than continuous wheat (Smika, 1970).

Green-manures

Long-term experiments have not shown any advantage in replacing fallow by green-manure crops in regions with less than 375 mm. rainfall. Thus, in long-term experiments in Montana, where annual precipitation ranged from 290–375 mm, green-manure crops of sweet-clover had no beneficial effects on dry-land wheat yields (Army & Hide, 1959). The negative results may be ascribed to the kind of crop used for turning under (frequently a cereal), to the soil-moisture depletion by the growing crop, to the drying out of the soil due to inversion when ploughing under the green-manure, and, finally, to the decomposition of the mass of organic matter.

Green-manures were found economically feasible for dryland cropping where at least 450–500 mm of rain can be expected (Army & Hide, 1959).

In trials in Nebraska, on a loess soil, a number of types of rotation were compared. The highest maize yields were obtained from rotations including two years of sweet-clover (Duley, 1960). Oats and wheat yields were also increased in a legume rotation as compared with a non-legume rotation as indicated in Table XXVII.

TABLE XXVII

Effect of Type of Rotation on Yields (kg/ha) of Oats and Wheat (Duley, 1960)

	Oats	*Wheat*
After two years of sweet-clover	2175	4080
In a maize–oats–wheat rotation	1 196	1842

In summary it may be said that a fallow–crop system, without soil-improv ing crops or practices, will reduce the organic matter content of the soil far more rapidly than does a continuous cropping system, and will eventually result in lower yields. The time required to reach this condition depends on the original level of soil fertility, the net quantity of nutrients removed annually, management practices (such as burning stubble, stubble-mulch farming, etc.), and the amount of precipitation.

In a cereal farming system, an increase of 100 per cent in yield on fallowed land, to compensate for the loss of a cropping year, is not a precondition for making fallow economically justified. Stability of production, distribution of labour, and economy in weed control, may justify fallow even under conditions of lesser response.

Where precipitation is too low for successful crop production, fallow is indispensable. Where crops can be grown with a reasonable prospect of success, fallow is optional; its value is determined by other factors in addition to the degree of crop response, and the areas under fallow will vary from year to year—depending *inter alia* on economic factors and soil moisture conditions. It is therefore far easier to decide in which regions fallow is essential than where its use is optional.

SUMMER-RAINFALL REGIONS IN THE LOW LATITUDES

In the semi-arid zone of Africa lying to the south of the Sahara, the two principal crops—sorghum and millets—are extensively grown on freshly cleared brush. They have proven to be extremely sensitive to soil depletion in organic matter. After a relatively short period of continuous sorghum or millet, yields become too low even for primitive farming, and the land has to be abandoned.

The rapid degradation of the soil resulting from cereal monoculture is an almost irreversible process under the conditions of primitive agriculture employed in this region. The exhausted soil, abandoned by the farmer, is covered only sparsely with vegetation. The process of regeneration is delayed by overgrazing and brush fires, so that it takes many years to return to the original level of soil fertility.

Non-cultivated fallow alternating with cereal production is therefore incapable of maintaining an acceptable level of production after brush-clearing. The situation is no better when groundnuts, another important crop of the region, are sown continuously. Yields decline rapidly, and even mineral fertilizers are unable to maintain a satisfactory fertility level.

Alternating a groundnut crop with two or more years of fallow (non-cultivated) alleviates the situation without, however, providing a radical solution, as illustrated by the example given in Table XXVIII (cf. Gillier, 1960).

TABLE XXVIII

Effect of Fallow on Yields of Groundnuts (average rainfall 421·8 mm)

	Yields of unshelled groundnuts	
Duration of fallow (years)	*kg/ha*	*Increase in percentage of control*
On depleted soil (control)	355	—
2	440	+24
3	610	+72
6	640	+80

It is clear that simply alternating groundnuts with periods of fallow does not allow for the production of high yields. Even with mineral fertilizers, yields are increased by only 270 kg/ha. Further, the respite provided by fallow in the depletion process is very transitory: after two years of cropping, even following six successive years of fallow, the drop in yield is considerable.

It has been found that an alternation of crops is the first essential to maintaining a reasonable level of fertility. Short cycles of sorghum–groundnuts, with light doses of mineral fertilizers, alternating with fallow, have been found not only capable of maintaining the original level of soil fertility, but even of gradually increasing yields.

Thus alternation of crops, with short periods of cropping alternating with fallow periods of sufficient length, and the use of mineral fertilizers at moderate doses, have given the possibility of economic production within the framework of traditional agriculture of this region. But the inclusion of green-manures in traditional farming has not been found to be justified.

Crop Rotation Under Irrigation

Establishing a permanent crop rotation for irrigated crops is far more complicated than for rain-fed crops, as the variety of different crops that can be grown is far greater. These can include grain-crops, forage and pasture crops, industrial crops, and vegetable and ornamental crops, whose economic justification is far more subject to fluctuation than the cereal crops which form the backbone of the rain-fed rotation.

Maintaining soil fertility and sanitation is far more difficult under irrigation than under rain-feeding, for reasons that have been discussed in Chapter XI (pp. 345–8).

Additional factors that have to be considered in establishing a crop rotation are the amounts of water available, their seasonal availability, their quality, and costs; the availability of labour and of specialized skills, the distance from markets, and many other local factors, have also to be considered. For these reasons, whilst it was possible to define prototype crop rotations (pp. 462–73) for rain-fed crops in the semi-arid regions, this cannot be done for irrigated crop rotations in the arid areas. At most, certain basic requirements and desiderata can be outlined:

1. *Permanency and Flexibility*

The rotation scheme must have permanency, and at the same time sufficient flexibility to allow for changes in the choice of crops when this becomes necessary as a result of fluctuations in the economic climate. This flexibility can be achieved by devising a crop rotation that is based not on a succession of crops, but on *groups* of crops with similar requirements and similar effects on the following crops.

2. *Periodic Drying of the Soil*

It is well known that a continuous succession of irrigated crops usually leads to a gradual decline in yield levels. Drying the soil at regular intervals is a most effective measure to counteract this tendency, for reasons explained in Chapter XI (pp. 342–4). Simply leaving the soil fallow is not a satisfactory solution as only the top-soil dries out, leaving the lower layers moist for very long periods. An excellent solution is to include a cereal, particularly wheat, as the last crop to be sown in the rotation. The wheat in this case has a sanitary function to perform, as it does not share the root diseases of the principal intensive crops grown. The cereal enables a thorough drying out of the soil and a steep reduction in the number of pathogens. Deep tillage, after harvest, leaves the soil in huge clods, which are then baked by the sun for several months, during which time the rhizomes of perennial weeds are killed. The clods disintegrate with the first rains, thereby providing a well-aerated soil and excellent tilth.

3. *Maintaining Soil Fertility*

In addition to the liberal use of chemical fertilizers, the turning under of organic manure, and green-manures grown as catch crops, as well as the considerable amounts of residues that are produced by crops under irrigation, will enable the maintenance of a satisfactory level of fertility.

4. *Long Time-intervals before Sowing Certain Crops Again*

For some crops, such as sugar-beets, that are very sensitive to certain diseases, a fairly long interval should be provided, before they are sown again in the same field.

5. *Choosing the Right Sequence of Crops*

(a) The continuous sowing of grain-crops (whether cereals or pulses) should be avoided as far as possible.
(b) Leguminous crops should not be sown one after the other, even when one crop is to be harvested for seed and the other for forage.
(c) Crops with common root-diseases should not be sown in succession to each other.
(d) The best 'preceding crops' should be reserved for the most remunerative cash-crops. Crops with the most favourable effects on the subsequent crops are, in order of merit:

1. Perennial pastures, lucerne;
2. Legumes for green-manure, soilage, silage, or hay; and
3. Heavily manured vegetable crops.

Prototype of an Irrigated Crop Rotation

The reasons why standard prototypes, suitable for a wide variety of conditions, cannot be proposed for irrigated crop rotations, have been outlined above. A single example will therefore be given of a crop rotation designed for specific conditions, but in which all the principles outlined above have been incorporated. This is a five-year rotation that was developed specifically for mixed farming in the coastal plain of Israel (Arnon, 1961):

Year	*Autumn–Winter*	*Spring–Summer*
First	Winter pulses (chick-peas, peas for canning)	Land preparation and manuring
Second	Sugar-beets	
	Autumn vegetables	Spring and summer vegetables
Third	Leguminous crop for hay, silage, or green-manure	Maize or sorghum for grain or silage
Fourth	Fallow	Cotton and or Peanuts
Fifth	Wheat	Deep-ploughing

In each five-year cycle are included: one soil-drying crop followed by deep-ploughing, one manured crop, two nitrogen-producing crops, and two successive dry summers. In the cash-crops, there is always the possibility of alternate choices according to the economic prospects of the crops. Crops other than those mentioned can easily be included in the framework of this rotation, without causing its breakdown or even requiring basic modifications.

Mixed Cropping

Mixed cropping is a characteristic of primitive agriculture dating from antiquity. It is a method which attempts to make the most of the potentialities of the environment. By planting together a succession of crops with varying planting and harvesting times and growth habits, plant nutrients in different soil layers are better exploited and light-energy is more effectively intercepted. Plants of the same species compete more intensively with each other than do plants of different species—mainly because of the latters' differences in the root systems and periods of peak water requirements, so that a limited water-supply is used more efficiently in a mixed-cropping system than in pure stands. The risks due to diseases, pests and climatic factors, are reduced and also better distributed in mixed stands, while weeds are more effectively smothered. When primitive varieties are used, the total yield from a given area under mixed cropping may be greater than from pure stands (Baldy, 1963).

A notable example of mixed cropping, practised by a primitive people, is the maize–beans–squash complex, which originated in Mexico during the period 1500–900 B.C. It was so successful that it spread out from Mexico and became the basis of practically all prehistoric Indian agriculture in all parts of America.

The complex of three crops exploited the soil and light-energy most effectively: the beans climbed on the maize stalks, exposing their leaves to the light without excessive shading of the maize leaves; the squash grew prostrate on the ground and choked out weed growth. The mixed cropping also produced a highly-balanced diet: the maize supplied most of the carbohydrates and certain amino-acids in which beans are deficient; the beans supplied the bulk of the protein ,and also phosphorus, iron, and the vitamins riboflavin and nicotinic acid. The squashes added calories and an increment of fat (Stakman *et al.*, 1967).

Intercropping is still practised on a large scale in the Mediterranean region, in Africa, and in India. In East Africa, it is exceptional for peasants to sow pure stands. A. C. Evans (1960) doubts whether it will be possible to introduce rotational systems of agriculture based on pure stands so long as the hoe is the main agricultural implement. He found that intercropping maize or sorghum with groundnuts generally gave higher overall crop production per unit area than growing these crops in pure stands. This was found to hold true in two areas of contrasting fertility as well as under conditions of contrasting rainfall; namely low, irregularly distributed rainfall and favourable rainfall.

The combinations castorbean–groundnuts and castorbean–soybean also usually showed an overall gain in production per hectare. When severe attacks of mirids reduced castor yields drastically, the associated crop gave a compensating yield increase (Evans & Sreedharan, 1962). Intercropping of cotton, both rain-fed and irrigated, with a wide variety of crops, in varying proportions, is widely practised in India. Common combinations are of cotton with sorghum, pulses, sesame, cucurbits, groundnuts, or castorbeans. Usually, one or two rows of the associated crop alternate with 8–10 rows of cotton. The mixed cropping generally gives a larger total yield than the pure crop of each does in the aggregate; it provides an insurance against complete crop failure, and it reduces soil erosion—particularly if one of the two associated crops has a trailing habit. In an investigation on root-rot (*Rhizoctonia* sp.) conducted in the Punjab, it was found that cotton, when intercropped with beans (*Phaseolus aconitifolius*), was much less damaged by the disease than when grown in a pure stand (Sawhney & Sikka, 1960).

With very few exceptions, particularly among forage and pasture crops, modern agriculture is based on pure stands. Weed control, whether mechanical or chemical, as well as pest and disease control, are hindered in mixed stands. Rational fertilizer applications are usually not possible to a mixture of different crops. However, probably the main limitation to mixed cropping in modern farming is that efficient mechanical harvesting of a mixed crop is not possible.

References

ARMY, T. J., BOND, J. J. & DOREN, C. E. VAN (1959). Precipitation yield relationships in dryland wheat production on medium- to fine-textured soils of the southern High Plains. *Agron. J.*, **51**, 721–4.

—— & HIDE, J. C. (1959). Effects of green-manure crops on dryland production in the Great Plains area of Montana. *Agron. J.*, **51**, 196–8.

ARNON, I. (1961). [*Field Crops—Principles and Practice*—in Hebrew] Hassadeh, Tel-Aviv, 2 vols, 867 pp., illustr.

BALDY, C. (1963). Cultures associées et productivité de l'eau. *Ann. Agron.*, **14**(4), 489–534.

CALLAGHAN, A. R. & MILLINGTON, A. J. (1956). *The Wheat Industry in Australia.* Angus & Robertson, Sydney, 486 pp., illustr.

CURL, E. A. (1963). Control of plant diseases by crop rotation. *Bot. Rev.*, **29**, 413–79.

DONALD, C. M. (1960). The impact of cheap nitrogen. *J. Aust. Inst. Agric. Sci.*, **26**, 319–38.

—— & WILLIAMS, C. H. (1954). Fertility and productivity of podzolic soil as influenced by Subterranean Clover and superphosphate. *Aust. J. Agric. Res.*, **5**, 664–87.

DULEY, F. L. (1960). Yields in different cropping systems and fertilizer tests under stubble mulching and plowing in eastern Nebraska. *Res. Bull. Nebr. Agric. Exp. Stn*, **190**, 53 pp.

EVANS, A. C. (1960). Studies of intercropping. I. Maize or sorghum with groundnuts. *E. Afr. Agric. J.*, **26**, 1–10.

—— & SREEDHARAN, A. (1962). Studies of intercropping. II. Castor-bean with groundnuts or soya-bean. *E. Afr. Agric. J.*, **28**, 7–8.

EVANS, C. E. & LEMON, E. R. (1957). Conserving soil moisture. pp. 340–59 in *Soil: The Yearbook of Agriculture*. U.S. Department of Agriculture, Washington D.C., xiii + 784 pp., illustr.

GILLIER, P. (1960). La reconstitution et le maintien de la fertilité des sols du Sénégal et le problème des jachères. *Oléagineux,* **15**, 637–43, 699–704.

HARPER, H. J. (1959). 65 years of continuous wheat. *Bull. Oklahoma Agric. Exp. Stn*, *B*-**531**, 38 pp., illustr.

LAUDE, H. H., HOBBS, J. A., SMITH, F. W., HEYNE, E. G., CLAPP, A. L. & ZAHNLEY, J. W. (1955). Growing wheat in Kansas. *Bull. Kan. Agric. Exp. Stn*, **370**, 36 pp., illustr.

LITTLEJOHN, L. (1956). Some aspects of soil fertility in Cyprus. *Emp. J. Expt. Agric.*, **14**, 123–34.

LUEBS, R. E. & LAAG, A. E. (1964). Tillage and nitrogen for dry-land grain in a winter rainfall climate. *Bull. Calif. Agric. Exp. Stn*, **805**, 34 pp., illustr.

LYON, T. L. (1936). The residual effects of some leguminous crops. *Bull. Cornell Univ. Agric. Exp. Stn*, **645**, 17 pp., illustr.

MATHEWS, O. R. (1951). Place of summer fallow in the agriculture of the Western States. *Circ. U.S. Dep. Agric.*, **887**, 1–17.

—— & ARMY, T. J. (1960). Moisture storage on fallowed wheatland in the Great Plains. *Proc. Soil Sci. Soc. Am.*, **24**, 414–18.

PATRICK, A. Z. & TOUSSON, T. A. (1965). Plant residues in relation to biological control. pp. 440–57 in *Ecology of Soil-Borne Plant Pathogens* (Ed. K. F. Baker & W. C. Sydner) Univ. of California Press, Berkeley, 571 pp., illustr.

PENDLETON, J. W. (1966). Increasing water-use efficiency by crop management. pp. 236–58 in *Plant Environment and Efficient Water Use* (Ed. W. H. Pierre, D. Kirkham, J. Pesek, and R. Shaw) Amer. Soc. Agronomy and Soil Sci. Soc. Amer., Madison, Wisconsin, xii + 295 pp., illustr.

PHILLIPS, L. G. (1963). Continuous Cropping with Peanuts and Cotton. *C.S.I.R.O. Div. Land Res. and Regional Survey*, Ann. Rep. 1962–63, p. 25.

RAMIG, R. E. & SMIKA, D. E. (1964). Fallow–wheat–sorghum: an excellent rotation for dryland in Central Nebraska. *Bull. Nebr. Agric. Exp. Stn.*, **483**, 11 pp.

RICHARDS, L. A. & RICHARDS, S. J. (1957). Soil moisture. pp. 49–60 in *Soil: The Year-book of Agriculture*. U.S. Department of Agriculture, Washington D.C., xiii + 784 pp., illustr.

SAWHNEY, K. & SIKKA, S. M. (1960). Agronomy. pp. 106–63 in *Cotton in India*. Examiner Press, Bombay, vol. I, viii + 339 pp., illustr.

SMIKA, D. E. (1970). Summer fallow for dryland winter wheat in the semi-arid Great Plains. *Agron. J.*, **62**, 15–17.

STAKMAN, E. C., BRADFIELD, R. & MANGELSDORF, P. C. (1967). *Campaigns Against Hunger*. Harvard Univ. Press, Cambridge, Massachusetts, xvi + 328 pp., illustr.

STAPLE, W. J. (1964). Dryland agriculture and water conservation. pp. 15–30 in *Research on Water*. Soil Sci. Soc. Am., Madison, Wisconsin, vi + 121 pp., illustr.

XVI

WEEDS AND THEIR CONTROL

WEEDS AND THE DAMAGE THEY CAUSE

Weeds have been defined in various ways, but the most appropriate definition for our purpose is: 'a weed is a plant that in a given situation is more detrimental to agriculture than beneficial'. This definition takes into account that weeds under certain circumstances can also be beneficial: they provide grazing in areas that cannot be cultivated, or during periods when the land is not producing a crop, at the same time protecting the land against erosion; certain weeds may serve as food in periods of scarcity; while many weeds have important medicinal uses (King, 1966).

It is common knowledge that weeds compete with crop plants for light, air, moisture and nutrients; but the full extent of the economic harm which they cause, and their direct effects on crop yields, are not generally realized. In arid climates the deprivation of water to the cultivated crop is frequently the most damaging aspect of weed competition—both in the growing crop and in the fallow period between crops,

The leaf area of a single plant of *Sinapis arvensis* (Wild Mustard, or Charlock) at blooming stage was found to be 7300 cm^2, as compared with 140 cm^2 for a single wheat plant with which it was growing in association (Pavlychenko, 1949). This disproportion in leaf area gives an indication of the relative demands of the weed and of the cultivated crop on soil moisture supplies, as well as on the amount of radiant energy intercepted by the weed, with its concomitant reduction of photosynthesis in the cultivated crop. It has been estimated that the amount of water saved by eliminating weeds in a maize field is equivalent to providing an entire irrigation at the time of maximum need (Mangelsdorf, 1966).

The effects of weed competition are mainly felt in the young crop; the much more rapid rate of growth of both the aerial parts and the roots of many weeds gives them a considerable advantage in depressing and even crowding out the crop plants among which they are growing. Weeds also increase harvesting costs, require costly cleaning of seeds, reduce water flow in irrigation and drainage channels, and increase fire hazards.

An additional undesirable effect is the poisonous nature of certain weeds, such as Darnel (*Lolium temulentum*), a weed that is widely associated with wheat. Flour prepared from grain contaminated with 0·5 per cent of Darnel seeds is toxic (King, 1966). Many weeds in natural pastures are poisonous to stock, as for example *Halogeton glomeratas* in the semi-arid range-lands of the United States; other weeds, such as *Allium* spp. (onions, etc.), taint the dairy produce of grazing cattle. Finally, many weeds serve as hosts for plant diseases, insects and nematodes, and are frequently major factors in the epidemic spread of pathogens, virus diseases and pests. Probably the best-known example is the role of Common Barberry (*Berberis vulgaris*) as an alternative host to the Stem-rust of wheat.

Weeds are naturally endowed with a number of characteristics which increase their survival rate and their ability to compete with crop plants. Their vitality frequently helps them to overcome the ill-effects of herbicides. They produce enormous quantities of seeds. In many species the seeds germinate only when conditions favourable to the survival of the seedlings are assured. In others, the seeds have various periods of dormancy: only a fraction of the total quantity of seeds germinates at a given time, so that seed reserves are always available in the soil, even if one or more generations are destroyed by natural causes or by the efforts of the cultivator.

Weed Control and Stages of Development

Efficient weed-control implies a knowledge of the life-cycle of the different weed species and of their special characteristics. Their rate of growth differs at different stages of development, and their vulnerability to weed-control measures varies accordingly. Plants are most vulnerable to herbicides during periods of rapid growth, and after they have depleted their food reserves (Åberg, 1964). From the point of view of weed control, three stages are of significance:

(a) *Seedling Stage*. The young plants are weak, their root system is not yet developed, and their total leaf area is limited. This is an ideal stage for weed control; mechanical methods such as harrowing, disking and cultivating, are especially efficient and economical during this period.

Selective control is especially effective only if the weeds are at this stage, whilst the cultivated plants are already more developed. Or alternatively, the weed seedlings can be selectively destroyed before the crop plant has emerged above the soil surface. An example of the considerable variations in suscepti-

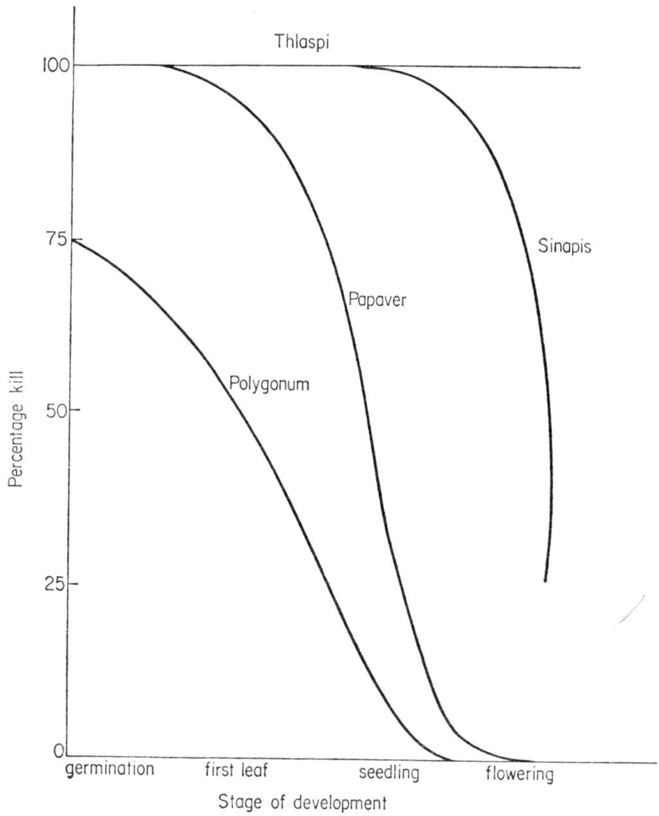

Figure 90. Effect of spraying four weed species with 0·2 per cent MCPA at different stages of development. *Thlaspi* (pennycuess) was completely killed at all stages. (From Blackman *et al.* (1949): by courtesy of Professor G. E. Blackman.)

bility to a herbicide, between species and at different stages of development within a species, is shown in Figure 90.

(b) *Vegetative Stage.* The growth rate is accelerated in the vegetative stage, the leaf area increases rapidly, and the root system is already extensive. This stage is not favourable for mechanical control, as even uprooted weeds are frequently capable of re-establishing themselves under favourable weather conditions.

The large and actively-growing leaf area increases the lethal effect of herbicides but, in order to be effective, high rates are required, which may cause excessive damage to the cultivated crops. Recovery of the crop may be hastened by fertilization, in particular with nitrogenous fertilizers. In some cases, e.g. with certain thistles such as *Cirsium arvense*, the most appropriate time for spraying with auxin herbicides is immediately before flowering, as at this time the amount of reserve carbohydrate is at its lowest level (Åberg, 1964).

(c) *Reproductive Stage*. Growth practically ceases in most weed species in the reproductive stage, and the assimilatory activity of the leaves decreases. This is generally not a favourable stage for effective weed control. However, the apparent absence of damage to the plant by late spraying, cannot be the only criterion: the chemical may be transferred to the seeds, and become active when the seeds germinate, causing damage to the seedlings (Åberg, 1964).

Spraying at the time of flowering may cause many abnormalities in crop plants, e.g. in maize, where flower malformations may be induced, with a resultant reduction in grain yield (Åberg, 1964).

Kinds of Weeds

Weeds, encompassing an enormous variety of species, can be classified in a number of ways. With the advent of selective weed-killers, the most purposeful grouping of weeds is probably into two main groups: narrow-leafed weeds and broad-leafed weeds.

NARROW-LEAFED WEEDS

Narrow-leafed weeds are generally monocotyledons, and include Gramineae, Cyperaceae, and Juncaceae. This group includes annuals and perennials. The annuals are particularly troublesome in cereal production, because the similar characteristics of weed and crop make selective control very difficult. *Avena sterilis*, *Phalaris paradoxa*, and *Lolium temulentum*, in cool-season crops, and *Setaria* spp. and *Panicum* spp., in warm-season crops, are examples of this group. Among the narrow-leafed perennials are some of the most ubiquitous weeds of the warm arid and semi-arid regions: Bermuda Grass (*Cynodon dactylon*), Johnson Grass (*Sorghum halepense*), and nutgrasses (*Cyperus* spp.). Their competition with the cultivated crop, mainly for water, is severe in dry habitats. However, it is only when the land is put under irrigation that the potentialities of the perennial weeds to compete with and depress the cultivated crop become fully apparent—hence the need for their eradication as the first step before irrigation is started.

BROAD-LEAFED WEEDS

Most of the annual weeds of the arid regions have a life-cycle adjusted to seasonal precipitation, and include ephemerals whose life-cycle is completed within a few weeks. However, the most pernicious annual weeds are those whose life-cycle has become adjusted to that of an annual crop, germinating together with the crop—their most depressing effect being in retarding the development of the young seedlings. They benefit from any practices carried out for the benefit of the cultivated crop, and scatter their seeds before or at the time of harvesting the crop.

In the dry regions there are many types of xeric weeds, such as thistles and related plants.

Among the broad-leafed perennials are also some of the most pernicious and ubiquitous weeds of the arid and semi-arid regions, such as Field Bindweed (*Convolvulus arvensis*) and Canada Thistle (*Cirsium arvense*).

WOODY PLANTS

A large number of woody plants, including many shrubs and trees, are detrimental mainly in the grazing lands of the arid regions, where they compete strongly with more palatable and nutritious forage plants. A characteristic example is the group of mesquites (*Prosopis* spp.).

PARASITIC WEEDS

The dodders (*Cuscuta* spp.) are a widely distributed genus, capable of both sexual and vegetative reproduction, and parasitizing a large number of crop plants.

Broomrapes (*Orobanche* spp.) are fairly widespread in semi-arid regions of the world. They are damaging mainly to a few crops, including certain legumes, vegetables and tobacco.

Witchweeds (*Striga* spp.) are restricted to the hotter arid regions, in which they parasitize mainly the summer cereals, including maize, sorghum, Sudan Grass and rice. The plants produce enormous numbers of extremely minute seeds.

METHODS OF WEED CONTROL

In the course of centuries, a number of mechanical, ecological, and chemical methods of weed control have evolved. Weeds are, however, so aggressive, persistent, and ubiquitous, that if control of them is to be effective, all available methods have to be integrated into a planned programme of continuous, long-range efforts to reduce the weed population. It is therefore not surprising that weed control involves some of the highest costs in crop production (Ennis *et al.*, 1963). In order to prevent the accumulation of chemical residues in the soil to a dangerous level, and in order to avoid causing shifts in weed populations, it is recommended to rotate different herbicides whenever this is feasible, and to combine chemical weed-control with appropriate cultural practices (ibid.).

CAUSES AND PREVENTION

Farmers are frequently responsible in the first place for causing the weed infestation of their fields. The most obvious way in which weeds are introduced is by the use of contaminated seed-supplies. Hence, the importance of seed laws to safeguard the farmer against unscrupulous or negligent seed suppliers, and of seed certification schemes to ensure proper supervision of seed production and distribution.

The use of manure, especially that of sheep that have grazed in weedy areas, unless properly decomposed, can contribute enormous amounts of seeds to the fields.

The introduction of combine harvesting has also favoured the spread of weed seeds. In the past, cereals were harvested by hand or by binders and thrashing was concentrated in a single place, so that weed seeds could be collected and destroyed. Nowadays, however, the combine-harvester does a preliminary cleaning job, separates a large proportion of weed seeds, and scatters them in the field behind the combine.

Irrigation and drainage ditchbanks, uncultivated patches, fence rows, weedy banks of roadsides, and country lanes, are also permanent sources of weed reinfestation.

Inappropriate tillage methods may help to divide and disperse the roots and rhizomes of perennial weeds. Delays in tillage, in mowing, or in the harvesting of crops, may provide the weeds with an additional opportunity to disperse their seeds. Controlling the weeds in one crop, and neglecting weed control in the next, gives a new start to weed infestation. It is estimated that if only 1 per cent of a typical stand of weeds survive control methods, it is capable of producing over 100 million new weed-seeds per hectare (Day, 1966).

MECHANICAL CONTROL

Even with the effective modern chemical methods available, mechanical methods of weed control are still frequently justified, as they may be less expensive and less dangerous to neighbouring crops or orchards. However, even intensive cultivations, planned for effective weed-control, cannot destroy in a single year more than about 50 per cent of the viable seeds. It is estimated that a 10-year control programme will reduce a typical weed-seed population of 500 millions per hectare, to about half-a-million (Day, 1966).

Hoeing and Hand-weeding

Hoeing and hand-weeding are probably the oldest agricultural occupations, and the back-breaking work involved still forms a considerable and inevitable part of the human effort involved in primitive agriculture. Hand-weeding rice was found effective in increasing yields by 45 per cent (International Rice Research Institute, 1966).

Even with the most modern methods of weed control, individual weeds or patches of weeds are apt to escape unscathed, and it is then usually justified to hoe or pull them—either because of their direct effect on the crop, as they impede mechanical harvest, or in order to eliminate a source of reinfestation. For example, the question was raised as to whether it is economical to hand-pull large weeds that have been missed in thinning sugar-beets and at the first hoeing. In an experiment carried out by Brickey (1966), the results indicated in Table XXIX were obtained:

TABLE XXIX

Effect of Weeds on Sugar-beet Yields

	in tonnes beets/ha	*Loss/ha*
No weeds	86	–
1 weed per 6 beets or 57 weeds/100 m²	79	$91·67
1 weed per 3·3 beets or 100 weeds/100 m²	74	$151·80

Cost of pulling weeds: $58·00 per ha.

Tillage

Even after the advent of chemical weed-killers, tillage methods are still, for many situations, the most effective and economical means of weed control. Ploughing, cultivating and harrowing, make possible weed control before sowing the crop, and there are appropriate tools that can do a satisfactory job even after its emergence. The younger and smaller the weeds are, the more efficient and economical will be their control.

The role of tillage in weed control consists of:

(a) Provoking the germination of weed seeds, which can then be easily destroyed by mechanical or chemical means;
(b) Bringing roots or stolons to the surface, where they will dry out under the sun;
(c) Repeated cultivation, thereby depleting the food reserves of the plant; and
(d) Uprooting or smothering the weeds with soil.

The tillage may consist in cultivating a field between cropping periods (clean tillage), in passing through a sown field with special equipment that destroys the weeds without severely damaging the crop (blind tillage, accomplished by harrows, rotary hoes, finger weeders, etc.), or in intertillage of row-crops.

The disturbance of the soil surface stimulates many weeds to germinate (King, 1966). Therefore, the first tillage operation usually stimulates the reappearance of a new wave of weeds, and needs to be followed by a second operation for their destruction. The implement chosen should be the one that under the given circumstances—soil condition, weed size, age of crop, etc.—

will destroy weeds efficiently at the least cost to the farmer. Tillage operations should be as superficial as possible, and adjusted to the minimum depth needed for effective weed control. Deeper tillage is more expensive, brings a new supply of weed seeds to the surface, and in intertillage damages the roots of the cultivated crop unnecessarily.

A special method for controlling deep-rooted perennial weeds is repeated cultivation at regular intervals, over a period of one or more years, until all underground buds have been destroyed. This is achieved by consistently preventing top-growth, so that the root reserves are exhausted and the weed is starved. However, the underground parts of hardy perennial weeds contain enormous food reserves, enabling them to survive frequent cultivations for years (Day, 1966). The method is therefore time-consuming, expensive, and has an unfavourable effect on soil structure. Any let-up in cultivations, or even a single excessively long interval between cultivations, may undo all the previous work. With the advent of chemical herbicides, it is doubtful whether this laborious method is justified, except under exceptional circumstances.

Tillage Combined with Irrigation

One of the most effective means of weed control is a pre-irrigation, before sowing, with a quantity of water sufficient to wet the soil to the full depth of the future root system. Then cultivating the field superficially before or during sowing (if a combined tillage and sowing operation is carried out) will destroy the weeds that have germinated following irrigation. The seeds of the crop are sown in moist soil at a depth that is sufficient to ensure germination without additional irrigation.

The surface soil dries out, effectively preventing germination of additional weeds until the next irrigation. In view of the earlier irrigation in depth, this next irrigation can be delayed for two months and more after sowing, so that the crop is well advanced when the next wave of weeds appears, making selective control easy.

Mowing

Mowing annual weeds at flowering time is a method of preventing seed propagation rather than an efficient method of weed eradication. Frequent and regular mowings of perennial weeds exhaust the food reserves of the underground storage organs, weaken the plants, and may occasionally lead to their death when environmental conditions become unfavourable. A similar effect is achieved by frequent and close grazing.

ECOLOGICAL METHODS

Crop Competition

Some crops such as rank-growing forage crops are able to compete successfully with weeds; their inclusion in the crop rotation will help to weaken if not to eliminate certain weed types, especially if the weeds start active growth at a time when the cultivated crop is already well developed (for example, *Cynodon dactylon* in a winter cereal). The choice of crop varieties, such as commercial hybrids, that have high seed viability, rapid germination, early emergence, seedling vigour, and rapid growth, also contributes to weed control. Weeds that have been weakened by competition are the more easily killed by selective herbicides. On the other hand, varieties with little vegetative development, such as the new dwarf varieties of wheat, are more susceptible to weed competition than are the leafy, high-growing varieties that they are replacing.

One of the traditional methods used for combating weeds was sowing a thick stand of the crop. Heavy seed-rates gave the crop a distinct competitive advantage over the weed population in the field. However in dry regions, where a relatively low plant population is one of the major means of adjustment to a limited soil-moisture supply, this method of weed control was self-defeating. On the other hand, if low sowing-rates were used, the additional space was taken up by weeds, again defeating the farmer's purpose. With the advent of selective herbicides, this dilemma has been generally overcome.

Smothering

Flooding. Flooding as a means of weed control is frequently used in rice production. Generally it is inefficient in achieving full control with other crops, as many weed seeds can withstand flooding for long periods, and the method is extremely wasteful of water. Undesirable effects on the soil and on raising the water-table may also result.

Mulches. Thick mulches of straw or compost, and, more recently, dark-coloured plastic mulches, can smother perennial weeds and prevent the germination of annual weeds. These methods are expensive, and usually are justified only on relatively limited areas for high-return crops.

Fire

Fire already served Palaeolithic Man to clear away unwanted growth, stimulate annual plants, and aid in food collecting. Burning is a method of improving production of grass in many primitive economies, and in modern range management controlled burning is an important tool for controlling brush in range-lands; it is also usually the first essential step in range reseeding and general improvement. The burning usually has to be followed by the application of herbicides to control resprouting from the stumps and prevent the development of new seedlings. Burning in preparation for tilling and planting is accepted practice in many agricultural economies (Sauer, 1952).

Woody plants that are of inferior value for grazing, may become established in grasslands which are not submitted to burning. In the Great Plains of North

America, the cessation of burning has contributed to the conversion of millions of hectares of prairies into mesquite jungles (Steward, 1956).

In connection with what may be termed 'flame cultivation', it may be noted that in cotton, for example, the stalk hardens much earlier than do the stems of the weeds that infest the crop. This makes it possible to find an appropriate time when a flame will destroy the weeds selectively, without harming the cotton. The operation depends for its success on the weeds being very small and much less developed than the cotton plants. The cotton plants should be at least 15–20 cm high and the stem 5 mm thick (Tupper & Matthews, 1962). The 'flame-weeder' has to be well adjusted, and the proper speed used, for the flaming to be effective.

The burning destroys only young weeds; after each irrigation or rainfall, a new 'wave' of weeds appears. Flaming can be repeated a number of times, 3–5 days after irrigation, until the appearance of the first bolls.

Crop Rotation

A succession of different crops facilitates weed control. Certain weeds are almost obligatory associates of specific crop plants, and if these are sown continuously, effective weed control may become impossible.

For example, *Lolium temulentum* (Darnel) is found in all cereal grain samples in deposits of carbonized plants, and is also found in abundance in ancient Egyptian grain (Helbaek, 1958). To this day, it infests wheat crops in temperate and in semi-arid regions. It is most difficult to separate the seeds of the weed from the cereal grains, and control in the growing crop is also extremely difficult.

Weeds whose growth and propagation is favoured by a certain crop may be weakened or checked by an appropriate succeeding crop. Leguminous crops for hay, grown during the rainy season, with their favourable nitrogen and moisture régime, encourage the rapid spread of perennial weeds such as Bermuda Grass (*Cynodon dactylon*). If the following crop is a closely-sown cereal, competition for nitrogen, moisture, and light, will greatly weaken the grass, which may be further damaged by close grazing of the stubble. Annual weeds which thrive in the winter cereal are easily destroyed during a winter fallow preceding the summer crop. Many other examples could be given.

Biological Control

Weed-eating Insects. Biological control of weeds by insects has had a few spectacular successes: in Australia the control of cactus (*Opuntia* spp., which by 1925 had infested 24 million hectares), by the moth *Cactoblastis cactorum*, is probably the best known. Klamath-weed or St John's-wort (*Hypericum perforatum*), an introduction from Europe which infests the ranges in the arid west of the United States, has been controlled by a beetle (*Chrysolina gemel-

lata). The beetle destroys the foliage, thereby impairing the ability of the plant to maintain the extensive root system that is essential for surviving summer drought (Huffaker, 1957).

Weeds introduced from foreign countries have frequently been the most difficult to control by conventional means, and have been the most susceptible to biological control, by an insect which is also an alien (Huffaker, 1957). The reason for the limited scope for biological control by insects is its serious potential danger. The need to find an insect that is specific in its choice of host—so that crop plants are not endangered by introducing the new pest—is imperative. The introduced insect must have considerable adaptability to a new environment, so that it can multiply rapidly and control the weed effectively.

Catch-crops and Trap-crops. These are used for the control of parasitic weeds, such as the witchweeds. A catch-crop that is host to the parasite is sown and ploughed under before the parasite produces seeds. A crop plant that is not a host to the weed but induces it to germinate, is called a trap-crop.

Grazing. Geese are frequently used for selective grazing in cotton. Provided they are properly managed, young geese will graze grasses and sedges in the cotton, without damaging the crop.

Sheep graze plants close to the ground, and continuous grazing may therefore considerably weaken certain perennial weeds and prevent others from producing seed. Goats can control brush in natural range.

CHEMICAL CONTROL

Chemicals for selective weed control have been in use since the end of the nineteenth century. Solutions of copper and iron salts, ammonium sulphate, potassium salts, and diluted sulphuric acid, have been used for destroying broad-leafed weeds in cereals. Kainite and calcium cyanamide dusts were also used for this purpose. These are all caustic herbicides, which burn the tissues of plants. The most widely used before the advent of the new hormonal weed-killers, was the nitrophenol DNOC (called Sinox in the U.S.A.), introduced in 1932—see p. 492. The selective action of all these herbicides is due mainly to morphological differences between the cereals and the broad-leafed weeds that are destroyed—see pp. 496–7.

The development of the selective organic herbicides is probably the most significant advance in agriculture of the last two decades. A whole array of new techniques and chemicals—low-volume spraying, improved surfactants, the precision application of herbicides to the soil, and granular formulations; new approaches such as pre-sowing and pre-emergence weed control; the discovery of new herbicides that kill grasses selectively, and of chemicals that are converted into herbicides by the plant or in the soil—have supplied the farmer with means for selective control of weeds in most crops, even including those that were considered to be most susceptible to the hormonal herbicides.

ABSORPTION OF HERBICIDES BY THE PLANT

Herbicides penetrate into plants mainly through the leaves or roots.

Foliar Absorption

The herbicides may act by *contact* or by *translocation*. In the former case, the foliage is killed, whilst in the latter case the herbicide moves throughout the plant, reaching young growing-points and moving down to the roots. The herbicides may penetrate the leaves partly through the stomata, but usually do so mainly through the cuticle. Stomatal penetration would enable rapid penetration by the herbicides, but is relatively ineffective in the hot and dry regions of irrigation agriculture because of stomatal closure and rapid drying of the spray (Crafts, 1964).

Whilst the herbicide salts are polar,* their acids, esters, etc., are non-polar, and will therefore tend to be absorbed more rapidly through the leaf cuticle, which is composed of non-polar materials. The absorption of the polar herbicides is improved by the addition of wetting agents, which may thereby increase the herbicide's toxicity considerably (Staniforth & Loomis, 1949), but at the same time may reduce its selectivity (Klingman, 1961). Absorption rates of herbicides are increased by high temperatures, and within biologically supportable limits the rate is doubled with each increase of 10°C. (ibid).

Absorption Through Roots

Herbicides can be applied to the surface of the soil and then incorporated into the upper soil layer by a light cultivation, by rain or by irrigation. They may also be injected below the soil surface. The chemicals are absorbed by the roots, polar herbicides being more rapidly absorbed than non-polar herbicides. However, non-polar substances may be converted in the soil quite rapidly into polar substances: e.g. the non-polar 2, 4-D ester may be converted successively into the 2, 4-D acid and then into a 2, 4-D salt. The sequence of events following the application of a herbicide acting through the soil is shown in Figure 91 (Holly, 1964).

* 'Polar' ions or molecules have regions that are 'electrically positive' and regions that are 'electrically negative'. Non-polar molecules do not possess strongly positive and strongly negative areas within the molecule. Most organic substances are non-polar. Polar compounds (such as most herbicidal salts) tend to be soluble in polar solvents (such as water) and insoluble in non-polar solvents, such as oils and most organic solvents (Klingman, 1961).

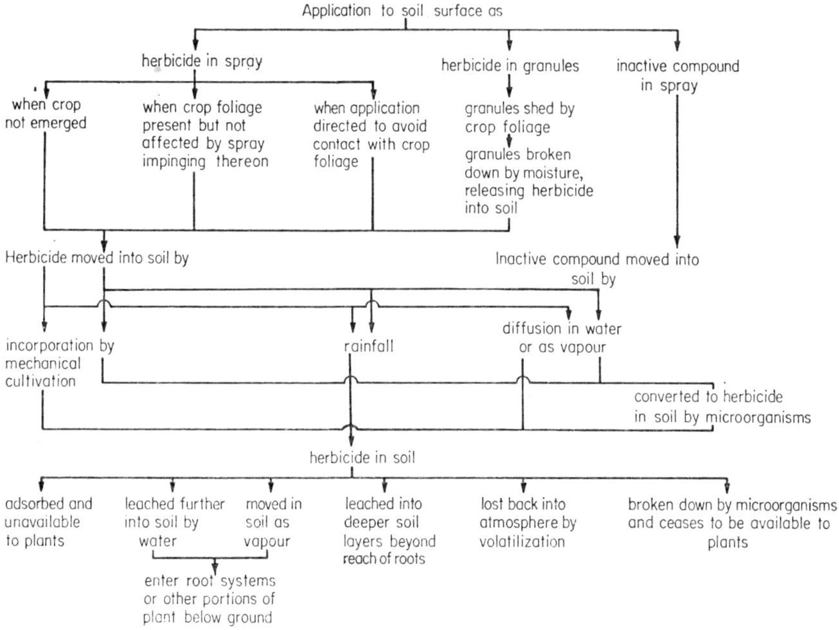

Figure 91. Schema showing the sequence of events following application of herbicides acting through the soil. (After Holly (1964): by courtesy of Academic Press, London.)

TRANSLOCATION OF HERBICIDES THROUGH THE PLANT

Herbicides that are absorbed by the leaves are translocated in the phloem towards other parts of the plant, moving in the general direction of the roots. If the herbicide is too toxic, as for example when it is applied at an excessive rate, the phloem is killed, thereby obstructing the translocation of the herbicide throughout the plant and reducing its effectiveness. Even if the top of the plant dies, the roots may still remain unaffected and growth may subsequently be renewed from underground buds, etc.

By contrast, herbicides that are absorbed through the roots are transported upwards through the xylem, together with plant nutrients, in the transpiration stream. As the xylem consists mainly of non-living tissue, it is not damaged by toxic substances. Even if the roots are killed, the translocation of the herbicide may continue. Translocation from the leaves may also occur downwards, through the xylem, in plants that are suffering from a water deficit (Kennedy & Crafts, 1927).

KINDS OF HERBICIDES

The number and variety of herbicides have increased enormously in recent years; some disappear after a relatively short period of use, to be replaced by other, more effective compounds. New compounds are constantly being added to the list. Up to the present, about thirty groups of herbicides have been investigated, including several thousands of active compounds, of which about one hundred are already in commercial use (Day, 1966).

It is interesting to note that most of the current weed control is actually being achieved with a very small number of chemicals. In 1964, three basic chemicals—2,4-D, Simazine, and Diuron, and their homologues and analogues—accounted for 75 per cent of all weed control by chemicals. Only two fundamental physiological effects on plants are involved in the action of these herbicides: 2,4-D causes abnormal plant growth, while Simazine and Diuron interfere with photosynthesis, and, for this reason, their herbicidal effectiveness depends to a large degree on light intensity (Overbeek, 1964). For almost the entire remaining 25 per cent weed control by chemicals, six more herbicides were used: Dinoseb (DNOB), Naptalam (NPA), Dalapon, Randox (CDAA) Chloropropham (CIPC), and Amitrole (AT) (ibid.).

It is not within the scope of this book to present a full account of all the herbicides that have so far been developed; only the most important groups of chemicals involved, and the principal herbicides, will be mentioned here.

Herbicidal Oils

Diesel oils and discarded lubricating oils have been used for general weed control on waste ground. Light distillates, with a medium boiling range and containing about 25 per cent of unsaturated and aromatic hydrocarbons, are used for selective control of weeds in umbelliferous crops, which have an inherent resistance to these oils (King, 1966). The selective action of these oils is based on their damaging the cellular membranes in susceptible plants, whilst those of resistant plants are not damaged (Ashton *et al.*, 1961). These oils are, however, bulky and relatively expensive, while they frequently taint the crop with an oily taste.

Phenols

Nitrophenols were the first organic chemicals to be used for selective weed control. As early as 1932, the sodium salt of 2-methyl-4,6-dinitrophenol became widely used in agriculture under the trade names of DNOC (Europe) and Sinox (U.S.A.) (Brian, 1964).

The toxic effect of the phenol herbicides appears to be due to the prevention of the generation of adenosine triphosphate (ATP, needed for all biological activities that require energy), thereby preventing growth (Overbeek, 1964).

Auxin-type, Growth-regulating Chemicals as Weed-killers

Certain synthetic auxins have the same general physiological action as the natural plant auxin 3-indoleacetic acid (IAA) on cell elongation in shoot tissue, but are not rapidly inactivated in plant tissues by the enzyme that destroys IAA.

Phenoxy Compounds. At certain concentrations, substituted phenoxy acetic acids can act as selective herbicides; they were the first and still are the most widely-used hormonal weed-killers. The best known are:

2,4-Dichlorophenoxyacetic acid (2,4-D)
2-Methyl 4-chlorophenoxyacetic acid (MCPA)
2,4,5-Trichlorophenoxyacetic acid (2,4,5-T)

2,4-D and MCPA are used mainly for the control of broad-leafed weeds in cereals, and 2,4,5-T is employed for the killing of undesirable woody species on which it is more effective than 2,4-D. Frequently, mixtures of 2,4-D and 2,4,5-T are used.

The phenoxy compounds are quickly absorbed and translocated throughout the plant both through the leaves and through the roots; sensitive plants may react within 10–12 hours after being sprayed. Though death may occur only weeks later, the weeds cease to compete with the crop plant shortly after being sprayed. The affected individuals show typical symptoms: grotesque, prostrate plants with malformed leaves and twisted stems. These are the result of disorganized cell division and deranged metabolic processes; they appear to be caused by an excess production of RNA that is induced by the 2,4-D and results in hormonal imbalance (Overbeek, 1964).

The phenoxy compounds are rapidly broken down by soil micro-organisms, so that there is no significant danger of build-up to toxic levels in the soil.

MCPA is slightly more selective than 2,4-D, and is therefore preferred for crop varieties that are relatively more susceptible to 2,4-D. Accounts of these and many growth-regulating substances are given by Audus (1959).

The esters of 2,4-D penetrate plant tissues somewhat more rapidly than the amine salts. The former are therefore preferred when the weather is not stable. They are somewhat more toxic to the crop than are the amine salts, and are therefore less selective. The 2,4-D compounds are also available commercially in granular form.

Phenoxy Butyric Acids. MCPB is less toxic than MCPA to clovers, cereals and peas, and yet is as toxic as MCPA to certain weeds (King, 1966).

Substituted Aryloxy Alkanols. The principal herbicide in this group, called 'Sesone', is sodium (2,4-dichlorophenoxy) ethyl sulphate. It is not readily absorbed when applied as a surface spray and does not injure plant leaves but is transformed into 2,4-D by microbial or chemical action in the soil, making it useful as a pre-emergence spray (King, 1966). In basic soils, it is the microbial action which predominates over the chemical (Klingman, 1961).

Phenylcarbamates. This group of herbicides interferes with the normal process of cell division in a way similar to that of colchicine. Plant roots are particularly susceptible and are rapidly killed (Crafts, 1961). For this reason, these herbicides are applied mainly through the soil.

The first herbicide of this group to be widely used, IPC (isopropyl N-phenyl carbamate), was rapidly replaced by CIPC (isopropyl-N-(3-chlorophenyl) carbamate), which is more toxic than the former to grassy weeds, and is more persistant under conditions of high temperature. Its phytotoxic effect is due to the inhibition of cell division in the root-tip (Overbeek, 1964).

Eptam (S-ethyl dipropylthiocarbamate) is used as a pre-emergence herbicide against grassy weeds and for the control of *Cyperus rotundus*. The chemical breaks down in moist soil, producing toxic vapours that penetrate plant roots.

Aliphatic Acids. The sodium salt of trichloroacetic acid (TCA) is used in a number of crops as a pre-emergence herbicide for the control of grassy weeds. It is fairly persistent in dry soil, but breaks down rapidly in moist, warm soils. It is very water-soluble, and is therefore easily leached from the surface layer of the soil. TCA is broken down by resistant species, whilst it accumulates in susceptible species.

Dalapon (2,2-dichloropropionic acid) is a strong acid that persists in the plant for weeks without undergoing any breakdown. It is slow-acting, has no acute toxic action, has little effect on the growth of the roots, but inhibits growth of the stem apex. Its action appears to be due to the breakdown of proteins into toxic products. Possibly, selectivity is due to differences between plant species in detoxifying the breakdown products resulting from the action of Dalapon (Overbeek, 1964).

Dalapon is very effective for the control of perennial grasses, such as *Cynodon dactylon*. It can be applied either to the foliage or to the soil, and is rapidly transported in the phloem and xylem (Overbeek, 1964).

Amino-triazoles. Amitrole, or AT (3-amino-1,2,4 triazole), interferes with glycine metabolism, thereby blocking the synthesis of purines and interfering with RNA synthesis (Overbeek, 1964). It is readily absorbed and rapidly translocated in the plant, causing marked chlorosis.

Triazines. These affect the Hill reaction, and hence photosynthesis, leading to the starvation of the plant. It has been shown that the addition of glucose to the leaves of a treated plant can serve as an 'antidote' to lethal doses of Simazine (Moreland *et al.*, 1959).

Simazine (2-chloro-4,6-bis(cthylamino)-s-triazine) is used as a pre-emergence herbicide (especially in maize) and, at higher rates, as a soil sterilant.

Atrazine (2-chloro-4-ethylamino-6-isopropylamino-s-triazine), although introduced several years after Simazine, is already of equal and possibly greater importance as a herbicide. It is predominantly used in maize, but has also found applications in a number of other crops, such as sorghum. In contrast to Simazine, Atrazine is effective through the leaves of plants, so that it can also be used as an early post-emergence herbicide in suitable crop plants, whereas Simazine can only be applied pre-emergence.

For this reason, and also because of its somewhat greater water solubility, Atrazine is a more effective herbicide under dry soil conditions than is Simazine (Birk & Roadhouse, 1962).

A number of other s-triazine herbicides have been developed, with different toxicity spectrums, such as Propazine (2-chloro-4,6-bis(isopropylamino)-1,3,5-triazine), Simetone (2-methoxy-4,6-bis(ethylamino)-s-triazine, and Prometone (2-methoxy-4,6-bis(isopropylamino-s-triazine).

Phenyl-substituted Ureas. Wilting is a first symptom of the toxicity of substituted ureas (Wort, 1965). Monuron or CMU (3-(p-chlorophenyl)-1-1-dimethyl-urea) and Diuron or DCMU (3-(3,4-dichlorophenyl)1-1-dimethyl urea) are extremely toxic at very low concentrations, which is an important characteristic for soil-applied herbicides. Both are used for selective pre-emergence treatments, and as general herbicides mainly against annual dicotyledons. Diuron is less soluble than Monuron, and is therefore less liable to be leached out of the surface layer of soil. Both herbicides are easily absorbed by the roots and quickly transported to all parts of the plant; they interfere with photosynthesis by inactivating chlorophyll.

These herbicides are broken down in the soil by the activity of micro-organisms. The rate of breakdown depends mainly on a favourable soil-moisture régime and on the presence of organic matter, and is generally not rapid.

The best-known herbicide of the chloracetamide group is Randox (2-chloro-N,N diallylacetamide). The chloroacetamides serve mainly as pre-emergence herbicides in a number of crops. They are toxic at very low concentrations to the germinating seeds of a large number of weeds, especially grassy weeds. They are, however, effective only in the earliest stages of germination, by inhibiting cell division in the root-tip (Overbeek, 1964).

Bipyridils. A large number of bipyridilium compounds have been tested in recent years, of which two, Diquat (1-1-ethylene-2,2-dipyridilium dibromide) and Paraquat (1,1-dimethyl-4,4-dipyridilium dichloride), have proved to be the most effective. The two are very similar in their biological properties, but Diquat is somewhat the more toxic to broad-leafed weeds, as is Paraquat to grassy weeds (Brian, 1964).

The bipyridils are contact herbicides, non-toxic to animals, that at very low concentrations kill weeds within a few hours after spraying, and are especially active under conditions of bright sunlight. They are rapidly absorbed and translocated to all parts of the plant, killing and drying the leaves. In contact with the soil, they are rapidly adsorbed by the soil colloids and inactivated, so that there is no danger of toxic residues from them.

The bipyridils can be sprayed on small weed seedlings that appear before a susceptible crop emerges above the soil. The crop will not be adversely affected, as the herbicide will have become inactivated by the soil before the crop emerges.

It has been found that Diquat effectively controls a dodder, *Cuscuta trifolii*, which is a major parasitic weed of lucerne in the Mediterranean region (Boon, 1965).

SELECTIVITY OF HERBICIDES

The selective action of herbicides is based on characteristics of the herbicide or the way it is applied, in combination with differences between specific morphological and physiological characteristics of the crop plant involved and the weeds it is desired to control. Selectivity is not absolute, the degree of selectivity depending on the dosage, the formulation, the method of application, environmental conditions, and the stage of plant growth (Day, 1966).

Morphological Basis of Selectivity

The vital meristems of broad-leafed plants are located at the tips of the shoots and in leaf axils, and are therefore relatively exposed to the action of herbicides. By contrast, the growing-points of cereals are located at the base of the plant and are protected by the leaf-sheath. Occasionally they are even situated below the soil's surface.

Cereal leaves have a waxy cuticle that is finely corrugated, so that there is no intimate contact between the toxic herbicide and the leaf epidermis (Figure 92). The leaves are also narrow and more or less erect, so that the amounts of herbicide that come in contact with leaves are relatively small.

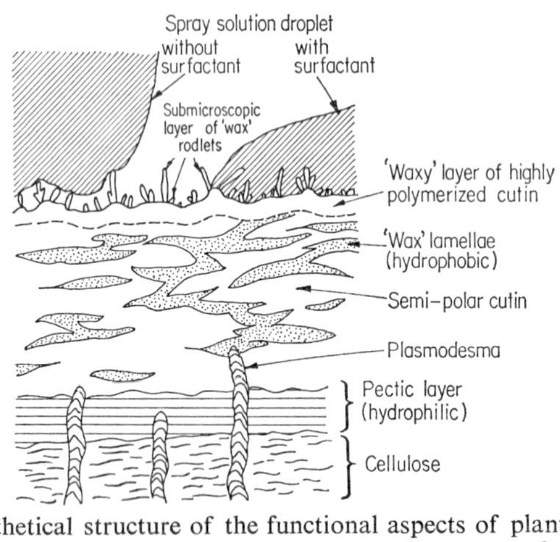

Figure 92. Hypothetical structure of the functional aspects of plant cuticle. The waxy rodlets of leaves having 'bloom' may prevent contact of a spray droplet with the leaf surface. (After Fried & Morris (1967): by courtesy of the Agricultural Experiment Station, Oregon State University.)

Broad-leafed weeds, as their name implies, have wide foliar surfaces and horizontal leaves that absorb large amounts of herbicide; the epidermis is smooth, so that the spray spreads in a thin film and is in intimate contact with the epidermis.

Differences in the thickness of the cuticle influence the amounts of herbicide absorbed by the leaves; plants with thick cuticles will be less susceptible to a herbicide than those with thinner cuticles. The number, size and location, of stomata vary for different plant species. As the stomata are one of the means of entry of the herbicide into the leaf, it follows that the differences in stomata between species will affect their relative susceptibility to herbicides.

The emergent stem of the plant is not easily wettable in most species; it also loosens the soil as the seedling emerges, so that effective contact with the herbicide in solution is minimal (Hartley, 1964). This is one reason for the absence of damage to the emerging seedlings of the crop plants, as they pass through the soil layer in which soil-applied herbicides are concentrated, whilst the weed seedlings germinate in the upper soil layer in which the herbicide is concentrated, and absorb the chemical through their roots.

Physiological Basis of Selectivity

Not all the mechanisms of the selectivity of action of herbicides have been elucidated. Differences in soil permeability, in metabolism, in the enzyme system, and in polarity, may be involved. By affecting one or more of these systems, certain essential biochemical processes may be hindered or over-stimulated, causing imbalance or dislocation of vital processes in the plant.

The simplest selective mechanism is in operation when a physical or biochemical process prevents the translocation of a herbicide in one species, giving it an advantage over plants in which the chemical is rapidly translocated to all parts of the plant. Alternatively, resistant plants metabolize the compound into an inactive form more rapidly than do sensitive plants.

2,4-D reacts with plant tissue substrate to form 2,4-D protein complexes, which have no toxic effects. In 2,4-D-resistant plants, the rate of formation of these complexes is very rapid, whilst it is slow in 2,4-D-susceptible plants (Butts & Fang, 1956).

Different enzyme systems may be involved in different plants, resulting in selectivity of herbicides. For example, certain weeds produce enzymes that convert homologues of 2,4-D such as 2-(2,4-DB) or Butaxone (4-(2,4-dichlorophenoxy) butyric acid), which have no herbicidal activity, into 2,4-D to which the weeds succumb, whilst other plants such as lucerne do not have this mechanism and so remain unaffected.

The opposite process is exemplified by the effect of Simazine, a very potent herbicide, which is absorbed by maize without ill-effect; the maize plant has a detoxification mechanism that hydrolizes the Simazine into an inoffensive compound (Gysin, 1960). Plants lacking this mechanism are killed by Simazine.

Dalapon inhibits pantothenic acid synthesis, a vital element in plant metabolism. There are marked differences in the ability of different plant species to withstand this disruptive action of Dalapon (Ashton *et al.*, 1961).

METHODS OF APPLICATION

Methods of application have been devised which ensure that the weeds will be contacted by the herbicide, whilst only minimal amounts reach the crop plants.

The sprays may be directed below the crop's foliage, so that only the stems of the crop plant are wetted, whilst the foliage of the weeds absorbs most of the application. Special shields may be used, which protect the crop plant from the spray.

Herbicides in granular form will roll off the crop's foliage and reach the soil. Rain or irrigation will move the herbicide superficially into the soil, where it will be absorbed by the shallow roots of the weeds.

In pre-emergence application, the location of the herbicide nearer to the site of the germinating weeds than to the seeds of the crop plant ensures the killing of the former, with a minimum of damage to the latter.

An improvement in selectivity can be achieved by coating the seed with a material such as activated carbon that adsorbs the herbicide, or alternatively by placing a layer of absorbant material just above the seed-row (Ripper, 1956).

Pre-sowing Treatments

Pre-sowing treatments may be applied to kill weeds that have emerged after seed-bed preparation has been completed, when actual sowing has been delayed for example because of weather conditions. By spraying with an appropriate herbicide, the need for additional tillage operations can be avoided.

The principal application of pre-sowing treatments in semi-arid regions is their use for controlling weeds during the fallow period between crops—the so-called chemical fallow. Their great advantage is in the possibility of controlling weeds in the early stages, or even in preventing their emergence. Tillage is too dependent on weather and soil conditions to be as effective as chemical weed-control in this respect.

The mechanical incorporation of the herbicide into the soil may have as objectives:

(a) *Placement of the Herbicide for Maximum Effectiveness.* The optimal depth in this case will depend mainly on the site of uptake of the herbicide, e.g. the triazines are mainly absorbed through the roots, many carbamates are taken up through the coleoptiles of gramineous plants, and EPTC penetrates through the bulblets of Nut-grass (*Cyperus rotundus*). In the last case, it was found that incorporation by disking had the additional advantage of tearing the chains of dormant bulblets and inducing their germination (Horowitz, 1965).

(b) *Protecting Herbicides against Loss by Volatilization and Photodecomposition.* A very thin protecting cover of soil is usually sufficient to achieve this aim (Hartley, 1964).

Pre-emergence Treatments

The herbicide is sprayed on the soil surface at the time of sowing or shortly after, and so is concentrated in the very shallow soil-layer in which the seeds of most annuals germinate. The seeds of the crop are planted at a lower depth; the crop may have tolerance to the herbicide or else, by the time of emergence of its seedlings, the weed-killer should have become decomposed and no longer toxic. The weeds near the soil surface having been killed, the field may remain clean until it is disturbed mechanically. The herbicidal effect may also persist for a long time after emergence.

Post-emergence Treatments

Herbicides applied in the standing crop must be truly selective. However, undesirable effects on the crop plants can be avoided by directing the spray so as to reduce to a minimum the amounts of herbicide reaching the crop plants, and by using adapted shields for protecting the rows of crop plants.

FORMULATIONS

Sprays have a number of advantages over dusts—in particular their greater efficacy and the reduced danger of drift. In certain countries the aerial application of dusts of the hormonal herbicides is forbidden by law. With low volume spraying, the main disadvantage of sprays—the need for transporting and spraying large volume of water—has been eliminated.

Many herbicides are now sold in granular form for application to the soil, either within a growing crop or for pre-emergence treatment. The solid forms are particularly advantageous where long-residual action in the soil is desired. They are also easier to apply and are frequently more selective than sprays. However, little is yet known of the possible differential response of granular forms in comparison with sprays under various environmental conditions, which can influence the degree of activation of the herbicide, its effectiveness, and its persistence. For example, EPTC was less persistent in the liquid form under extreme drought conditions than was the granular formulation. By contrast, the spray formulation of Amiben was more effective in the control of weeds under dry conditions than was the granular formulation (Cardenas & Santelmann, 1966).

FACTORS AFFECTING THE PERFORMANCE OF HERBICIDES

Climatic Conditions

The susceptibility of a species to a herbicide depends to a large degree on environmental factors. Any factor that increases herbicidal retention, penetration and absorption by the weeds, will probably increase effectiveness (Hammerton, 1967).

The temperature prevailing after treatment has a considerable effect on the length of time between application and death of the weeds. In general, warm temperatures promote penetration. In particular, the auxin-like herbicides are extremely slow-acting at temperatures below 20°C; their efficiency is also reduced at temperatures above 30°C (Helgeson, 1957). Information on optimum temperatures of application is extremely important in the hot, dry regions.

Low relative humidities reduce the permeability of the plant tissues to the penetration of the herbicides, and high air humidity has the opposite effect. In tests with the diethanolamine salt of maleic hydrazide, it was found that, at 100 per cent relative humidity, absorption was four times as rapid as at 50 per cent relative humidity (Smith *et al.*, 1959).

The combination of too-high temperatures and low humidity is detrimental to the absorption of herbicides as not only does the cuticle tend to become thicker and less penetrable but spray drops dry more rapidly under these conditions (Foy *et al.*, 1967). Light stimulates the opening of the stomata and increases the production of photosynthates which promote the systemic movement of the herbicide in the phloem (ibid.).

Rain shortly before application increases the hydration and turgor of the tissues, thereby increasing leaf wettability and in consequence herbicidal susceptibility (Hammerton, 1967).

Rain occurring shortly after spraying may wash the chemicals off the plants before they have had time to take effect. The minimum time-interval necessary between spraying and rainfall depends on the chemical constitution: esters require no more than half an hour without rain to ensure successful application, whilst amines require about six hours, and the metallic salts about twelve hours. It is generally not recommended to spray plants that are wet with dew or from recent rain or irrigation, as the herbicide may then be excessively diluted. However, when dusts are used, foliage that has been wetted gives better adherence and improves the efficiency of the dust.

The main disadvantage of pre-emergence applications of herbicides is their dependence on the weather following application. Rainfall, beyond a certain amount, may leach the herbicide from the effective superficial soil-layer, and may even endanger the seedlings of the crop plant.

Shade, under certain circumstances, may affect the effectiveness of herbicide application. Competition for light, following treatment with herbicides may be important in preventing the recovery of weeds that have been weakened by herbicides (Hammerton, 1967).

Granular formulations of CIPC, uniformly distributed on the soil, and applied before dodder (*Cuscuta* sp.) was wrapped around the host plant, were found to provide excellent control of this parasitic weed in the field. Partial shade from the crop at the time of application increased the effectiveness of the CIPC. Control persisted beyond the expected soil-life of CIPC, because shade from the crop plants inhibited dodder development and the dry soil surface reduced dodder germination (Dawson, 1966).

Edaphic Conditions

For many herbicides, the most important path of entry is through the roots. When herbicides are applied to the soil there is, however, no direct contact with the active surface of the weeds to be destroyed. The chemicals must reach their target by diffusion and water flow—processes which are extremely dependent on weather and soil properties. This explains the considerable variability in the performance of soil-applied herbicides, and the reason for their higher reliability in irrigation agriculture (Hartley, 1964).

Adsorption of Herbicides by Soil. The adsorptive capacity of the soil for herbicides depends on the quantity and nature of its colloids. The fraction of the herbicide that is adsorbed by the soil is either permanently or temporarily inactivated. Hence, the efficiency of a treatment tends to be in inverse proportion to the adsorptive capacity of the soil for the herbicide concerned. The rates of application of herbicides have to be adjusted accordingly; higher rates for effective weed control are needed in a heavy clay than in a sandy soil, and still higher rates are required in soils with a high organic-matter content.

In general, as the adsorption of herbicides on the clay particles is an exothermic process, increased temperatures result in reduced adsorption by clay and enhanced effect of the herbicide (Harris & Warren, 1964). For example, the phytotoxicity of pyrazone, applied to early sowings of sugar-beets in September, is greater than at later sowings, when temperatures are much lower (Horowitz, 1966).

Losses of Herbicides. When the herbicides are surface-applied to dry soils, the very high temperatures of the soil surface under the strong sunshine in the arid regions can cause considerable losses by volatilization and possibly photodecomposition (Hartley, 1964).

VOLATILIZATION

The rate of volatilization differs for different herbicides—a factor to be taken into account when choosing the appropriate herbicide. For example, the ester of 2,4-D is approximately 20 000 times more volatile than Monuron

(Klingman, 1961). At a temperature of 60°C, losses by volatilization after twenty four hours of exposure were 90 per cent for Atrazine and Propazine, 30 per cent for Simazine, and negligible for Prometone (Kearny *et al.*, 1964).

It is of interest to note that many of the volatile herbicides are adsorbed more strongly by a dry soil than by a moist one; when sprayed on a moist soil, they have to be incorporated into the soil immediately after application in order to avoid the loss of a considerable proportion of the chemical (Horowitz, 1966). By contrast, they can be sprayed on a dry surface with negligible loss; Atrazine, for example, can retain its effectiveness for several weeks under dry conditions, whilst it is less effective if sprayed on already moistened soil shortly before the weeds germinate. Similar results were obtained with CDED: losses from a wet soil surface at 44°C were found to be much more rapid than from a dry soil (Taylorson, 1966).

This does not, however, apply to all herbicides: under conditions of high temperature and soil moisture, soil persistence of NPA(N-1-naphthylphthalamic acid), Amiben (3-amino-2,5-dichlorobenzoic acid), and DCPA or Dacthal (dimethyl ester of tetrachloroterephthalic acid), was found to vary inversely with their volatility (Cardenas & Santelman, 1966).

PHOTODECOMPOSITION

Many trials have shown that solar radiation, and in particular ultra-violet rays, can degrade many herbicides and attenuate their biological effects Jordan *et al.*, 1965). However, these effects are less marked in the field than in the laboratory, probably because the micro-relief of the soil and the adsorption process provide some degree of protection against the radiation effects (Horowitz, 1966.)

As long as the herbicide remains on the surface of the soil, it is not only subject to degradation, but it also has little or no phytotoxic effect. For this reason, a herbicide with low solubility, such as Simazine, is ineffective in arid regions without irrigation (Hartley, 1964). More soluble herbicides, such as Atrazine, can at least be washed into the soil by a light shower. Conversely, in regions of high rainfall, soluble herbicides can be leached downwards to a depth at which they become either dangerous to the crop and therefore less selective, or ineffective when leached beyond the root-zone. Disappointing results with herbicides applied to the soil have therefore been frequently reported in dry regions in the absence of rains or, conversely, after excessive rains.

If a herbicide, immediately after application, is made to penetrate the soil for a few millimetres, as is possible by applying a light irrigation with sprinklers, not only are losses considerably reduced, but the chemical is concentrated in the zone in which it will do the most damage to germinating weeds, and the least to the cultivated crop. This irrigation will, incidentally, also cause the weed seeds in the surface layer of the soil to germinate. For these reasons, soil-applied herbicides (in particular for pre-emergence treatments) are most widely used in areas in which intensive irrigation agriculture is practised and where labour costs are high (Hartley, 1964).

IRRIGATION METHODS

It has been shown that CIPC, NPA, EPTC, and Sesone are more effective with sprinkler irrigation than when furrow-irrigated, whilst the effectiveness of Diuron was least affected by the method of irrigation (Menges, 1963). The improved toxicity of the herbicides was due to reduced volatilization following their incorporation into the soil by water, and to greater effectiveness due to their being brought into the area in which weed seeds were germinating (Jordan *et al.*, 1963). Soil incorporation has improved the weed control by some herbicides, but may occasionally reduce the effectiveness of certain herbicides if too much dilution occurs in the area of the weed seeds (Ashton & Dunster, 1961).

The incorporation into the soil of volatile herbicides such as EPTC and Dichlobenil (2,6-dichlorobenzonitrile) usually increases their toxicity to weeds (Jordan *et al.*, 1963).

For most herbicides, overhead sprinkler irrigation is more effective than mechanical incorporation, probably because of the former's more uniform incorporation and better penetration (Figure 93; Menges, 1963). When the source of activating water is from above, the downward movement of water carries the herbicide by gravity flow, resulting in a better and more controlled distribution of the herbicide in the soil. With furrow irrigation, water flow is sidewards and upwards, and there is a tendency for the herbicide to accumulate at the surface of the soil (Cardenas & Santelman, 1966).

Problems Resulting from the Use of Herbicides

PERSISTENCE OF HERBICIDES IN THE SOIL

Herbicides in the soil may be destroyed by micro-organisms, may decompose chemically or by photodecomposition, may be inactivated by adsorption on the soil colloids, or may be lost by leaching or volatilization (Klingman, 1961). The principal factor involved in the degradation of the herbicides is *biological*. Decomposition of herbicides to subtoxic levels by soil micro-organisms may require a few days or up to a year or more.

In general, well-managed irrigated soils are warm, moist, fertile, and well-aerated—conditions that are conducive to the rapid decomposition of organic herbicides. By contrast, in unirrigated agriculture, during the extended periods when the soil is dry, microbiological decomposition may cease completely (Figure 94).

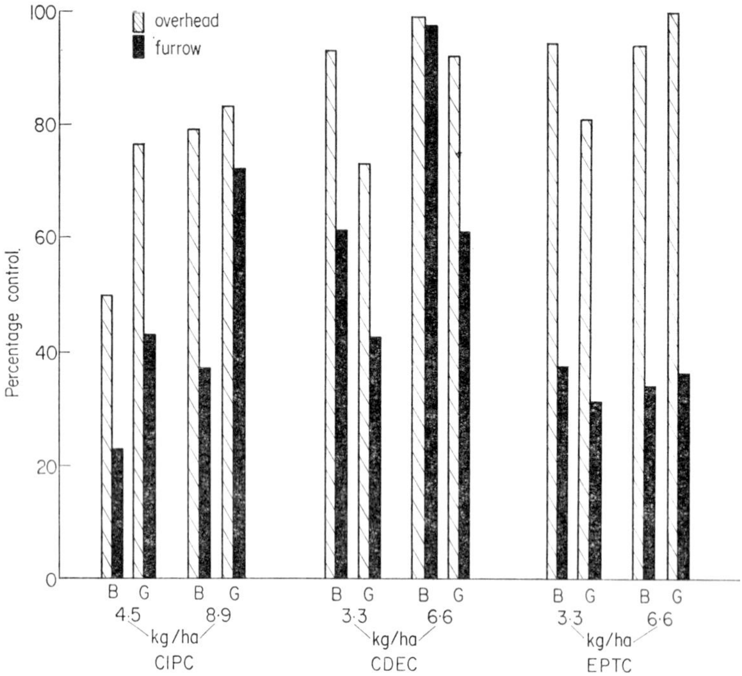

Figure 93. Effect of pre-emergence herbicides as influenced by irrigation method.
B: broad-leafed weeds
G: grass weeds (the soil was a silty clay-loam)
(After Menges (1963): by courtesy of the Weed Science Society of America.)

The persistence of herbicides in the soil varies considerably according to their chemical composition. Paraquat and Diquat are almost immediately decomposed. Most of the phenoxy group of herbicides are rapidly decomposed in the soil, whilst Simazine and Diuron are very persistent. Endothal (disodium 3,6-endoxyhexahydorophthalate), applied as a pre-emergence treatment for sugar-beets, remains herbicidal for 2–3 weeks only; it is possible to sow cotton approximately six weeks after a soil application of Dalapon plus 2,4-D. By contrast, Atrazine applied at the beginning of the rainy season in a winter rainfall region, may effectively control weeds throughout the winter months, and even throughout the growing period of sorghum that is sown in the following spring (Horowitz, 1966).

The problem of the persistence of herbicides in the soil has two contradictory aspects: persistence is desirable in order to ensure a weed-free environment for as long as possible, but it is undesirable if it interferes with the sowing of a susceptible crop following the crop for which it had been used. In addition, the build-up of residues may reach toxic levels for most, if not all, crops.

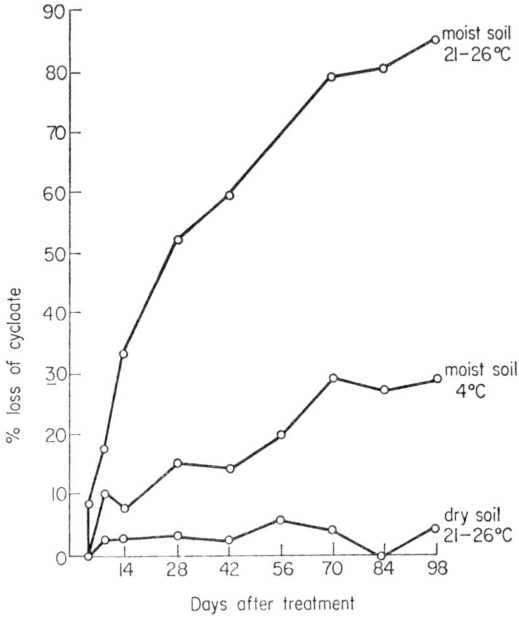

Figure 94. Effect of soil moisture and temperature on losses of herbicide from the soil. A thiocarbamate (cycloate), at the rate of 6·8 kg/ha, was incorporated 7·5 cm deep in a loam soil. (From Gray & Weierich (1968): by courtesy of the authors.)

Other effects, such as those on soil micro-organisms, have been discussed in Chapter XI (*see* pp. 341–2).

The detrimental effects of herbicide residues are usually more serious in arid than in humid regions, and in rain-fed agriculture more than under irrigation. The erratic and unpredictable rainfall of the semi-arid regions makes it difficult to forecast the time that will be required for the breakdown of a herbicide to innocuous levels. The long dry periods may permit persistence of herbicidal activity, which may then interfere with cropping schedules.

In Nebraska, phytotoxic concentrations of Simazine and Atrazine were found to have accumulated in the subsoil 16 months after treatment (Burnside *et al.*, 1963). 2,3,6-TBA (2,3,6-trichlorobenzoic acid) was found to destroy soybeans five seasons after application. Monuron reduced oat yields in the third growing-season after its application (Burnside *et al.*, 1965). Though less dangerous in irrigated agriculture, toxic build-ups may still occur there. In California and Arizona it was found that annual applications of Diuron or Monuron for eight consecutive years on irrigated cotton had not resulted in a build-up of residues to a level that would be toxic to the cotton. However, cereals were adversely affected after a single year's application (Arle *et al.*, 1965).

The build-up of herbicide residues to a toxic level can be at least partly overcome by dilution (tillage, irrigation), accelerating breakdown (adding organic manures or maintaining a favourable soil moisture régime), or sowing a non-susceptible crop, and, mainly, by careful use of the herbicides themselves through accurate dosage and calibration, appropriate choice of herbicides, avoiding the continuous use of a single type of herbicide, etc.

INDUCED RESISTANCE TO HERBICIDES

Unlike insects, weeds have, up to the present, only rarely developed strains that are resistant to the chemicals which are used for their eradication. The main reason is probably that the development of resistance is an evolutionary process, requiring a minimum number of generations; this is achieved in a far shorter time by most insects than by plants, which normally produce at most only one generation a year (Day, 1966). However, Longchamp (1965) reported that, in fields of maize which had been regularly treated with Simazine from 1957, plants of *Panicum crus-galli* had become resistant to doses of 9 kg/ha of active material, whilst originally, complete control of this weed was achieved with only 2 kg/ha. Moreover, after some years of treatment with Dalapon, M. Horowitz (personal communication) has been able to isolate several strains of *Cynodon dactylon* which differed in their sensitivity to that herbicide.

EFFECT OF HERBICIDES ON YIELD OF CROP PLANTS

In addition to the indirect effect of the herbicides on the cultivated crop, as a result of the removal of weed competition, a certain amount of herbicide usually reaches the crop plant, and its direct effect on the crop is therefore of importance. These concentrations are naturally sub-lethal even in the case of sensitive crop plants. It has been found in many cases that, at these concentrations, many herbicides have a stimulating effect on growth, and may improve yields independently of the effect of weed removal (Wort, 1965).

EFFECTS OF HERBICIDES ON PLANT COMPOSITION

Herbicides can affect the composition of crop plants in a number of ways:

1. By causing changes in metabolism, resulting in quantitative changes in the amounts of some of the metabolites produced.
2. The formation of unusual compounds may occur, following the action of certain herbicides.
3. Some herbicides may cause alterations in the physical properties of the cell-walls and in the permeability of cell-membranes, and, as a result, water and nutrient uptake may be influenced (Brian, 1964).
4. Residues of the herbicide itself, or its derivatives, may remain in the agricultural product.

Many herbicides have been found to influence plants' contents of sugar, starch and protein, usually with the result of improved quality (Wort, 1965). In particular, a favourable effect on the protein content of the grain in wheat, barley and maize, following the application of 2,4-D, has frequently been reported (ibid.). Moisture content is generally increased following herbicide application (ibid.). On the other hand, oil content is usually decreased, as for example in the seeds of linseed (Dunham, 1951).

CHANGES IN CROP MANAGEMENT FOLLOWING MODERN WEED-CONTROL PRACTICES

The introduction of new and effective, selective weed-control measures, has had a considerable impact on traditional management practices.

A new criterion in the choice of varieties is their relative susceptibility to the standard herbicides. Varieties such as dwarf wheats, which could not have held their own against broad-leafed weeds, can now be sown safely. When pre-emergence treatments are anticipated, the seed-bed has to be prepared accordingly; direct seeding can replace planting, and sowing to a final stand can replace high rates of sowing with subsequent thinning. In general, rates of sowing and distances between rows can be reduced when the farmer is certain of a relatively weed-free stand. Depth of seeding has to be adjusted to the depth of herbicide placement.

The need for sowing companion crops with legumes, in order to help with their establishment, has frequently become superfluous.

TILLAGE

Tillage operations can be considerably reduced by the appropriate use of herbicides, and 'minimum tillage' has become a practical proposition (see pp. 434–8).

As a result of numerous studies on methods of utilizing herbicides in semi-arid areas for weed control during fallow periods, 'chemical fallowing' (Plate 34) is in many cases replacing the traditional clean-tilled fallow and even the newer methods of mechanical stubble-mulch farming (Wiese *et al.*, 1967). With chemical control of weeds in fallow, yields were identical to those obtained from conventional tillage (Baker *et al.*, 1956). Chemical fallow greatly increases the crop residues that are available for wind-erosion control. With an initial amount of residue after harvest of approximately $2\frac{1}{2}$ tonnes/ha, after 11 months chemical fallow left 80 per cent of this, sweep tillage left 66 per cent, and disking left less than 3 per cent (Wiese & Army, 1958).

Chemical fallow reduces tillage operations, not only during the fallow period, but also subsequently in the growing crop. For example, when Propazine was applied to wheat stubble, it was found that three tillage operations during the fallow period were eliminated, as was one in the following sorghum crop. Soil moisture storage and sorghum yields were the same as those obtained by traditional tillage during fallow and inter-row cultivation of the sorghum crop (Wiese *et al.*, 1967).

Chemical fallow has also been shown to reduce costs of production. When Atrazine was used for weed control in the fallow after wheat and a subsequent sorghum crop, net returns were markedly increased over those obtained with mechanical weed-control methods (Phillips, 1964). Following chemical fallowing, mechanical harvesting has been facilitated in weed-free crops such as cotton, and is much improved in efficiency in other crops.

The increasing importance of pre-emergence selective weed-control confers an additional advantage to overhead irrigation, as compared with surface methods of irrigation. It has been shown that the efficiency of pre-emergence application of several weed-killers is greatly increased when the application is followed by overhead irrigation as compared with furrow irrigation. As already explained on page 503, the source of activating water is from above, the downward movement of water carries the herbicide by gravity flow, resulting in a better and more controlled distribution of the herbicide in the soil. With furrow irrigation, water flow is sidewards and upwards, and there is a tendency for the herbicide to accumulate at the surface of the soil (Cardenas & Santelman, 1966).

References

Åberg, E. (1964). Susceptibility: factors in the plant modifying the response of a given species to treatment. pp. 401–22 in *The Physiology and Biochemistry of Herbicides* (Ed. L. J. Audus) Academic Press, London, xix + 555 pp., illustr.

Arle, H. F., Miller, J. H. & Sheets, T. J. (1965). Disappearance of herbicides from irrigated soils. *Weeds*, **13**, 56–60.

Ashton, F. M. & Dunster, K. (1961). The herbicidal effect of EPTC, CDEC and CDAA on *Echinochloa crusgalli* with various depths of soil incorporation. *Weeds*, **9**, 312–17.

——, Harvey, W. A. & Foy, C. L. (1961). Principles of selective weed control. *Circ. Calif. Agric. Exp. Stn.*, **505**, 20 pp., illustr.

Audus, L. J. (1959). *Plant Growth Substances* (Plant Science Monographs, Ed. N. Polunin) Leonard Hill, London, and Interscience, New York, xxii + 553 pp., illustr.

Baker, L. O., Krall, J. L., Aeshein, T. S. & Hartman, T. T. (1956). Chemical summer fallow in Montana. *Down to Earth*, **11**(4), 21–23.

Birk, L. A. & Roadhouse, F. E. B. (1962). Penetration of and persistence in soil of the herbicide Atrazine. *Can. J. Pl. Sci.*, **55**, 21–7.

Blackman, G. E., Holly, K. & Roberts, H. A. (1949). The comparative toxicity of phytocidal substances. *Symp. Soc. Exp. Biology*, **3**, 283–317.

Boon, W. R. (1965). Diquat and Paraquat—new agricultural tools. *Chem. Ind. Rev.*, **19**, 782–8.

Brian, R. C. (1964). The classification of herbicides and types of toxicity. pp. 1–37 in *The Physiology and Biochemistry of Herbicides* (Ed. L. J. Audus) Academic Press, London, xix + 555 pp., illustr.

BRICKEY, J. (1966). How much does one weed cost? *Crystal-Ized Facts, Denver*, **20**(1), 5–6.

BURNSIDE, O. C., FENSTER, C. R. & WICKS, G. A. (1963). Dissipation and leaching of Monuron, Simazine and Atrazine in Nebraska soils. *Weeds*, **11**, 209–13.

——, WICKS, G. A. & FENSTER, C. R. (1965). Herbicide longevity in Nebraska soils. *Weeds*, **13**, 277–8.

BUTTS, J. S. & FANG, S. C. (1956). Tracer studies on the mechanism of action of hormonal herbicides. *Conf. Radioactive Isotopes in Agric.*, East Lansing, Michigan, pp. 209–14 (cited by L. J. King (1966)).

CARDENAS, J. & SANTELMANN, P. W. (1966). Influence of irrigation and formulation on activity of NPA, Amiben and DCPA. *Weeds*, **14**, 309–13.

CRAFTS, A. S. (1961). *Translocation in Plants.* Holt, Rinehart & Winston, New York, x + 182 pp.

—— (1964). Herbicide behaviour in the plant. pp. 75–100 in *The Physiology and Biochemistry of Herbicides* (Ed. L. J. Audus) Academic Press, London, xix + 555 pp., illustr.

DAWSON, J. H. (1966). Factors affecting dodder control with granular CIPC. *Weeds*, **14**, 255–8.

DAY, B. E. (1966). The scientific basis of weed control. pp. 102–44 in *Scientific Aspects of Pest Control.* Nat. Acad. of Sci., Washington, xi + 470 pp., illustr.

DUNHAM, R. S. (1951). Differential responses in crop plants. Pp. 195–205 in *Plant Growth Substances* (Ed. F. Skoog) Univ. of Wisconsin Press, Madison, xiv + 476 pp., illustr.

ENNIS, W. B., SHAW, W. C., DANIELSON, L. L., KLINGMAN, D. L. & TIMMONS, F. L. (1963). Impact of chemical weed control on farm management practices. *Adv. Agron.*, **15**, 161–210.

FOY, C. L., WHITWORTH, J. W., MUZIK, T. J. & CURRIER, H. B. (1967). The penetration, absorption and translocation of herbicides. pp. 3–62 in *Environmental and Other Factors in the Response of Plants to Herbicides* (Ed. V. H. Fried & O. R. Morris) *Tech. Bull. Corvallis Agric. Exp. Stn*, **100**, 130 pp.

FRIED, V. H. & MORRIS, O. R. (1967). Environmental and other factors in the response of plants to herbicides. *Tech. Bull. Corvallis Agric. Exp. Stn*, **100**, 130 pp.

GRAY, R. A. & WEIERICH, A. J. (1968). Behavior and persistence of thiocarbamate herbicides in soils under different environmental conditions. *Proc. 9th British Weed Control Conference, Brighton*, Vol. I, pp. 94–101.

GYSIN, H. (1960). Chemistry and herbicidal properties of Triazine derivatives. *Adv. Pest Control Res.*, **3**, 289–358.

HAMMERTON, J. L. (1967). Environmental factors and susceptibility to herbicides. *Weeds*, **15**, 330–6.

HARRIS, C. I. & WARREN, G. F. (1964). Absorption and desorption of herbicides by soil. *Weeds*, **12**, 120–6.

HARTLEY, G. S. (1964). Herbicide behaviour in the soil. pp. 111–61 in *The Physiology and Biochemistry of Herbicides* (Ed. L. J. Audus) Academic Press, London, xix + 555 pp., illustr.

HELBAEK, H. (1958). Plant economy in ancient Lachish. pp. 309–17 in *Lachish IV: The Bronze Age.* Wellcome-Marston Archaeological Expedition, 351 pp., illustr.

HELGESON, E. A. (1957). *Methods of Weed Control.* F.A.O. Rome, x + 189 pp., illustr.

HOLLY, K. (1964). Herbicide selectivity in relation to formulation and application methods. pp. 423–64 in *The Physiology and Biochemistry of Herbicides* (Ed. J. L. Audus) Academic Press, London, xix + 555 pp., illustr.

HOROWITZ, M. (1965). Data on the biology and chemical control of nutsedge (*Cyperus rotundus*) in Israel. *Pans* (*c*), **11**, 389–416.

—— (1966). A rapid bioassay for PEBC and its application in volatilization and adsorption studies. *Weed Res.*, **6**, 1–21.

HUFFAKER, C. B. (1957). Fundamentals of biological control of weeds. *Hilgardia*, **27**(3), 101–57.

INTERNATIONAL RICE RESEARCH INSTITUTE (1966). Methods of weed control. *I.R.R.I. Reporter*, **2**, (2), 1–4.

JORDAN, L. S., DAY, B. E. & CLERK, W. A. (1963). Effect of incorporation and method of irrigation on pre-emergence herbicides. *Weeds*, **11**, 157–60.

JORDAN, L. S., MANN, J. D. & DAY, B. E. (1965). Effects of ultraviolet light on herbicides. *Weeds*, **13**, 43–46.

KEARNY, P. C., SHEETS, T. J. & SMITH, J. W. (1964). Volatility of seven s-triazines. *Weeds*, **12**, 83–87.

KENNEDY, P. B. & CRAFTS, A. S. (1927). The application of physiological methods to weed control. *Pl. Physiol.*, **2**, 503–6.

KING, L. J. (1966). *Weeds of the World: Biology and Control* (Plant Science Monographs, Ed. N. Polunin) Leonard Hill, London, and Interscience, New York, xxxii + 526 pp., illustr.

KLINGMAN, G. C. (1961). *Weed Control as a Science.* John Wiley, New York, viii + 421 pp., illustr.

LONGCHAMP, R. (1965). La lutte biologique contre les mauvaises herbes. pp. 56–61 in *La Lutte Biologique.* Rapport Général de la Commission d'Etudes, Délégation de la Recherche Scientifique, Paris.

MANGELSDORF, P. C. (1966). Genetic potentials for increasing yields of food crops and animals. pp. 66–71 in *Prospects of the World Food Supply—A Symposium.* National Academy of Science, Washington D.C., 84 pp., illustr.

MENGES, R. M. (1963). Effects of overhead and furrow irrigation on performance of pre-emergence herbicides. *Weeds*, **11**, 72–76.

MORELAND, D. E., GENTNER, W. A., HILTON, J. L. & HILL, K. L. (1959). Studies on the mechanism of herbicidal action of 2-chloro-4,6 bis (Ethylamino)-s-triazine. *Pl. Physiol.*, **34**, 432–5.

OVERBEEK, J. VAN (1964). Survey of mechanisms of herbicide action. pp. 387–400 in *The Physiology and Biochemistry of Herbicides* (Ed. J. L. Audus) Academic Press, London, xix +555 pp., illustr.

PAVLYCHENKO, T. K. (1949). Plant competition and weed control. *Agric. Inst. Rev.* 4, 142–5.

PHILLIPS, W. M. (1964). A new technique for controlling weeds in sorghum in a wheat–sorghum–fallow rotation in the great plains. *Weeds*, **12**, 42–4.

RIPPER, W. E. (1956). A new method of selective weed control for related plants in particular broad-leaf weeds in beet. *Proc. 3rd British Weed Control Conference, Blackpool*, Vol. I, pp. 225–32.

SAUER, C. O. (1952). *Agricultural Origins and Dispersals.* Bowman Memorial Lecture, Ser. 2. American Geographical Society, New York, 110 pp.

SMITH, A. E., ZUKEL, J.W., STONE, G. M. & RIDDELL, J. A. (1959). Factors affecting the performance of maleic hydrazide. *J. Agric. Fd Chem.*, **7**, 341–4.

STANIFORTH, D. W. & LOOMIS, W. E. (1949). Surface action in 2,4-D sprays. *Science*, **109**, 628–9.

STEWARD, J. H. (1956). Cultural evolution. *Sci. Am.*, **194**(5), 69–83.

TAYLORSON, R. B. (1966). Influence of temperature and other factors on loss of CDEC from soil surfaces. *Weeds*, **14**, 293–6.

TUPPER, G. & MATTHEWS, E. J. (1962). Mechanized weed control in cotton. *Ark. Fm Res.*, **11**(1), 5–7.

VAN OVERBEEK, J. *See* OVERBEEK, J. VAN.

WIESE, A. F. & ARMY, T. J. (1958). Effect of tillage and chemical weed control practices on soil moisture storage and losses. *Agron. J.*, **50**, 465–8.

——, BURNETT, E. & BOX, J. E. (1967). Chemical fallow in dryland cropping sequences. *Agron. J.*, **59**, 175–7.

WORT, D. J. (1965). The effect of herbicides on the chemical composition of crop plants. *Ier. Col. CICRA. Congr. Int. Antiparisitaires, Naples*, pp. 3–13.

XVII

INSECT PESTS

Adaptations to Aridity

Adaptation to life under arid conditions has not necessitated any drastic modifications in insects (Pradhan, 1957). Insects in these areas can therefore not be sharply differentiated from those of the non-arid regions.

COLOUR

The colours of most typical desert insects are either those of the desert itself—ochre, brown, or sandy-grey—or else they are predominantly black. The desert colours characterize those insects that are relished by predators, the black or blackish insects being generally distasteful.

AESTIVATION

Many insect species of the arid regions are capable of remaining in a state of dormancy for extensive periods during periods of dryness, excessive heat, or hot winds. This adaptation is very similar to that of the ephemeral plants. Actually, insects and plants appear together and conclude their life-cycles in the very short period during which favourable conditions occur. This may also occur to insects living on or in cultivated species—for example, *Sesamia* larvae parasitizing dry-land sorghum will become dormant in summer within the stalks, whilst on irrigated sorghum dormancy occurs much later (M. Vermesh, personal communication).

Sometimes it is the eggs that remain dormant, even for several seasons, until sufficient rainfall occurs. One example of this is the locust *Schistocerca gregaria.*

UNDERGROUND EXISTENCE

Many insects of the arid regions spend much or all of their time underground, or under stones, where they find a more favourable microclimate than in the open.

PROTECTION OF EGGS, LARVAE AND PUPAE

Examples of adaptation to the harsh environment of the desert are egg-pods, larval cases, and pupal cocoons, which are constructed of materials that give effective protection.

BEHAVIOURAL ADAPTATIONS

Many of the insect species of the desert are carnivorous and capable of storing their prey for their progeny, as a supply to be used during adverse periods. The phytophagous insects have adjusted their life-cycles to those of the ephemeral plants of the desert, or accumulate stores of plant food, or feed on a wide variety of plants.

Some insects are physiologically so well adapted to arid and semi-arid conditions that humid conditions are actually harmful to them. Examples of this phenomenon are the chinch-bug *Blissus leucopterus* and the Pale Western Cutworm (*Porosagrotis orthogonia*). The former is a major pest of the gramineous crops, which under humid conditions is held in check by a Fungus (*Sporotrichum globuliferum*). The intensity of infestation of the latter pest depends on the amount of rainfall in the preceding summer: when precipitation is less than a certain amount, a heavy outbreak occurs.

LACK OF STABILITY OF DESERT ENVIRONMENTS

Ecological theory states that difficult and extreme environments such as deserts, to which only a few species can adapt themselves, are characterized by lack of stability, whilst complex environments, with many interacting species, have a relatively high degree of built-in stability. A typical example is the extreme fluctuations of locust populations, from 'very low population level to virtual population explosions' (Cole, 1966).

EFFECTS OF TEMPERATURE

The warm season of the semi-arid and arid regions is usually characterized by high temperatures and low air humidity during the daytime, with a considerable drop in temperature during the night.

High temperatures, if not excessive, accelerate the developmental cycles of insects, and in certain cases the fluctuations in temperature may even increase the effect of high temperature. Hence, there tends to be a larger number of generations per year in these regions than in those with temperate climates. For example, the optimum temperature for aphid reproduction is about 26°C and their life-cycle is shortened and their activity increases as temperatures increase.

Hot weather stimulates the aphids to move frequently during the daytime, as the plants wilt or their palatability decreases. This increases the rapidity of spread of viruses of which they are the vectors. However, still higher temperatures have adverse effects on the aphids, and at 32° C they cease to infest crop plants and virus spread becomes negligible. Such hot weather also decreases the infectivity of individual vectors (Broadbent, 1967).

Effects of Irrigation on Insect Populations in Arid Regions

EFFECTS ON NUMBER OF SPECIES AND ON POPULATION LEVELS

In an arid environment, irrigation causes profound changes in the nature of the vegetation, both cultivated and spontaneous, as well as modifications of the microclimate, which are bound to affect the nature and number of the insect population. The insects are favoured by the continuous supply of fresh food and the possibility of surviving, between crops, on the weeds of irrigated lands. Intensification of agriculture in arid and semi-arid zones therefore inevitably causes an enormous increase in the numbers of insect species attacking crops—many of them autochthonous or anyway native, whereas others are imported.

The predators and parasites of autochthonous pests are usually also present, whilst imported pests are frequently separated from their natural enemies.

A few years after irrigation was initiated in a desert area of Arizona, seventy three insect species were identified as pests of the cultivated crops (McKinney, 1939). And in addition to increasing the number of different species, irrigation is also conducive to enormous increases in insect population numbers.

In arid regions, the rainy season, which is favourable to the breeding of insects, is relatively short. During the dry season, adverse physical and biological conditions cause high mortality; indeed many insect populations are hardly able to maintain themselves under these conditions (Joyce, 1955). Irrigation extends the period that is favourable for breeding, and at the same time reduces or completely eliminates conditions which check the increase in numbers of insect populations. In the absence of natural checks or effective control, epidemics may break out.

In Arizona (U.S.A.), wide variations in the degree of infestation by the Two-spotted Spider-mite, *Tetranychus bimaculatus*, were observed between plots of lucerne receiving different amounts of water. The plants grown under the most favourable moisture régime had the most mites, and infestations were far lower in the relatively 'dry' plots.

The influence of the different irrigation schedules upon mite population is believed to be both ecological and nutritional. The sparse growth of the lucerne grown under relatively dry conditions allowed the sunlight to penetrate to the ground surface, and as a result air temperatures 15 cm above the ground were markedly higher and relative humidities were very much lower in the dry plots than in those more frequently irrigated. The hardened, sparse growth of the lucerne grown under conditions of moisture stress was apparently nutritionally less favourable for the mites than the lush growth of the well-watered lucerne (Butler, 1955).

Irrigation in Israel has brought about drastic changes in the lepidopterous pest complex. *Chilo simplex*, *C. pulverosellus* and *C. hypenalis*, have practically disappeared. By contrast, species not previously recorded, such as *Chilo agamenon* and *Ostrinia nubilalis*, have become important pests (Rivnay, 1963). The continuous supply of fresh food made possible by irrigation, at a time when temperatures are favourable for the rapid evolution of insects, engenders an increase in the number of generations produced in a year.

Lixus junci, an autochthonous insect in Israel, became a major pest with the extension of irrigated stock beets, producing three generations a year instead of the single generation in spring that was the case before irrigation was introduced. *Heliothis armigera* and *Ostrinia nubilalis* became important pests of maize only after it was grown extensively under irrigation.

The transition from relatively harmless insects into major crop pests was not a gradual process, but occurred explosively as soon as a continuous supply of fresh food became available (M. Vermesh, personal communication, 1968).

On the other hand, crops are sometimes less damaged by insect attacks when irrigated than when grown under a drier régime. For example, Ranger lucerne was more tolerant to leafhoppers when an adequate water-supply was available than when it was grown under conditions of water-stress (Wilson *et al.*, 1955). Sprinkler irrigation may reduce a thrips population by about 50 per cent (Smith, 1957).

ARID REGIONS AS BARRIERS TO THE SPREAD OF INSECT PESTS

Arid regions can form a very effective barrier to the spread of certain insects. For example, the Pink Bollworm (*Platyedra gossypiella*) has been established in Egypt for almost half-a-century, but it did not cross the Sahara and infest the cotton fields of West Africa. That this was not due to adverse ecological conditions in the West African countries is indicated by the fact that, after the Pink Bollworm was introduced with infested seed samples sent by post, it spread rapidly in the countries to the south and west of the Sahara (Vayssière, 1954).

Canals that cross desert regions serve as routes for the infestation of irrigated areas within the desert by insects which, because of lack of adaptation to dry conditions, could not have reached such areas otherwise. An example is the Cucumber Beetle (*Diabrotica balteata*), which has spread widely in irrigated areas of California, along the irrigation canals (McFarlane *et al.*, 1957).

The development of irrigated 'oases' in the desert creates bridges which facilitate the migration of pests across the desert. The Beet Leafhopper

(*Circulifer tenellus*), the carrier of the curly-top virus disease, migrates long distances over desert areas by this means (Carter, 1961).

DRYLANDS AS RESERVOIRS OF POTENTIAL INSECT PESTS

The drylands can also serve as reservoirs, from which insect infestations may develop and then come to damage adjacent irrigated areas. Following the introduction of irrigation in the geographically isolated Salt River Valley of Arizona, certain insect species that originally subsisted on wild native host-plants increased in abundance and became adapted to the cultivated crops, whilst many other species were introduced by natural dispersion or by artificial carriage (McKinney, 1939). Many insects migrate over long distances, from their normal habitats in arid lands, to neighbouring irrigated or semi-arid rain-fed cultivated fields.

During the dry season, many moths migrate at night in mass flights from arid habitats in which they breed during the favourable parts of the year, to irrigated or higher-rainfall areas (Uvarov, 1962). Examples of this phenomenon are the Cotton Leafworm (*Alabama argillacea*) and *Agrotis ypsilon*. The adults of the former come annually, probably from arid parts of Central America, to the southern United States, where they produce several generations which cause considerable damage to the cotton crop, before finally disappearing (Fenjves, 1950). The latter insect, a serious pest in Israel, immigrates annually from unknown sources, multiplies in cultivated crops, and then disappears (Bernard, 1954).

A typical case of an insect for which certain desert regions are the natural habitat, but which may become a major pest of cultivated areas far beyond the fringe of the desert, is the Desert Locust (*Schistocerca gregaria*) (Plate 35). This Locust requires moist sand in which to lay its eggs, sufficient green, tender food for the young hoppers, and succulent green food or high air humidity for reproduction (Bodenheimer *et al.*, 1930). These requirements are certainly not typical for desert conditions, but they are met in ephemeral ecological islands. Therefore, the survival of the Desert Locust is dependent on the ability of the insect to move from one favourable habitat to another—a nomadic habit very similar to that of many desert mammals.

Land development schemes in the desert, based on irrigation, may easily change precarious and ephemeral ecological islands into more or less permanently favourable breeding grounds for the locusts (Uvarov, 1954), thereby enabling the pest to become still more dangerous than in the past. The Rocky Mountain Grasshopper (*Melanoplus mexicanus*) normally produces a single generation annually in the deserts of Arizona. The extensive cultivation of lucerne under irrigation has, however, made it possible for the Grasshopper to produce several generations a year, and it can now only be kept in check by an expensive chemical control programme (ibid.). The effect of land development in the desert on the increased incidence of this locust is probably not an isolated example, but an indication that irrigation schemes in the desert may well create new, unanticipated entomological problems (ibid.).

Many other insects, feeding on the native vegetation of arid and semi-arid regions, have become important pests of cultivated crops. A few examples are the Wheat Thrips (*Haplothrips tritici*), which is abundant in wild perennial grasses in Kazakhstan (Uvarov, 1962), the Sorghum Midge (*Contarinia sorghicola*), a pest that causes severe damage to sorghum in arid regions and is closely related to a genus that occurs on wild sorghum and allied species (Barnes, 1954), the Cotton-boll Weevil (*Anthonomus grandis*), which has spread into the cotton-growing regions of the United States from desert regions of Mexico in which its host plants are wild cotton species (Uvarov, 1962), and the Beet Leafhopper (*Circulifer tenellus*), which is the vector of curly-top virus in sugar-beets, having migrated to the fields of the cultivated crop in the southern United States from the desert where it feeds on sagebrush (*Artemisia*) and salt-bushes (Lawson *et al.*, 1951).

PREFERENTIAL FEEDING ON IRRIGATED CROPS

In general, insect pests are more attracted to the succulent growth of irrigated crops than to the hardened, sparse growth that is characteristic of plants grown under conditions of moisture stress

For example, irrigated cotton was more heavily infested by three species of mirids—the Cotton Fleahopper (*Psallus seriatus*), the Rapid Plant-bug (*Adelphocoris rapidus*), and the Tarnished Plant-bug (*Lygus lineolaris*)—than was unirrigated cotton (Adkisson, 1957). Infestation by the stem-borer of wheat (and rice), *Sesamia inferens*, in India, was far greater on irrigated than on unirrigated wheat (Chowdhury & Sharma, 1960). Irrigated wheat was found to be more attractive than non-irrigated wheat to the Hessian Fly (*Mayetiola destructor*) and to the frit-fly *Oscinella frit* (Pavlov, 1959).

METHODS OF INSECT CONTROL

COLLECTION AND DESTRUCTION

In many countries with primitive agriculture, especially those with an overabundance of labour, the collecting and destroying of caterpillars and eggs is still practised on a large scale with crops such as cotton, e.g. for the control of *Prodenia*.

CULTURAL PRACTICES

Until the advent of insecticides, cultural methods were practically the only ones available to the farmer for active insect control. These include time of sowing, sanitation, crop rotation, tillage, sowing trap-crops, strip cropping, etc. Even at present, appropriate cultural methods should not be neglected, and their contribution to keeping the harmful insects at a low level of incidence can be considerable.

In Israel, early spring sowing of sorghum is the principal and most effective method of avoiding disastrous damage by the Sorghum Shoot-fly (*Atherigona exigua*). Conversely, a delay until June in the sowing of maize is an essential precaution for the prevention of 'dwarfness', a virus disease propagated by an insect vector.

The destruction of weeds, which serve as alternate hosts to certain insects, considerably reduces the sources of infestation of the cultivated crop.

Preparing a good seed-bed and providing favourable conditions for rapid and even germination, is an effective method for reducing the damage caused to the young seedlings by soil-inhabiting insects.

Chopping and ploughing-under the stalks and residues of the cotton crop, has been found to destroy 85 per cent of the larvae of the Pink Bollworm (Fife *et al.*, 1963).

Ploughing to a depth of 15–20 cm is completely effective in controlling *Syringopais temperatella*, one of the most damaging pests of wheat in the Near East.

PREVENTIVE CONTROL

The classical example of preventive control is that of locusts, of which two species—the Red Locust (*Nomadaeris septemfaseiata*) and the Migratory Locust (*Locusta migratoria*)—used to cause periodical devastations in Africa and Asia. The foci of infestation of these pests are in relatively restricted areas. By maintaining a constant watch over these potential outbreak areas, it has been possible to reduce the dangers of large-scale invasions considerably (Uvarov, 1962).

The preventive control of the Desert Locust (*Schistocerca gregaria*) is more difficult, because this Locust is nomadic and has no permanently located outbreak areas (Uvarov, 1962). Preventive control, in this case, is based on international supervision of the seasonal movements of swarms in the areas most frequently infested, so that effective control operations can be planned at an early stage.

USE OF RESISTANT VARIETIES

Breeding of plants for insect resistance has received far less attention in the past than has breeding for disease resistance. In general, the insect-host relationship is far less specific and more complex than is the case with pathogens (Beck, 1965). Knowledge of the mechanisms which determine plant resistance is still fragmentary.

Notwithstanding these limitations, a few varieties that are resistant to insects have been developed: varieties of maize resistant to the Corn-borer (*Ostrinia nubilalis*), varieties of wheat resistant to the Hessian Fly (*Mayetiola destructor*) (Painter, 1958), and jassid-resistant varieties of cotton, are a few examples.

Resistance in each case is due to one or more mechanisms such as the following:

Antibiosis. A constituent present in the resistant variety is lethal to the insect or adversely affects its life-cycle or fertility.

Preference. Certain plant characteristics such as hardness, hairiness, and leaf density, may repel insects; or they may prevent oviposition, feeding, or the establishment of the insect (Moseman, 1966). *Gossypium tomentosum* has no nectaries, and attracts far fewer insects than do other cotton species (Anon., 1960). Resistance to jassids is based on the hairiness of cotton plants.

Field Resistance. This is the ability of the plant to recover rapidly without excessive damage, after an attack by insects.

INSECTICIDES

The insecticides in use in the late 1920s were derived mainly from plants, and included nicotine, pyrethrum, hellebore, quassia and derris. A few inorganic compounds were also used: arsenicals—extremely effective stomach poisons with high mammalian toxicity—and the less dangerous fluorides.

During the Second World War and the following years, a number of completely new and extremely effective insecticides were developed. The first of these new contact poisons, which penetrate into the insect through the cuticle or trachea, entering the nervous system or the blood system, was DDT. Though it had been synthetized as early as 1874, its efficacity as an insecticide was discovered only in 1939. It was then noted that it was not only extremely effective in killing a wide range of insects, but that its residual effect lasted for up to one month under field conditions. In the period following these discoveries concerning DDT, a large number of additional compounds were produced and introduced into agricultural practice.

The most widely used insecticides are: (a) *chlorinated hydrocarbons* (organochlorines), containing carbon, hydrogen, and chloride (e.g. DDT, Dieldrin, Aldrin, Endrin, Toxaphene, Lindane, Methoxychlor, Chlordane, and Heptachlor); (b) *organic phosphorus compounds*, containing phosphorus, sulphur, oxygen, carbon, and hydrogen (e.g. Parathion, Malathion, Phosdrin, and Tepp (tetraethyl pyrophosphate) (Weisner, 1963); and (c) *carbamates* (a number of carbamic acid esters have been found to be effective insecticides, killing insects by cholinesterase inhibition). The carbamates are highly toxic to aphids, thrips, and granary weevils, the most widely used being Dimethan and Sevin (Carbaryl). The latter is a naphthyl carbamate with low mammalian toxicity, and has proved very effective in controlling many insects of field-crops (O'Brien, 1967).

Systemic Insecticides

Systemic insecticides can be applied in a number of ways (Ridgway, 1967):

To the Seed and in the Furrow. These are the most prevalent methods. The insecticides can be applied directly to the seed, and treated seed is already available commercially. A more recent development is in-furrow applications of such systemic insecticides as Disulfoton and Phorate in granular formulations applied with the seed. In-furrow applications appear to be less phytotoxic than seed treatments, at equal rates per hectare.

Foliar Applications. When applied to the leaves, systemic insecticides penetrate the leaves; but as translocation is limited, the systemic action is localized.

Side-dress Applications. The uptake of systemic insecticides from the soil depends on a number of factors, including moisture conditions, and is therefore not very effective.

Stem Applications. Such applications of systemic insecticides are much more effective than side-dress application; for example, 22 gm/ha of Cl-47031 applied to the stems of cotton gave as good control of Cotton Fleahopper and spider mites as 230 gm/ha applied to the soil. A tractor-mounted rotary-brush applicator for treating the stems of plants has been developed.

Antifeedants

Antifeedants are chemicals that inhibit feeding: insects that remain on the treated foliage eventually starve to death without eating the leaves. An example of an antifeedant is Brestan (triphenyl tin acetate) which, when used as a fungicide, inhibits the feeding of insects on the treated foliage (Ascher & Rones, 1964). Antifeedants have been reported to be effective against the Cotton Leafworm (*Spodoptera littoralis*) (Ascher & Nissim, 1964).

The main advantage of antifeedants is that they do not harm the parasites and predators of insect pests. Their main limitation is that they are effective only against surface-chewing insects (Wright, 1967).

BIOLOGICAL CONTROL

Biological control is defined as 'the direct and indirect use of living organisms to control pests and to reduce damage below economic levels' (Chant, 1966). The living organisms employed are insects or pathogens.

A large percentage of insect species prey on other insects, and there is no doubt that the main factor preventing a 'population explosion' of an insect species is its natural enemies. These fall into two main groups: the *predators*, which are usually larger and stronger than their victims, and the *parasites*, which live on or in the bodies of insects that are much larger and stronger than themselves (Plate 36).

The individual predator attacks and destroys numerous individuals of the species on which it preys. Predators are usually active and agile, but reproduce more slowly than their prey. Therefore, the prey are usually capable of outbreeding their predators (Cole, 1966). By contrast, the parasite is not very active, has a short life-cycle, and usually preys on a single individual victim.

Not all natural enemies of insects are beneficial to Man. The natural enemies of harmful insects have their own predators and parasites which are therefore, *ipso facto*, harmful from Man's point of view. These, in turn, also have their natural enemies, which are therefore beneficial.

The relationships between insects, their natural enemies, and the hyperparasites, are extremely complicated, and it is easy to understand how Man's intervention can readily upset the very delicate system of checks and balances occurring in Nature between these components of a biological system (Figure 95).

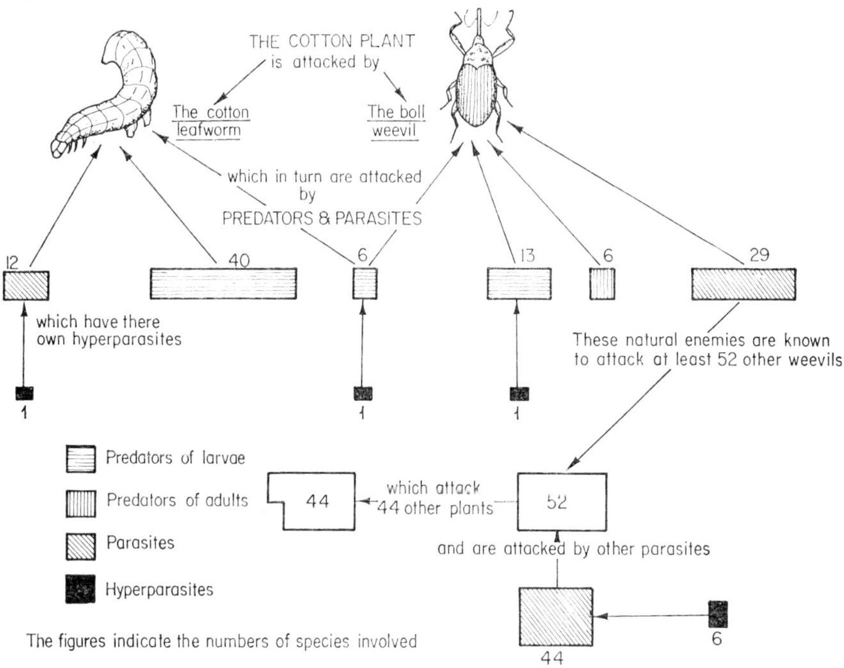

Figure 95. Schema showing some of the interrelationships of helpful and harmful parasites associated with the Cotton-boll Weevil and the Cotton Leafworm. (After Pierce (1912): by courtesy of the U.S. Department of Agriculture.)

There are two basic approaches in the use of parasites and predators for biological control (Chant, 1966):

(a) *Conservation:* increasing the effectiveness of natural enemies by influencing the environment, especially by minimizing the effects of factors that reduce the populations of natural enemies.
(b) *Introduction:* based on the search for new natural enemies in other regions and usually countries, their rearing and subsequent release. This method has proved its effectiveness especially in the control of harmful insects that had been introduced from foreign countries. As a result of their disassociation from the natural enemies of their native habitat, these may become pests.

The advantages of the biological methods of insect control are (Moseman, 1966): (a) the system, once established, is relatively stable and self-perpetuating; (b) it does not have the undesirable side-effects of chemical control; and (c) it is relatively inexpensive in comparison with the conventional chemical methods.

Microbial Insecticides

Microbial insecticides are suspensions of micro-organisms and/or their products that are specific to certain insects and cause their death. The study of this type of control has expanded considerably in recent years, because of the problems of insect resistance, the emergence of secondary pests, and toxic residues that are becoming increasingly acute following the use of conventional insecticides (Tanada, 1967).

The number of successes in this field is still small, and is limited to pathogens with resistant stages that retain their viability for a long time and can be applied with conventional dusters or sprayers. The two most important successes in field crops are the use of *Bacillus thuringiensis* for the control of the European Corn-borer (*Ostrinia nubilalis*) and other lepidopterous insects, and of nuclear-polyhedrosis viruses against the Lucerne Caterpillar (*Colias philodice eurytheme*).

Most of the attempts to use Fungi, Protozoa, and nematodes, have met with only very limited success up to the present (Tanada, 1967).

Much work has been devoted in recent years to substances which are toxic to insects, and which are formed by micro-organisms that could be applied in pest control. Such toxins are produced mainly by Bacteria and Fungi, though Protozoa and viruses may also be a potential source.

PARA-BIOLOGICAL CONTROL

The introduction of sterile insects into the natural population has been successfully accomplished in a number of cases. The method involves the rearing, sterilization and release, of insects in numbers sufficiently large to overwhelm the natural population (Moseman, 1966). The sterilization is achieved by irradiation, chemosterilants or hybridization.

Chemosterilants. These are chemicals that are administered orally or by contact and produce irreversible sterility in insects, without marked adverse effects on their mating behaviour or viability.

The application of chemosterilants to insect control is still in its early stages, but it is considered to be most promising, with a potential 99 per cent control of an insect species compared with the 90 per cent control that is expected with conventional insecticides (Kilgore, 1967).

Lethal Genes. Methods of genetic selection used to improve plants and domestic animals may serve to introduce lethal genes or characteristics with low survival value into insect populations (Moseman, 1966). No successes under field conditions due to this approach have, however, yet been recorded.

Traps

Light, sound, or chemicals, may serve as lures for attracting insects to traps. Thousands of insect species, in particular nocturnal fliers, are attracted to light, and many types of traps have been devised. The traps generally consist of a lamp mounted above a collection device in which the insects are killed by a suitable volatile insecticide. Whilst conspicuous successes in insect control by these means in the field have not yet been recorded, the method may be of special interest when combined with the application of chemosterilants to the captured insects. Following capture and treatment, the insects are freed to rejoin the natural population (Henneberry & Howland, 1966).

Electromagnetic Energy

Radio Frequency (RF). Electric fields have been successfully used for the control of insects attacking stored grain, which cause considerable damage in the warm dry regions. Complete mortality of adult Rice Weevils (*Calandra oryzae*) has been achieved by using RF treatments for a few seconds to raise the grain temperature to about 38°C. As temperatures of this order would not normally be lethal to insects, even if applied for longer periods, it is supposed that the insects selectively absorb energy at a faster rate than the grain which they are infesting (Nelson, 1967).

Infra-red Energy. This has been studied for the possible control of the Pea Weevil (*Bruchus pisorum*) in stored pulses, and of grain weevils in stored rice (Tilton & Schroeder, 1963).

Ionized Radiation. For insect control in grain, this is also receiving much attention (Cornwell, 1966).

Sex-attractants

These may provide a means of luring insects to their deaths, and the possibility exists that some of these attractants may be produced synthetically

(Moseman, 1966). The use of caged virgin females in black light-traps has proved to be very effective with certain insect species. This method can be combined with chemosterilant baits, so that sterile males are released into the population (Nelson & Seubert, 1966). Some insect species have been attracted by recordings of their own mating songs (Belton, 1962). However, none of these methods has yet been sufficiently developed for practical application (Shorey & Gaston, 1967).

Photoperiodic Effects

It has been observed that exposures to light as brief as a photoflash, during a critical time in the normal dark period, can upset the life-cycle of an insect. This opens up new possibilities of insect control, by forcing insects to bypass their normal diapause under field conditions. Daily exposures of a few minutes to red and infra-red light were also found to inhibit the normal development of certain insects (Barker *et al.*, 1964).

Promising results in the laboratory have been obtained by using light to control diapause of the Pink Bollworm (*Platyedra gossypiella*) and of the European Corn-borer (Nelson, 1967).

Ultrasound

Experiments have shown that bat-like ultrasound in the field reduced infestation of sweet corn by the European Corn-borer by more than 50 per cent, indicating that this approach may have great potentialities (Belton, 1962).

INTEGRATED CONTROL

For a few years, great successes were achieved with the new insecticides; but gradually their shortcomings revealed themselves. Optimistic forecasts, predicting the complete extermination of major pests, proved to be premature. Certain insects were effectively controlled, to be replaced by others which had previously been minor pests. For example, the chemical control of jassids (*Emploasca* spp.) on cotton in the Sudan, resulted in an increase of the Whitefly (*Bemisia gossypiperda*); when the Whitefly was controlled by an insecticide, damage due to Cotton Bollworm (*Heliothis armigera*) increased markedly (Joyce, 1955). It was shown that three consecutive sprayings with one of the chlorinated hydrocarbons were sufficient to annihilate all the natural enemies of these pests.

Strains resistant to the new insecticides developed within a relatively short period, by an evolutionary process, in more than 200 insect species (Ordish, 1967). Meanwhile public awareness of the danger to human health of the residues of certain insecticides, especially after their concentration through 'food-chains', has increased. Preventive chemical control (before the appearance of the pest) and routine control (treating 'by the calendar') proved to be not only expensive but also frequently ineffective and even dangerous.

A phenomenon known as 'resurgence' also became increasingly frequent (Chant, 1966). When insect populations increase steeply, stresses due to severe competition cause a marked reduction in reproduction. The obverse is also true: when the population density of a pest is reduced to a low level by an effective chemical-control programme, reproduction rates may increase to such an intensity that a population explosion of epidemic proportions occurs. This process is facilitated if the chemicals used have also decimated the natural enemies. It is only when population density is reduced below a certain, extremely low level, that the decline may continue until complete extinction occurs; but it is very unusual to achieve this by the use of pesticides alone.

On the other hand, it is well known that the natural enemies of an insect species are not capable of controlling it completely (this could lead to their own extinction) or even continuously. Normally, they are able to maintain the numbers of the harmful insect within certain limits, so that it does not cause appreciable economic damage (Biliotti *et al.*, 1962). Under circumstances which favour a very rapid increase of the harmful insect, however, its natural enemies are no longer able to cope with the population explosion, and the intervention of Man, using appropriate control techniques, becomes necessary.

The fundamental premise on which integrated control is based, is that complete eradication of a pest is neither necessary nor even desirable. The development of an integrated control system requires the continued existence of pest populations, albeit at a low density. This is known as the 'dirty-field principle', and is an essential condition for the permanence of a balanced system of pests and natural enemies. The presence of the insect pest, when maintained at a certain level, does not cause economic damage (Chant, 1966).

One aspect of integrated control is 'adjusting the spray programme, so as to cause minimum damage to the natural enemies of the pest that is being controlled' (Ordish, 1967).

The lowest population density at which the rate of damage to the crop exceeds the cost of controlling the pest is called the *economic injury level*; it is different for different crops, areas, seasons, and cost fluctuations. The *economic threshold* is the pest population density at which control measures have to be initiated in order to prevent the pest population from increasing to the economic injury level. The economic threshold has to be lower than the economic injury level in order to permit sufficient time for control measures to be applied and to take effect (Figure 96).

When a pest population, responding to favourable environmental conditions, approaches the economic injury level, insecticides are usually the only control method that can act quickly enough to prevent a population build-up that will cause serious damage to the crop (Smith & Bosch, 1967). For

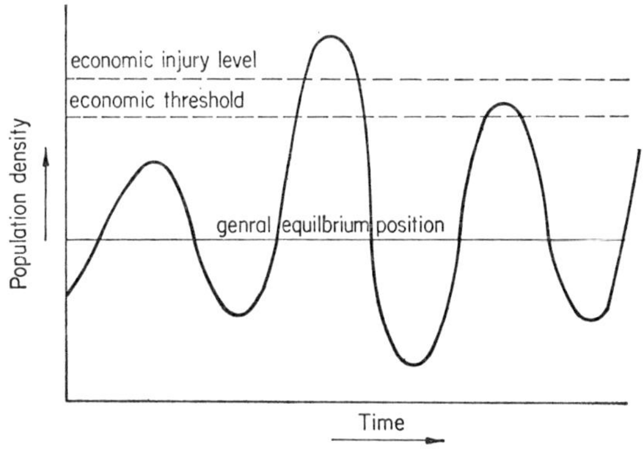

Figure 96. Schematic graph of a theoretical arthropod population over a period of time, indicating its general equilibrium position, economic injury level, and economic threshold. (After Stern (1965): by courtesy of the Food and Agriculture Organization of the United Nations.)

example, the bollworm *Heliothis zea* attacks, among other crops, lucerne (alfalfa), cotton, and sweet corn, in California. In lucerne, the population never reaches a density which causes economic damage. In cotton, the economic threshold is four worms per 100 plants, and cotton fields usually require one or two sprayings during the season to control this bollworm. In sweet corn, the market demands, for human consumption, ears that are entirely free from worms, so that the economic threshold in this case is at zero level. For this reason, insecticides may have to be applied at three-day intervals. The economic threshold may be somewhat higher when demand considerably exceeds supply (Stern, 1965).

Knowledge of the economic threshold for each crop and pest, at which insect damage can no longer be tolerated, is essential to the success of an integrated control system; otherwise, matters may get out of control and serious damage be caused to the crop. However, it is not easy to determine these thresholds, because of the large number of factors involved and the everchanging economic and environmental conditions (Chant, 1966). An understanding of the interactions in the ecosystem is also essential (Moseman, 1966). Refraining from panicking and applying chemical control measures prematurely, requires a high degree of self-control on the part of the farmer, and confidencc in the advice and instructions given by the research and extension workers. Precipitate action will result in a disruption of the whole programme.

When once the pest is reduced to an acceptable level, chemical control is stopped. The proper timing of chemical control reduces the danger of replacing one pest with another, of creating resistance to pesticides, and of eradicating beneficial insects; at the same time, it makes possible considerable savings, by drastically reducing the number of sprayings required. Chemical control of insect pests will continue to be an essential element in any control programme.

The chemical pesticides to be used in an integrated control system should preferably have the following characteristics (Cole, 1966):

(a) Lack of residual effect; by breaking down rapidly into harmless substances, the selective pressure for the pests to evolve resistant strains is reduced to a minimum, and the problem of residues toxic to humans and disruptive of the ecosystem is also avoided.
(b) High specificity, so that only the pest which is under attack is controlled. This will avoid harm to the natural enemies of the pest.

These characteristics are the diametric opposites of those of the insecticides that were previously considered the most successful, such as DDT, which has a broad spectrum of effectiveness and long persistence (Moseman, 1966).

Integrated control also refers to other planned combinations of control methods, such as 'harmonizing techniques in an organized way into a multi-faceted flexible system' (Chant, 1966). For example, pesticides are very efficient when the insect population is high, and inefficient when it is low. The release of sterile males is inefficient at high population densities and efficient at low densities. Accordingly we see the value of 'integrated control', which combines the chemical and biological approaches (Ordish, 1967).

An example of successful integrated control is that of the Spotted Alfalfa Aphid (*Therioaphis maculata*), a pest which was introduced into California from the Middle East and threatened the entire lucerne industry. It was reduced to minor status, within six years, by the introduction of two parasites, the development of resistant varieties, harvesting by strip cutting methods to preserve predator and parasite populations, and the occasional use of a selective systemic insecticide (Stern, 1965).

Another example is that of the control of the pests of cotton in the Canate Valley of Peru (Barducci, 1965). The early cotton pests in this region—Leafworm (*Anomis texana*), Bud Weevil (*Anthonomus vestitus*), Cotton Aphid (*Aphis gossypii*), Minor Bollworm (*Moscinca peruella*), White Scale (*Hemichionaspis minor*), and Cotton Stainer (*Dysdercus peruvianus*)—were controlled by natural predators and parasites, hand collection of insects and damaged squares, and spraying with arsenical insecticides and nicotine sulphate. In 1939, considerable damage was caused by another bollworm, *Heliothis virescens*. Attempts to control this pest by insecticides caused a build-up of *Aphis gossypii*. The problem became worse from season to season and culminated in crop failure in 1949.

A new policy was subsequently adopted, using early-maturity varieties, crop rotation, and fixed dates of planting. The growing of ratoon cotton was prohibited. Synthetic organic insecticides (DDT, BHC, and Toxaphene) were used and gave good control during the first three years. Yields were good. However, as the effectiveness of the above-mentioned insecticides declined, they were replaced by others, including Aldrin, Dieldrin, Endrin, and Parathion. The rates used increased progressively, the sprays were applied at shorter intervals, and mixtures of insecticides were used. After a short period, leafrollers (*Argyrotaenia sphaleropa* and *Platynota* spp.), the Major Leafworm (*Pseudoplusia rogationis*), the Mealy-bug (*Pseudococcus citri*), and the Leaf Perforator (*Bucculatrix thurberiella*), became serious pests. The situation continued to deteriorate, and culminated in a complete crop failure in 1956. It was then decided to adopt an integrated control programme: synthetic organic insecticides were almost completely banned, being replaced by arsenicals and plant-derived insecticides, tillage of dry soil to destroy pupae of *Heliothis*, and other adapted cultural practices—among which repopulation with predators and parasites imported from other regions, and destruction of crop residues, were the principal measures adopted. After seven years of integrated control, the insect problems were overcome, and yields were on the increase.

Application of Pesticides

FORMULATIONS

Dusts

The active chemical is combined with an inert water-repellent carrier (talc, pyrophyllite, etc.); special materials may also be added in order to improve adhesion of the dust to the foliage.

Dusting has a number of advantages: no water is required, equipment is cheaper and lighter than that required for spraying, and less labour is required for application. The dusts disperse well and reach the underside of the leaves. The advantages are, however, generally outweighed by a number of inconveniences: the danger of 'drift' is greater than with spraying, so that the dusting can only be carried out during windless periods, while for dusting to be fully effective, it is desirable that the plants be wet with dew or rain; yet the dusts are easily washed off by rain or overhead irrigation. At equal amounts of active material, sprays are usually more effective than dusts.

Granulated Pesticides

Granules are formed by combining the active chemical with a coarse-grained carrier, such as calcium carbonate. They can be used against soil-inhabiting insects, in which case they are broadcast on the soil surface and harrowed in. They can also be combined with baits, in which case they are left on the soil surface. Granulated insecticides also have a special use in the control of insects that bore into the stems of cereals and so are well protected against sprays and dusts. The granules, when broadcast over the crop, roll down the leaves and lodge within the leaf-sheaths, where they form toxic deposits which kill the stem-borers.

Wettable Powders

Wetting agents may be added to the inert carriers of the pesticides, so that they can be applied in water suspensions. Colloids may also be added to prevent the rapid settling of the particles.

Emulsifiable Solutions

The active ingredients may be mixed with appropriate solvents such as oils, alcohol, or xylene, and emulsifiers may be added to improve the dispersion in water. Before application, the solution is diluted in water. Emulsifiable solutions have the advantage of being less likely than wettable powders to obstruct the nozzles of sprays.

Aqueous pesticide mixtures, when applied in full sunshine, may burn the crop's foliage, the droplets acting as lenses on the leaves. It is therefore recommended in hot climates to spray in the early morning or late afternoon (Wyniger, 1962).

METHODS OF APPLICATION

Dusts

The simplest method of applying dusts, that is frequently employed in underdeveloped regions, is to use coarse-meshed bags or sacks, which are gently shaken over the plants. Equipment for dusting is available in many forms: from simple, hand-operated dusters, to engine-driven equipment. The latter can be of the knapsack type, carried on the back of the operator, or they can be animal-drawn or tractor-mounted. In the dusting machines, an airstream is generated which blows the dust on to the crops.

Sprays

High-volume Sprays. In this method, high pressure is used to force the liquid pesticide through appropriate nozzles. High-volume sprays give very good coverage but require large amounts of water—as much as 600 to 800 litres per hectare.

Low-volume sprays. The low-volume method requires the same amounts of active ingredients per unit area as the high-volume method, but instead of using water as the carrier for the spray, the liquid is dispersed by the action of an airsteam directed against a jet of the liquid. The airsteam can be produced by a fan, propellor, or turbine fan. As little as 50 1/ha of spray are needed, and it may be economical to use organic solvents, such as kerosene or fuel oil, instead of water (Martin, 1964).

The earlier low-volume spraying machines used small volumes of high-pressure air, which provided a very fine spray with little carry or penetration. By using large volumes of relatively low-pressure air, this defect was overcome, but the machines required were too expensive and cumbersome for large-scale farm use on field-crops (Holmes, 1960). Recently, a very efficient and relatively simple hydraulic low-volume sprayer was developed.

For small farms, especially in the developing countries, small motorized, low-volume knapsack sprayers are available. Low-volume sprays are more subject to drift than high-volume sprays, and the operation can therefore be carried out only in calm weather. Certain pests, such as red spider and aphids, require complete wetting of the leaves for their effective control, and low-volume sprays therefore cannot be used in such cases (Wyniger, 1962).

Aircraft

The development of aeroplane dusters and sprayers has provided an entirely new dimension to pest control. The farmer is no longer dependent on soil conditions when he needs to control pests, diseases, or weeds, and pesticides can therefore be applied at the time when they are most effective. The possibility of covering large areas in a very short period facilitates the containment of outbreaks and prevents their epidemic spread; it also facilitates community action instead of control by individual farmers, which is frequently ineffective if neighbouring fields are left untreated.

Aircraft spraying is not as thorough as ground spraying and is also more dependent on weather conditions, but it generally gives adequate control. Drift is one of the main hazards of aircraft application.

Helicopters have better manœuverability, and are therefore more suitable, than propeller-driven aircraft for the treatment of small or inaccessible areas. Ferrying time for the former is usually lower, but speed of coverage is far less. The down-draught of the rotor makes the helicopter more effective in aerial spraying, and it gives better penetration of dusts in dense foliage, than the propeller-driven aeroplane.

The principal disadvantages of helicopters are their high initial cost and restricted load-carrying capacity.

REFFERENCES

ADKISSON, P. L. (1957). Influence of irrigation and fertilizer on populations of three species of mirids attacking cotton. *Pl. Prot. Bull. F.A.O.*, **6**, 33–6.

ANON. (1960). Cottons that starve insects. *Agric. Res.*, **9**(5), 3–4.

ASCHER, K. R. S. & NISSIM, S. (1964). Organotin compounds and their potential use in insect control. *Wld Rev. Pest Control*, **3**(4), 188–211.

—— & RONES, G. (1964). Fungicide has residual effect on larval feeding. *Int. Pest Control*, **6**(3), 6–8.

BARDUCCI, T. BOZA. (1965). *Experiencas Sobre el Empleo del Control Biológico y de los Metodos de Control Integrado de las Plagas del Algonodera en el Valle de Canaete-Peru.* Seminario del Comité Consultivo Int. del Algodon, Washington D.C.

BARKER, R. J., COHEN, C. F. & MAYER, A. (1964). Photoflashes: A potential new tool for control of insect populations. *Science*, **145**, 1195.

BARNES, H. F. (1954). *The Sorghum Midge Problem.* Rep. Sixth Commonw. Ent. Conf., Lond., pp. 101–3.

BECK, S. D. (1965). Resistance of plants to insects. *Ann. Rev. Ent.*, **10**, 207–32.

BELTON, P. (1962). Effects of sound on insect behaviour. *Proc. Ent. Soc. Manitoba*, **18**, 1.

BERNARD, F. (1954). Rôle des insectes speciaux dans les terrains du Sahara. pp 104–11 in *Biology of Deserts* (Ed. J. L. Cloudsley-Thompson) Inst. of Biology, London, iv + 224 pp., illustr.

BILIOTTI, E., GRISON, P. & MILAIRE, H. (1962). La lutte biologique. *Bull. Techn. Inf. Ing. Ser. Agric.*, **167**, 143–62.

BODENHEIMER, F. S., REICH, K., SEGAL, N. & FRÄNKEL, G. (1930). *Studien zur Epidemiologie, Oekologie und Physiologie der Afrikanischen Wanderheuschrecke* Schistocerca gregaria, *Forsk.* Paul Parey, Berlin, 124 pp., illustr.

BOZA BARDUCCI, T. *See* BARDUCCI, T. BOZA.

BROADBENT, L. (1967). The influence of climate and weather on the incidence of plant virus diseases. pp. 99–104 in *Weather and Agriculture* (Ed. J. H. Taylor) Pergamon Press, London, xi + 225 pp., illustr.

BUTLER, G. D. JR. (1955). The effect of alfalfa irrigation treatments on the two-spotted mite in alfalfa. *J. Econ. Ent.*, **48**, 221–2.

CARTER, W. (1961). Ecological aspects of plant virus transmission. *Ann. Rev. Ent.*, **6**, 347–70.

CHANT, D. A. (1966). Integrated control systems. pp. 193–218 in *Scientific Aspects of Pest Control.* J. Wash. Acad. Sci., xi + 470 pp., illustr.

CHOWDHURY, S. & SHARMA, R. G. (1960). Effect of methods and frequency of irrigation on the incidence of stem-borer of wheat (*Sesamia inferens* Wlk.) at Irrigation Research Farm. *Indian J. Agron.*, **4**, 264–8.

COLE, L. C. (1966). The complexity of pest control in the environment. pp. 13–25 in *Scientific Aspects of Pest Control.* J. Wash. Acad. Sci., xi + 470 pp., illustr.

CORNWELL, P. B. (1966). *The Entomology of Radiation Disinfestation of Grain—A Collection of Original Research Papers.* Macmillan (Pergamon), New York, xx + 236 pp., illustr.

FENJVES, P. (1950). Einige Probleme der angewandten Entomologie in Venezuela. *Mitt. Schweiz Ent. Ges.*, **23**, 135–54.

FIFE, L. C., ROBERTSON, O. T. & GRAHAM, H. M. (1963). Winter carryover of the pink Bollworm under various cultural practices in central Texas. *J. Econ. Ent.*, **56**, 172–5.

HENNEBERRY, T. J. & HOWLAND, A. F. (1966). Response of male cabbage loopers to black-light with or without the presence of the female sex pheromone. *J. Econ. Ent.*, **59**, 623–6.

HOLMES, E. (1960). The contribution of crop protection to increased world food supplies. *Outlook on Agriculture*, **11**(3), 23–38.

JOYCE, R. J. (1955). Cotton spraying in the Sudan Gezira. II. Entomological problems arising from spraying. *Pl. Prot. Bull. F.A.O.*, **3**, 97–103.

KILGORE, W. (1967). Chemosterilants. pp. 197–239 in *Pest Control* (Ed. W. Kilgore & R. L. Doutt) Academic Press, New York, xii + 477 pp., illustr.

LAWSON, F. R., CHAMBERLIN, J. C. & YORK, G. T. (1951). Dissemination of the beet leafhopper in California. *Tech. Bull. U.S. Dept Agric.*, **1030**, 59 pp., illustr.

MCFARLANE, N. L., AYERS, R. S. & WINWRIGHT, G. L. (1957). California desert agriculture. *Circ. Calif. Agric. Exp. Stn*, **464**, 66 pp., illustr.

MCKINNEY, K. B. (1939). Common insects attacking sugar-beets and vegetable crops in the Salt River Valley of Arizona. *J. Econ. Ent.*, **32**, 808–10.

MARTIN, H. (1964). *The Scientific Principles of Crop Protection.* Edward Arnold, London, viii + 376 pp., illustr.

MOSEMAN, A. H. (1966). Pest control—its role in the United States economy and in the world. pp. 26–38 in *Scientific Aspects of Pest Control.* Nat. Acad. of Sci., Washington, xi + 470 pp., illustr.

NELSON, S. O. (1967). Electromagnetic energy. pp. 89–145 in *Pest Control* (Ed. W. W. Kilgore & R. L. Doutt) Academic Press, New York, xii + 477 pp., illustr.

—— & SEUBERT, J. L. (1966). Electromagnetic and sonic energy for pest control. pp. 135–66 in *Scientific Aspects of Pest Control.* Nat. Acad. of Sci., Washington D.C., xi + 470 pp., illustr.

O'BRIEN, R. D. (1967). *Insecticides, Action and Metabolism.* Academic Press, New York, xi + 332 pp., illustr.

ORDISH, O. (1967). *Biological Methods in Crop Pest Control.* Constable, London, 242 pp., illustr.

PAINTER, R. H. (1958). Resistance of plants to insects. *Ann. Rev. Ent.*, **3**, 267–90.

PAVLOV, I. F. (1959). [The influence of shelter-bets, irrigation and perennial grasses on the frequency of the Hessian Fly and fruit-fly (*Diptera*) *Cecidomycidae* and *Chloropidae*—in Russian] *Ent. Obozr.*, **38**, 326–40.

PIERCE, W. D. (1912). The insect enemies of the cotton boll weevil. *U.S. Dept. Agric. Bur. Ent. Bull.* **100**, 99 pp., illustr.

PRADHAN, S. (1957). The ecology of arid zone insects (excluding locusts and grasshoppers) *Arid Zone Res.*, **8**, 199–240.

RIDGWAY, R. L. (1967). Research on systemic insecticides and methods of applying them for control of cotton insects. *Cot. Grow. Rev.*, **44**(1), 39–50.

RIVNAY, E. (1963). Present status of lepidopterous pests of maize and other gramineous crops in Israel. *Pl. Prot. Bull. F.A.O.*, **11**, 1–3.

SHOREY, H. H. & GASTON, L. K. (1967). Pheromones. pp. 241–65 in *Pest Control* (Ed. W. W. Kilgore & R. L. Doutt) Academic Press, New York, xii + 477 pp., illustr.

SMITH, R. F. (1957). The effect of irrigation on insects and other arthopods. *Irrig. Eng. Maintenance*, **7**, 20–1.

—— & BOSCH, R. VAN DEN (1967). Integrated control. pp. 295–340 in *Pest Control* (Ed. W. W. Kilgore & R. L. Doutt) Academic Press, New York, xii + 477 pp., illustr.

STERN, W. M. (1965). Significance of the economic threshold in integrated pest control. *Proc. F.A.O. Symp. Integrated Pest Control, Rome*, **2**, 41–56.

TANADA, Y. (1967). Microbial pesticides. pp. 31–81 in *Pest Control* (Ed. W. W. Kilgore & R. L. Doutt) Academic Press, New York, xii + 477 pp., illustr.

TILTON, E. W. & SCHROEDER, H. W. (1963). Some effects of infra-red irradiation on the mortality of immature insects in kernels of rough rice. *J. Econ. Ent.*, **56**, 727–30.

UVAROV, B. P. (1954). The desert locust and its environment. pp. 85–89 in *Biology of Deserts* (Ed. J. L. Cloudsley-Thompson) Inst. of Biology, London, iv + 224 pp., illustr.

—— (1962). Development of arid lands and its ecological effects on their insect fauna. *Arid Zone Res.*, **18**, 235–48.

VAYSSIÈRE, P. E. L. (1954). Les relations entre les zones désertiques et la population des parasites des plantes. pp. 80–4 in *Biology of Deserts* (Ed. J. L. Cloudsley-Thompson) Inst. of Biology, London, iv + 224 pp., illustr.

WEISNER, J. B. (1963). Use of pesticides. *Residue Rev.*, **6**, 1–22.

WILSON, M. C., DAVIS, R. L. & WILLIAMS, G. G. (1955). Multiple effects of leaf-hopper infestation on irrigated and non-irrigated alfalfa. *J. Econ. Ent.*, **48**, 323–6.

WRIGHT, D. J., JR. (1967). Antifeedants. pp. 287–93 in *Pest Control* (Ed. W. W. Kilgore & R. L. Doutt) Academic Press, New York, xii + 477 pp., illustr.

WYNIGER, R. (1962). *Pests of Crops in Warm Climates and their Control.* Verlag f. Recht u. Gesellschaft, Bâle, 555 pp., illustr.

XVIII

PLANT DISEASES

Plant Diseases and Aridity

The conditions that are prevalent in arid and semi-arid regions have varying effects on different plant diseases. There are two main categories of pathogens, from the point of view of their development under arid conditions (Rotem & Palti, 1968):

(a) *Pathogens Well Adapted to Surviving Dry Spells or Seasons.* These are generally weak or only moderately strong parasites, and include species of *Alternaria* and powdery mildews. Irrespective of weather conditions, these diseases manage to develop epidemically every year if susceptible hosts are present, and are not markedly dependent on microclimatic conditions. The main influence of irrigation on these diseases is indirect, being due to its effect on the host plants and the possibility of growing them throughout the year. Techniques of irrigation have little influence on the intensity of these diseases.

(b) *Pathogens with High-moisture Affinities.* Pathogens belonging to this group, such as rusts, downy mildews, and *Phytophthora* spp., are generally strong pathogens, whose development is limited to specific seasons. They are greatly affected by microclimatic effects, such as dew and shade, and by methods of crop management, such as irrigation, planting density, etc. In the dry regions, conditions for a build-up of the massive quantities of inoculum required to produce epidemics are generally sub-optimal for these pathogens. As only small amounts of inoculum survive the dry spells, a fairly long period of time is required to achieve a build-up of inoculum to epidemic proportions; hence the marked time-lag between the occurrence of environmental conditions favourable to the pathogen and the actual outbreak of the epidemic.

One advantage of the arid regions is the possibility of growing, under irrigation, disease-free seed during the dry period.

For this category of pathogens, irrigation has a considerable effect on disease intensity, both by increasing the predisposition of the host and by the creation of a favourable environment for the pathogen. Methods and frequency of irrigation, as well as timing of the water application, have marked effects on the severity of infection.

TEMPERATURES AND AIR HUMIDITY

In dry climates, two factors have an inhibitory effect on plant diseases; these are high temperatures and low air-humidity (Reichert, 1954).

Diseases that originate in cooler climates are adversely affected mainly by high temperatures. *Tilletia tritici* occurs in wheat only when rain and low temperatures prevail shortly after germination. Similarly, *Claviceps purpurea* is favoured by wet seasons with limited sunshine (Heald, 1933). *Puccinia graminis*, which is prevalent on wheat in lower Egypt, does not occur in the Sudan, because of the low air-humidity there (Vayssière, 1954).

Sclerotinia sclerotiorum, with an optimal temperature between 17 and 21°C, attacks plants during the winter, whilst summer crops remain healthy. Sunflower (*Helianthus annuus*), which is attacked by the disease in damp, cool climates, is practically immune when grown in a dry, hot climate (Reichert, 1954).

For pathogens originating in hot climates, it is the dryness of the air which is the limiting factor. For example, *Puccinia sorghi* and *Ustilago zeae* attack maize in Israel only when it is grown in the vicinity of the sea or under overhead irrigation. Other maize diseases, such as *Diplodia zeae*, *Sclerospora macrospora*, and *Physoderma zea-maydis*, occur in regions with frequent rains during the growing period but not in the arid regions.

In general, root and stem diseases are more prevalent than leaf diseases in arid regions (Reichert, 1954). For leaf diseases, caused by wind-borne fungi, commonly require rainfall and high air-humidity in order to spread and infect crops (McNew, 1966). There are, however, exceptions, as will be shown in the following paragraphs.

Arid-regions may also be a source of infection of important diseases. For example, black rust (*Puccinia graminis*) is a major disease of wheat in many regions. In moist, cold regions, low temperatures destroy the uredospores on wheat, so that the main remaining source of infection consists of the spores that develop on a secondary host, the barberry (various species of *Berberis*). By destroying the barberry bushes, an essential link in the life-cycle of the rust is eliminated, and rust epidemics can be prevented.

However, in warm arid regions, temperatures do not fall sufficiently to kill the uredospores produced on wheat and native grasses which continue to grow throughout the winter. These are potent sources of infestation—not only for the arid regions themselves, but also for adjacent wheat belts—causing epidemics when conditions are favourable, even in regions in which barberry bushes have been eliminated (Ordish, 1967).

SOIL MOISTURE

High soil-moisture and increased turgidity of the plants at the time of infection generally favour most air-borne diseases. However, the relationship between moisture régime and plant disease does not always follow a consistent pattern.

For example, moisture stress was found to increase susceptibility of sesame to *Alternaria* and *Corynespora* blights, probably as a result of induced senescence; the recovery of infected plants was also delayed, as compared with plants growing under conditions of adequate moisture supply (Culp & Thomas, 1964). Similarly, tobacco grown under conditions of limited moisture was found to be more susceptible to *Peronospora tabacina* than was tobacco grown with plentiful moisture. However, in this case it was found that, irrespective of the previous moisture régime, actual inoculation only occurred when the leaves were turgid, and only negligible infection occurred on flaccid leaves (Rotem *et al.*, 1968).

A xerophytic environment favours powdery mildews (*Erysiphe* spp.) and therefore plants grown under conditions of deficient moisture are more heavily attacked than are those growing under a more favourable moisture régime (Yarwood, 1949). Powdery Mildew (*Erysiphe polygoni*) was found to increase in severity on Red Clover when this was grown under conditions of limited moisture (Curl & Weaver, 1958).

Whilst root-rots due to *Rhizoctonia*, *Fusarium*, etc., may increase as a result of excess moisture, they are usually more harmful to a crop under conditions of water-stress than under optimal water conditions (Menzies, 1967).

Effects of Irrigation on Plant Diseases

In general, irrigation favours the incidence and spread of many crop diseases by its influence on macroclimates as well as on microclimates and on the soil moisture régime. It affects both the host and the pathogen. The rapid succession of crops (especially if monoculture is practised), the longer periods that the stomata are widely open, the increased succulence of the tissues as a result of favourable moisture supply and higher rates of nitrogen fertilization, and the increased plant populations necessary to make efficient use of the favourable growing conditions, are all contributory factors that increase the susceptibility of the crop and create more favourable conditions for infection.

The greater planting densities required in irrigation farming, and the more luxuriant vegetative growth resulting from a favourable moisture régime, ensure more shade and longer periods of high air-humidity under the leaf canopy and accordingly slower drying of the soil surface. These conditions are favourable for the infection of the lower leaves and root collars of field-crops, which effect is especially marked in the case of *Cercospora personata* on peanuts (Rotem & Palti, 1968). Occasionally, however, the improved vigour of the crop plant due to a favourable moisture régime will increase its resistance to disease.

By changing the microclimate and often to some extent also the macroclimate, irrigation affects the viability of air-borne spores and may favour spore disposal; it may also increase the speed with which the infection is spread. The interrelation between irrigation and plant diseases is complicated by the interactions of a number of factors affecting the time and extent of disease development in irrigated crops. Some of these factors are (Rotem & Palti, 1968):

(a) Weather: rain, air humidity, temperature, radiation;
(b) Pathogen characteristics: time of spore disposal, drought resistance of the spores, speed of infection;
(c) Method of irrigation used, frequency of irrigations, and time of day at which water is applied; and
(d) Cultural practices specific to irrigation farming.

The following examples illustrate the relationship between irrigation and plant disease:

Certain diseases, and in particular those caused by downy mildews, are especially favoured by the young and succulent growth promoted by irrigation. Examples include the increased incidence of downy mildew of peas and lucerne (Rotem & Palti, 1968), and of *Sclerospora graminicola* on pearl millet (Kenneth, 1966).

Sclerotium rolfsii, which attacks many crops, requires a combination of high temperatures (30–35°C) and favourable soil moisture conditions (Watkins, 1961). Normally absent in arid and semi-arid regions, it becomes a serious disease of summer crops when irrigation is introduced (Avizohar-Hershenson & Palti, 1962).

Irrigation can also affect crop diseases by its effect on soil temperature. For example, optimum temperatures for verticillium wilt of cotton (*Verticillium albo-atrum*) are 20 to 25°C; no damage occurs to the plant at temperatures above 33°C (Garber & Halisky, 1958). This can explain why irrigation, by reducing soil temperatures, promotes the development of root diseases such as verticillium wilt.

Irrigation may increase the incidence of virus diseases indirectly, through its effect on the vector responsible for spreading the disease.

The Beet Leafhopper (*Circulifer tenellus*) is the only known vector of curly-top disease affecting a number of important crops such as sugar-beet, beans, melons, tomatoes, and others. In regions with cold winters, this vector's successful hibernation, which is essential for a rapid build-up of the population subsequently, is prevented by the low temperatures (Carter, 1961), but in arid regions with mild winters, the vector hibernates on typical winter annuals. When irrigation is introduced into these areas, ideal conditions are offered for the rapid dissemination of the Leafhoppers in the green cultivated areas, from which the virus disease is spread to a number of crops (Lawson *et al.,* 1951).

When desert land is first developed for irrigation, a period of relative freedom from plant diseases is usually experienced. However, certain diseases, when once introduced, spread rapidly in these newly irrigated areas, especially if no proper crop rotation is practised. In the Columbia basin, in northwestern N. America, *Fusarium* root-rot of beans increased from a trace in the first year of irrigation, to serious damage in the third year, when beans were grown continuously (Menzies, 1954). Many pathogens are introduced with infected seeds, into regions in which the disease that they produce was previously non-existent.

Effect of Method of Irrigation on Plant Diseases

Natural precipitation generally favours the incidence of foliar diseases due to Fungi and Bacteria; it is therefore reasonable to assume that overhead irrigation will have similar effects. Certain authors state that, under semi-arid conditions, the use of a method of surface irrigation instead of sprinkling may contribute more to disease control than does any chemical treatment (Snyder *et al.*, 1965). However, experience has shown that the danger from sprinkler irrigation has been over-emphasized, at least for field crops. There are several important differences between natural precipitation and overhead irrigation that are of relevance to the incidence and spread of plant diseases. For many pathogens, infection depends on the length of time that the plants remain wet after inoculation. Precipitation occurs over relatively large areas, usually after a period of cloudy weather, lowered temperatures, and increased air-humidity —which conditions may continue for some time after the rain.

Overhead irrigation applies 'rain' suddenly, under conditions of high temperature and light intensity and low air-humidity. The irrigated area is an 'oasis', surrounded by hot, dry air. Shortly after irrigation ceases, the foliage of the irrigated crop becomes completely dry. For most plant diseases, free water must be present on the leaf surfaces for periods of not less than 6–12 hours, to permit infection to take place. This applies to many of the most important disease agents (Gottlieb, 1950). Secondary sporulation, which leads to the main damage to the crop, requires a considerably longer time (Menzies, 1967). Rainfall, which is effective in causing the spread of these diseases, is usually associated with far lower temperatures than overhead irrigation. The optimum temperatures for spore production, germination and infection, are usually far lower than those prevailing when irrigation is usually necessary (ibid.). Therefore, whereas overhead irrigation may be conducive to infection, the dry hot weather between irrigations may, in many cases, effectively preclude the survival and spread of the inoculum.

Whilst one should not overemphasize the dangers inherent in overhead irrigation, neither should they be overlooked. Sprinkling differs from other irrigation methods in wetting the foliage of the crop, washing off fungicide deposits, and dispersing inoculum of many Fungi. It therefore creates conditions under which the host is more prone to disease, while also favouring the dissemination of the agents of disease and reducing the effectiveness of fungicides at a crucial time.

Although overhead irrigation periods may be of shorter duration than rain, they may extend by a few hours the period of wetness resulting from dew, which also favours the spread of certain diseases. The dew period in arid zones is also greatly extended during irrigation periods (Schein, 1968).

Sprinkler irrigation also affects leaf temperature. In a study in Israel, it was found that when air temperatures were 24 and 29°C in the morning and at noon, respectively, the leaf temperature of unirrigated potatoes was 30 and 36°C, whilst that of the irrigated leaves was only 22°C. When irrigation was stopped, leaf temperatures began to rise, but remained below that of unirrigated plants for the whole day. This may be of considerable significance in dry climates, as the leaf temperatures are frequently critical for infection (Rotem & Palti, 1968).

Though it is less dangerous than rainfall, there is no doubt that overhead irrigation can favour the spread of certain diseases, and is therefore more dangerous in this respect than surface irrigation, as is shown by the following examples:

Bacterial leaf-diseases of field beans were found to be far more severe with sprinkling than with furrow irrigation (Menzies, 1954). In the arid Negev, *Phytophthora* blight on potatoes was found to develop profusely under overhead irrigation, whilst it did not occur with furrow irrigation (Rotem *et al.*, 1968). *Cercospora* leaf-spot (*C. personata*) on peanuts also reacted in the same way (Palti & Stettiner, 1959).

For most forage crops, it was found that leaf diseases were more severe under a high moisture régime with sprinkler irrigation than otherwise. The diseases which showed the greatest increase in severity under high-level irrigation were *Pseudopeziza* leaf-spot on intermediate White Clover, *Curvularia* leaf-spot on Ladino Clover, and *Pseudopeziza* leaf-spot on alfalfa (Curl & Weaver, 1958).

In view of the general advantages of sprinkler irrigation (*see* Chapter VIII), techniques must be adopted to minimize the undesirable effects on disease propagation. This can be done by:

(a) Avoiding excessive and unbalanced fertilizer applications which are conducive to luxuriant vegetative growth and increased succulence;
(b) Applying large amounts of water at relatively long intervals of time rather than small amounts frequently;
(c) Choosing the time of day when sprinkling is least likely to enable the pathogen to complete the infection process; and
(d) Allowing each irrigated strip to dry before irrigating the adjacent strip (Rotem & Palti, 1968).

It has been shown experimentally that numerous light sprinkler irrigations, applied at short intervals of time, aggravate the damage caused to groundnuts by pod rots (*Pythium* spp.) as compared with an equal total amount of water applied in heavier irrigations at longer intervals. Allowing the top-soil to dry for a few days before resuming irrigation, after the onset of the main blossoming period, was fairly effective in reducing the damage caused by pod rots (Frank, 1967).

Though surface irrigation is generally less conducive than overhead sprinkling to the spread of diseases, there are exceptions. Thus in the irrigated desert areas of southern California, stands of lucerne are often partly killed during the hot summer months, by flooding—an injury called 'scald'. Apparently, the roots of the plants are damaged by a combination of excessive soil moisture with high soil temperatures (Erwin *et al.*, 1959).

Disease Control

Disease control methods aim at preventing the build-up of inoculum, evading infection, and increasing the resistance of the crop to infection—or improving its ability to recover from infection.

USE OF CHEMICALS

Since the advent of Bordeaux mixture in the middle of the 19th century, many chemicals have been developed for the control of plant diseases. Since 1934, some fifty new fungicides and bactericides have been brought into wide use (McNew, 1966). Most of these are effective only as external prophylactics; they do not control the disease when once it is established, and they have no effect on pathogens within the vascular system of the plant.

The great difficulty in designing new plant protectants is the need for selective toxicity to the pathogens, without excessive phytotoxic effects. In this respect the organic fungicides, first developed shortly before the Second World War, are far superior to the inorganic copper, sulphur, mercury, and zinc compounds, which were available to farmers before the advent of the organics. The most widely used of the new compounds are the alkylene bisdithiocarbamates, such as Nabam, Zineb and Maneb. These are essentially inert materials—until they undergo decomposition at a controlled rate, which releases a toxicant that destroys fungous spores with which it comes in contact. Proper formulation increases the effectiveness of the fungicides (McNew, 1966).

No dramatic advances in plant disease control by application of chemicals have been made in recent years. In this respect, plant pathology lags behind entomology. However, several less conventional approaches to plant disease control have been investigated in recent years (Goodman, 1964).

USE OF CHEMOTHERAPEUTANTS

Systemic chemotherapeutants, capable of attacking pathogens within the plant, *a priori* appear to be the most efficient approach. However, the development of systemic fungicides has been up to the present, generally disappointing. Host and pathogen tend to have very similar biochemical systems; chemicals that are effective in suppressing the pathogen are usually either too phytotoxic or too expensive.

A new approach is the search for chemicals which, by their effect on plant metabolism, improve the host's resistance to, or recovery from, infection. Certain growth regulators have been found to be effective in this respect. Some of the toxins produced by pathogens interfere with the protein synthesis of the plant; it has been found possible to overcome this antimetabolic effect by adding the appropriate amino-acid (Woolley *et al.*, 1955).

USE OF ANTIBIOTICS

Antibiotics have been widely tested for the systemic prevention and eradication of plant diseases. Three antibiotics—streptomycin, griseofulvin and cycloheximide—are most promising, and have been found effective against 25 bacterial diseases and 50 fungal diseases of plants (King & Splitter, 1966). The artificial introduction of antibiotics or of inhibitory antagonists into the soil has, however, generally been unsuccessful—presumably because of adsorption by soil colloids and decomposition by the soil microflora (McNew, 1966).

In the field of *biological control*, some of the pathways that are being investigated are:

(a) The use of bacteriophages for the control of plant bacterial diseases;
(b) The use of bacterial species indigenous to a host plant, to suppress virulent bacterial pathogens; and
(c) Preinfecting certain host species with a mild virus pathogen to protect the host from subsequent infection by a more destructive virus pathogen (Loebenstein, 1963). It is possible that the use of killed or extracted virus protein may accomplish the same purpose. However, no practical control methods of this nature have yet been developed for field crops.

The use of natural enemies is still being explored. It is well known that certain Fungi are parasitized by other Fungi. Competition between Bacteria and *Fusarium oxysporum* resulted in the suppression of the Fungus as a result of biologically-induced nitrogen deficiency, as was evidenced by the fact that the addition of nitrogen cancelled the competitive effect (Marshall & Alexander, 1960). Parasitic Fungi, viruses, and predators, all attack soil nematodes (Boosalis & Mankau, 1965). However, no practical application of these findings has yet been accomplished.

Disease Prevention Methods

A basic difficulty in designing crop production methods which will minimize the dangers of infection by disease is that plant pathogens generally have environmental requirements very similar to those of their hosts, so that conditions conducive to high production by the crop are also, *a priori*, favourable to the pathogen.

However, many crop diseases can be avoided or kept in check by the use of certain preventive measures and suitable agronomic practices:

Eradication of Alternate Hosts

Eradication of species of barberry for the control of black stem-rust of wheat (*Puccinia graminis*) has already been mentioned (p. 535). It is estimated that, in the United States alone, the campaign for eradicating barberry reduced annual losses due to the disease to one-fifth of their original level (McNew, 1966). For the prevention of virus disease, the choice of sites in which the disease vector is absent, and the choice of sowing dates at which the vector is not yet active, are very effective preventive measures. Moreover, chemical control of the vectors became practicable after the introduction of systemic insecticides. Whilst the control affected by systemic insecticides is not absolute, as the vector insect may transmit the disease before the insecticide has taken effect, the losses from virus diseases can be considerably reduced by their use.

SEEDING METHODS

The aim here is to improve air movement and promote the rapid drying of foliage by such means as check-planting of sugar-beets instead of having solid rows; this has helped to retard the spread of *Cercospora beticola* (McNew, 1966).

NUTRITION

The quantity and balance of the nutrients supplied to the crop may have a considerable bearing on its susceptibility to disease. This aspect has been dealt with in detail in Chapter XII (*see especially* pp. 404–5).

Soil-inhabiting Pathogens

Because of the large volume of chemicals that would necessarily be involved and the rapid adsorption of them by soil colloids and organic matter, chemical treatment of soils for the control of soil-borne diseases is too expensive for routine use in the field. Therefore, biological measures, such as crop rotation and the stimulation of antibiosis, are widely used for the prevention of build-up of inoculum potential.

CROP ROTATION

By maintaining a sufficiently long interval between susceptible crops, during which non-susceptible crops are grown, the population of soil pathogens is contained within reasonable levels. For highly specialized parasites, this method is fairly effective; others, which can lead a saprophytic existence, may survive for very long periods between susceptible crops.

ANTIBIOSIS

Some sequences of non-susceptible crops have a drastic effect on soil pathogens, such as cannot be attributed solely to the absence of a suitable host. These effects may be due to root secretions of organic acids, amino-acids, or carbohydrates, that influence the balance among micro-organisms in the rhizosphere; this influence may be exerted by direct toxic effects on the pathogen, by stimulating antibiosis, or by increasing the numbers of antagonistic micro-organisms (McNew, 1966). The soil abounds in powerful antagonists that compete with parasites or poison soil-borne plant pathogens. They are selective for the pathogens they attack, and are also selective for the kinds of organic materials upon which they thrive (Snyder, 1960). Legumes can reduce Take-all disease in wheat (*Ophiobolus graminis*) by fixing soil nitrogen and thereby starving the Fungus (Garrett, 1965). The crop residues of Sweet-clover, grown in rotation with cotton, reduce infection by Texas Root-rot of cotton (*Phymatotrichum omnivorum*) presumably by stimulating the activities of Bacteria that are antagonistic to the pathogen (Lyle *et al.*, 1948).

Another method of achieving the same effect is the ploughing under of a green-manure crop. Where *Phymatotrichum* Root-rot of cotton is serious, the turning under of a pea crop for green-manure makes it possible to harvest a crop of cotton. This is now a standard practice for combating Root-rot of cotton in Arizona (Snyder, 1960). The leguminous herb *Crotalaria spectabilis* has a suppressant action on root-rot nematodes (*Meloidogyne* spp.) (McNew, 1966).

The effects of crop residues vary according to the kind of residues and the micro-organisms involved. In general, oat and lucerne stems reduced the severity of root diseases of wheat, whilst wheat and barley residues increased their severity (Curl, 1963).

Oat residues reduced stem-rot severity of groundnut seedlings due to *Sclerotium rolfsii*, as compared with groundnut or clover residues. In the oat-residue soil, the activity of *Trichoderma*, and its potential as a biological control agent of *S. rolfsii*, were enhanced (Mixon & Curl, 1968).

Huber *et al.* (1965) found that crop residues which increased the rate of nitrification reduced the severity of *Fusarium* and *Rhizoctonia* infection, whilst residues that inhibited nitrification increased disease severity.

Breeding for Disease Resistance

The use of disease-resistant varieties is without doubt the most effective and economical method of disease control, especially for those diseases for which no effective chemical control methods have yet been developed. In many cases, the continued cultivation of a crop in a certain area became possible only after resistant varieties had been developed.

NATURE OF DISEASE RESISTANCE

Genes for resistance to disease have been found in many wild species related to our cultivated crops, and effective plant-breeding methods have been developed for transferring these genes to cultivated varieties within a relatively short time. The basic difference between breeding for disease resistance and breeding for other physiological or morphological characters of the crop plant is that, in the latter case, the plant breeder is concerned with a single biological entity, whilst in the former case he must work within two biological systems and their interactions, namely the host plant and the pathogen. Many of the pathogens have a considerable number of physiological races, each capable of attacking one or more varieties of the host plant. The black stem-rust of wheat, for example, has over 200 recognized parasite races.

BREAKDOWN OF DISEASE RESISTANCE

New virulent races may develop by hybridization, and may build up to epiphytotic proportions within three to five years, leading to the collapse of resistance of established varieties. On the other hand, the breeding of resistant varieties takes about ten years, even with easy-to-handle crops, whilst their life expectancy does not exceed 5 to 15 years (McNew, 1966). The use of pure, homozygous crop varieties—bred to almost absolute uniformity—is also a major factor facilitating the rapid breakdown of disease resistance, by increasing the potential for adjustment and rapid propagation of new virulent pathogenic races.

The work carried out in recent years at the International Maize and Wheat Improvement Center, located at Chapingo, Mexico, is an illustration of the need for a continuous, dynamic plant-breeding programme in order to cope with the ever-changing spectrum of prevalent virulent races of cereal rusts (Stakman *et al.*, 1967).

In the autumn of 1948, four outstanding varieties of wheat that had been bred at Chapingo were first released to farmers. These varieties derived their rust resistance from two sources: the 'Hope'-type, due to a Yaroslav emmer parent, and the 'Kenya' type, probably due to a *Triticum turgidum* parent. As a result of their spectacular success, by 1951 no less than 70 per cent of the total wheat acreage in Mexico was sown with one or other of these four superior wheat varieties. Stem-rust was apparently under control. However, in the summer of 1950, race 15B, a hitherto non-virulent race of stem-rust in the northern U.S.A., suddenly spread like wildfire in the northern U.S.A. and southern Canada, ruining late fields of varieties whose near-immunity to rust for more than a decade had finally broken down. Race 15B was carried into Mexico in the autumn of 1950, and, the following year, completely ruined late-sown crops of the varieties with 'Hope'-type resistance. In the summer of 1953, race 139, another rare race which had hitherto shown only limited and weak parasitism, caused the breakdown of the varieties with 'Kenya'-type resistance.

In the meantime, new varieties had been bred that combined 'Hope' and 'Kenya' types of resistance. This was, however, matched by another major shift in the population of stem-rust races, with the suddenly increased virulence in 1953 of the two races 29 and 48.

The continuous breeding programme, based on hybridization between varieties of very diverse parentage, and encompassing some 50 000 varieties and hybrid lines in the breeding nursery, was able to make more resistant varieties available as the changes occurred, thereby averting a catastrophe.

FACTORS INVOLVED IN DISEASE RESISTANCE

Genetically controlled resistance may be due to physiological, morphological, or anatomical, characteristics of the plant.

Physiological Factors

Resistance may be due to the nature of the reaction of the cells of the host plant to the invasion of the pathogen. Obligatory parasites, such as the wheat rusts, cannot maintain themselves except on living plant tissues. If there is full adaptation of the pathogen to the host, the disease will establish itself and spread rapidly. In the field, the whole spectrum of host-parasite relationships may occur from absolute incompatibility—which confers immunity on the host—to complete compatibility, in which case the parasite will completely invade the plant.

In some cases, hypersensitivity of the host to the parasite may provide an effective defence mechanism. Thus certain varieties of wheat are so sensitive to infection that the site of invasion is immediately surrounded by necrotic tissue, which effectively prevents growth and sporulation of the Fungus. A similar reaction also restricts the development of some virus diseases (McNew, 1966).

Chemical Constituents in the Plant

The presence of anthocyanin in the plant tissues frequently confers some degree of disease resistance. Resistance to rust is related to the amount of phenols present in wheat tissues.

In linseed it has been found that varieties which have the same types of protein as the Fungus causing Linseed Rust (*Melampspora lini*) are susceptible to the disease, whereas those in which this type of protein is absent are resistant (Flor, 1956).

Morphological Factors

Disease resistance may be due to certain morphological characters, such as hairiness or a waxy covering of the vegetative parts. Certain pathogens penetrate into the plant tissues through the stomata, the size of which, and the length of time during which they are open, have relevance to the susceptibility of the plant to infection. In wheat, for example, it was found that the plants most susceptible to rust were those in which the stomata are wide open during the early morning hours, at a time when the plants are wet with dew and conditions for infection are optimal (Hart, 1949).

The dwarf wheats, in addition to their increased ability to resist lodging, are also less prone to infection by stem-rust, because their stems are better protected by the leaf-sheaths than are those of the long-strawed varieties.

Disease Avoidance

Many varieties are not attacked by certain diseases because the stage of growth at which the plant is most susceptible to attack does not coincide with the periods during which the pathogen is most virulent. Early maturing varieties of wheat, for example, may thereby escape severe damage from rust epidemics.

NEW APPROACHES TO BREEDING FOR DISEASE RESISTANCE

The rapid breakdown of resistance in varieties in which attempts have been made to build up specific resistance by adding genetic sources of resistance to the known races of a particular disease, and which are therefore always one step behind a new physiological race of the pathogen, has led plant breeders to search for new approaches to the problem. One such approach is to attempt to breed varieties with multigenic resistance, on the assumption that it is highly improbable that a new race of pathogen would be able to acquire, by a random breeding process, several genes for virulence, which at one and the same time would enable it to overcome a host with a wide spectrum of resistance.

This approach replaces the concept of plant resistance or immunity with that of disease tolerance. As specific resistance to a given race of pathogen is more absolute than multigenic resistance to a broad spectrum of races, disease-tolerant plants are not immune to the disease, but can resist massive infection and produce a reasonable crop.

REFERENCES

AVIZOHAR-HERSHENSON, Z. & PALTI, J. (1962). *Sclerotium rolfsii* in Israel. *Pl. Dis. Reptr*, **46**, 145–9.

BOOSALIS, M. G. & MANKAU, R. (1965). Parasitism and predation of soil micro-organisms. pp. 374–89 in *Ecology of Soil-borne Plant Pathogens* (Ed. K. F. Baker & W. C. Snyder) Univ. of Calif. Press, Berkeley, 571 pp., illustr.

CARTER, W. (1961). Ecological aspects of plant virus transmission. *Ann. Rev. Entomol.*, **6**, 347–70.

CULP, T. W. & THOMAS, C. A. (1964). *Alternaria* and *Corynespora* blights of sesame in Mississippi. *Pl. Dis. Reptr*, **48**, 608–9.

CURL, E. A. (1963). Control of plant disease by crop rotation. *Bot. Rev.*, **29**(4), 413–79.

—— & WEAVER, H. A. (1958). Diseases of forage crops under sprinkler irrigation in the south-east. *Pl. Dis. Reptr*, **42**, 637–44.

ERWIN, D. C., KENNEDY, B. W. & LEHMAN, W. F. (1959). Xylem necrosis and root-rot of alfalfa associated with excessive irrigation and high temperatures. *Phytopathology*, **49**, 572–8.

FLOR, H. H. (1956). Genic systems in flax and flax rust. *Adv. Genetics*, **8**, 29–52.

FRANK, Z. R. (1967). Effect of irrigation procedure on *Pythium* rot of groundnut pods. *Pl. Dis. Reptr*, **51**, 414–16.

GARBER, R. H. & HALISKY, P. M. (1958). Influence of soil temperature on *Verticillium* wilt of cotton. *Phytopathology*, **48**, 393.

GARRETT, S. D. (1965). Toward biological control of soil-borne plant pathogens. pp. 4–16 in *Ecology of Soil-borne Plant Pathogens* (Ed. K. F. Baker & W. C. Snyder) Univ. of Calif. Press, Berkeley, 571 pp., illustr.

GOODMAN, R. N. (1964). Trends in plant protection against micro-organisms. pp. 248–55 in *Global Impacts of Applied Microbiology* (Ed. M. P. Starr) J. Wiley, New York, 572 pp.

GOTTLIEB, D. (1950). The physiology of spore germination in fungi. *Bot. Rev.*, **16**, 229–57.

HART, H. (1949). Nature and variability of disease resistance in plants. *Ann. Rev. Microbiol.*, **3**, 289–316.

HEALD, F. D. (1933). *Manual of Plant Diseases.* McGraw-Hill, New York, xii+ 953 pp., illustr.

HUBER, D. M., WATSON, R. D. & STEINER, G. W. (1965). Crop residues and plant disease. *Soil Sci.*, **100**, 302–8.

KENNETH, R. (1966). Studies on downy mildew diseases caused by *Sclerospora graminicola* (Sacc.)/Shroet. and *S. sorghi* Weston & Uppal. *Scripta Hierosolymitona*, **18**, 143–72.

KING, N. B. & SPLITTER, E. J. (1966). Antibiotics. *Agric. Sci. Rev.*, **4**(4), 1–9.

LAWSON, F. C., CHAMBERLIN, J. C. & YORK, G. T. (1951). Dissemination of the beet leafhopper in California. *Tech. Bull. U.S. Dept Agric.*, **1030**, 59 pp., illustr.

LOEBENSTEIN, G. (1963). Further evidence on systemic resistance induced by localized necrotic virus infections in plants. *Phytopathology*, **53**, 306–8.

LYLE, E. W., DUNLAP, A. A., HILL, H. O. & HARGROVE, R. D. (1948). Control of cotton root-rot by Sweet-clover in rotation. *Bull. Texas Agric. Exp. Stn*, **699**, 21 pp.

MCNEW, G. L. (1966). Progress in the battle against plant diseases. pp. 73–101 in *Scientific Aspects of Pest Control.* Nat. Acad of Sci., Washington D.C., xi + 470 pp., illustr.

MARSHALL, K. C. & ALEXANDER, M. (1960). Competition between soil Bacteria and *Fusarium. Pl. & Soil*, **12**, 143–53.

MENZIES, J. D. (1954). Effect of sprinkler irrigation in an arid climate on the spread of bacterial diseases of beans. *Phytopathology*, **44**, 553–6.

—— (1967). Plant diseases related to irrigation. pp. 1058–64 in *Irrigation of Agricultural Lands* (Ed. R. M. Hagan, R. H. Haise & T. W. Edminster) Agronomy Series; Amer. Soc. Agron., Madison, Wisconsin, **11**, xxxi + 1180 pp., illustr.

MIXON, A. C. & CURL, E. A. (1968). Influence of plant residues on *Sclerotium rolfsii* and inhibitory soil micro-organisms. *Crop Sci.*, **7**, 641–3.

ORDISH, O. (1967). *Biological Methods in Crop Pest Control.* Constable, London, 242 pp., illustr.

PALTI, J. & STETTINER, M. (1959). [*Pests and Diseases of Field and Garden Crops in the Huleh Valley 1955–58*—in Hebrew.] Bull. of the Local Council, Upper Galilee, iii + 29 pp., illustr.

REICHERT, I. (1954). The occurrence of plant diseases in arid climates and their agricultural significance. Pp. 68–75 in *Biology of Deserts* (Ed. J. L. Cloudsley-Thompson) Inst. of Biology, London, iv + 224 pp., illustr.

ROTEM, J., COHEN, Y. & SPIEGEL, S. (1968). Effect of soil moisture on the predisposition of tobacco to *Peronospora tabacina. Pl. Dis. Reptr*, **52**, 310–13.

—— & PALTI, J. (1968). Epidemiology of some field crop diseases under semi-arid conditions with special reference to irrigation effects. p. 167 in *First Internat. Congr. Pl. Pathol.*, London, 225 pp.

SCHEIN, R. D. (1968). Planning in plant disease problems. pp. 141–51 in *Agroclimatological Methods.* Proc. Reading Symp., UNESCO, Paris, 392 pp., illustr.

SNYDER, W. C. (1960). Antagonism as a plant disease control principle. pp. 127–36 in *Biological and Chemical Control of Plant and Animal Pests* (Ed. L. P. Reitz) Am. Assoc. Adv. Sci., Washington D.C., xii + 273 pp., illustr.

——, GROGAN, R. G., BARDIN, R. & SCHROTH, M. N. (1965). Overhead irrigation encourages wet-weather plant diseases. *Calif. Agric.*, **19**(5), 11.

STAKMAN, E. C., BRADFIELD, R. & MANGELSDORF, P. C. (1967). *Campaigns Against Hunger*. Harvard Univ. Press, Cambridge, Massachusetts, xvi + 328 pp., illustr.

VAYSSIÈRE, P. E. L. (1954). Les relations entre les zones desertiques et la pullulation des parasites des plantes. pp. 80–4 in *Biology of Deserts* (Ed. J. L. Cloudsley-Thompson) Institute of Biology, London, iv + 224 pp., illustr.

WATKINS, G. M. (1961). Physiology of *Sclerotium rolfsii*, with emphasis on parasitism. *Phytopathology*, **51**, 110–13.

WOOLLEY, D. W., SCHAFFNER, G. & BRAUN, A. C. (1955). Studies on the structure of the phytopathogenic toxin of *Pseudomonas tabaci*. *J. Biol. Chem.*, **215**, 485–93.

YARWOOD, C. E. (1949). Effect of soil moisture and mineral nutrient concentration on the development of bean powdery mildews. *Phytopathology*, **39**, 780–8.

XIX

AGRICULTURAL DEVELOPMENT IN DRY REGIONS

Within the arid zone is found the whole range of *per caput* incomes: from less than $200 for countries such as India (U.S. Department of Agriculture, 1965) up to $3 665 for California (U.S. Department of Commerce, 1968). Some of the most developed and flourishing agricultures can be found in the dry regions, but very large parts of the world's arid and semi-arid regions are still underdeveloped or even undeveloped.

Transforming traditional agriculture into modern farming, by applying scientific knowledge and the experience gained in the developed arid regions, involves social and economic problems of great complexity. The interrelationships between these and technological progress in agricultural production are briefly reviewed in this chapter.

Role of Agriculture in Developing Countries

A formidable task faces the developing nations. The 1960s have been called a decade of development. However, the most important social and economic change that has taken place in the underdeveloped countries is a *population explosion* that makes it much more difficult than formerly for them to solve their economic problems. The flow of financial resources from developed countries is not increasing, but has levelled off; there is also a reduction in private long-term investment, due to a slowing down of development, to internal political instability, and to boundary conflicts.

The demand in the developed countries for the traditional export commodities of the developing countries has also fallen off, whilst outlets for new exports have not developed sufficiently. The considerable income-gap between developed and underdeveloped countries is widening, and food production *per caput* is declining. Two-thirds of the people in underdeveloped countries suffer from malnutrition or undernutrition; disease and illiteracy, in addition to certain social customs, generally handicap any programme aimed at increasing productivity in agriculture (Myrdal, 1965).

There are many opinions as to the 'preconditions' which are essential if a 'take-off' into 'self-sustaining growth' is to be achieved by developing countries, but two essential facts are generally acknowledged:

(a) *Agriculture has a crucial role to play, and 'take-off' in agriculture is the first essential step.* The economic development of low-income countries predicates a considerable increase in their production of agricultural commodities. Rapid population growth and rising incomes create increased demands for food and clothing. The only alternative to increasing local production is imports for consumption, which then compete for scarce foreign exchange with the requirements for investment goods needed for the development of industry and agriculture. Developing agriculture is therefore an essential prelude to industrial growth (Kellogg, 1962). In addition to improving the existing nutritional levels in both quantity and quality—and providing for a rapidly increasing population—agriculture must make substantial contributions which will make possible national economic growth on a wide front (Mellor, 1962). For this it is necessary for agriculture to

(i) produce export crops as a source of foreign currency,
(ii) provide raw materials for industrial processing,
(iii) increase efficiency, so as to free labour for the expansion of industrial and other economic activities, and
(iv) provide a source of savings which can be invested in industrial development.

The failure to develop agricultural production potentials will therefore impede economic progress in general, and industrialization in particular.

(b) *These requirements cannot be achieved by reliance on traditional agriculture.* In a detailed analysis of the factors involved in achieving a transformation of traditional agriculture, Schultz (1964) reached the conclusion that traditional agriculture, as such, is in a state of equilibrium, in which the traditional factors of production available are used as efficiently as possible under the prevailing circumstances.

Traditional agriculture cannot contribute much to the overall development of a country, because its production capacity is low and most of the income is used for feeding the food-producers themselves. Schultz (1964) further states that subsistence agriculture is characteristically poor 'because the factors on which the economy is dependent are not capable of producing more under the circumstances'. Simply increasing the input of traditional factors of production with the existing state of the 'art' gives too low a rate of return to induce further investment. Hence, a breakthrough is possible only by providing 'improvements in the quality of the inputs'—namely, new agricultural inputs with a relatively high pay-off. Virtually all these 'new' inputs of potential promise must come from *outside* traditional agriculture—improved varieties, fertilizers, equipment, pesticides, etc.—and their success depends on their being used efficiently.

Christensen & Yee (1964) estimate the minimum required increase of agricultural productivity that must be achieved by developing countries to enter the take-off stage of economic development, at 4–5 per cent per year. How difficult this is to achieve is illustrated by the fact that in Japan, with its spectacular success in increasing output per unit area, the long-term annual rate of yield increase has usually ranged between 1·0 and 1·5 per cent per year (Brown, 1965); but that it can be achieved is shown by one of the most spectacular success stories of recent years—the development of agriculture in Mexico. That country has had a large population explosion, with an annual increase of over 3 per cent. Up to 1955, Mexico was largely dependent for food on American aid; by 1965, however, it had become entirely self-supporting in this respect. During the decade 1955–65, Mexico increased its wheat production fourfold and doubled its maize production (Baade, 1965).

Dependence of Agricultural Development on Other Economic Aspects

The essential role of agriculture in providing the means for a take-off into self-sustained growth has been discussed above. Agriculture cannot, however, develop in isolation; its own further development is also dependent on the expansion of non-farming activities.

Development of agriculture consists in the gradual transformation of subsistence agriculture to production for the market and the purchase of production inputs from other sectors of the economy. A pre-condition is the existence of a non-agricultural population with a reasonably high standard of living. For this reason, industrial development has a profound effect on agriculture. By making possible a rapid increase in incomes, it creates expanded markets for agricultural produce. It causes changes in the social structure and norms, which in turn influence the values and behaviour of farmers and have, furthermore, an impact on farm technology (Hoselitz, 1962). Agricultural and industrial development are therefore highly interdependent on, and complementary to, each other.

Modern farming is also dependent on a supply of industrial products, machinery, chemicals, packing materials, etc. Agriculture is most successful in regions in which industry and other resources, such as mining, tourism, etc., promote the development of an adequate infrastructure of roads, commerce, social services, and amenities, and moreover share in their costs.

Industries that process agricultural products, such as oil expressing and refining, milling of cereals and pulses, textile production, sugar production, food preservation, etc., make possible the diversification of agricultural production. Modern food-processing also makes a considerable contribution to the economy of the country by reducing waste and preventing seasonal gluts.

The integration of industries in the rural areas has additional benefits. It reduces the mass emigration of the rural population to the towns and, as a corollary, improves the standards of services provided to the rural population in the matters of health, education, culture, etc. Local markets for agricultural products increase the profitability to the farmers, while the production of raw materials for local industries diversifies farming; moreover, seasonal work in industry may balance seasonal work in agriculture, providing fuller employment throughout the year.

Requirements for Agricultural Development

Modern agriculture is highly dynamic and flexible. It responds rapidly to changes in demand for farm products by replacing one crop with another and by adjusting the technologies to new situations. It is characterized by continuously increasing productivity in terms of yields or output per worker.

Transforming traditional agriculture into modern agriculture is accompanied by profound changes in the factors of production and in their relative importance. If efforts to transform agriculture are to be successful, technology has to be improved in a wide variety of fields. All this has to be accompanied by a complex infrastructure which serves agricultural production and which is able to provide the necessary supplies and facilities for both production and marketing. The farmer must feel that his efforts will be worthwhile and that their outcome will be in accordance with his wants. Incentives have to be provided to motivate the farmer to change his methods of production—notwithstanding the risks involved—in such items as appropriate pricing, credit, land reform, and other measures.

Research has to provide the necessary know-how; education is essential to make the farmer receptive to new ideas and capable of applying new technologies; extension services are needed to provide the link between research and the farmer.

New social forms and structural changes in rural society are also needed, to enable the farmer to cope with the new complexities with which he will be increasingly faced, as he moves from traditional to modern agriculture.

Factors of Production

LAND AND WATER

When the far-sighted Cambridge don Thomas R. Malthus in 1798 first proposed his thesis that population, if unchecked, increases geometrically, while subsistence increases only arithmetically, there were still vast areas in the world which could be developed relatively easily. This is now no longer the case, while the actual cultivated area of the world is far from providing the calculated minimum adequate diet for its total human population (Stamp, 1953).

Agricultural development is possible in one of two directions, which are not, however, mutually exclusive: bringing more land under cultivation, or increasing productivity per unit of land. The first alternative is possible with traditional methods alone, whereas the second is entirely dependent on technological change.

The area of potential arable land on the earth is more than three times the area actually harvested in any given year, yet the area of good new land that can be developed relatively easily has become very small. A very large proportion of the potentially arable land is in the arid or semi-arid regions.

For most of this land, productivity can be increased only by supplying water; hence, the two factors of production—land and water—have to be considered together.

Quality of Land

The inherent fertility of the land, whilst important, is not a prerequisite for increasing productivity; for relatively poor land can be made highly productive.

Schultz (1964) illustrates this with two sets of comparisons: in Mexico, agricultural production has been increasing at the annual rate of 7·1 per cent; the country has been modernizing its industry and agriculture concurrently. By contrast, in Chile agricultural production has increased only 1·6 per cent per year, and in Argentina and Uruguay it has stagnated; and yet in Uruguay, in particular, land and environment have an agricultural potential similar to that of California. In Japan, agricultural production has increased recently at the rate of 4–6 per cent per year, but in India it has done so at an annual rate of only 1–2 per cent. Total production per hectare in Japan is eight times that in India. However, the land in India is naturally of a better quality, and three times as much land *per caput* is irrigated in India as in Japan.

Water Development Potential

All available and potential water resources will have to be developed to meet the demands of a not-too-distant future.

If a water source is available, the development and settlement of deserts involve fewer technical difficulties than in the cases of jungles, marshy areas, or mountainous regions.

During the 19th century, large irrigation projects were executed, mainly in Egypt, India, U.S.A., and what is now U.S.S.R., increasing the total irrigated area of the world about fivefold—from 8 million hectares to 49 million hectares. Expansion of irrigated areas continued in the present century, so that at present the area irrigated is estimated at over 160 million ha, of which 150 million ha are in the arid lands (Gulhati & Smith, 1967). A further expansion of the irrigated areas in arid regions is still possible. Thus in the dry parts of India, it is estimated that not more than 10 per cent of the water available in the rivers is utilized. In the middle-Niger, an area of West Africa larger than that irrigated in Egypt could be developed. In Latin America, the area that could be developed is estimated at around five million hectares (Dickson, 1956). Irrigated areas developed in recent years in the Soviet Republics of Central Asia now cover five million hectares, and it is planned to increase the irrigated areas by another 15 million hectares in the course of the next 10 to 15 years (United Nations, 1963). It is estimated that, by the end of this century, after the numerous development plans in different parts of the world that are in various stages of planning and development will be completed, the world's total irrigated area may exceed 200 million hectares (Jewitt, 1966).

There are still very great amounts of unused water in the rivers of the world. The Irrawaddy, Salween and Mekong, of south-eastern Asia; the Yangtze and the Yellow Rivers of China; the Ob, Yenisei and Lena, of the U.S.S.R.; the Niger, Congo and Zambezi, in Africa; the great Alaskan and Canadian rivers in North America; the Amazon, Orinoco and Parana Rivers in South America, are all examples of potential water sources which could irrigate enormous tracts of land.

Even in great rivers, such as the Nile and the Tigris-Euphrates, that have been used for irrigation from time immemorial, 40 per cent or more of their mean flows are discharged without being used for irrigation (White, 1962).

Shortcomings of Many Irrigation Schemes

There are many irrigation projects around the developing world which are not giving reasonable yields and are not leading to the anticipated significant intensification of production (Burke-Knapp, 1966).

Most irrigation projects are concerned with major irrigation structures: diversion dams, storage reservoirs, canals, etc. These indeed provide water; but without a coordinated development plan providing the necessary infrastructure, and without the inputs required for the effective use of the water, these projects cannot be fully productive. They can even cause yields to decline after a relatively short period. Irrigated agriculture, to be fully effective, must be capital-intensive; other inputs have to be used intensively and at a high level of technology. Unfortunately, these aspects are frequently neglected. The authorities are concerned mainly with the prestige derived from spectacular large-scale structures; the irrigation engineers are concerned exclusively with the diversion, storage, and conveyance, of water, and usually have little understanding of the effective application and use of the water on the individual farm (U.S. President's Science Advisory Committee, 1967).

It would be a great fallacy to suppose that irrigation *per se* is equivalent to intensive agriculture and is capable of increasing productivity and raising the standard of living when applied under primitive conditions. Successful irrigation involves far more than elucidating engineering problems and applying water to the land. It becomes an effective tool for increased production, maintained on a high level, only when applied in conjunction with changes in crops and varieties and when combined with the most intensive and up-to-date cultural practices.

Irrigation facilities used for traditional farming are seldom adequate for producing high-yielding crops and for double cropping.

The need for adapting farming practice to the potentialities of irrigation farming can best be illustrated by the disappointing results obtained in a large number of irrigation projects in developing countries, notwithstanding substantial investments. One example is that of the Office du Niger in the Mali Republic. After thirty years of experience, yield levels were still low: average yields were 1600 kg/ha for rice and 700 kg for seed-cotton; even where yield levels had been previously raised to 2000 kg for rice and 1000 kg for cotton, the yields showed a tendency to decline. The only source of production increases was the development of additional land for irrigation. In 1958 a change of policy was adopted by the Office du Niger, namely, to stop the expansion of the irrigated area and to concentrate on improving cultural practices (Wilde & McLoughlin, 1967).

The greatest agricultural investments in the Near East have been for the exploration and development of water resources. And yet, the overall contribution of irrigation to agricultural output has been small. Poor irrigation techniques and inadequate drainage have caused salinity problems (F.A.O., 1966).

Lack of provision of technical and financial assistance for field levelling and field drainage was responsible for disappointing results of irrigation schemes in Iraq (Stippler & Darwish, 1966).

Six institutional bottlenecks in the Near East prevented the full impact of irrigation schemes on agricultural production (F.A.O., 1966). These were:

1. Lack of incentive to investment in view of the agrarian structure and systems of land tenure.
2. Shortage of institutional credit; more than 85 per cent of all farm credit is provided by moneylenders who charge exorbitant interest, the farmers

having to sell their products immediately after harvest at prices dictated by their creditors.

3. Inadequate supplies of fertilizers, insecticides, etc.; supplies are either not available locally at the right time, or are not in sufficient quantities or at reasonable prices.
4. Poor marketing facilities, lack of storage, and no effective guaranteed prices for basic commodities.
5. Inadequate extension service; the present ratio is 1 to 8000 farmers, in contrast to a minimum desirable level of 1 to 1000, and the standards of training of extension workers are also unsatisfactory.
6. Absence or inadequacy of farmers' organizations.

There is general agreement, among both Indian and foreign experts, that disappointing results from irrigation projects in India were due to the lack of relation between water inputs and other inputs, and the lack of effective coordination between the departments dealing with irrigation, agriculture and agrarian reform. A contributing factor was the inappropriate size of land holdings. But the main reason was that areas under irrigation were increased without concern for intensification of production methods (Mason, 1966).

The improved control and use of water on the individual farm should be complementary to the major irrigation projects. Enormous amounts of water are usually lost during conveyance from the major source of the works to the individual fields. These are not only a waste, but contribute to raising the levels of ground-water and salinity.

The general tendency is to over-irrigate, especially if the water is subsidized. This tendency can be avoided only if information is available on the most efficient way to use water, and if a field service is organized to advise the farmer on when and how to irrigate.

In addition to large-scale irrigation projects, water resources can be developed at the level of the individual farm or village. Low-lift pumps and tubewells can, in the aggregate, make considerable contributions to the development of a region.

In Pakistan, in the course of the Second Plan, 25 000 private tubewell installations were made by farmers (Bucha, 1968). These not only provided much added water, but made a substantial contribution to overall productivity by reducing the level of the ground-water in the region and by leaching salts out of the root-zone (Ghulam, 1965).

In Italy, a large number of small dams were built to make use of the run-off from eroded hilly areas. Approximately 3000 small artificial lakes were created, providing for local irrigation of over 100 000 ha of land. Yields of a wide range of crops were increased by as much as 400 per cent, improving the standard of living of the population in these dry, hilly regions (Ruperti, 1963).

Potential Contribution of Irrigation to Agricultural Development

Irrigation projects on a national scale have broad social goals of raising the productivity and standard of living of farmers, and making a major contribution to the expansion and stabilization of the economy of a region.

In the *arid regions*, large-scale irrigation projects are usually concerned with bringing into cultivation huge tracts of land that were completely barren in the past. They thereby make possible intensive settlement of the population from other areas, and in addition facilitate the transformation of local nomads into sedentary farmers.

In the *semi-arid regions*, in which agriculture was dependent on rain-fed crops, the situation is entirely different. In these regions, practically all land that is not too steep, rocky or shallow, is generally already under cultivation—including a not inconsiderable proportion of lands on which, under rational land-usage methods, cultivation would be considered as unjustified owing to soil erosion hazards.

These areas are characterized by alternating wet and dry seasons and by considerable variability in the amount and distribution of their rainfall. The erratic climate and relatively low rainfall leave very little scope for diversification of crops or for intensification of production; for the choice of crops adapted to these conditions is limited, and yield levels are as erratic as is the rainfall.

Under these conditions crops, such as cotton, that require high investments in labour or cash, cannot be remunerative, except in very restricted, favoured areas; moreover industrial crops, such as sugar-beets, are not economically justified because an industry, already handicapped by a working season of only 2–3 months, cannot possibly maintain itself when the supply of raw material is also subject to great fluctuations due to unpredictable climatic conditions.

Future prosperity and economic stability of the semi-arid regions therefore depend largely on the development of water resources to expand irrigation agriculture.

Objectives of Irrigation in Semi-arid Regions

Though the seasonal rainfall pattern in each semi-arid region is generally uniform, this is far from being the case in the *amount* of rainfall. One can distinguish, roughly, between areas with an annual rainfall of less than 300 mm, in which crop production based on natural precipitation is precarious, if not impossible, and the higher-rainfall areas in which regular yields of a few adapted crops are more or less assured.

In the low-rainfall areas there is a possibility of choice of objectives: (a) supplementary irrigation aimed at assuring a more or less constant level of yields of traditional crops in lieu of the alternating crop failures and relative successes which are typical for the area or (b) full irrigation aimed at the general intensification of crop production. Shimshi *et al.* (1957) have calculated, on the basis of long-term meteorological records and the results of irrigation experiments, that for the northern Negev in Israel (250 mm average annual precipitation) the prospects of obtaining an economic yield of wheat (80 per cent of the maximum yield obtainable when water is not a limiting factor) occur about once in ten years. Supplementary irrigation at the rate of 100 mm, given before sowing, ensures success in one year out of two; if in addition the farmer has at his disposal a reserve of water of about 50–60 mm, to cover deficits arising from insufficient rainfall up to earing, the probability of success becomes 95 per cent.

Whilst this in itself is a great advance, there is no essential difference between crop production with supplementary irrigation in the low-rainfall area and rain-fed production in the higher-rainfall regions—except that production costs are appreciably increased in the former instance. Supplementary irrigation in the low-rainfall area cannot therefore appreciably raise the standard of living, unless the farm unit per family is relatively large—which is usually not possible where the farm population is dense.

The main importance of irrigation, is not as a supplement to rainfall, but as a cultural method enabling full control of the moisture supply to the crop, that makes possible a more intensive succession of crops, a greater variety of crops, and higher yield levels. Replacing a rain-fed crop, such as wheat, with an irrigated crop such as cotton, makes it possible to treble and even quadruple the net income from a unit area.

The development of irrigation also has an impact on the development of non-agricultural sectors of the economy, such as commerce, services and transportation; it contributes to the balance of payments and provides a market for the industrial production of agricultural equipment and supplies.

The classic example of a successful large-scale development scheme based on irrigation is the Gezira Project in the Sudan. This scheme was established on a clay plain of 0·4 million ha, lying between the White and Blue Niles. The flat land is extremely well suited to surface irrigation. Environmental conditions are very favourable for growing long staple cotton, a cash-crop of high value.

In the case of the Gezira Project, large-scale development was preceded by pilot-scale tests based on research carried out by an efficient organization especially set up for the purpose. The social controls imposed on the settlers were designed to protect them against antisocial actions but not to suppress individual initiative. The controls consisted of the planned family holding, tenancy rather than ownership, prescribed rotation, and a prohibition on mortgage and fragmentation. Organized services such as a research station and seed farms, supply of fertilizers and pesticides, availability of machinery for cultivation and pest control, as well as supervised credit and marketing, made it easier for the individual to obtain maximum returns. Nonetheless each settler's returns depended on his own initiative and effort. At present, the Gezira Scheme contributes over 50 per cent of the total revenue of the Sudan Government (U.S. President's Science Advisory Committee, 1967).

LABOUR

Agriculture in the developing countries employs nearly two-thirds of their population but its share in the gross national product is less than one-third. Production per man in agriculture is therefore only about one-fourth of what it is on the average in other activities, based on pecuniary values (O.E.C.D., 1967*a*).

For the next generation, at least, it is expected that in countries in which at present 60 to 80 per cent of the population is occupied in agriculture, farm population will increase faster than the opportunities for absorption by industry and services. Even if birth control measures are successful, their influence will be felt only in the next generation; 'the future workers are already born or will soon be born' (Myrdal, 1965). This will mean that costs for labour are likely to remain low, relative to those for capital (Bachman & Christensen, 1967). Any attempt to reduce farm labour before alternative opportunities are available is socially unjustifiable and will only increase the number of unemployed in the towns.

Agricultural development must therefore be aimed, in the early stages, at creating a *labour-intensive* technology producing high yields. This is exactly the opposite aim to the situation encountered in developed countries, which have had to solve the problems of the continuously decreasing quantities of manpower available to agriculture.

Japan was very successful in improving labour-intensive agriculture with very little land per agricultural worker, and the pattern developed by the Japanese may have considerable relevance to other developing countries during a certain stage of their development. It should, however, be noted that even in Japan, with its high labour output in agriculture as compared with many other Asian countries, productivity per man in agriculture is far lower than that of industry or modern services.

In a comparison between labour utilization on selected farms engaged in similar types of farming in India and in Japan, it was found that 50 per cent more labour per land unit is employed in Japan, where labour utilization is twice as high and operating expenses are eight times as high as in India. The cost of fertilizers alone is sixteen times as high, and the gross production is ten times as high, on Japanese farms as on Indian farms (Mellor, 1962).

Labour-intensive agriculture is not an objective in itself. Farming on the deep alluvial soils in the valleys of China, is one of the most 'intensive' types practised anywhere in the world, but is associated with unceasing drudgery and an extremely low standard of living.

The very large agricultural population of the developing countries constitutes an enormous reservoir of manpower for the development of other sectors of their economy. The transition from traditional to modern farming sooner or later involves problems of reallocation of this labour force.

The achievement of an acceptable standard of living for the farmer and the rural labour force requires a reduction in the number of people employed in agriculture, both in absolute and relative figures. The improvement of the national economy depends on industrialization and expansion of services, which must draw on the reserves of manpower available to agriculture. The difference in quality of this manpower and that of developed countries can diminish rapidly if education and training are adequate. The difference in wage levels may remain very big for a long time to come (O.E.C.D., 1967*a*), giving the developing countries an initial advantage in proceeding with industrialization.

The size of the agricultural population at each stage of development is determined by the rate of increase of the non-agricultural sources of employment. After industry and services have taken the labour they require, what remains stays in agriculture (Malassis, 1966). Therefore, the rate of reduction of the percentage of people engaged in agriculture is a fair index of economic progress.

Japan, Greece, Israel, Mexico and Taiwan, are examples of countries in which increased productivity of agriculture created an economic surplus and freed labour for other activities, thereby enabling industrial growth to take place and structural changes to occur in the national economies (Christensen & Yee, 1964).

CAPITAL

Capital is needed at the national level for the financing of infrastructure projects—such as ones of electric power, transportation, irrigation and flood control—which do not as a rule attract private investors.

The World Bank, in some twenty years of operation, has provided large amounts of capital for loans and credits in indirect support of the agriculture of the developing countries. However, only in recent years has the Bank become aware that its earlier efforts did not take sufficient account of the numerous human and institutional problems that impeded the effective use of capital resources. Experience has shown that capital can be applied rapidly and effectively only when the other elements of the agrarian structure are adequate—in particular, extensive land reform and efficient agricultural administration (Arnon, 1968). For this reason, many countries are not able to absorb much capital for agricultural development at the present time (Burke-Knapp, 1966).

In traditional agriculture the main inputs are labour and land. Even other inputs, such as seeds tools, and draught animals, are frequently farm-produced.

With modern agriculture, at the farm level capital largely substitutes for labour and land, two-thirds of the total inputs being represented by capital. The chemicals, machines, improved seeds, fuel, etc., required for increasing productivity and efficiency, are purchased from sources outside the farming sector (Heady, 1962). Capital in most developing countries is extremely scarce —hence the need to utilize it in a way that ensures maximum increases in production potential.

The intensive use of capital for replacing labour in agriculture may be uneconomic and anti-social as long as labour is underemployed, the cost of labour is low, and alternative sources of employment are insufficient.

On the other hand, the application of modern technology entails the reduction of the labour employed in agriculture. As general economical development and industrialization of a country proceeds, the labour force available to agriculture decreases. Yet agricultural production can increase substantially notwithstanding an ever-decreasing labour force, provided more capital is invested and the level of knowledge improves (O.E.C.D., 1967*a*).

An illustration is provided by the change of techniques adopted in Mexican agriculture since 1945, which has resulted in the use of less land and labour, and more capital equipment and supplies. Agricultural crop output in 1945–49 was about 60 per cent greater than in 1925–29. In producing this 60 per cent greater output, only 16 per cent more labour and 23 per cent more land were used, but over 300 per cent more capital, equipment and supplies (More, 1955).

Table XXX gives an example of the differences in costs of labour and capital between developed and less-developed countries (Bachman & Christensen, 1967).

TABLE XXX

Comparative Costs of Labour and Fertilizers in the U.S.A. and India

	U.S.	*India*
Daily farm wage rates ($)	6·60	0·29
Cost of ammonium nitrate (cents per kg)	0·30	0·38
Kg of ammonium nitrate equivalent in value to one day of labour	22·0	0·80

The capital requirements for the transformation of agriculture and the growth of industry are formidable. In developing countries, agriculture is frequently the only major potential source of this capital. Success depends on agriculture becoming capable of making capital investments out of its own revenues, for sustained growth and development; this capital then becomes a substitute for labour.

Technological Factors

There will certainly be new land brought into cultivation and, often, irrigation, the acreage depending mainly on the financial and human resources required for its development. Most of the required increase in production will, however, have to come from increasing productivity of the land already under cultivation and, in arid regions, irrigated.

At the lowest level of production—subsistence hand-tool agriculture—less than the equivalent of 300 kg of cereal is produced *per caput*. If productivity is increased by between 300 and 500 kg *per caput*, a surplus is created that can be marketed. 500 kg *per caput* is a turning-point at which it becomes possible to replace, in part, hand-tool labour by draught animals. Over 750 kg *per caput* is a crucial level above which farmers can keep domestic animals, such as pigs and poultry, that compete with humans for cereals and other foods. From this point onwards, progress towards modern farming technology is possible (Clark, 1968).

Many factors can contribute to an increase in crop yields: improved husbandry, improved varieties, changes in crops, the use of commercial fertilizers, better seeds, crop protection, improved irrigation techniques, food technology and mechanization.

IMPROVED HUSBANDRY

Many countries, in the early stages of development, may increase yields by improving certain simple cultural practices, that do not require expensive purchased inputs. More careful tillage, better stands of plants, timely sowing and weeding, can all contribute to improved yields.

There are, however, serious limitations to these possibilities. The climatic pattern of most dry regions, characterized by the sharp demarcation between wet and dry seasons and the relatively short wet season, creates an uneven work-load throughout the year, with high seasonal labour peaks.

Certain desirable agronomic practices, such as timely sowing, may conflict with these peak labour requirements of traditional subsistence crops, and therefore be unfeasible in practice. This may account for slipshod and hurried methods of soil preparation, sowing and weeding.

In rain-fed crops, simple practices for conserving soil moisture may increase yields in the semi-arid regions and may even prevent crop failures. In northern Nigeria, yields of rain-fed cotton, grown in traditional agriculture, are around 165 to 220 kg/ha of cottonseed. Experimental work indicated that, by using improved varieties and fertilization, yields could be increased to a maximum of 660 kg/ha. More recent research has shown that mulching, in addition to improved fertilizers and varieties, increases the percolation rate of water into the soil, and thereby further increases yields considerably. These relatively high levels were practically independent of variations in rainfall; thus in the course of five years with distinctly different rainfall patterns, yields per annum of 2400 kg/ha or more were obtained (Hutchinson & Wilman, 1965). Tie-ridging can also increase cotton yields significantly by conserving and storing water from the early rains. As a result, the cotton plants are better able to survive the dry period which usually occurs after the early rains (Wilde & McLoughlin, 1967).

IMPROVED CROP ROTATIONS

The merits of integrating crop production and animal husbandry have been discussed in Chapter XV, and the spectacular effects on productivity of including in the rotation leguminous crops for pasture or forage in Israel and Australia have been described (pp. 463–4). Other improved cropping practices are also possible as a first step in transforming traditional agriculture, and as a framework in which the response of crops to other factors affecting production is increased.

However, by themselves, improved husbandry practices have limited effects on yield, and cannot bring about a marked expansion of production. Even an improved crop rotation without fertilizer for the legume will be largely ineffective. And yet, in the early stages of development, improvements in husbandry are extremely important, as they build up traditional agriculture to a point where a qualitative change becomes possible. They increase the farmers' ability to buy more costly inputs such as fertilizers and insecticides, and enhance the effects of these.

ANIMAL HUSBANDRY

The wisdom of allocating land and other resources to the production of forage crops, and their conversion into food by animals, has often been questioned for countries with heavy population pressure in relation to the available land and water resources.

Forage plants are considered to be energetically inefficient (Figure 97). It has been estimated that 20 calories of animal feed are required to produce approximately 1 gram of protein. The conversion rate from calories of forage to calories of animal food is approximately 7:1, so that 1 gram of animal protein requires 140 calories of forage crops. If by breeding and improved management techniques it became possible to improve the conversion rate, it would still stand at 6:1 (O.E.C.D., 1967*a*). The relative inefficiency of forage crops is assumed to be the main reason why crops are grown for direct human

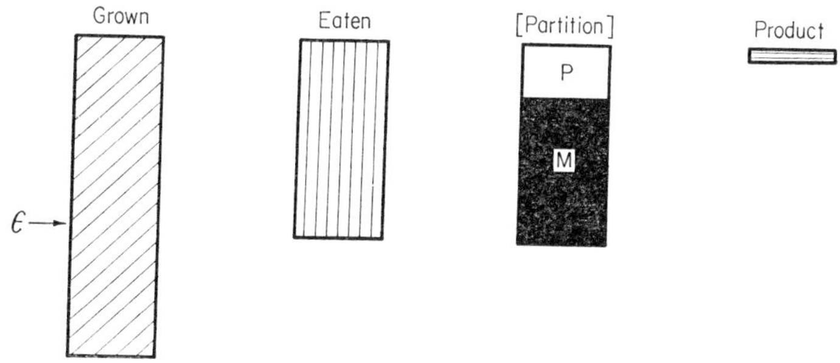

Figure 97. Energy flow in a forage-beef production system. (From Raymond (1968): by courtesy of the American Society of Agronomy.) Only 60 per cent of the 'forage energy' grown (E) is in fact eaten by the animals and about three-quarters of the energy consumed is used for maintenance (P). Of the remaining energy used for production (M), only one-quarter approximately is available in the final product.

consumption in the densely populated dry regions, whether the local agriculture is based on irrigation or on natural rainfall.

Before deciding that forage production is a wasteful process, it should be remembered that the production of animal feed is the only way of utilizing large areas of permanent grass, in particular in low-rainfall areas. Animals also eat crop residues that would otherwise be wasted. The effect on soil fertility of including forage crops in the rotation, and hence the effect of such inclusion on overall productivity of the land, must also be stressed. Intensive livestock production is frequently a prerequisite to raising the standard of living of large sections of the agricultural population.

IMPROVED VARIETIES

The sowing of improved varieties of the crops traditionally grown is, for the individual farmer, probably the cheapest of all the means available to him for increasing production. As cereals are the main source of nutrients for the populations of the developing countries, increased yields will depend to a large extent on the results of efforts to produce and propagate high-yielding adapted varieties of wheat, maize, rice and sorghum.

In Pakistan, the adoption of high-yielding and fertilizer-responsive varieties of wheat and rice, developed in Mexico and the Philipines by the Rockefeller and Ford Foundations, proved to be the main catalyst of changes in agriculture (Bucha, 1968).

CHANGES IN CROPS

As traditional farming evolves, an increasingly large proportion of its income must be earmarked for sale in order to pay for the services and inputs on which increased productivity depends.

Changes in the types of crops grown, in particular changes from food crops for home consumption to cash crops, also make it possible to mobilize labour more intensively. However, such changes are not easily accepted by subsistence farmers, and will certainly not occur spontaneously. They require a number of preconditions—in particular, government initiative in extension, marketing, financing, etc. For example, high-value crops for export can, as a rule, be developed only when local markets are capable of absorbing part of the produce.

As the productivity of the land and of the labour engaged in subsistence production improves, the farmer will show more readiness to devote some of the resources available to him to the production of cash-crops.

FERTILIZERS

In the course of twenty years (1946 to 1966), world fertilizer consumption increased by 543 per cent, while the cultivated acreage increased by less than 20 per cent. In 1966–67, world consumption of all fertilizers totalled nearly 48·3 million tonnes, an increase of 9·6 per cent over 1965–66; world consumption of fertilizers was then actually increasing at a faster rate than world production, which was increasing by 7·9 per cent per annum.

Most of the world's fertilizers are used in the already developed countries. In the early 1960s, Europe, North America and Japan, used nearly 90 per cent of the world's supply of all fertilizers on less than 40 per cent of the world's arable land, whilst Asia, U.S.S.R., Africa and South America, with 66 per cent of the world's arable land, used only 33 per cent of the total supply of fertilizers. However, some of the most striking increases in fertilizer use in recent years are occurring in the developing countries (O.E.C.D., 1967*b*).

Much of the rise in crop yields in industrialized countries can be attributed to increased use of fertilizers. For example, it has been estimated that half of the rise in crop production per unit area in the United States since 1930 is due to additional fertilizers (Durost & Barton, 1960). In India, increased fertilizer use was estimated to be responsible for 4·6 million tons out of the 11·2 million tons increase in food production during the second Five-year Plan (F.A.O., 1967). It is, however, difficult to separate out the effects of other improved practices which, in addition to making their own specific contribution to yield increases, greatly enhance the response of the crop to fertilizer applications.

Over all the regions and countries of the fertilizer programme of the Freedom from Hunger Campaign of F.A.O., nitrogen had positive effects in 97 per cent of the locations, phosphorus in 90 per cent, and potassium in 85 per cent (Couston, 1967). The most successful fertilizer treatment increased yields by an average of 60 per cent and was profitable in over 90 per cent of all locations. If the other production inputs had also been introduced, the overall effect on productivity would have been even greater (Olson, 1968). The value/cost ratio averaged from 4 to 8, indicating that the investment in properly used fertilizers is highly profitable.

In the early stages of development of traditional farming, though plant nutrients are the major limiting factor (if water-supply is adequate), the yield increases due to fertilizers are, of necessity, limited. Marginal productivity of increments in fertilizer is low because other essential production factors that interact with fertilizers are not present.

Limited knowledge of the most effective way of using fertilizers may also reduce their impact. Factors such as optimal rates, the best timing of application, the most suitable nutrient carrier, and combinations of fertilizers in the right proportion, are highly relevant to the overall effects obtained. Faulty or negligent application techniques, such as uneven distribution of the fertilizer, may also reduce their effectiveness.

In semi-arid regions, in particular, inappropriate use of fertilizers may stimulate vegetative growth at the expense of the yield of the economic product.

There must be a sufficient margin between returns and cost, and this is estimated as five to one initially for fertilizers (Nelson, 1967).

All these problems fade as knowledge concerning fertilizer application increases as a result of research, and know-how of the farmers improves through education and extension (Mellor, 1962).

Increased use of fertilizers is already important in the early stages of development, for practical and for psychological reasons: practical, because the returns are quick and little capital is required; the use of fertilizers is probably the most responsive single factor for increased yields per hectare or per unit of water. The psychological reasons are that few inputs have such strikingly visible effects on the crop. The fertilizer itself is a tangible input, so that the relation between cause and effect is most evident. For these reasons, fertilizer application is considered a 'lead' practice which predisposes the farmer to adopt other improved practices. Every improvement in varieties and in management practice that increases yields, also increases fertilizer requirements. Increased amounts are needed to make possible the potential yield increase due to improved practices, and to replace the additional nutrients removed from the soil.

Fertilizers bear such a highly complementary relationship to other yield-increasing practices, that the amounts of fertilizers used per hectare of land have been found to be a reliable index of progress in the adoption of yield-increasing technologies in general. Williams and Couston (1962) report a 0·87 coefficient of correlation between fertilizer consumption and grain yields in forty countries. In Japan, fertilizer is a major expense item, accounting for 30 per cent of all cash expenditures for agricultural production (Christensen &

Stevens, 1962). Generally, the countries with low levels of fertilizer use and yield-value* of crop production are the relatively undeveloped countries; high levels of fertilizer use and yield-value of crop production characterize countries with a modern, highly productive agriculture and an efficient industrial sector (Williams & Couston, 1962).

REQUIREMENTS OF FERTILIZERS

It has been estimated that an average annual growth-rate of 2–6 per cent in total food production in the developing countries by 1980, would require an increase in the consumption of fertilizers amounting to four times the level of consumption of 1965. An annual increase in food production of 3·10 per cent would require a sixfold increase in fertilizer requirements. Thus the consumption of fertilizers must increase at a much faster rate than agricultural production itself (O.E.C.D., 1967*b*). As a result, imports of fertilizers by developing countries have been rising fast in recent years; the need to concentrate on yield per unit area and per unit of water will certainly involve a continuation of rising imports of fertilizers.

Increases in fertilizer use pose considerable problems for countries that are dependent on fertilizer imports, and that often have already big deficits in their balance of payments and a shortage of hard currency.

SEED TECHNOLOGY

The provision of good seeds has two objectives: to ensure that the farmer will sow the variety that is most suitable for his conditions, and, so far as possible, to guarantee that he will get a full and uniform stand of plants.

The individual farmer, even in advanced countries, is usually not well-equipped to maintain the purity and genetic identity of the varieties which he grows. With hybrid varieties, whose importance in a number of widely-grown crops is increasing, it is of course out of the question for the individual farmer to produce his own seed.

In addition to the need for maintaining the genetic identity of improved varieties, a certified source of seed will ensure that the farmer obtains sound seed, with a high percentage of germination and emergence, and free from seed-borne diseases and noxious weeds. A further advantage will accrue if the seed is already treated against seed- and soil-borne diseases and soil insects. A striking example of what centralized control of seed disinfection can achieve is the seed-treatment campaign against bacterial blight of cotton in northern Nigeria, which was extremely effective in controlling the disease at very low cost to the farmer.

* The yield-value index of crop production per hectare is obtained by multiplying the production of each recorded crop by its regional price-weight, aggregating these values, and then dividing by the total number of hectares of crops (Williams & Couston, 1962).

The supply of certified seed in advanced countries is usually assured by private firms or cooperative enterprises. In developing countries this may be an important function to be carried out by an appropriate service in the Ministry of Agriculture.

PEST CONTROL

The direct loss of agricultural products as a result of the combined effects of pests, diseases and weeds, cannot be estimated accurately. For the United States, in spite of advanced farm technology, it is estimated that the annual loss from these causes is at least 30 per cent. The overall loss in developing countries is certainly greater (Ennis *et al.*, 1967).

Pest problems are generally greatly intensified as the level of fertilizers, water, and other production inputs, increases (Ennis *et al.*, 1967). General pest control is another leading practice that is fairly easily adopted by farmers, mainly because the disappearance of insects and other pests after treatment is spectacular and the good results are immediately evident (Central Treaty Organization, 1962). However, the initiative for pest and disease control cannot, as a rule, be left to the individual farmer—especially if he is still lacking in experience of modern technology. Firstly, he may fail to identify the pest or disease and be unaware of the appropriate control methods. Secondly, even if he does effectively control an outbreak in his own field, he will frequently have wasted his efforts and money if his neighbours neglect to do the same.

There is accordingly need for the creation and maintenance of an efficient crop protection service as one of the first essentials of agricultural development. This service will have to forecast insect invasions and disease epidemics whenever possible, advise the individual farmer on appropriate measures, and organize joint large-scale control measures whenever the need is indicated. Locust control is one example.

The control of the desert locust is based on intensive campaigns aiming at the destruction of the locusts in their breeding grounds, and thereby preventing large-scale damage in the cultivated regions bordering on the desert. For certain species of locust, the ratio of the extent of the area in which damage occurs to the outbreak areas is not known, but their migration routes are known, so that control is based on frontier defence—attacking the locusts, wherever possible, outside crop areas (Gunn, 1960).

MECHANIZATION

The tractor is frequently considered to be the symbol of agricultural development, and yet agricultural progress does not necessarily depend on immediate mechanization. True, agriculture based on abundant cheap labour doing back-breaking work perpetuates a low standard of living. Premature mechanization, however, is not only socially wrong but can cause considerable damage when practised indiscriminately. Agricultural machinery requires relatively large capital investments, and the main justification for its use in developed countries is to save labour. In most underdeveloped countries, capital is the scarce resource and labour is in ample supply, at least in their early stages of development.

At first, much can be done to increase the productivity of labour by the use or improvement of simple, traditional tools and implements and the partial replacement of manual labour by animal draught. For some regions this alone may constitute an important technological improvement.

Replacement of wooden by steel ploughs, of steel-rimmed by rubber-tyred wheels, and of sickles by scythes, for example, can be major technical improvements. The development of simple drills in order to change from broadcast to row planting in certain crops, has manifest advantages. In India, drill sowing of Ragi (*Eleusine coracana*) increased yields by approximately 20 per cent as compared with broadcast sowing (Patil, 1963).

As the economy invests more and more in industrial growth, which draws off farm labour, the proportion of farm workers to total population decreases, and the need to increase the productivity of the individual farm worker becomes progressively more acute. As a result, the need and justification for mechanization increase. There is, however, no clear-cut demarcation between an era of hand labour and animal traction, and an era of mechanization.

In certain cases, mechanization at an early stage of development may be fully justified as the only means of rapidly expanding the area under cultivation. In particular, mechanization may be an essential complement to irrigation for developing large tracts of arid land.

An example of large-scale irrigation development based on mechanization is the rapid expansion of cotton and wheat production in northern Mexico since the Second World War, which has accounted for a substantial part of the phenomenal growth of agricultural production of Mexico (McPherson & Johnston, 1967).

At first sight, the statement: 'It is not necessarily inefficient for the cultivator of India to plow his field half a dozen times with a stick, while a moldboard plow would do it in one operation and save labour, because the farmer would have no alternative use for his labour' (Heady, 1962) may appear quite logical. This, however, does not always hold true. High yields may depend on the timeliness with which certain operations are carried out. In particular, sowing dates, enabling the most effective use to be made of rainfall, may be critical for certain crops in certain regions. Land preparation for sowing in time may be possible on large areas only if tillage can be mechanized. There is, however, a danger inherent in partial mechanization of agriculture in that labour bottlenecks may be created. For example, mechanical tillage may encourage the sowing of larger areas of cotton than can be effectively kept weed-free, when weeding is done by hand.

A quality of work may be achieved by mechanical tillage that is not possible with animal draught, and certainly not with manual labour. For example, in the Near East, ploughing with the traditional plough only scratches the soil surface. As a result, the Wheat-fly (*Syringopais temperatella*) causes enormous damage and reduces yields considerably. Mechanical tillage, which turns the soil to a somewhat greater depth, is fully effective in destroying the larvae of the pest.

Extensive use of agricultural machinery may therefore be justified even at the labour-intensive stage, when it makes possible large land expansion for field operations that cannot be performed manually or with traditional implements, and where it produces large increases in yields.

Some Problems of Mechanization

In developing countries, a number of factors operate which tend to make mechanization costs excessive. These factors include the need for foreign exchange, difficulties with maintenance and repair, the need for thorough land-clearance, and inefficient operation and utilization (Wilde & McLoughlin, 1967). The economic justification for mechanization will depend on the efficiency with which tractors and machinery can be operated, and the extent to which output can be increased by increasing yield and/or by extending cultivated areas. The efficiency of operation can be improved by providing contract services to farmers; by pooling individually owned land into larger holdings, by organized settlement schemes (in particular irrigation schemes), and by having large state or collective farms.

Experience has shown that groups of farms for maintaining tractor services have advantages over government hire-services. The latter are notoriously inefficient, presumably because of the lack of personal incentive, whereas farm-groups can consolidate stretches for cultivation, and therefore less travelling time to and from work is involved and fuller use of tractors throughout the year is possible (Mettrick, 1968).

Whether the farmers should maintain the services cooperatively, or should encourage private *entrepreneurs* to provide the service, depends very much on local conditions—mainly on the possibility of cooperation and the technical know-how available. Private tractor contracting can be profitable to the *entrepreneur* and advantageous to the farmers providing there is sufficient demand, sufficient experience in operating the machinery, and the number of tractors justifies the existence of adequate repair and servicing facilities.

Increased Output

The introduction of mechanized equipment cannot be justified economically unless production is intensified to pay for increased costs. Expenditure cannot become modern so long as income remains primitive! Therefore, mechanization must be accompanied by improved cultural practices.

Mechanization of Small Farms

Whilst economies of scale are not critical for many improved cultural practices, such as the use of better varieties, fertilizers, weeding, etc., this does not generally hold true for machinery.

In addition to the various organizational forms discussed above, which by pooling mechanized equipment make it economically available to relatively small farm units, much progress has been made in recent years in adapting modern farm machinery to the needs of the small family farm. Japan pioneered in this field, but West European countries have also made important contributions.

COMPLEMENTARY NATURE OF TECHNOLOGICAL FACTORS

Agricultural progress cannot, as a rule, be piecemeal. There is little point in introducing improved varieties if they are unable to develop their potentialities owing to lack of nutrients; there is no justification in adopting practices aimed at producing what might become a bumper crop, if disease prevention and pest control are not carried out and a large part of the crop is thereby lost. Living standards will hardly rise if weed control is ineffective because it has to be carried out by back-breaking manual labour.

For these reasons, single-practice programmes, such as introducing irrigation, applying fertilizers, using good seeds, controlling pests, etc., usually give poor results (Kellogg, 1962).

An illustrative example of the inefficiency of fertilizers, when used on varieties that are not adapted to high levels of fertility, is provided by Wellhausen (1967). In Mexico, native varieties of wheat were susceptible to black stem-rust. When fertilizers were applied to these varieties, the profuse tillering and heavy growth of foliage created a favourable microclimate for the germination of rust spores. As a result, heavy infections and epidemic spread of rust caused complete crop failures. By contrast, in unfertilized stands, the sun penetrated the thin stands, causing dew to dry up early in the morning; rust germination under these conditions was limited, and a build-up of the disease to epidemic proportions was thereby avoided. Yields under these conditions were low—around 750 kg/ha—but fairly constant. Conversely, an example of the need for fertilizer application before improved varieties can give highly increased yields is described by Kellogg (1962). In fertilizer trials on maize in India, the increases obtained over unfertilized local varieties were of 1290 kg/ha from hybrid seed alone, 1110 kg/ha from the application of fertilizer alone, and 3480 kg/ha from combination of the two.

Another aspect that leads one to argue against single practices, even when effective, is that at best they provide only small increases in yield over traditional farming and therefore have very little impact. Increases of 10 or even 20 per cent over an average grain yield of 500–600 kg/ha that is usual for traditional farming are smaller than the normal seasonal variations in yield due to climate, and will therefore not even be attributed by the farmer to the improved practice. It is also doubtful whether the increase is sufficient to justify the cost of the input required. In the early stages of development, it is almost essential that spectacular increases of at least double or triple the normal yield be obtained. This is usually not possible with a single practice, but may be reasonably expected with an appropriate 'package' (Figure 98).

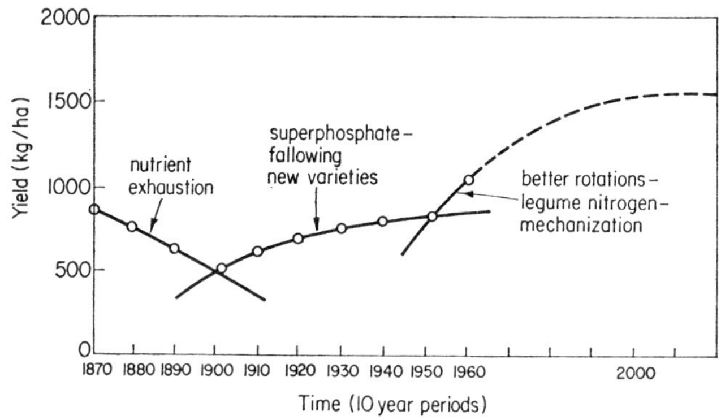

Figure 98. Mean decennial yields of wheat in Australia during the past hundred years. (After Donald, 1964.)

Certain techniques, when applied in combination, can give spectacular results in a very short time. A combination of improved variety, appropriate fertilization, adjusted plant population, efficient weed-control, and plant protection, can give increases in yield that range up to several hundred per cent. The expenditure required from the farmer for inputs is apt to be very low in relation to the additional yield produced—provided only that the prices charged for fertilizers and pesticides are not inflated as a result of deliberate policy, dependence on unscrupulous middle-men, unrealistic distribution costs, or other man-made factors that disrupt the cost-ratio of crop and input factors.

Infrastructure and Services

Facilities will be needed to produce, distribute, and supervise the use of, the materials required for improved agricultural production. Good seed of improved varieties must be grown, cleaned, and distributed, by special organizations set up for the purpose. Government control of the quality of fertilizers, feedstuffs and, in particular, pesticides and fungicides, is essential if the farmer is not to be cheated and the efforts of extension to introduce improvements are to be made ineffective.

New markets, marketing techniques, and a transportation system, will have to be developed. An efficient marketing system refers not only to the disposal of farm products, but must ensure that the farmer is able to purchase the inputs he requires when and where he needs them.

In developing countries, fluctuations in yields have had disastrous consequences in the past; these could be overcome, at least partially, by adequate transport and storage facilities. The availability of these facilities is not only an essential precondition to change in agriculture, but also provides a stimulus to farmers to effect such change.

In Latin America, storage service is provided to the farmer by commercial banks which own an extensive network of warehouses that are located in the principal producing and consuming centres of the country. These banks finance agricultural activities and supply credit for the purchase of agricultural inputs. Grain and cotton are stored by the banks, and secured loans are given (Salazar, 1968).

Public works such as access roads, drainage facilities and flood control, are incidentally an effective outlet for the use of underemployed agricultural workers.

Human Factors

Successful development of agriculture is frequently dependent on an understanding of the society within which it is to take place, and in particular of this society's system of values and constraints (Wilde & McLoughlin, 1967).

THE FARMER

In advanced countries the farmer is a man of business, with a fairly high degree of business acumen and a positive attitude to agricultural science. He is on the lookout for improvements which will bring him higher profits. He has his eye on the market, and he has the necessary skill to adjust his farm business to changing conditions.

In contrast, in the undeveloped and developing countries, agriculture is carried on by farmers who are very small producers and who produce more for subsistence than for the market. They have a low standard of education and a restricted mental horizon. The *per caput* annual income of these farmers is commonly between U.S. $80 and $150. In addition to these handicaps, the soil is usually exhausted and the land resources of the individual farmers are limited. Crop failures do not simply mean a reduced income, but may result in outright starvation. It is small wonder, then, that these farmers hesitate to adopt new and untried practices (Anderson, 1958).

As the farmers of advanced countries differ from the farmers of undeveloped and developing countries, so the agricultural extension problems also differ. In an advanced country the problem is how to communicate effectively the results of research to a farmer who is mentally prepared to accept new practices that will give him higher returns. In contrast, in the undeveloped and developing countries, one must not only solve the problem of how to communicate information but also that of how to motivate the farm operator to accept technological change and use improved practices to his advantage.

Preliminary data from India show that many farmers are unable to absorb the knowledge and the investment that are readily made available to them. The limiting factors appear to be the ability and skill in handling investment (Bose, 1961).

In a survey which included 200 cultivators in West Pakistan, Sturt (1965) found that, in spite of the resistance to change that is often attributed to farmers in developing countries, over a third of the farmers made changes that were recommended by the extension service: in cropping patterns (34 per cent), in varieties planted (29 per cent) in use of fertilizer (20 per cent), in kinds of farm implements used (2 per cent), and in the use of livestock (15 per cent). These were all changes that were considered to be within the means and ability of the average farmer. Most of the changes that took place required little capital expenditure, but some changes were not made even if they required no cash expenditure, when the farmers were not motivated to make changes. The primary motivation of the cultivators for making changes was found to be to have more food to eat. Little or no interest was shown in better housing, household ornaments, or entertainment, or in investing in land improvements or equipment. These became factors of motivation only after the primary need for sufficient food was satisfied.

In all development programmes the human factor must therefore be given full consideration.

SOCIAL FACTORS

Of equal importance to the human element in influencing the adoption of new farm practices is the social framework, which must also be taken into consideration in planning development programmes.

There are certain communities and groups that do not require much persuasion to adopt new techniques, and in which the innovator enjoys respect and prestige, while there are others, and they probably form the majority, in which tradition and established customs are conducive to resistance to change; in spite of all manner of persuasions and inducements, the latter communities seem inert and indifferent. They ignore the opportunities to increase their production and income even if opportunities are available.

The Indian sociologist, Nair (1961), found that the tacit assumption made in planning, 'that given equal opportunity, financial incentive, and resources, all communities will respond, and respond similarly, in their productive attitude to work, aspirations, and habits', is not valid. She gives numerous examples to show that the social factor ultimately determines whether the resources are properly used, or not, for increasing production. 'A community's attitude to work can be a more decisive determinant for raising productivity in Indian agriculture than material resources, or for that matter even technology.' Nair continues, '. . . in planning for the farming community it is apparent that there cannot be any economics in isolation from sociology. There are many causal relationships and connections between purely economic factors and social and cultural conditions which cann ot be ignored or excluded from economic analysis and planning.' Her conclusion is: 'Development will not become a self-generating process with its own momentum unless the value system of the community, and the social structure containing it, are first altered and adjusted to be in harmony with the socio-economic objectives of planning.'

This brings us to the question of how the general problem should be tackled, in order to bring about changes in the habits and mental horizons of the rural people, which also predicate changes in the socio-cultural systems. An accepted premise is that it is necessary to adapt to the social framework of the farmers without forcing changes upon them, and that any attempt to replace traditional leaders, or to approach the farmers directly rather than through the traditional leaders, is doomed to failure (Admoni, 1963).

There is no doubt that the readiness of farmers to adopt new ideas and practices is influenced by the nature of the leadership and the degree of control it exercises over their community. Reliance on the 'traditional leaders' may be justified as a short-term, tactical approach; but normally a strengthening of their traditional stranglehold—economic, moral, and social—on their people will be self-defeating in that it will reinforce the conservative, anti-change tendencies. These last are very strong in any case, and will stifle the initiative of the more progressively minded members of the clan and village. Bose (1963) has a case when he states 'that it is necessary to change the entire cultural pattern of an underdeveloped society before we can hope for permanent technological changes.' Dumont (1962) also stresses the evil influence of the gerontocracy of the villages—in this case in Africa. By making the strongest economic power the preserve of the chiefs with their large families, many primitive societies entrust the levers of progress to the oldest, who are often the least receptive to modern techniques.

In most undeveloped and developing countries, the adults are emotionally committed to primitive agricultural techniques and to tribal customs to a degree which seriously hampers or even blocks progress and offers a serious danger of 'counter education' (Chatelain, 1963).

Reliance on, and strengthening of, the existing social patterns will also, in most cases, mean a perpetuation of the degraded role played by woman in primitive societies. In most of the undeveloped and developing countries, girls and women are an essential part of the farm labour force. Their vocational training is, however, usually neglected, as a result of established social patterns and prejudices (Chatelain, 1963). The status of women is far from identical in the different countries, and, to quote the recommendation of F.A.O. in this connection: 'Traditional ways of thought must be taken into consideration and can only be modified with caution.' But modified they must be, for sooner or later the day will come when the peasant woman will become conscious of her slavery and of her exhausting and unattractive life; then she will flee to the city and the men will follow suit. In some European countries the villages stand empty because they failed to do the only thing that was capable of keeping the women on the land, namely, provide them with vocational training. The lesson is worth studying.

The need for extending advisory work to women is a prerequisite for achieving the ideal of 'dual responsibility for the homestead', or husband–wife teams as a factor for progress.

SEDENTARIZATION

A social change more extreme than any other, and specific to the arid regions, is the sedentarization of nomads as a prerequisite to irrigation development of deserts. Millions of nomads have been settled in recent years in the Asian parts of the Soviet Union and in Iran, Iraq, and Egypt (White, 1962), but much must still be learnt on how to transform nomads into sedentary farmers—a way of life not only alien to them, but also deeply despised by them.

INCENTIVES

Adopting new practices, with all the uncertainty and risks that this involves, can be critical for farmers who produce barely enough for survival. In particular, the semi-arid regions are areas of high risk, due to erratic rainfall.

The rate at which farmers in a traditional framework will adopt a new practice will depend on its profitability, with due allowance for risk and uncertainty (Schultz, 1964).

Incentives must therefore aim at improving profitability and reducing risks, at least for a more or less protracted transition period. Experience in developing countries has shown that increases in agricultural productivity can be obtained in a relatively short period if incentives that are favourable to the farmer are provided (Mason, 1966). The principal incentives are prices, production credit, tax structure and policy, and land-tenure arrangements.

Prices

Schultz (1966) considers a system of efficient prices as one of the basic economic requirements for the development of agriculture. He distinguishes three sets of prices: of farm products, of agricultural inputs, and of consumer goods and services that farmers have to buy.

Prices of Farm Products. In subsistence farming, only a small fraction of farm production occasionally reaches the market, so that the prices of farm commodities have almost no impact on crop production. However, prices can be used to change cropping patterns, first by encouraging a shift from low to high yields, and subsequently from low-value to high-value crops. The unfortunate tendency in most developing countries is to base pricing more on the need for low food costs in cities than on the production incentives needed by farmers, thereby impairing the effectiveness of agriculture in contributing to the overall development of the economy (U.S. President's Science Advisory Committee, 1967).

Prices of Agricultural Inputs. For modernizing production, farmers must buy fertilizers, chemicals for the control of pests, diseases and weeds, and tools, equipment, machines and fuel. In most undeveloped and developing countries the prices of these inputs are excessively high. A balance must be maintained between the prices the farmer obtains for his products and the cost of the inputs he is being encouraged to buy and make.

Prices of Consumer Goods and Services. The prices of the consumer goods and services that farmers buy are the key to the purchasing power of the net income earned by the farmers, hence their importance (Schultz, 1966). In most developing countries these prices have been rising relative to the prices that the farmers obtain for their products.

The *stability* of the three sets of prices described above is also of considerable importance.

Credit

The high costs of credit is one of the main reasons why subsistence farmers do not adopt more rapidly improved practices which require cash outlays.

In a survey carried out in the CENTO countries (Central Treaty Organization, 1962), it was found that the main reason given by farmers for not using fertilizers was money. They had no money of their own to invest, and if they borrowed money they had to pay often double the amount after the crop was harvested. Interest rates range from 60 to 250 per cent!

Experience in developing countries has shown the absolute necessity of providing short-term credit at reasonable cost to farmers buying inputs such as fertilizers, insecticides, etc., even when the prices of the inputs are subsidized. The credit must be available at the time when the inputs are purchased, and must be extended until the crop is marketed.

An Agricultural Development Bank and cooperative credit societies are useful institutions for supplying credit to farmers. In view of its importance, agricultural credit is the second largest category, after irrigation, of World Bank lending to the developing countries. Experience has shown that lending to credit institutions, to be effective, must be accompanied by extensive technical assistance in order to develop sound management and operating procedures. The effective use of credit is also dependent on sound technical advice to farmers in guiding their operations (Burke-Knapp, 1966). Centralized agricultural credit is usually provided in developing countries by agricultural banks, generally sponsored by governments, which provide most of the capital. In some cases, the credit is linked with the adoption of certain improved practices or farm development, in which case the bank operates in close collaboration with the extension service.

For the decentralization of agricultural credit, the most effective instrument is the cooperative, provided the farmer can be educated to accept and support this type of institution. Cooperative purchase of inputs will reduce their costs, while cooperative sale of farm produce will protect the farmer from pressures to sell at low prices—especially if his products are perishable. Quality control and improved packing made possible by cooperative marketing will also have a favourable effect on his income. When such cooperatives also handle credit, they have satisfactory guarantees that the credit will be repaid.

LAND REFORM

Land reform may be concerned with problems of land tenure and ownership or with the excessive fragmentation and dispersal of holdings. Both aspects may be major obstacles to agricultural development.

Most undeveloped and developing countries have problems of land tenure: traditional land ownership and tenure patterns (such as absentee landlords); and short-term leases and share-cropping on unfavourable terms to the farmer, hamper agricultural development.

It has become increasingly clear that the adoption of improved techniques is not possible unless the farmer operates land which he owns or holds securely, and from which he obtains an equitable share of the produce. If he is a tenant who retains only a small proportion of his yield, and all investments conducive to higher yields have to be made at his own expense, he has no incentive to adopt improved practices. Land reform is therefore an essential precondition to agricultural progress. It is not easy to achieve, because of the political forces which frequently oppose it, and because of the enormous technical planning involved. In Mexico, land reform was initiated following the revolution of 1910, and has not yet been completed.

Land reform may have different objectives. It may aim at changing tenant farmers into owners of their land without basically affecting the pattern of land distribution, as in India. Alternatively, it may aim at breaking up existing large estates into small viable farm units as in the United Arab Republic. The basic policy may be to establish large collective farms as in the communist countries, or large state farms as in certain newly independent countries in Africa. Where land reforms have been enacted, they have frequently created new problems, replacing those that they succeeded in solving. For example, where ownership passed from large landowners to potential farmers who did not have the necessary managerial and farming skills or the funds needed to develop their newly acquired land, overall productivity was reduced. In many cases, land reform resulted in frustrations and the return to larger landholdings (Sola, 1968).

The problem of land reform is probably most acute when irrigation is being introduced in semi-arid regions, where rain-fed agriculture is traditionally practised and pressure on the land is extreme. The existing land-ownership pattern may then be the greatest obstacle to irrigation development. If the land is divided into very small holdings, and the parcels of land of the individual farmer are widely separated, no modern irrigation farming is possible before these holdings are consolidated. On the other hand, the promise of development made possible by irrigation may be the lever to achieve the necessary land reform.

By contrast, advantages of the desert regions that are to be developed by irrigation, are that the present population is sparse, land as distinct from water rights has little value, and problems of tenure are not critical. In areas of limited rainfall, land reform is almost meaningless unless it also gives title to the use of the available irrigation water.

Size of Farms

A problem closely related to that of land tenure is the size of the individual farm. In most undeveloped and developing countries, the land is either excessively split up into very small producing units that are difficult to farm economically, or it is concentrated in the hands of a few large landowners.

Evidence from many countries indicates that economic progress in agriculture is possible under a great variety of farm-size conditions (Bachman &

Christensen, 1967), and that the size of farms *per se* is neither a prerequisite nor an obstacle to agricultural development.

Under the prevailing social and economic conditions of many of the developing countries, large holdings may be a factor retarding modernization and intensification. If large landowners lack initiative, ability, or capital, to develop their land, large units are also a handicap to progress.

Technical know-how and managerial ability are very scarce in developing countries, and this is one reason why small farms may be more efficient than large ones. Small farms also favour labour-intensive forms of production. Farm size tends to grow with the decline in labour supply relative to available land and capital. The application of improved technology makes the use of larger units of machinery and equipment profitable (Bachman & Christensen, 1967). On the average, each person in the farm population of the developed countries will have about twenty times as much land as those in the undeveloped and developing countries. Compared with the Far East alone, it would be fifty times as much (O.E.C.D., 1967*a*). Four hundred hectares are considered the minimum area required for economic operation of irrigated farms in California (Gregor, 1959). The investment in pumping equipment, soil-levelling and conditioning, and recurrent investment in improved farm equipment, make large units a precondition for economic operation.

Farm size has social, political, and security, aspects in addition to technical and economical considerations. In many developing countries, governments are interested in setting up large state farms, with the most modern equipment, for reasons of prestige or political conviction, or as a short-cut from subsistence agriculture to modern farming. On the other hand, very strong emotional and social factors militate against the abolition of small farms, and the problem therefore arises of making small farms viable in the age of mechanization. Many improved cultural practices that can transform traditional agriculture—such as better-yielding varieties, use of fertilizers, improved water management, etc.—do not involve economies of scale and can be used equally rationally on small and large farms. For the relatively few functions that do require large-scale investment and organization, the problems can be overcome by the provision of cooperative specialized services and/or government or private contractual services (*see* Mechanization, pp. 569 *et seq.*).

The minimum size of a farm should be dictated by a balance between productivity and efficiency on the one hand, and an acceptable standard of living on the other. There is much evidence to show that excessive emphasis on size, and the exclusive use of large units, also has drawbacks and dangers—particularly if such units are centrally directed. Rigid bureaucratic control, lack of adaptability, and rigid organizational forms, generally characterize this type of farming unit. By contrast, a mixed structure of units of various sizes appears to be the most promising in the developmental stages of agricultural expansion (Raup, 1967). Large farms and small farms are not mutually exclusive, and can well exist within the same region, and even in proximity to each other.

In Mexico, the land reform led to the adoption and establishment of two different categories of landholding: 'ejdos' or common lands, and small private holdings. It was thought that the two types of land-use together would ensure a better utilization of the land (Osoyo, 1968). A formula which has many merits is that of a large state- or corporation-owned farm, surrounded by relatively small producing units. Among farmers who still require a long period of education towards cooperation, a neighbouring large farm can play an important role. The fully equipped and mechanized large farm can provide contractual services (machinery, pest control, etc.) to the smaller units and also facilities for processing, grading, packing, and marketing, their produce. More important still, the large farm should be the first to adopt improved cultural practices, and can then serve as a guide and demonstration area as well as a help in influencing the small farmer to accept the same methods.

Another approach that has been proposed is to consolidate a large number of small holdings by forming large agricultural corporations in which the landowners are represented by stocks (Ozal, 1963).

Supporting Services

Agricultural productivity has increased most in countries that have invested sums in agricultural research and the education of the farmer (Christensen & Stevens, 1962).

The remarkable success of farmers in Japan, in increasing productivity per unit area and per farm worker, is ascribed by Schultz (1964) first and foremost to the public investments in agricultural research and in schooling of the farming community. The ability of the farmers to adopt new practices was thereby enhanced, even when the latter were complex and difficult (ibid.).

AGRICULTURAL RESEARCH

Role of Agricultural Research

The mission of agricultural research has been defined by Aldrich (1966) as:

1. To apply all possible sources of scientific discovery to the solution of the technical and practical problems of agriculture;
2. To engage in basic research where the lack of fundamental knowledge may impede progress; and
3. To solve the specific problems with which agriculture is faced.

The relative emphasis placed on each of these objectives depends mainly on the stage of development of agriculture in each individual country and on the country's economic requirements.

Increased productivity is usually the first demand to be made on agricultural research in a developing country, during a period when agricultural methods are still primitive, the population is increasing rapidly, and demand elasticities are still at levels which justify and allow increased agricultural output (Heady, 1961).

For instance in Israel, immediately after independence, the first objective of agricultural policy was to make the country self-supporting in food production. The research programme was therefore geared mainly to the problem of increasing yields of the principal food-crops, and of increasing the productivity of animal husbandry in supplying milk and other dairy products, meat, and eggs. As the result of an intensive research programme, yields were increased three- to tenfold and more: yields of grain under dryland conditions rose from 600 to 5000 kg and more per ha, and, under irrigation, from 3000 to over 10 000 kg. The previous primitive breeds of cattle and sheep were replaced by locally developed or selected breeds; milk production increased from 800–1 500 kg to 6000 kg/cow/year, placing Israel in the top rank of dairy producers in the world.

The financial value of Mexico's increased wheat and maize production that is a direct result of research, is the equivalent of an annual interest of 400 per cent on the total money spent on the research which was required during the period from 1943 to 1962. The return on the wheat programme alone was at least 800 per cent (Ardito-Barletta, 1962).

Imported Research or 'Own' Research

A tremendous amount of agricultural research work is being carried out in all parts of the world (Arnon, 1968). The need for research in developing countries can therefore be legitimately questioned. In a country which is struggling to establish a sound economy, which lacks trained personnel, and in which the agricultural population is still primitive, research may appear to be a luxury which can be ill afforded. It may well be asked whether elementary logic does not compel a developing country to concentrate on disseminating and applying knowledge already available in other countries, and which can be found in the numerous textbooks, bulletins, and journals, that are devoted to agricultural progress—knowledge which is increasing daily at such a pace that it is almost impossible to keep abreast of developments. In other words, should the available limited personnel with adequate training be devoted to extension instead of research?

This apparent logic is, however, a dangerous fallacy. As a policy it would be self-defeating. A direct transfer of technology, developed largely in the temperate zones for temperate crops, to the subtropical and tropical regions in which most of the undeveloped and developing countries are situated, is not feasible, and has even been the cause of considerable damage in certain cases. Basic principles can be established anywhere in the world, but their application to a specific environment requires local research teams working under local conditions.

Certain practical results of applied research can, nevertheless, be transferred from humid temperate zones to the dry regions without the differences in climate being an insurmountable obstacle—provided the results are properly adapted to the new environment.

Let us consider plant breeding as an example. Thousands of new, improved varieties of the principal crops are developed and released yearly in the world. One cannot, however, simply go shopping for improved crop varieties—an Introduction Service must be established and the promising introductions tested for adaptability, disease-resistance, and technological suitability, according to modern scientific procedures. Even such a simple 'application' of scientific achievements developed elsewhere is therefore far from simple. In addition, introduction of new varieties of crops, if carried out without the necessary scientific safeguards, may cause untold damage by introducing pests or diseases that were previously non-existent in the country.

Even the best varieties bred are practically useless unless placed under appropriate ecological conditions of nutrient and water requirements, day-length, crop sequence, and weed, pest, and disease control, while many other factors may need elucidation and due consideration. These cannot be determined by reading textbooks published in foreign countries, but frequently require appropriate research under local conditions.

However good imported varieties may be, they will in time generally show some weakness which seriously limits their usefulness. This can usually be overcome by an appropriate breeding programme, carried out by local scientists working under local conditions.

Sooner or later, emergency situations will develop: an unexpected invasion of insects, an unexplained epidemic of disease, or one of the other numerous emergency situations which normally appear in progressive agriculture. These cannot be tackled simply by consulting textbooks or journals, but require teams of research workers who are experienced in their respective professions and fully acquainted with local conditions.

It would be a mistake to conclude from the foregoing that the duties of research workers in a developing economy are limited to 'trouble shooting'. The research worker must pioneer new developments, think ahead of farmers and planners, and spearhead agricultural progress. Research workers have to be constantly alert to the opportunities provided by new information resulting from basic research carried out elsewhere, and by new technological advances. It would, of course, amount to criminal neglect not to take advantage of every step forward that is made in agricultural knowledge in any part of the world;

but even the evaluation of promising leads in the scientific literature requires people trained for research, who can judge what is desirable to test and adapt to local conditions, and who are able to carry out this work. As a result of increasing specialization, usually only workers engaged in research in a given field are able to keep abreast of developments, and to evaluate correctly the potential importance and applicability, to their own specific conditions of the results of basic research carried out elsewhere.

Thus an essential element in the transformation of traditional agriculture is an appropriate agricultural research programme and the necessary infrastructure needed to carry it out. Research on production factors such as those mentioned above, that may have a rapid and considerable impact on yields, does not require expensive or sophisticated basic research in the early stages. It is usually well within the capabilities of developing countries, provided the minimum research infrastructure is available. This includes a network of experimental stations with testing facilities, and a competent staff which can carry research to the point of final application.

Research findings do not automatically transform themselves into agricultural practices; even in the best of cases there is usually a considerable time-lag between the development of a new technique and its application. Moreover, frequently, considerable resistance has to be overcome before a new practice is accepted by farmers.

EXTENSION

It is a truism to state that close links between research and the farming community are essential if research findings are to contribute to agricultural progress. For this, a strong extension service is needed. The extension worker is the first to try out, under normal farming practice, the new ideas resulting from research, to investigate their economic justification, and to adapt them to the realistic necessities of agricultural production; this he does in relation to different types of farming and to varying environmental conditions. Reciprocally he brings the problems of the farmer to the research worker, and may even contribute to their solution.

It is generally agreed that the objectives of extension work in agriculture are not only to bring about an improvement in farming through the application of science and technology, but also to promote the social, cultural, recreational, intellectual, and spiritual, life of the rural people. There is also general agreement, especially in developing countries, that whilst this advisory service has to adapt itself to the existing social framework of the farmers, it must also be active in promoting change towards a more progressive social framework as a prerequisite for technological change.

Penders (1956) correctly stresses the need to ensure participation of the farmers themselves in the extension work, which includes formulating policies and requirements, planning the programme, and carrying it out. The contribution of well-chosen local leaders can be considerable. This is possible not only in developed countries, but can, with suitable encouragement, be adopted also in developing countries.

LIAISON PROBLEMS BETWEEN RESEARCH AND EXTENSION

While the need for close cooperation between research and extension appears to be axiomatic, its achievement in most countries is the exception rather than the rule. Extension work and research are usually organized in different services, and there exists a general tendency towards separation of interests and even alienation between the two. This may easily lead to a situation in which the extension worker, instead of serving as a link between research and the farmer, becomes, on the contrary, an obstacle between them. Lack of contacts between research and extension workers easily leads to conflicting conceptions and opposing instructions to farmers, to the detriment and confusion of these last.

Practically everybody is agreed that such a situation is incompatible with the needs of the agricultural community and must be avoided—nowhere more vigorously than in developing countries. Lip service to this idea is, however, not sufficient; nor can the problems be solved only by contacts at the individual level.

The two services of research and extension, though they have a common goal, have basic differences in their methods of work and objectives, and therefore require distinct and separate administrative machinery in order to ensure their efficient functioning. This can, however, strengthen the centrifugal tendencies, which can be counteracted by the following means:

(1) Setting up a *common organizational framework* for the two administrative units, thereby assuring maximum co-operation between the two and enabling a fair amount of movement of personnel between research and extension—what the French aptly call 'ambivalence du personnel'.

The policy and programme of the two services should be formulated jointly, whilst the implementation of policy remains the separate responsibility of each service at least so far as its own aspect is concerned.

(2) Reducing the differences in the 'atmosphere' in which research and extension work are carried out—differences which make them drift apart even further—by:

(a) Having the leading extension workers participate in establishing the research programme, determining priorities, etc.

(b) Co-opting into the research programme extension workers who have suitable qualifications. This especially applies to regional work and can also serve for demonstration purposes.

(c) Locating regional extension services, wherever possible, at the regional experimental station.

(3) Organizing, in common, certain units which serve both services—such as libraries, central laboratories for the diagnosis of fertilizer requirements, and others for disease identification, etc.

Strategy of Development

The crucial role of agricultural development in overall economic growth is widely recognized, but there is still little agreement as to the most appropriate strategy for modernizing agriculture in developing countries (Johnston, 1964).

The complementary nature of agricultural techniques, and the futility of attempting to introduce them one at a time, has been discussed above. In addition, the adoption of technological improvements is in its entirety dependent on many social, economic, and political, factors.

Relatively simple changes 'may lead to events which wrench social, economic and political relationships lying deep in the structure of society. For example, they may project millions of people out of subsistence living into a monetized economy for the first time' (Krebs, 1964). A single innovation may require a fundamental transformation. For instance, if animal- or tractor-drawn equipment is introduced to overcome a labour bottleneck, a whole train of innovations may become necessary. The use of implements may make necessary the complete clearing of land, removing all stones and tree-stumps. This makes the traditional farming methods uneconomical; other methods, such as crop rotation, manuring, and fertilization, become essential. If livestock are maintained for traction and manure production, a forage crop has to be introduced into the crop rotation, etc. (Wilde & McLoughlin, 1967).

Because of these highly important interrelationships among various factors, the effectiveness of a development programme will depend on the ability to find an appropriate combination of simultaneous changes.

It is a characteristic of undeveloped and developing countries 'that nearly everything seems to need doing at once' (Kellogg, 1962). But equally characteristic are the limitations in human resources and capital, making it impossible to tackle everything at the same time.

Satisfactory agricultural progress cannot be piecemeal; it must proceed in carefully planned stages that take into account potential resources as well as limitations due to environment, economic factors, and human resources. Therefore, agricultural development first and foremost requires a determination of priorities. Certain factors must be applied simultaneously, in order to interact concurrently if they are to have an impact on development; others have to be added in sequence as the need arises. The priorities of action are dependent on the stages of agricultural growth. One can roughly distinguish three stages, as follows:

FIRST STAGE—IMPROVING TRADITIONAL AGRICULTURE

First, the preconditions for development have to be provided; of these, changes in the land-tenure system should receive first priority. If the area to be developed is arid, water resources must be developed and the whole infrastructure providing water to the individual farm must be established. Improved transportation facilities may make it possible to bring more land under cultivation and irrigation.

At the farm level itself, priority should be given to methods that do not require heavy cash investments, but that give large returns by increasing the efficiency and productivity of the existing labour-intensive agriculture. Efforts will aim at introducing high-yielding and disease-resistant varieties, at providing high-quality seed, at improving seed-bed preparation and sowing techniques, at proposing more rational crop sequences and diversification of crops, at providing improved tools and equipment, and at recommending adequate levels of fertilization.

These inputs and potentialities have the following characteristics: (a) they come from outside traditional farming; (b) they depend on government initiative and help; and (c) the complementary nature of the various factors requires the simultaneous occurrence of a number of developments (Mellor, 1962).

Although the cash outlays required by the individual farmer are relatively modest at this stage, he is incapable of providing them himself; therefore these inputs have to be financed by credit advances to the farmers.

SECOND STAGE—COMMERCIAL AGRICULTURE

As a result of the factors brought into play in the first stage, production increases: total productivity increases of from fourfold to eightfold are possible in favoured areas, while from twofold to fourfold increases averaged over large areas are considered conservative (Mellor, 1962). An increasingly larger proportion of farm output is sold, and agriculture becomes capable of making a positive contribution to the overall economic development of the country. It is during this stage that conditions are most favourable for the promotion of co-operative frameworks.

THIRD STAGE—INDUSTRIALIZED AGRICULTURE

As a result of industrial development, the non-farm population that is dependent on purchased food becomes bigger, and the demand for agricultural products rises rapidly. Labour becomes gradually more scarce and costly; by contrast, capital becomes more abundant. As a result, agriculture becomes more and more mechanized; specialization becomes prevalent, and farm size increases as holdings become consolidated.

PACKAGE PROGRAMMES

An interesting approach to the large-scale introduction of improved practices into traditional farming is the 'Package Programme' carried out in a number of districts of India (Malone, 1965). More than one million farmers are already taking part in the plan, comprising about two per cent of India's farming population (Anon., 1966). In the fourth Five-year Plan period, India intended to extend the 'package' principle to more than 100 of the country's 325 agricultural districts.

The basic concept underlying the programme is that agricultural progress will be more rapid, and the adoption of improved practices will be more effective, if:

(1) A 'package' of complementary, improved production practices, locally adjusted to climate soil and irrigation conditions, is established by specialists.

(2) The 'package' deal is applied across the whole farming community, by helping whole groups of farmers in each village to make a break with tradition, adopting the package of improved practices. These include clean seeds treated against seed-borne diseases, equipment for improving seed-bed preparation, fertilizers, plant protection measures, and effective irrigation practices. The 'package' is adapted to each locality, but always consists of a combination of interacting practices (Anon., 1966).

(3) The technical supplies (seeds, fertilizers, pesticides and implements) needed for the execution of the programme, and the necessary credits to finance implementation of the plans, are made available to the villages in time and in sufficient quantity, together with a number of supporting services that are required for this purpose: these include suitable transport and marketing arrangements, adequate storage facilities, seed- and soil-testing laboratories, workshops, and credit institutions.

(4) A general educational programme clarifying the benefits to be derived from the package plan, including demonstration plots, is executed. Each demonstration is carried out on two plots in a farmer's field; on one of these plots the 'package' of improved practices is applied, whereas the other is farmed in the traditional way as a kind of control.

(5) A simple farm plan is worked out with each farmer participant, indicating the crops he intends to grow and the supplies he will require.

The 'Package Programme' was started in 1960, in seven selected districts, with support from the Ford Foundation, which provided financial help and technical guidance. Malone (1965) studied the effect of the Programme in the Tanjone District, in which almost 63000 farms participated. The Programme in this District was carried out by three extension officers and twenty community workers for each extension area, consisting of about 90 villages and 83000 farms. Energetic extension workers achieved an enrolment in the plan of 50–70 per cent of the farmers within two years.

Participation included all sizes of farms, with tenants as well as owners participating. In the three-year period following the introduction of the plan, the tonnage of fertilizer used in the Tanjone District increased by 90 per cent; the application of fertilizers on Package Programme farms was double the average for the Tanjone District. The input of 'new' practices (fertilizers, plant protection materials, etc.) increased by 250 per cent. More livestock feed was purchased than before. Labour input increased by about 20 per cent on the same acreage, as a result of employing improved practices and diversification of crops. Rice yields increased by 19 per cent (about 450 kg/ha); this modest increase was due mainly to the small response to fertilizer application of the rice varieties used at the time. More minor field crops, vegetables, fruit and milk, were produced. On the average, there was one dollar of net gain for every dollar spent.

The implementation of the package plan has shown that farmers take the first steps away from tradition more quickly and more successfully if they do it as a group, together with their neighbours.

Planning

The planning of agricultural development must be an integral part of overall national planning, in order to ensure a balanced expansion of the various sectors of the national economy and the integration of agriculture in its overall development. This requires weighing investments in agricultural development against other development potentials. For example, the need to allocate water to activities other than agriculture, such as industrial or municipal uses, discussed in Chapter VI (see p. 166), may have a considerable bearing on the planning of agricultural development.

A central organization is required that has the responsibility for planning, and then assigning the execution of the programme to the agencies involved.

Planning of agriculture is necessary on three levels: at the national level (macro-planning), at the level of the individual farm-units (micro-planning), and at the intermediate level of the region (Weitz & Rokach, 1968).

At the *national level*, a basic requirement for agricultural development is a clear definition of long- and short-term objectives at this level, with the assignment of priorities in accordance with overall needs. An overall survey of resources provides the essential raw material for planning and the determination of priorities.

Information on the estimated increase in population, on the annual increase in the gross national product, on the annual increase in *per caput* consumption, and on the projected rate of industrial development, is essential to the planning process. Assessments have to be made on the internal and foreign market prospects for agricultural products. A survey of the basic natural resources—land, labour, water and climate—and of the prospects of increased productivity per unit of land, water and labour, are also prerequisites of planning. The inventory of resources must also provide information on human population in terms of size, distribution, educational level and health standards.

Production targets for all agricultural commodities, based on local requirements and export possibilities, have then to be established.

Linear programming, in which the production and marketing factors pertaining to each branch of production are weighted and calculated to produce an integrated programme, is being increasingly applied to agricultural planning.

Planning at the *level of the individual farm* is concerned with finding the most appropriate farm structure that combines profitability for the individual farmer with a maximum possible contribution to the national economy.

Planning at the *regional level* is not to be considered as a small-scale application of the national plan. It has the dual function of adjusting the national plan to the conditions encountered at the micro-level, and of coordinating agricultural development with other sectors of the economy in the region. It is at the regional level that 'comprehensive rural planning' becomes effective. This is defined as 'the integrated development of agriculture, industry, and services, within a rural area, with reference to all aspects of rural life, i.e. economic, social, organizational and environmental' (Weitz & Rokach, 1968). A well-defined region, the size and nature of which permits integrated planning of all sectors, appears to be the natural unit for comprehensive rural planning. The region itself may be defined on the basis of economic criteria, on the basis of ethnic or cultural communities, or on common physical and ecological features, such as a river basin, which constitute a unit for development purposes.

SOIL SURVEYS

A prerequisite for planning at all three levels is a survey of the country's soils.

A reconnaissance survey leading to a map on scales of from 1:30 000 to 1:100 000 or more makes possible a first, superficial appraisal of the agricultural potential and the delineation of areas that appear promising for development. Then follow detailed surveys on scales of from 1:1 000 to 1:20 000, which make possible land-classification and the planning of irrigation and settlement schemes.

After the soils have been identified and classified, a land-capability classification is made. A number of capability classes are established, based on an evaluation of important soil characteristics such as depth, texture, slope, permeability, water-holding capacity and inherent fertility. Vegetation maps provide additional information. In arid regions, the main aim of soil survey studies is to determine the irrigation potential (United Nations, 1963).

THE TIME-ELEMENT IN PLANNING

The time-element is important for successful planning. More, perhaps, than in any other economic field, time plays a great role in agricultural development, and must be realistically appraised in planning. Time is required to produce improved inputs, such as new varieties, or to learn how to use existing inputs to the best advantage. Time is needed to influence the farmers to adopt new plans and methods, and to create the necessary infrastructure and the markets that will have to absorb increased production (Gillie, 1967).

CONTROL AND DIRECTION OF PRODUCTION

Whilst it is neither desirable nor practicable to oblige the individual farmer to produce according to planning on a national scale, it is possible to use a number of inducements whereby government can direct production. These include:

Price policy, whereby minimum prices are guaranteed for certain commodities.

Subsidies to decrease the cost of certain inputs, such as fertilizers, foodstuffs, protection chemicals, and water, as well as for specific development projects.

Quotas are allocated for certain commodities; production above these quotas is not entitled to subsidies.

Loans are made on easy terms to farmers who wish to produce favoured commodities. 'Oriented' credits are a major instrument for the implementation of government plans.

Marketing boards are established, one for each of the principal commodites; they comprise representatives of the producers, consumers and government, and are an effective tool for translating the official agricultural plan into practice. The boards establish regional and individual quotas, determine guaranteed minimum prices, and regulate the supply to the markets (Figure 99).

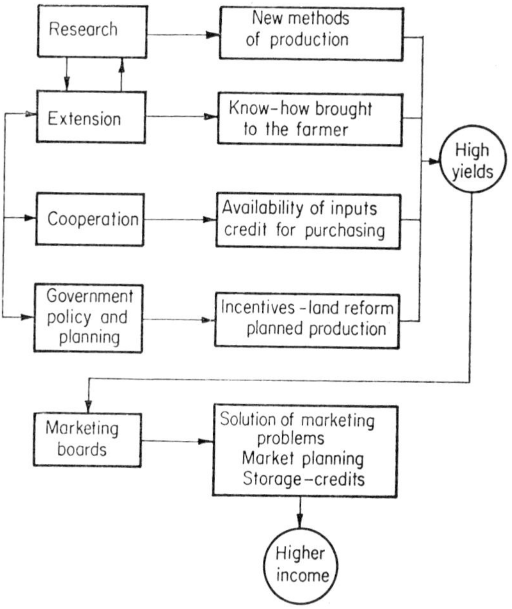

Figure 99. Schema showing chain of events leading to higher income to farmer.

References

ADMONI, Y. (1963). Adapting extension methods to different social groups in Israel. Paper presented at the Rehovoth Conference on *Comprehensive Planning of Agriculture in Developing Countries*. Rehovoth, Israel, 11 pp. (mimeographed).

ALDRICH, D. G. (1966). Agricultural research, a key to understanding our environment. pp. 50–3 in *Symposium on Research in Agriculture*. U.S. Dept. Agric. Res. Serv., Washington D.C., iv + 112 pp.

ANDERSON, W. (1958). Technological developments in world agricultural production, *J. Fm Econ.*, **40**, 1470–81.

ANON. (1966). India's package program raises food output. *Wld Fmg*, **8**(6), 28–36.

ARDITO-BARLETTA, N. (1962). *Cost and Social Return to Agricultural Research in Mexico*. Ph.D. dissertation, Univ. of Chicago, cited by Schultz, T. W. (1967).

ARNON, I. (1968). *Organisation and Administration of Agricultural Research*. Elsevier, Amsterdam–London–New York, xii + 342 pp., illustr.

BAADE, F. (1965). Agricultural problems of Turkey, Pakistan, and India. pp. 24–37 in *Conference on Asian Development: Towards a strategy for development co-operation*, University Press, Rotterdam, 104 pp.

BACHMAN, K. L. & CHRISTENSEN, R. P. (1967). The economics of farm size. pp. 234–54 in *Agricultural Development and Economic Growth* (Ed. H. M. Southworth & B. F. Johnston) Cornell Univ. Press, Ithaca, xv +608 pp., illustr.

BOSE, S. P. (1961). Characteristics of farmers who adopt agricultural practices in Indian villages. *Rural Sociology*, **26**, 138–46.

—— (1963). General review of problems of agricultural education and extension in developing countries. Paper presented at the Rehovoth Conference on *Comprehensive Planning of Agriculture in Developing Countries.* Rehovoth, Israel, 9 pp. (mimeographed).

BROWN, L. R. (1965). *Increasing World Food Output.* U.S. Dept. Agric. Foreign Agric. Econ. Rep., **25**, 140 pp., illustr.

BUCHA, M. K. B. (1968). Agricultural development in Pakistan. pp. 41–50 in *Strategy for the Conquest of Hunger.* Rockefeller Foundation, New York, 131 pp.

BURKE-KNAPP, J. (1966). The role of international agencies in aiding in world food production. pp. 11–17 in *Proc. Symp. on World Food Supply,* Nat. Acad. Sci., Washington D.C.

CENTRAL TREATY ORGANIZATION (1962). *Travelling Seminar for Increased Agricultural production.* Regional tour report, Office of U.S. Coordinator CENTO Affairs, Ankara, 160 pp.

CHATELAIN, R. (1963). Main problems of agricultural training at village level. Paper presented at the Rehovoth Conference on *Comprehensive Planning of Agriculture in Developing Countries.* Rehovoth, Israel, 10 pp. (mimeographed).

CHRISTENSEN, R. P. & STEVENS, R. D. (1962). Putting science to work to improve world agriculture. pp. 158–76 in *Food: One Tool in International Economic Development* (Ed. E. O. Haroldsen) Iowa State Univ. Press, Ames, x + 419 pp., illustr.

—— & YEE, H. T. (1964). The role of agricultural productivity in economic development. *J. Fm Econ.*, **46**, 1051–61.

CLARK, C. (1968). Health, population, and agricultural growth. Plenary lecture 2, International Seminar on *Change in Agriculture*, Reading, England, 15 pp., illustr. (mimeographed).

COUSTON, J. W. (1967). *Physical and Economic Summary of Trial and Demonstration Results.* FFHC Fertilizer Program 1961/62–1964/65, F.A.O., Rome, 116 pp. (mimeographed).

DE SOLA, F. *See* SOLA, F. DE.

DE WILDE, J. C. *See* WILDE, J. C. DE.

DICKSON, B. T. (1956). The challenge of arid lands research and development for the benefit of mankind. pp. 47–66 in *The Future of Arid Lands* (Ed. G. F. White) Amer. Assoc, Adv. Sci., Washington D.C., ix + 453 pp., illustr.

DONALD, C. M. (1964). Phosphorus in Australian agriculture. *J. Aust. Inst. Agric. Sci.*, **30**, 75–105.

DUMONT, R. (1962). Accelerating African agricultural development. *Impact Sci. Soc.*, **12**, 231–53.

DUROST, D. D. & BARTON, G. T. (1960). Changing sources of farm output. *Prod. Res. Rep. U.S. Dept. Agric.*, **2**, 3–12.

ENNIS, W. B., JANSON, L. L., ELLIS, I. T. & NEWSOM, L. D. (1967). Inputs for insecticides. *U.S. President's Science Advisory Committee, Washington*, **3**, 130–75.

F.A.O. (1966). *Indicative World Plan.* F.A.O., Rome, 672 pp.

—— (1967). *Fertilizers.* Ann. Rev. Wld Production, Consumption and Trade. F.A.O., Rome, 205 pp. (mimeographed).

GHULAM, M. (1965). Private tubewell development and cropping patterns in West Pakistan. *Pakistan Dev.*, **5**, 1–53.

GILLIE, F. B. (1967). *Basic Thinking in Regional Planning.* Mouton, Paris, 95 pp.

GREGOR, H. F. (1959). Push to the desert. *Science*, **129**, 1329–39.

GULHATI, N. D. & SMITH, W. C. (1967). Irrigated agriculture: an historical review. pp. 3–11 in *Irrigation of Agricultural Lands* (Ed. R. M. Hagan, H. R. Haise & T. W. Edminster) Agronomy Series, Amer. Soc. Agron., Madison, Wisconsin, II, xxxi + 1180 pp., illustr.

GUNN, D. L. (1960). The biological background of locust control. *Ann. Rev. Ent.*, **5**, 279–99.

HEADY, E. O. (1961). Public purpose in agricultural research and education. *J. Fm Econ.*, **43**, 566–81.

—— (1962). Research and economic development needs, opportunites and problems. pp. 1–31 in *Food: One Tool in International Economic Development* (Ed. E. O. Haroldsen) Iowa State Univ. Press, Ames, x + 419 pp,, illustr.

HOSELITZ, B. F. (1962). Agriculture in industrial development. pp. 125–47 in *Food: One Tool in International Economic Development* (Ed. E. O. Haroldsen) Iowa State Univ. Press, Ames, x + 419 pp., illustr.

HUTCHINSON, J. B. & WILMAN, D. (1965). The strategy of food production. *Discovery, London*, **26**, 20–5.

JEWITT, T. N. (1966). Soils of arid lands. pp. 103–24 in *Arid Lands: A Geographical Appraisal* (Ed. E. S. Halls) UNESCO, Paris, xviii + 461 pp., illustr.

JOHNSTON, B. F. (1964). The choice of measures for increasing agriculture production: a survey of possibilities in East Africa. *Trop. Agric.*, **41**, 91–113.

KELLOGG, C. E. (1962). Using agriculture resources for economic development in underdeveloped countries. pp. 148–57 in *Food: One Tool in International Economic Development* (Ed. E. O. Haroldsen) Iowa State Univ. Press, Ames, x + 419 pp., illustr.

KREBS, W. A. W. (1964). The developing countries new frontiers for research. pp. 46–51 in *Proc. 17th Nat. Conf. Administration in Research.* Univ. of Colorado, Denver.

MCPHERSON, W. W. & JOHNSTON, B. F. (1967). Distinctive features of agricultural development in the tropics. pp. 180–233 in *Agricultural Development and Economic Growth* (Ed. H. M. Southworth & B. F. Johnston). Cornell Univ. Press, Ithaca, xv + 608 pp., illustr.

MALASSIS, L. (1966). Role of agriculture within the context of general economic growth. pp. 31–36 in *Higher Education in Agriculture.* O.E.C.D., Paris, 170 pp.

MALONE, C. C. (1965). Some responses of rice farmers to the package program in Tanjore District, India. *J. Fm Econ.*, **47**, 256–69.

MASON, E. S. (1966). *Economic Development in India and Pakistan.* Harvard Cent. Internat. Affairs, Cambridge, Massachusetts, 67 pp.

MELLOR, J. W. (1962). Increasing agricultural production in early stages of economic development. pp. 219–43 in *Food: One Tool in International Economic Development* (Ed. E. O. Haroldsen) Iowa State Univ. Press, Ames, x + 419 pp., illustr.

METTRICK, H. (1968). Mechanization of peasant agriculture in East Africa. Case Study 12e, International Seminar on *Change in Agriculture*, Reading, England, 10 pp. (mimeographed).

MORE, C. A. (1955). Agricultural development in Mexico. *J. Fm Econ.*, **37**, 72–80.

MYRDAL, G. (1965). *The 1965 McDougall Memorial Lecture.* F.A.O., Rome, 213 pp.

NAIR, K. (1961). *Blossoms in the Desert.* Duckworth, London, 200 pp., illustr.

NELSON, L. R. (1967). Fertilizer requirements for increased food needs. *The World Food Problem.* U.S. President's Science Advisory Committee, Washington D.C., **3**, 95–118.

O.E.C.D. (1967*a*). *The Food Problem of Developing Countries.* O.E.C.D., Paris, 114 pp.

—— (1967*b*). *Supply and Demand Prospects for Chemical Fertilizers in the Developing Countries.* O.E.C.D., Paris, 206 pp., illustr.

OLSON, R. A. (1968). The fertilizer programme of the freedom from hunger campaign. Case Study, International Seminar on *Change in Agriculture*, Reading, England, 56 pp. (mimeographed).

OSOYO, R. (1968). Mexico: from food deficits to sufficiency. pp. 6–15 in *Strategy for the Conquest of Hunger.* Rockefeller Foundation, New York, 131 pp.

OZAL, K. (1963). A review of irrigation development in Turkey. Paper C/261, U.N. Conference on *Application of Science and Technology for the Benefit of the Less Developed Areas*, Geneva, vol. 3, viii + 309 pp.

PATIL, N. P. (1963). Economics of drill sowing over broadcasting. *Report Frm Man. Res. Center Agric. Stn Hebbal, India.*

PENDERS, J. M. A. (1956). *Methods and Programme Planning in Rural Extension.* H. Veenman, Wageningen, 324 pp., illustr.

RAUP, H. M. (1967). Land reform in agricultural development and economic growth. pp. 267–97 in *Agricultural Development and Economic Growth* (Ed. H. M. Southworth & B. F. Johnston) Cornell Univ. Press, Ithaca, xv + 608 pp., illustr.

RAYMOND, W. F. (1968). Efficiency of land-use of ruminant production, pp. 25–33 in *Forage Economics—Quality* (Ed. C. M. Harrison) *ASA Spec. Publ.* No 13,v + 144 pp., illustr.

RUPERTI, R. (1963). The part played by artificial lakes in transforming hill farm irrigation in arid zones. Paper C/501, U.N. Conference on *Application of Science and Technology for the Benefit of the Less Developed Areas*, Geneva, vol. 3, viii + 309 pp.

SALAZAR, J. M. (1968). Latin America: the need for increased agricultural production. pp. 51–61 in *Strategy for the Conquest of Hunger.* Rockefeller Foundation, New York, 131 pp.

SCHULTZ, T. W. (1964). *Transforming Traditional Agriculture.* Yale University Press, New Haven, Conn., 212 pp.

—— (1966). Increasing world food supplies: the economic requirements. pp. 18–23 in *Proc. Symp. World Food Supply*, Nat Acad. Sci., Washington D.C., xiv + 212 pp.

—— (1967). Education and research in rural development. pp. 391–402 in *Rural Development in Tropical Latin America* (Ed. K. L. Turk & L. V. Crowder) New York State College of Agriculture, Ithaca, xx + 446 pp., illustr.

SHIMSHI, D., GAIRON, S. & RUBIN, I. (1957). [*Experiments in wheat irrigation in the northern Negev*—in Hebrew] Prelim. Rep. Agric. Res. Stn Rehovoth, Israel, **167**, 61 pp., illustr. (mimeographed).

SOLA, F. DE (1968). Change and opportunity in Central America. pp. 23–9 in *Strategy for the Conquest of Hunger.* Rockefeller Foundation, New York, 131 pp.

STAMP, L. D. (1953). *Land for Tomorrow: The Underdeveloped World.* Faber, London, 186 pp.

STIPPLER, H. H. & DARWISH, M. J. (1966). *Land Tenure and Land Utilization in Shakha* 8, Dujaila Project 1965/66, Baghdad, Minist. Agr. Rep., Govt. of Iraq—Ford Foundation.

STURT, D. W. (1965). Producer response to technological change in W. Pakistan. *J. Fm Econ.*, **47**, 625–33.

UNITED NATIONS (1963). *Science and Technology for Development.* United Nations, New York, vol. 3, viii + 309 pp.

U.S. DEPARTMENT of AGRICULTURE (1965). *Changes in Agriculture in 26 Developing Nations 1948–1963.* Foreign Agric. Econ. Rep. **27**,

U.S. DEPARTMENT OF COMMERCE (1968). *Statistical Abstract of the United States.* U.S. Govt Printing Office, Washington D.C., xii + 1034 pp., illustr.

U.S. PRESIDENT'S SCIENCE ADVISORY COMMITTEE (1967). *The World Food Problem.* Rep. Panel World Food Supply, Washington D.C., vol. 3, xxii + 332 pp., illustr.

WEITZ, R. & ROKACH, A. (1968). *Agricultural Development: Planning and Implementation* (Israel Case Study) D. Reidel, Holland, xix + 404 pp.

WELLHAUSEN, E. J. (1967). Opportunities for crop improvement. pp. 237–55 in *Rural Development in Tropical Latin America* (Ed. K. L. Turk & L. V. Crowder) New York State College of Agriculture, Ithaca, vol. 2, xii + 446 pp., illustr.

WHITE, G. F. (1962). *The Changing Role of Water in Arid Lands.* Univ. Arizona Press, Tucson, 15 pp.

WILDE, J. C. DE & MCLOUGHLIN, P. F. M. (1967). *Experiences with Agricultural Development in Tropical Africa.* Johns Hopkins Press, Baltimore, vol. 2, xii + 446 pp., illustr.

WILLIAMS, M. S. & COUSTON, J. W. (1962). *Crop Production Levels and Fertilizer Use.* F.A.O., Rome, 48 pp., illustr.

SUBJECT INDEX

For Latin names index, see pp. 631–34

LATIN NAMES INDEX

AUTHOR INDEX